THE FAUNA OF BRITISH INDIA

INCLUDING

CEYLON AND BURMA

THE FAUNA OF BRITISH INDIA

INCLUDING

CEYLON AND BURMA.

ODONATA.

IN THREE VOLUMES

VOL. I

BY

LT.-COL. F. C. FRASER, I.M.S.

GYAN

Published by

Gyan Publishing House
5, Ansari Road
Daryaganj, New Delhi-110002
Phone: 011-47034999, 9811692060
E-mail: books@gyanbooks.com

Distribution Network
gyanbooks.com
India, USA, Canada, UK, Australia

ISBN: 978-81-212-0211-4 (Set)
978-81-212-0248-0 (HB)
First Published, 1933

2nd Impression 2024

Printed at: Gyan Press, Delhi.

THE FAUNA OF BRITISH INDIA INCLUDING CEYLON AND BURMA (VOL. I)

Author: F. C. FRASER

THE FAUNA OF BRITISH INDIA,

INCLUDING

CEYLON AND BURMA.

Published under the authority of the Secretary of State for India in Council.

EDITED BY LIEUT.-COL. J. STEPHENSON, C.I.E., M.B., D.Sc., F.R.S., I.M.S. (*ret.*).

ODONATA.

VOL. I.

BY

LT.-COL. F. C. FRASER, I.M.S.

LONDON:

TAYLOR AND FRANCIS, RED LION COURT, FLEET STREET.

March, 1933.

CONTENTS.

PREFACE.

THE first mention we find of Indian Dragonflies is the description of *Neurobasis chinensis* by Linnæus in 1758: this, however, from a non-Indian specimen. Thereafter, Johannsen followed with the description of *Rhyothemis variegata* in 1768, and Drury with some half-dozen descriptions in 1773. Twenty years later Fabricius, the pupil of Linnæus, published in his 'Entomologia systematica' (1792–94; Suppl. 1798) a further eight descriptions. We find no further mention of the Indian fauna until more than forty years later (1842), when Rambur published his 'Histoire naturelle des Insectes, Névroptères,' which contained the descriptions of over twenty species found within Indian limits. It was, however, during the life of the late Baron Edmond de Selys-Longchamps that the greatest advance was made in the study of the Indian Dragonfly fauna. From 1850 until 1890 he published a long series of papers dealing with the Dragonfly fauna of the world, and the descriptions of more than half of the Indian species are to be sought for in these works. At his death, in 1890, a number of workers on the Order made their appearance, among whom must be mentioned especially the late Dr. F. Ris and Mr. E. B. Williamson, along with Dr. F. Laidlaw, the two latter writing especially on the Indian fauna. Williamson contributed two valuable papers on the Indian AGRIIDÆ and GOMPHIDÆ, whilst Laidlaw published a series of papers in the 'Records of the Indian Museum' dealing especially with the Indian GOMPHIDÆ, AGRIIDÆ, and several subfamilies of the CŒNAGRIIDÆ. The present writer drew his inspiration from the two latter writers, and began a series of papers on the same subject in 1918 in the 'Journal of the Bombay Natural History Society' which have now run into forty parts, and which have formed the basis of the present work. The first few of these papers were written during the Great War and, owing to lack of references and paucity of material, cannot be said to have been of a high order; they were, however, intended to stimulate an interest in the subject, as the Journal had a wide circulation amongst Indian readers. They had the desired effect, for through their agency the writer got into communication with a number of collectors who during the following years continued to supply him with endless material. After the War had terminated, the writer was able to pay visits to well-known European collections, and to make a prolonged study of the rich material in the British and Paris Museums. Access was also obtained on several occasions to the MacLachlan collection through the courtesy of the nephew of that great naturalist, by whose care the collection is now preserved.

I have myself made extensive collections in the Eastern and Western Ghats of India, the Darjeeling district of Bengal, and in parts of Sind, Ceylon, and Burma, and these collections have been enormously enhanced by gifts of specimens made by a number of willing helpers. Many thanks are especially due to Mr. T. Bainbrigge Fletcher, Imperial Entomologist to the Government of India, who since 1918 has sent me a vast amount of material and notes, and kept me constantly informed of all publications which have appeared on the ODONATA. Without his help in these and other ways much of the present volume would have been difficult to write. I am also indebted to Col. F. Wall, I.M.S., who made very extensive collections for me in Burma and Ceylon, as well as supplying me with data about the species captured. Mr. H. V. O'Donel collected for me in the Duars, finding several new species; Lt.-Col. A. Frere collected in the Palni Hills; Mr. C. M. Inglis sent me many collections from Sikkim and Thibet; and a few collections were also sent me from the Indian Museum and the Forest Research Institute, Dehra Dun. To all these tireless helpers I now take this opportunity of tendering my sincere thanks and acknowledgements for the assistance they have rendered in the compilation of this work.

To the Committee of the Bombay Natural History Society I have to express my thanks for permission to use the many figures and plates which appeared in the pages of the Journal illustrating the papers referred to, and also for their offer on loan of the blocks.

To the Authorities and Trustees of the Indian and British Museums I am especially indebted for the loan of types and the opportunity of studying material otherwise inaccessible. To Mr. Kenneth Morton of Edinburgh, Dr. F. Laidlaw of Devon, and Mr. E. B. Williamson of Indiana, U.S.A., I wish to express my thanks for the valuable advice they have given me from time to time, throwing light in dark places, and tendering willingly any assistance which I asked for. I owe a debt of gratitude to the late Dr. F. Ris of Rheinau, who was the greatest worker of our time on the Order, and who was always ready to give his advice when asked for; his valuable opinions will be greatly missed by all students of Odonatology; his aptitude for throwing light on obscure points was almost uncanny, and was an ample testimony to his wide knowledge of the Order.

There are about 500 species of Dragonflies, included in about 90 genera, known to occur within Indian limits, including Burma and Ceylon, and of these, 39 genera and 168 species and subspecies, belonging to the Suborder ZYGOPTERA, are described in this volume.

F. C. FRASER.

GLOSSARY.

Paratype.—Strictly, any one of the rest of the series collected with, and at the same time as, the type, or forming part of the series from which the type is described ; in a wider sense (as usually in these volumes), specimens lodged in accredited public or reference collections for the use of students, and diagnosed by approved specialists as similar in all respects to the type ; or shortly, representative specimens lodged in public or well-known collections.

Allotype.—The first specimen of the opposite sex described (R. J. Tillyard, 'Insects of Australia and New Zealand,' p. 510 (1926)).

Teneral.—Immature or freshly emerged specimens which have not yet become fully chitinized or attained their adult colouring.

Lilaceous.—Colour of lilac blossom ; pale purplish.

Pruinescence.—A powdery white excrescence developing on various parts of the body in adult dragonflies, usually of the male sex.

Pulverulent.
Pruinosed. } —Coated in part or wholly with pruinescence.

Enfumed.—Smoky tint ; pale blackish-brown ; like a smoked glass.

Petiolated.—Stalked like a leaf.

Braced.—With a strong oblique, short, transverse nervure, usually extending from the under side of the proximal end of the pterostigma to the radius.

Termen.—The posterior border of wings.

Apical.—Tip of wing when applied to the wings ; the part most distal from the thorax when applied to the abdominal segments.

Basal.—The part of the wing nearest the thorax when applied to the wings ; the part nearest the thorax when applied to the abdominal segments.

Clypeus.—The face, or sclerite, of the head, situated between the frons above and the labrum below, the latter being attached to it.

Anteclypeus.—The portion of the clypeus to which the labrum is attached ; the lower or most anterior portion of the clypeus.

Postclypeus.—The portion of the clypeus lying between the frons above and the anteclypeus below.

Frons.—The forehead, or upper portion of head lying in front of the eyes, and often forming a prominent shelf-like ridge at the upper part of face.

Occiput.—The back of the head; its size and shape depend on the extent of the space between the eyes, and, when these meet, it exists as a mere tiny triangular area behind.

Labrum.—The upper lip.

Labium.—The lower lip.

Epistome.—The postclypeus.

Carina.—A ridge, more or less defined, running along the mid-dorsum of thorax and abdomen.

Wedgewood blue.—Pale greenish-blue.

Cambridge blue.—Pale azure blue.

Obsolete.—(Lat. *obsolesco*, go out of use, decay.) Becoming atrophied or almost invisible; fading away or obscured when applied to coloured markings.

Ocellus.—A simple eye as contrasted with the compound eye. All dragonflies possess three simple eyes usually arranged in a triangle on the vertex in front of and between the eyes.

SYSTEMATIC INDEX.

ODONATA (Paraneuroptera).

INTRODUCTION.

The ODONATA comprise the Dragonflies of popular entomology. Although less known than butterflies or moths, they form a conspicuous feature of the average Indian landscape and are common to all parts, even the most arid areas. The beauty and elegance of their flight, their gay colours, and their countless numbers, especially during the post-monsoon months when the flighting season is in progress, force their presence on the tardiest observer.

There are few insects with which they can be confused, but very occasionally specimens of ant-lions and may-flies are received which have been sent under the mistaken impression that they are dragonflies. From these insects dragonflies may be readily distinguished by having only minute and almost invisible antennæ, these organs being conspicuous objects in the other groups just mentioned.

Dragonflies are popularly supposed to sting, but in reality they are perfectly harmless insects, and the forceps which adorn the end of the abdomen are merely accessory sexual appendages; beyond their great powers of flight they have no defensive appliances.

Economically they are of great importance in destroying noxious flies and mosquitoes, as well as the smaller moths which are regarded as pests. Life in the tropics would soon become unbearable were it not for the beneficent work of vast numbers of dragonflies acting as scavengers of the atmosphere. Some species which take to the wing only after dark or at dusk live entirely on mosquitoes.

They have but few enemies if we except themselves, as they are much given to cannibalism, both in the larval and imaginal stages. Certain species, such as *Orthetrum sabina*, live almost exclusively upon their own kind, haunting the margins of tanks and lakes, where they make a rich living off the smaller fry. Among their enemies must be reckoned birds, lizards, fishes, frogs, and spiders. Birds take a heavy toll of these insects during their emergence, when their weak flight makes them fall an easy prey to swallows, martins, minas, and king-crows. I have at times watched numbers of newly emerged dragonflies rising from their watery element, and not a single one appeared to escape from the swallows and martins skimming overhead. Lizards not uncommonly attack those species which are given to alighting on the ground. Fishes and frogs feed on them both during the larval and early imaginal stages. Tillyard mentions the finding of as many as thirty-five undigested heads of dragonflies in the stomach of a single trout. I have frequently observed fish rising for dragonflies as the latter skimmed the surface of the water, and have also noted frogs acting in the same way. A rare specimen which I had knocked into the water was snapped up in this way before I could retrieve it.

Spiders and mantids play but a small part in this respect, and it is extremely rare to find a dragonfly caught up in a spider's web. Lastly, Tillyard mentions that large numbers of the smaller species are caught by the Giant Sundew of Australia (*Drosera binata*), while I have occasionally noted the same circumstance in the case of *Ischnura delicata* and the common Indian Sundew, *Drosera peltata*.

A few parasites are known to attack dragonflies, either as larvæ or as the imago. Several species of small Hymenoptera (Tetragrammidæ and Myrmaridæ) deposit their eggs in those of dragonflies. A small nematode is not uncommonly found in the gut of the imago during the process of cleaning specimens, whilst gregarines sometimes occupy the same situation in the larvæ.

Few examples of symbiosis are known in regard to these insects. A small red mite (a Hydrachnid) is commonly found clinging to the wings, legs, and thorax of certain species, more particularly *Ceriagrion coromandelianum* of the Indian fauna, and species of *Sympetrum* of temperate climates; these attach themselves to the insects at the time of emergence as a means of dispersal to other localities; it is unknown why certain species only are thus employed as living aeroplanes, but the fact remains. Another form of symbiosis is met with in certain larvæ which become coated with dense masses of algæ, which serve to disguise them effectually from their enemies and prey alike.

Like many other insects, dragonflies are much given to migratory flights, though the habit appears to be confined to but a few species. Among Indian forms must be mentioned *Pseudagrion decorum* and *P. microcephalum*, *Pantala flavescens*, *Anax parthenope*, *Hemianax ephippiger*, *Aciagrion occidentale*, and *Onychargia atrocyana*. The larger species depend on their own powers of flight for these journeys, but the smaller and weaker species rely entirely on upper air-currents. Migratory flight is usually in a direction from east to west, but during October and November vast numbers of the smaller species may be seen flighting northwards up the West Coast, their emergence appearing to coincide with the draining of paddy-lands prior to the harvest.

The dragonfly collector will soon find that the greater part, at least, of his collecting has to be carried on in the neighbourhood of water. As an order, these insects cling with remarkable tenacity to their watery birth-places, and although on emergence they wander afar, they return to their former homes for purposes of breeding; males, indeed, rarely depart from such places, but await the return of the females over ponds and streams. It is this circumstance which accounts for the fact that the number of males taken by the collector vastly exceeds the number of females. In a few rare cases where mating takes place in the forest or away from water, this order of preponderance is, of course, reversed; such a case is that of *Anaciæschna martini*.

Some species form isolated colonies which may be found inhabiting the same localities over a space of many decades. This applies equally well to species which breed in rivers, lakes, or marshes, but more particularly to riverine species, as tanks, lakes, and marshes are apt to dry up in times of drought, the fauna therein perishing *en masse*. Old colonies thus die out, and an entirely new group of species may be found replacing them in succeeding years. Rivers which dry up during a portion of the year are rarely of much use from a dragonfly collector's point of view, as the species, especially the rarer forms, fail to get established there.

As dragonflies breed in water, it happens that the number of species found bears a direct ratio to the rainfall or water-supplies of any particular district. So far as India is concerned, the Western Ghats, the southern slopes of the Himalayas, and the wet tracts of Burma and Ceylon are by far the richest in species. The drier zones are not entirely barren, and in a few cases have their own distinctive fauna.

In spite of their great powers of flight, dragonflies are peculiarly local, and this applies to all the rarer and more uncommon forms in particular, some of which will be found confined to a very short stretch of river or a small patch of marshy lands in the forest. Thus *Chloroneura apicalis* has been found only on a short stretch of the Cauvery River at Fraserpet, Coorg, barely half a mile in extent. The same applies to *Platysticta deccanensis*, a species similarly confined to a small stream in the Cochin forests, or again to *Mortonagrion varralli*, confined to a few acres about the Vihar Lake near Bombay, and many other instances can be cited. Prolonged search for these species in many other localities has been quite fruitless.

In addition to their local distribution, dragonflies are very seasonal, this again applying especially to the more uncommon forms, which are usually single-brooded and appear for but a short space of a few weeks. So far as India is concerned, the best months for collecting are May and June and September to the first half of November, that is, during the months which precede and follow the south-west monsoon. Common species are not only widely distributed, but occur nearly the whole year round. Quite a number of species are strictly confined to certain altitudes, a zone lying between 2,000 and 3,500 feet yielding the richest fauna. The hot plains appear to be inimical to these, and to act as a barrier to their spread; thus isolated mountain groups have developed a peculiar fauna of their own, resembling islands in this respect.

The Indian Odonate fauna has borrowed largely from Malaysia and, to a less extent, from the Palæarctic, but it is rich in entogenic groups, the origin of some of which is decidedly obscure. The chain of the Western Ghats has a rich fauna, many forms of which are entirely restricted to it. Ceylon is equally rich in this respect, but shows some affinities to the fauna of Southern India. In the north-eastern provinces—Bengal, Assam, and Bihar—the Malaysian affinities become pronounced, and increase further towards Lower Burma, which is purely Malaysian. A few Ethiopian and Australian forms have made their way to India; examples of such are *Zygonyx torrida isis* from the former and *Hemicordulia asiatica* from the latter.

Methods of Collecting and Preservation.

Breeding from Larvæ.—From the collector's point of view this is not very satisfactory, as a dragonfly after emergence takes several days to develop its full colours and become fully chitinized, that is, hardened. Thus a newly fledged insect will not only be poor in colouring, but will also shrivel up after death and make an unsightly specimen. Even keeping for a few days with artificial feeding does not improve them much in this respect. There are, however, some species which are only to be obtained in this manner—probably far more than we are aware of, as larvæ of many species are frequently dredged up, the imagines of which have never been seen in that particular neighbourhood. Other species depart into the heart of the jungle directly after emergence, and are never seen from that time onwards; this applies particularly to the females and to the Gomphines.

The aquatic life of larvæ varies greatly in length, that of the genus *Lestes* being probably the shortest—about three months in all. The larger species take from one to three years to reach maturity, a time long enough to tax the patience of the most patient collector. It is best to keep only those larvæ which are nearly or fully grown. Riverine larvæ are the most difficult to rear, as they need running water in which to develop. Mention has already been made of their cannibalistic propensities, so that it is very necessary to keep them apart, or the numbers will be found to diminish with alarming rapidity. For the most part they feed on their own kind, but tadpoles, small fishes, and the larvæ of may-flies or caddis-flies form a suitable diet.

As regards *the Imagines,* they must be netted in the same way as butterflies, and no special remarks need be made on this method. It is, however, to be noted that the eyesight of a dragonfly is remarkably acute, so that a stroke made head-on will very frequently miss its object, whereas a quick following stroke will usually secure the insect. When settled, it is best to strike from below, as their sight is less acute from that angle.

For the guidance of the collector the following habitats may be specially mentioned :—

Rocky beds of mountain streams, especially those winding through heavy jungle, since these afford the only open spaces in which they can sun themselves. Here some species will be found patrolling up and down on a very limited beat ; others work over a greater distance. Some species will be found resting on rocks in mid-stream, or perched on herbage beside the stream. Others, particularly the small Protostictas and Platystictas, will be found lurking in dark hollows amongst boulders in the bed or sides of the streams ; from such situations it will be necessary to flush them out by gently inserting the net into their supposed hiding-places.

Weedy banks of tanks, ponds or lakes, or in the shallows of drying streams. Here the smaller species are to be found, and, on account of their minute size and slow movements, they must be sought for carefully ; many species may be taken by sweeping the net amongst rank herbage in such situations.

Hovering over tanks or roads. One will need a fund of patience to net such species, but sooner or later they may come within striking-distance. I have employed dust-shot from a ·22 Winchester rifle to bring down the larger species when flying at an altitude out of reach of the net, and very little damage has been done to specimens which could only have been secured in this way.

In the dark undergrowth of jungles. Many of the smaller genera, such as *Copera*, *Platycnemis*, and *Mortonagrion*, are only found in such habitats. With careful beating they will be noticed making their way through bushes or dense herbage, and may be persuaded to come out into the open by shepherding them with the handle of the net.

Nearly all these species possess what are known as *recognition marks*, i. e., a pale-coloured spot on the terminal segments of the abdomen, apparently intended as a guide for the opposite sex. These minute spots of light are seen moving stealthily about in the undergrowth, and are just as useful a guide to the collector.

By careful beating, night-flying or crepuscular dragonflies may also be put up in deep jungle, especially in the neighbourhood of streams or marshes adjoining forest.

Lastly, many species prefer open country, and will be found resting on bare patches of ground in scrub country or jungle.

Specimens may be *killed* in an ordinary cyanide bottle, or put alive into triangular paper packets or ordinary envelopes and killed later on when they have got rid of the contents of the gut. This latter is of importance for the preservation of specimens, more particularly of their colours. In tropical countries these insects are particularly prone to decompose and drop to bits, as well as to lose their colours, so that early steps should be taken to preserve them. If put alive into packets, these will be found to be full of excreta at the end of the day. After killing the insect, the abdomen should be split up from the thorax to the end along the ventral surface, and the gut removed with a pair of small tweezers. With a little practice this manœuvre is easily accomplished in the larger species, but is a difficult procedure with the smallest, and may be neglected in these. After clearing out the gut, a pig's bristle should be run through the body by inserting it through the underside of the thorax between the second pair of legs, at which point it is cut cleanly off so as not to show. This strengthens the dried insect and prevents it breaking to pieces later, especially when relaxing. The smaller species especially must be bristled, as this not only strengthens them, but clears the abdomen of its contents and helps to preserve them. (Pig's bristles may be had from the Cawnpore tanneries in India.) After cleaning and bristling, rapid drying is of great service for preserving the natural colours, and for this purpose a drying-tin for cigars or cheroots is very useful and effective.

Specimens may now be *mounted* in store-boxes or cabinets, after having been set out and dried on appropriate setting-boards, after the manner of setting butterflies. Dragonflies may, however, be kept permanently in paper packets, a method which is convenient for examination, as, unlike butterflies, they stand much handling without damage. For this purpose specimens may be placed in triangular paper packets and these put in envelopes, three or more in each, according to the size of the insect, the name, date, locality, name of collector, and altitude, or any other data, being typed on the front of the envelope. These envelopes are stored in long tins made to accommodate 25, 50 or more, card-index fashion, so that any particular specimen may be found at a moment's notice. This method is particularly useful for a collector subjected to frequent transfers, as thousands of specimens may be confined in a small space, and suffer no damage during the frequent transits. The smaller tin boxes are nested into a large tin-lined trunk in which some preservative such as camphor or naphthaline is placed at the bottom. A catalogue of the collection is kept on the card-index system ; any other data may be entered on the cards.

Classification.

The Odonata fall naturally into two large suborders :—

A. The Zygoptera.—Dragonflies in which the eyes are well separated, and in which the fore- and hind-wings are approximately of the same shape and breadth.

B. The Anisoptera.—Dragonflies in which the eyes are usually confluent across the middle line or at the most very slightly separated, and in which the fore- and hind-wings are of variable shape and the hind usually considerably broader at the base than the fore.

To these some authors have added a third suborder, the Anisozygoptera, known from only a single species and a larva, of which latter the imago has not yet been discovered. Since, however, the characters of the only known species conform closely to the Zygoptera, it seems more convenient to include it in this suborder ; it is to be regarded as a link connecting the two suborders.

Several different schemes of classification have been adopted by specialists in this order, most recognizing five families. The suborder Zygoptera comprises only two of these families—the Cœnagriidæ, dealt with in this volume, and the Agriidæ.

So far as the Indian fauna is concerned, the Cœnagriidæ are further subdivided into seven subfamilies and the Agriidæ into five. Of the latter, two, viz., Philoganginæ and Caliphinæ, contain species which it has been found impossible to fit into other subfamilies on account of the obscurity of their relationships with other genera.

Morphologically the Dragonflies show some relationship to the Stone-flies (PERLARIA), and to the May-flies (PLECTOPTERA), but the relationship is more apparent than real, since when we come to study their life-history we find that they occupy a very isolated position.

In their life-history there are three stages—the egg, the

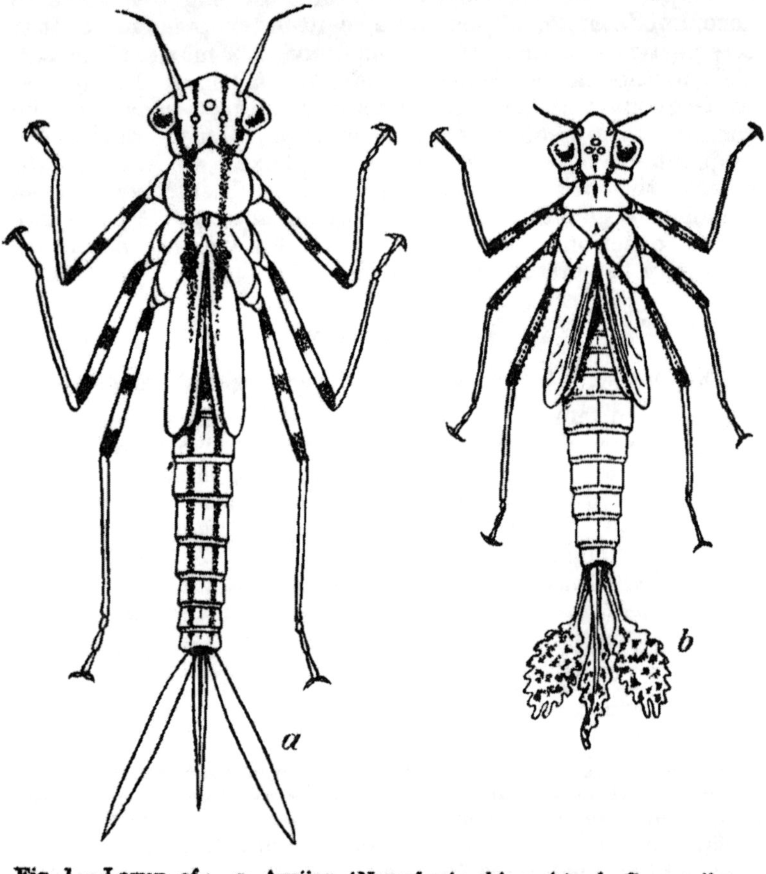

Fig. 1.—Larvæ of :—*a.* Agriine (*Neurobasis chinensis*) ; *b.* Cœnagriine (*Copera marginipes*).

larva, and the imago or perfect insect. Unlike the Butterflies, they have no pupal or resting stage, but the change from larval to imaginal life is direct, although with just as great a change in the structure at metamorphosis. They differ also from the Butterflies by being amphibiotic, that is, part of their life, the larval stage, is passed in a watery environment in which they breathe by means of tracheal gills; the second and shorter part is aerial, breathing then being carried on by means of thoracic and abdominal spiracles.

(1) *The Egg.*—This is round or fusiform, and pale yellow in colour. In the ZYGOPTERA the eggs are invariably deposited in the tissues of plants or in floating debris by means of a robust ovipositor; usually submerged or partially submerged objects are chosen, but quite a number of species

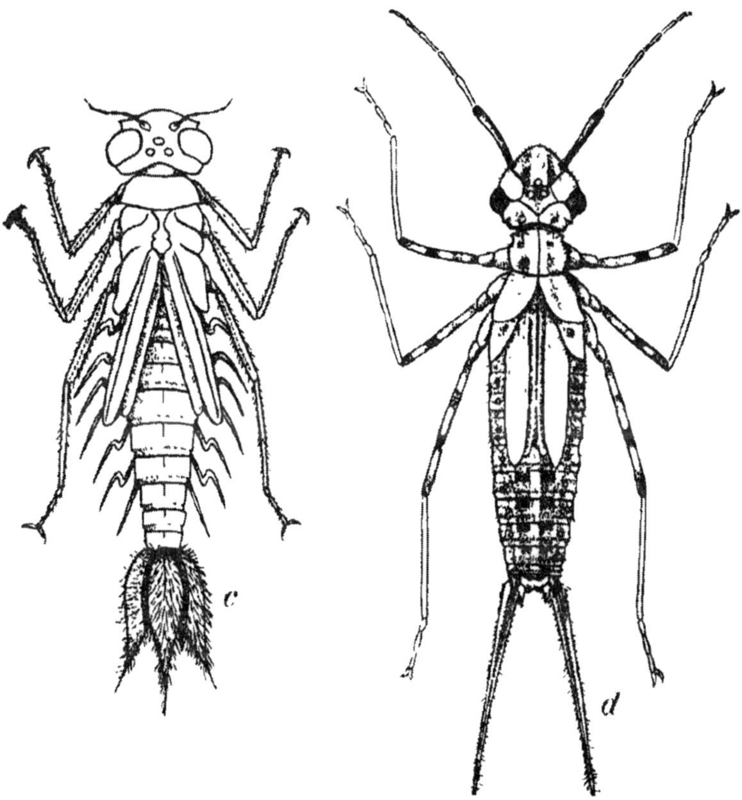

Fig. 1.—Larvæ of :—*c*. Epallaginine (*Indophæa fraseri*); *d*. Libellaginine (*Libellago lineata indica*).

oviposit in the tissues of plants, reeds, or stems of alders, etc., overhanging water, and sometimes at a height of several feet from the water's surface. In the ANISOPTERA the eggs are usually dropped straight into the water: a mass of eggs is exuded and stroked off by the insect rapidly skimming the water's surface, dipping its abdomen at intervals as each batch of eggs becomes ready for casting off. There are, however, exceptions, for we find some which oviposit in a similar manner to the ZYGOPTERA, whilst others deposit their eggs in the soil of gulleys, from which they are washed down later on by a spate of rain; a third, and more archaic, group deposit a mass of eggs on foliage overhanging the water, from whence the larvæ drop as soon as they hatch out.

(2) *The Larva.*—The egg hatches out as a tiny larvule enclosed in a thin membranous envelope which is to protect it until it finds its watery environment, at least in the case of those which hatch out of water. This envelope is soon cast off, and the free-swimming larva or *nymph* makes its appearance. The larvæ are of varied form, and are described below in more detail under their various genera or species, but they all agree in possessing a highly specialized labium. This organ lies flush beneath the head and thorax, and

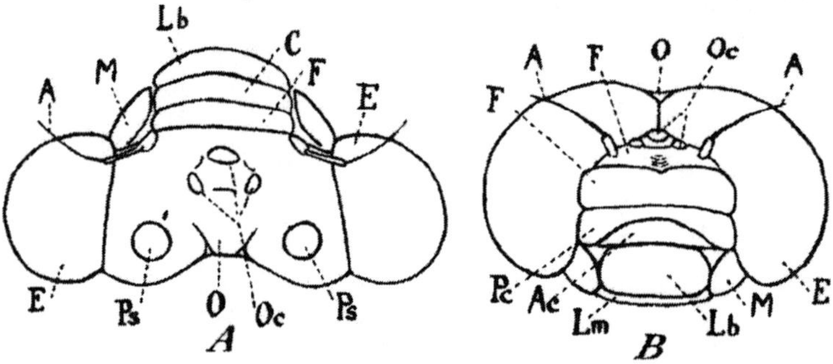

Fig. 2.—A. Head of a Zygopterous dragonfly seen from above. B. Head of an Anisopterous dragonfly seen from the front.

A, antennæ; *Ac*, anteclypeus; *C*, clypeus; *E*, eye; *F*, frons; *Lb*, labrum; *Lm*, labium; *M*, bases of mandibles; *O*, occiput *Oc*, ocelli; *Pc*, postclypeus; *Ps*, postocular spots (these latter not always present).

possesses, between the mentum and submentum, a hinge at which it is folded in two; it can be extended at will by opening out. It is known as the *mask*, and is employed in seizing its prey by means of the powerful hooks and jaws situated at its apex. The mask is variable in shape, and bears importantly on the question of classification. In the ZYGOPTERA the larvæ are usually elongate, cylindrical, and possess lamellar caudal gills to the number of two or three. In some genera (of ZYGOPTERA only) they possess also pairs of so-called abdominal gills, which are, however, employed solely for the purpose of anchoring themselves to the surface of rocks or stones in swiftly flowing torrential streams.

(3) *The Imago.*—The perfect insect is made up of a head, a prothorax, thorax, and abdomen, the thorax carrying a pair of membranous wings and three pairs of legs, the abdomen composed of ten segments, and carrying the genitalia of the male on the 2nd, 9th, and 10th segments, and appendages for seizing the female on the anal end of the 10th.

The Head.—This is composed mainly of two large compound eyes which are either fused broadly across the middle line, as in most ANISOPTERA, or merely meet at a point as in the more primitive ANISOPTERA, or are slightly or widely separated as in the ZYGOPTERA. In addition to the compound eyes, there are three tiny ocelli situated between and in front of the former, or grouped round a small eminence known as the *vesicle*; the antennæ are very short and inconspicuous, filiform, composed of four to seven segments; the mouth-parts formed for biting, the mandibles with strong incisor and molar teeth, the maxillæ with simple unjointed palp

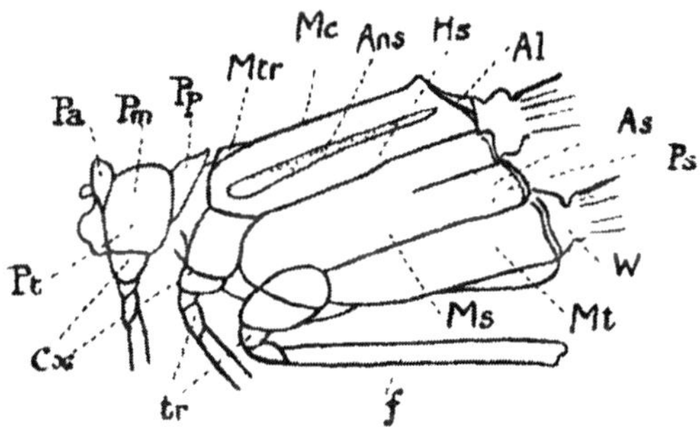

Fig. 3.—Prothorax and thorax of a Zygopterous dragonfly.

Pt, prothorax; *Pa*, anterior lobe of prothorax; *Pm*, middle lobe of prothorax; *Pp*, posterior lobe of prothorax; *Mtr*, mesothoracic triangle; *Mc*, mid-dorsal carinal suture or ridge; *Ans*, antehumeral coloured stripe; *Hs*, humeral suture; *Al*, ante-alar sinus; *As*, antero-lateral suture; *Ps*, postero-lateral suture; *W*, wing-bases; *Ms*, mesepimeron; *Mt*, metepimeron; *Cx*, coxæ; *tr*, trochanter; *f*, femur.

and robustly toothed internal lobe; the labium short and very broad, made up of a median and two lateral lobes, the former cleft in the ZYGOPTERA and more archaic types, the latter smaller than the median lobe in the ZYGOPTERA, larger in the ANISOPTERA and higher types, and made up of two segments, a basal and a large movable hook.

The Thorax.—Composed of a small *prothorax* which forms a neck, and a larger mass formed from the fused meso- and metathorax—the *synthorax*. By means of the prothorax the insect is able to move its head freely in all directions, mainly by the aid of its legs, which are used to push the head into various positions. The prothorax consists of an anterior lobe, collar-like and narrow, a middle lobe, tumid, and forming the greater part of the structure, and lastly, a posterior lobe, important for the purposes of classification, as its shape differs not only in the genera, but also in the species of individual genera ; this latter lobe is often armed with hooks or spines on its hinder border in the female, more rarely in the male also. The synthorax is very oblique, so that the anterior surface comes to be the dorsum, carrying back the wings and thrusting the legs forward beneath the head.

The *legs* are not used for walking, but are formed for clinging and for catching prey. For this latter purpose the legs are

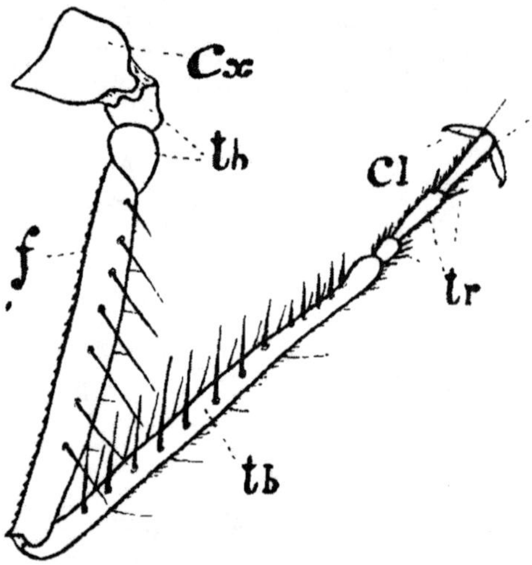

Fig. 4.—Leg of a Zygopterous dragonfly.
Cx, coxa ; *th*, trochanter ; *f*, femur ; *tb*, tibia ; *tr*, tarsus ; *Cl*, claw.

bunched together to form an open basket, the interstices of which are closed in by long spines situated on the femora and tibiæ. Coxæ short and robust ; trochanters slender, formed of two segments ; tibiæ in some genera keeled, in others much dilated ; tarsi made up of three segments, the terminal one ending in a pair of claws, which usually bear a minute spine near the end.

The *wings* during rest are held vertically over the thorax and abdomen as in the Zygoptera, or opened widely and horizontally as in *Philoganga* and the Anisoptera. In some genera, such as *Lestes*, the wings are held half open during

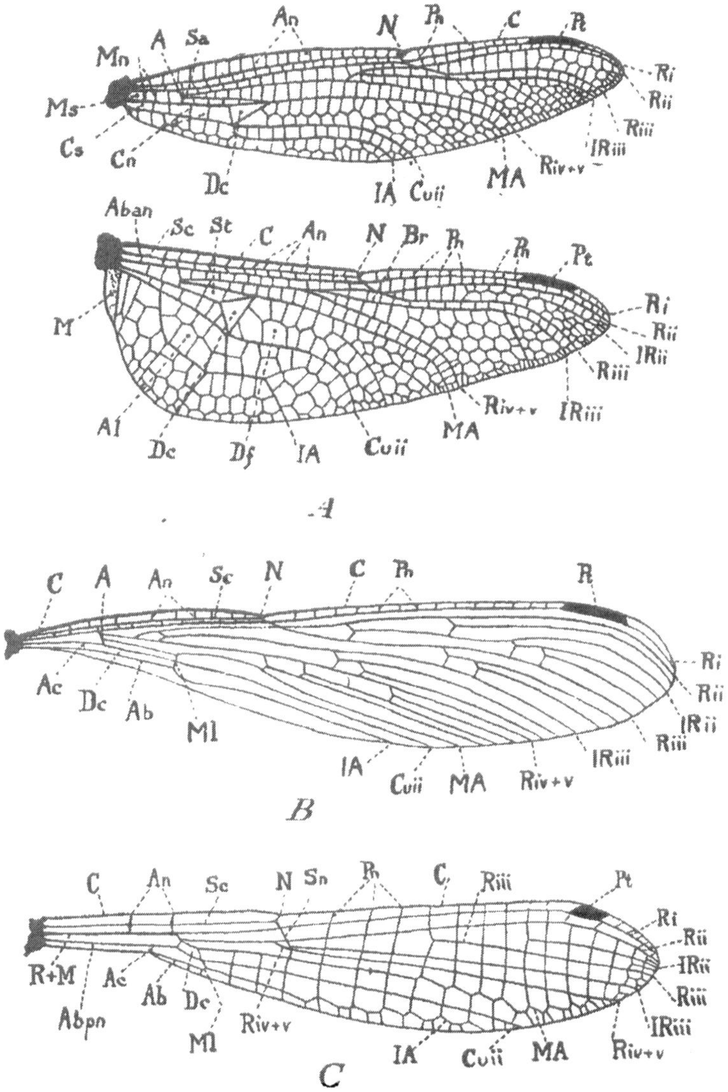

Fig. 5.—A. Fore- and hind-wings of an Anisopterous dragonfly. B. Fore- or hind-wing of a Zygopterous dragonfly of the family Agriidæ. C. The same of the family Cœnagriidæ.

For explanation of the lettering, see pp. 14, 15 and Notation Table; *Pt*, pterotigma. (In fig. B only the main and intercalated nervures have been shown, the short transverse nervure being omitted for the sake of clarity.)

rest. The membrane is hyaline or partly opaque, coloured or uncoloured, without hairs or scales, but has numerous fine spines on the under surface of the supporting ribs or nervures of the wings.

The *venation* is highly specialized. Several systems of notation are in vogue—the old Selysian system, which is more a nomenclature than a notation; the Comstock-Needham, which has largely superseded the former, and is still employed by several specialists in the order; lastly, the revised and more modern Tillyard notation, which seeks to interpret the homologies of the veins and bring their notation into line with that of other orders of insects. This latter is employed in these volumes; but, to avoid confusion, is compared in tabular form with the other two systems.

The classification of dragonflies has been largely built up on the variations of structure of the wings, and it is thus necessary to recognize certain parts and nervures (veins or ribs) in these organs. The important parts of the wings are :—

The *Costa* (*C*).—Anterior border of wing from base to apex.

The *Node* (*N*).—An artificial joint in the costa situated variably at or proximal to its middle.

The *Pterostigma*.—A thickened, chitinized, variably shaped cell situated on the costal margin near the apices of the wings. It may be absent in one or other wing, or in one of the sexes, or may differ in shape in fore- and hind-wings in the male, and lastly, may be braced, or not, at its proximal end.

The *Base*.—The side nearest the body of the insect. The wings are an elongated triangle, the third side being known as the hind or posterior margin.

Spaces.—Certain areas of the wing near the base have received names, and are noted in the diagram (fig. 5); these are important for classification.

Nervures.—These are the ribs or scaffolding on which the membrane is built up, and comprise the costa, subcosta, a many-branched radius, median, cubital, and anal nervures.

Discoidal Cell (*Dc*).—Near the base of each wing will be found a triangular or quadrilateral cell which acts as a keystone from which other nervures appear to radiate. This cell is triangular in the ANISOPTERA, and made up of two cells known as the *trigone* or *triangle* and the *hypertrigone* or *hypertriangle*, but is a single quadrilateral cell in the ZYGOPTERA, where it is known as the *quadrilateral*. It is here called the *discoidal cell*, as it functions similarly to that space found in the wings of Lepidoptera.

Subnode (*Sn*).—The oblique nervure descending from the node.

Antenodal Nervures (*An*) (antecubital nervures) and *Postnodal Nervures* (*Pn*) (postcubital nervures).—These are the nervures lying respectively between the costa and subcosta, and costa and radius on the proximal and distal sides of the node.

Median Space (*Ms*).—Sometimes called the *basal space*. It lies proximal to the arc, between it and the base of wing, and may be traversed by nervures or entire, a distinction of generic value in the ZYGOPTERA.

Submedian or Cubital Space (*Cs*).—Lies beneath and parallel to the median space between the base of wing and discoidal cell.

Anal Bridge (*Ab*).—A short nervure running parallel to the hinder border of the wing and passing below the discoidal cell, variable in the genera of the ZYGOPTERA, complete or incomplete, usually arising from the hinder border of the wing at, or proximal to, the level of the nervure *Ac*, and extending out to the medio-anal link or under side of discoidal cell.

Anal Crossing (*Ac*).—A short nervure situated at the base of the wing extending from *Cuii+IA* to the anal bridge or hinder border of the wing. Where the neuration is at all dense, this nervure will be indistinguishable from the general neuration.

NOTATION TABLE.

	Tillyard's.	Comstock-Needham's.	Selysian.
Costa	*C*	*C*	Costal nervure.
Subcosta	*Sc*	*Sc*	Subcostal nervure.
Radius median ...	*R+M*	*R+M*	Median nervure.
Radius	*Ri*	*Ri*	Median nervure.
Arculus vein	*Rs+M*	*M*	Arc.
Radial sector	*Rs*	*Mi–iii*	Superior sector of arc.
Branches and intercalated branches of radius.	*Rii*	*Mi*	Principal sector.
	IRii	*Mia*	Postnodal sector.
	Riii	*Mii*	Nodal sector.
	IRiii	*Rs*(*Ms*)	Subnodal sector.
	Riv+v	*Miii*	Median sector.
Anterior median .	*ṀA*	*Miv*	Inferior sector of arc.
Posterior median .		(Absent in the Odonata.)	
First cubitus	*Cui*	(Absent in the Odonata.)	
Second cubitus ...	*Cuii*	*Cui*	Superior sector of triangle.
First anal	*IA*	*Cuii*	Inferior sector of triangle.
Cubito-anal	*Cuii+IA*	*Cu*	Submedian nervure.
Second anal	*A′*	*A*	Postcostal nervure.

The Abdomen.—In the majority of the ZYGOPTERA the abdomen is elongate and cylindrical, and varies far less than it does in the other suborder, where it may be similar, or short and broad, flattened or tumid in parts, constricted or fusiform, etc. It consists invariably of ten segments, the anal appendages being attached to the last.

Genitalia.—Male: situated on the ventral surface of segment 2 and comprising the *lamina*, a hood-like structure closing in the genital sac nearest the thorax; the *hamules*, or hooks used for copulating with the female, an anterior and posterior pair being usually present, infinite in shape and variety; the *penis*, an intricate organ varying greatly in the families, genera, and even in the species of individual genera, in the ZYGOPTERA usually taking the form of a chitinized stem with a membranous apex which is bent strongly or folded back over the stem and furnished, or not, with long and curling branches; the *lobe*, a small flask-shaped receptacle at the apical border of the 2nd segment, in which the apex of the penis is tucked away when not in use, and which is also employed as a receptacle for the sperm.

The *primitive penis* is absent, but there is a gonopore at the junction of the 9th and 10th segments, from which the spermatozoa are at first passed into the lobe on segment 2 by the insect curling its abdomen underneath until the two segments are in juxtaposition. Copulation then takes place with the female by means of the functional penis.

Female: the organs consist, in all the ZYGOPTERA and some of the ANISOPTERA, of a robust *ovipositor* on the under side of segment 9, enclosed by a pair of *vulvar scales* or plates furnished with *styles* which are probably tactile organs.

Anal Appendages.—The ANISOPTERA possess a pair of superior and a single inferior anal appendage, but the ZYGOPTERA have in the male sex both paired superior and inferior appendages. These organs are employed in seizing the female by the head or prothorax prior to copulation, and are of infinitely varied design, length, and shape, furnishing a fascinating study in themselves to students of the order; whilst often of generic value, they are more useful as a specific character.

Suborder ZYGOPTERA.

Head transversely elongate; eyes more or less widely separated, usually more so than in the genus *Gomphus*; vesicle usually absent, ocelli lying free in middle of vertex; labium with middle and lateral lobes approximately equal, the mid-lobe deeply fissured; wings usually identical in shape, the hind dilated occasionally at their middles, never at the base; trigones represented by a four-sided irregular structure (hereafter called the *discoidal cell*), which may be squared, or obtusely or acutely pointed outwardly; pterostigma usually present, occasionally absent in one or all wings, variably shaped; antenodal nervures numerous or reduced to the two primitive pairs; wings usually held closed together when at rest; a pair of superior anal appendages at the end of 10th abdominal segment, and a pair of inferior anal appendages at the sides of anal orifice; female with well developed ovipositor somewhat similar to that of *Gynacantha*.

Larvæ variably shaped, possessing caudal gills, two to three in number, which show a great variety in shape and size, spine-like, vesiculated or paddle-shaped; breeding in still or running water, more often in the latter.

The Zygoptera are divided up into two large families distinguished by the number of antenodal nervures, as follows:—

1. Only two antenodal nervures: discoidal cell usually entire; arc midway between base of wing and node, or nearer node than base; usually small insects ... **Cœnagriidæ.**

2. More than two antenodal nervures, never less than five; discoidal cell usually traversed, but sometimes entire; arc nearer base of wing than node; larger insects .... **Agriidæ.**

Family CŒNAGRIIDÆ.

Dragonflies of small or medium size, with characters of the suborder ZYGOPTERA, of which they form by far the largest division.

Wings very narrow and of equal size and shape, the base of hind not more dilated than the fore, all four markedly petiolate; venation very simple, especially at base of wings, only two antenodal nervures, discoidal cell entire, elongate, squared or acutely pointed distally, often differing slightly in shape in the fore- and hind-wings, arc situated about midway between base of wing and node or nearer node than base. Abdomen long and attenuated.

Larvæ elongate, cylindrical, breathing by means of caudal gills, all active swimmers, living usually in weed. Breeding in still and running water.

Distribution.—Cosmopolitan. The Indian fauna is a very rich one, and belongs to seven subfamilies, of which the CŒNAGRIINÆ form the greatest bulk.

Key to the Subfamilies of the Cœnagriidæ.

1.	The nervure *IA* absent and *Cuii* markedly reduced	*PLATYSTICTINÆ*, p. 99.
	The nervure *IA* present, *Cuii* of normal length or reduced	2.
2.	No intercalated sectors at apical end of wing	3.
	A variable number of intercalated sectors at apical end of wing	4.
3.	Cells of wings mostly four-sided; discoidal cell with costal side about one-fifth shorter than lower side and with outer angle subacute	*PLATYCNEMINÆ*, [p. 150.
	Cells of wings mostly four-sided; discoidal cell with costal side of the same length as lower and with outer end of cell squared.	*PROTONEURINÆ*, [p. 209.
	Cells of wings mostly five-sided; discoidal cell with costal side much shorter than lower, and with outer end acutely pointed.	*CŒNAGRIINÆ*, p. 272.
4.	*Riv+v* and *IRiii* taking origin closer to node than arc; oblique vein absent; distal end of discoidal cell subacute or blunt	*MEGAPODAGRIINÆ*, [p. 88.
	Riv+v and *IRiii* taking origin closer to arc than node; an oblique vein present between *Riii* and *IRiii* about half-way between node and pterostigma; distal end of discoidal cell very acute.	5.

5.	*Cuii* at its origin from discoidal cell strongly arched towards the costa; moderately large species	SYNLESTINÆ, p. 19.
	Cuii at its origin only slightly arched towards the costa; rather small species	LESTINÆ, p. 27.

Subfamily SYNLESTINÆ Tillyard (1917).

Synlestinæ Tillyard, Biology of Dragonflies, p. 177 (1917); Laid., Rec. Ind. Mus. vol. xix, p. 187 (1920); Fras., J. Bombay Nat. Hist. Soc. vol. xxxiii (1929).

Dragonflies of medium size, considerably larger than the average Cœnagrionid, usually with green metallic head and body; resting with wings wide spread. Wings with a long petiole, hyaline or barred with blackish-brown, discoidal cell elongate, narrow, acutely angulated distally; sectors of arc arising just a little above middle of arc; *Riii* arising from *Rii* far distal to node; *Riv+v* arising from *Rs* at or a little proximal to the subnode; *Cuii* arched strongly costalwards at its origin; *Riv+v* and *IRiii*, except in the genus *Megalestes*, arising closer to node than to arc and closely convergent at their distal ends; an oblique vein present or absent between *Riii* and *IRiii* at about midway between node and pterostigma; pterostigma variable, usually elongate; intercalated sectors at apical end of wings usually present; anal bridge (*ab*) very short, but well developed in the Indian genera; anal field small. Abdomen very long and slim. Superior anal appendages of male forcipate, spined outwardly.

Larva.—Long and slender; pedicel of antennæ much elongated; caudal gills lamellate, lanceolate, oar-shaped; mask long, median lobe deeply and narrowly cleft, lateral lobes without setæ, with short robust movable hook, two apical robust teeth, and the biting border minutely dentate. Legs long and spider-like. Gizzard with a few robust teeth on each fold. Breeding in streams in montane or submontane areas.

Distribution.—Only two genera known from within Indian limits, viz., *Megalestes* Selys and *Orolestes* MacLach. This subfamily is represented in Australasia by *Synlestes*, in Africa by *Chlorolestes*, in Asia by the two Indian genera mentioned above.

Key to the Genera of the Synlestinæ.

1.	Wings stalked to the level of the middle of discoidal cell; *Riv+v* and *IRiii* arising nearer to the node than arc.....	2.
	Wings stalked only to the level of *ac* or a little proximal to it; *Riv+IRiii* arising nearer to the arc than node ...	3.

2.	Nervures at apex of wings curved strongly downwards towards posterior border of wing; wings of male usually barred black; African species only	CHLOROLESTES Selys.
	Nervures at apex of wings not curved strongly downwards towards posterior border of wing; wings of male hyaline; Australian species	SYNLESTES Selys.
3.	Body non-metallic; wings stalked to the level of *ac*; wings of male usually partly coloured; S. Asiatic species only.	OROLESTES MacLach., [p. 25.
	Body metallic green; wings stalked to slightly proximal to the level of *ac*; wings never coloured; S. Asiatic species only	MEGALESTES Selys, [p. 20.

Genus **MEGALESTES** Selys (1852). (Fig. 6.)

Megalestes Selys, Bull. Acad. Belg. (2) vol. xiii, p. 293 (1862); Kirby, Cat. Odon. p. 159 (1890); Munz, Mem. Amer. Ent. Soc. no. 3, p. 48 (1919); Laid., Rec. Ind. Mus. vol. xix, pp. 186, 187 (1920); Fras., J. Bombay Nat. Hist. Soc. vol. xxxiii, pp. 835, 836 (1929).

Comparatively large Cœnagriids with vivid metallic-green body, marked with pale yellow, and abdomen of great length. Wings long and narrow; pterostigma elongate, nearly three

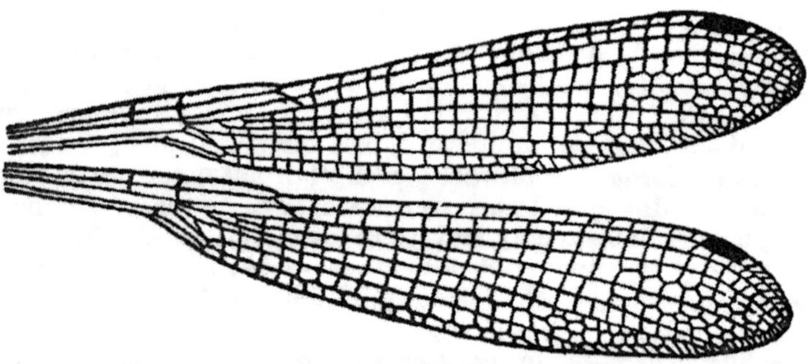

Fig. 6.—Wings of *Megalestes major* Selys, male.

times as long as broad, broader at middle than at either end, braced; postnodal nervures numerous; arc situated at the distal antenodal nervure; postcostal nervure (*ac*) arising much nearer the distal antenodal nervure, and petiolation of wings ceasing some distance before its junction with the anal bridge (*ab*), this point slightly distal to the level of the mid-point between the two antenodals in the fore-wing, and well distal to that point in the hind; discoidal cells very acutely pointed distally, the costal side not quite double the length of the basal in the fore-wing, and slightly more than double in the hind; *IRiii* and *Riv+v* arising much closer to arc than node, and running parallel to each other for some distance; oblique vein usually present between *Riii* and *IRiii*; *Riii* arising from *Rii* much nearer node than pterostigma; 3 or 4 intercalated sectors at the apical ends of wings.

Abdomen very long and slender, much longer than the wings: legs moderately long, hind femora extending to hinder end of thorax and armed with 10 to 12 robust spines; tibiæ with longer, finer spines; tarsal claw-hooks robust, situated at end of claws. Anal superior appendages forcipate; inferior rounded. Female with robust ovipositor, abdomen considerably shorter and stouter than in male.

Distribution.—N.E. INDIA, NEPAL, SIKKIM, BURMA, Indo-China, and S. China.

Genotype, *M. major* Selys.

Key to the Species of Megalestes.

Thorax black beneath; inferior anal appendages with two long spines	*irma* Fras., p. 23.
Thorax pale yellow beneath or pruinosed white; inferior anal appendages with only a single spine	*major* Selys, p. 21.

1. **Megalestes major** Selys. (Figs. 6 & 7.)

Megalestes major Selys, Bull. Acad. Belg. (2) vol. xiii, p. 293 (1862); Kirby, Cat. Odon. p. 159 (1890); Needham, Proc. U.S. Nat. Mus. vol. xxvi, pl. li, fig. 7 (1903); Laid., Rec. Ind. Mus. vol. xix, pp. 148, 185–7 (1920); Fras., J. Bombay Nat. Hist. Soc. vol. xxxiii, pp. 836, 837, fig. 1 (1929).

Male.—Abdomen 48–54 mm. Hind-wing 33–36 mm.

Head: labium yellow; labrum brilliant metallic emerald-green; bases of mandibles bright citron-yellow; anteclypeus black; postclypeus, frons, and vertex glossy metallic emerald-green, with a royal blue reflex when viewed from certain angles; occiput and whole area behind ocelli and eyes dull metallic green with a coppery reflex; eyes dark brown; antennæ black. *Prothorax* dull metallic emerald-green, pruinosed white in adults at sides and mid-dorsum; posterior lobe dull black, with a medial spot of yellow in tenerals. *Thorax* brilliant glossy metallic emerald-green on dorsum, antealar sinus, and laterally as far as the postero-lateral suture; mid-dorsal suture and lateral suture finely black, as also a narrow stripe along the anterior border of metepimeron, posterior to which the rest of the metepimeron bright yellow; the lower portion of mid-lateral suture also narrowly yellow, this colour broadening out below over trochanter; beneath entirely light yellow, with the lower part of sides and whole of under surface, in adults, pruinosed white so as to obscure the ground-colour.

Legs bright yellow as seen from behind, dull black or enfumed when viewed from the front; spines and tarsi black. *Wings* hyaline, uncoloured or palely enfumed, especially at apices; pterostigma variably yellow to brown in tenerals, black in adults, covering 2½ to 3½ cells, strongly braced; 16 to 21 postnodal nervures in fore-wings, 13 to 16 in the hind; other items of venation as for genus. *Abdomen* dull metallic green on dorsum and sides, more brilliantly metallic on segments 1 and 2; the sides of segments 1 and 2 and base of 3 pale yellow; intersegmental joints narrowly black as well as beneath; dorsum of segments 9 and 10 pruinosed white in adults. *Anal appendages* black; superiors slightly longer than segment 10, forcipate, the apices meeting or even over-

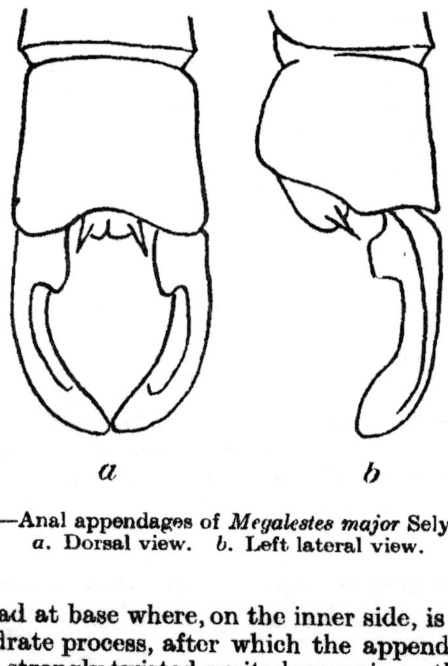

Fig. 7.—Anal appendages of *Megalestes major* Selys, male.
a. Dorsal view. *b*. Left lateral view.

lapping, broad at base where, on the inner side, is seen a short, robust, quadrate process, after which the appendix is slightly tapered and strongly twisted on its long axis, so that the outer side comes to look rather upward; a robust spine on the inner side at junction of apical and medial thirds; inferior appendages rudimentary, rounded, and with a moderately robust short spine directed upwards and analwards.

Female.—Abdomen 45–47 mm. Hind-wing 35–36 mm.

Anal appendages moderately short, about three-fourths the length of segment 10 in Assam material, and about half the length of that segment in a specimen from Dehra Dun, shortly conical, pointed, widely separated, black. Ovipositor dark blackish-brown, with a spot of yellow variably present laterally, very robust, extending to end of abdomen.

Distribution.—ASSAM, BENGAL, SIKKIM, NEPAL, and the N. PUNJAB Hills up to an altitude of about 8,000 ft. The specimens reported by Martin from Tonkin are a distinct race, or even species, and not *M. major*. Dehra Dun, 26. ix.; Shillong and Mawphlang, Assam, 10. x., Mangpu, Darjeeling District, 3,000 ft., 18. v., and 12. vii.; Muktesar, Kumaon Hills, 7,000 ft., 9. v.; Kakani, Nepal, 7,000 ft., 18. vii.

I found the larva, which has been described by Dr. Laidlaw (*l. c.*), in a sandy and gravelly stream at Ghum, near Darjeeling, 8000 ft., in the month of May, the water then being icy cold at that elevation, but the larva, nevertheless, full-grown. Lower down the hill slopes, at about 6000 ft., in a scrub jungle and hillside marsh, I found tenerals emerging from small brooklets which formed a network throughout the marsh.

Mr. T. Bainbrigge Fletcher has sent me the following notes on the occurrence of this species in Shillong:—"Moderately common amongst bushes below Kenilworth, 5000 ft., during the second half of August, 2 or 3 seen daily. Always solitary, although 2 or more may be seen in one locality, generally 100 yards up and down stream and about 20 yards from the stream, never more, and on bushes. Slow unsustained flight, settling with wings half expanded. None seen *in cop.* Specimens seen right up to the 20th October." Similar habits were observed by myself near Darjeeling, tenerals hiding up in dark jungle in company with *Caliphæa confusa*.

2. **Megalestes irma** Fraser. (Fig. 8.)

Megalestes irma Fras., J. Darjeeling Nat. Hist. Soc. pp. 32, 38 (1926); id., J. Bombay Nat. Hist. Soc. vol. xxxii, pp. 837, 838 (1929).

Male.—Abdomen 60–65 mm. Hind-wing 36–37 mm.

Head: labium yellow; labrum glossy metallic emerald-green; bases of mandibles bright yellow, the upper apical half glossy black; anteclypeus yellow, marked with two submedial blackish spots inclining to metallic; postclypeus, frons, and vertex glossy metallic emerald-green, with a peacock-blue reflex in certain lights; rest of head dull metallic green; eyes brown; antennæ black. *Prothorax* dull black, with two large rounded bosses and the mid-dorsum dull metallic green, and the whole of the posterior lobe except the extreme base, which is black, bright yellow; the lower part of the sides also yellow. *Thorax* brilliant emerald-green on dorsum and sides as far back as the hinder suture, the sutures finely black, a narrow medial stripe on each side of the mesepimeron bright chrome-yellow, as also the whole of metepimeron except a narrow irregular stripe of metallic green along its anterior border. Beneath black, pruinosed white. *Legs* wholly black except the trochanters and coxæ of the anterior pair, which are yellow. *Wings* hyaline, uncoloured ; pterostigma reddish-brown, strongly braced, covering 2 to 3 cells ; postnodal nervures to fore-wings 18–24, hind-wings 17–18 ; *ac* and *ab* in the hind-wing meeting at hinder border of wing, petiolation ending, however, before *ac* in the fore-wings. *Abdomen* much longer than in *M. major* and nearly twice the length of wings in some specimens, dull black, with a dull metallic green or coppery reflex ; segments 1 and 2 brighter glossier metallic green and rather broadly yellow on lower part of sides ; segments 9 and 10 pruinosed white on dorsum.

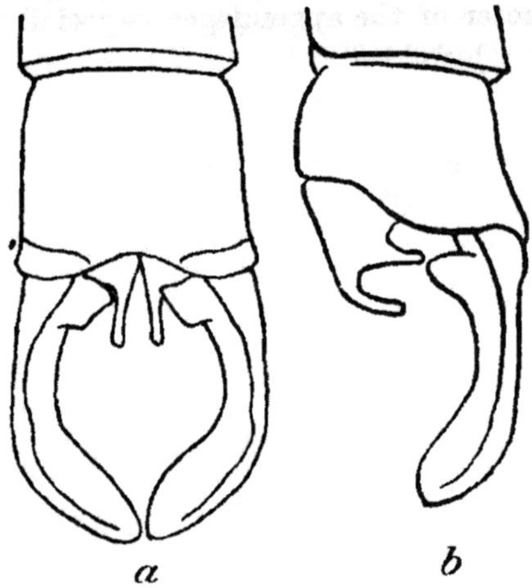

Fig. 8.—Anal appendages of *Megalestes irma* Fraser, male. *a*. Dorsal view. *b*. Left lateral view.

Anal appendages black, differing markedly from those of *M. major* ; superiors as long as segment 10, forcipate, strongly curved in to meet at apices, outer borders unspined ; no vestige of an inner medial spine as in *M. major* ; base thickened and with an inner blunt but prominent tongue-like spine, somewhat similar to that seen in *M. major*, but longer, curled inwards and downwards, and tipped with yellow ; inferiors differing still more, short, rounded, tumid masses with two strongly imbricated spines, one, short and robust, on the outer side, shaped like a rose-thorn ; the second, springing from the apex below, very long and curling strongly upwards, like the horn of a goat.

Female.—Abdomen 51 mm. Hind-wing 37 mm.

Similar to the male, but with shorter and more robust

abdomen. Yellow on abdomen more extensive, obscure basal rings on segments 3 to 6, and a ventro-lateral stripe on segments 8 and 9. *Anal appendages* shortly conical, black, pointed. Ovipositor robust, extending to end of abdomen, blackish-brown.

Distribution.—SIKKIM. *Type,* male and female, in the Darjeeling collection; paratype males in the Fraser collection, one of which goes to the British Museum. All specimens collected by Mr. Chas. M. Inglis at Gangtok, 5,000 to 6,000 ft., Sikkim, 29. v. 24, near the source of the Rani Khole River.

Differs in many respects from *M. major*, viz., by the black spot on bases of mandibles; by the great length of abdomen; by the greater number of postnodal nervures; by the black legs and black under surface of thorax (which latter forms an easy method for identification); and lastly, by the entirely different character of the appendages, especially the inferiors, which bear two spines instead of only one.

Genus **OROLESTES** MacLachlan (1895).

Orolestes MacLach., Ann. Mag. Nat. Hist. (6) vol. xvi, p. 21 (1895); Needham, Ent. News, vol. xxii, pp. 342–4, pl. xi, figs. 1–4 (1911); Munz, Mem. Amer. Ent. Soc. no. 3, p. 48 (1919); Laid., Rec. Ind. Mus. vol. xix, pp. 146, 148, 186 (1920); Fras., J. Bombay Nat. Hist. Soc. vol. xxxiii, p. 838 (1929).

Moderately large Cœnagriids with wings partly coloured opaque black; body metallic or non-metallic, and abdomen of great length. Wings long and narrow, partly opaque blackish-brown; pterostigma elongate, more than four times as long as broad, dilated at the middle, braced; postnodal nervures numerous; arc situated at the level of distal antenodal nervure; postcostal nervure (*ac*) situated much nearer the distal antenodal nervure, and petiolation ceasing at its level, that is, *ac* meets *ab* at the hinder margin of wing; discoidal cell broad, markedly acute distally, its hinder border longer than costal and thrice as long as the base; *IRiii* and *Riv+v* arising much closer to arc than to node, and running parallel to each other for some distance; oblique vein usually present between *Riii* and *IRiii*; *Riii* arising from *Rii* 8 cells after the node in fore-wings and 7 cells after in the hind; only 2 intercalated sectors at apical end of wings, between *Riii* and *Riv+v*. Abdomen slender, long, much longer than the wings; legs moderately long, hind femora extending to hinder end of thorax, spines of moderate length and similar to those of *Megalestes*; anal appendages very long, forcipate, inferiors short, triangular, without spines.

Distribution.—N.E. INDIA, Darjeeling district; Malaysia, Indo-China, and Formosa. (Female unknown.)

Genotype, *O. selysi* MacLach.

3. Orolestes selysi MacLachlan. (Fig. 9.)

Orolestes selysi MacLach., Ann. Mag. Nat. Hist. (6) vol. xvi, pp. 21–3 (1895); Mart., Mission Pavie, p. 18 (1911); Laid., Rec. Ind. Mus. vol. xix, pp. 148–50, 186 (1920); Fras., J. Bombay Nat. Hist. Soc. vol. xxxiii, pp. 838, 839, text-fig. 2 (1929).

Male.—Abdomen 57 mm. Hind-wing 36–39 mm.

Head: labium yellow; labrum olivaceous green, as also the bases of mandibles; rest of head black except the back of occiput and eyes, which are yellowish; eyes brown. *Prothorax* olivaceous green, clouded on the mid-dorsum; posterior lobe simple, rounded. *Thorax* bronzed green on the dorsum as far out as the hinder margin of humeral suture; an ill-

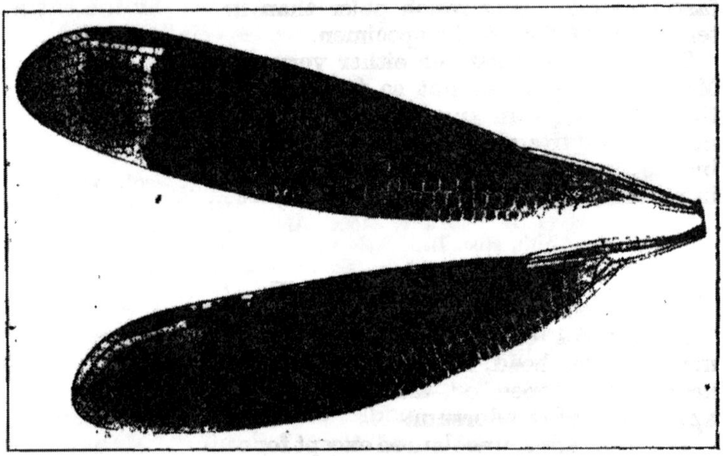

Fig. 9.—Wings of *Orolestes selysi* MacLach., male.

defined stripe pale olivaceous on either side of mid-dorsal carina; the sides pale olivaceous green or yellow, with a trace of the dark bronzed green on upper part of mid-lateral suture. *Legs* black, femora yellowish-brown on flexor surface. *Wings* hyaline at base and apex, blackish-brown between these two points with a steely blue reflex, this broad fascia with an oblique border proximally and a straight, slightly irregular border distally, where it ends about one cell proximal to the pterostigma and is edged narrowly with a milky opalescent white border; the cell-middles in the dark area paler; reticulation black; pterostigma dark brown between thick black nervures, 3·5 mm. in length; 21–24 postnodal nervures in fore-wing, 19–20 in the hind. *Abdomen* bronzed green on dorsum of segments 1 and 2, bronzed black on the remainder, but segments 3 and 4 with a medial bluish area and ringed with black at either end; segments 1 and 2 broadly olivaceous green laterally, narrowly so on segments 3 to 5 and basal half of 6; margins of segments 7 to 9 minutely spined apically; segment 10 slightly notched. *Anal appendages* black, the inferiors tipped with yellow; superiors half as long again as segment 10, slender, regularly curved and forcipate, so that the apices meet or overlap; outer border finely spined along apical half; a robust basal tooth on inner side and a dilatation which extends along inner border nearly to apex and ends in a broad, triangular, acute tooth; apices rounded, blunt; inferiors barely half the length of superiors, broad, triangular, with inner borders straight and closely apposed.

Distribution.—DARJEELING DISTRICT: two males in the MacLachlan collection in rather poor condition, the dark fascia in one being much paler than in the other, from the teneral condition of the specimen.

The species must be either very rare or very local, as Mr. Chas. Inglis has not so far rediscovered it, although all the other Selysian species from that area have fallen to his net. It is quite easily recognized from all other Indian Lestids by its coloured wings, these being hyaline and uncoloured in all others save for some slight or occasional enfumation.

Subfamily LESTINÆ. (Fig. 10.)

Dragonflies of small size, with metallic or non-metallic markings on head, body, and abdomen; resting with wings nearly wide open or, more rarely, with the wings closely apposed over the dorsum.

Wings hyaline, uncoloured except for occasional enfumation; with a long petiole, petiolation begins at the postcostal nervure (*ac*) about midway between the two antenodal nervures, *ac* meeting *ab* at the hinder border of the wing; discoidal cell elongate, narrow, similar or differing in the fore- and hind-wings, very acutely angulated at the distal end; sectors of arc arising from middle of arc; *Riii* arising from *Rii* well distal to the node; *Cuii* but slightly arched or not at all at its origin; *Riv+v* and *IRiii* arising much nearer arc than node and far proximal to the subnode; an oblique vein always present between *Riii* and *IRiii* at about midway between node and pterostigma; pterostigma variable, nearly always at least twice as long as wide, often more so; intercalated sectors usually present at apical end of wings: anal bridge (*ab*) always present and well developed; anal field small.

Abdomen long and slender; superior anal appendages of male forcipate, spined outwardly; inferior appendages variable.

Larva long, slender, abdomen cylindrical: caudal gills long, rounded at apex, paddle-shaped, secondary tracheæ arising

at right angles to the main stems ; mask with the lateral lobe greatly expanded, concave ; deeply and irregularly toothed ; setæ on middle lobe and movable hooks usually to the number of 5 and 6 respectively ; middle lobe simple, not fissured, movable hook of great length ; gizzard with 8 folds, 4 major and 4 minor, the former with a specialized dentition of 3 to 5 very large teeth and numerous smaller ones ; antennæ with elongate pedicel. Breeding in tanks and ponds, rarely, if ever, in running water.

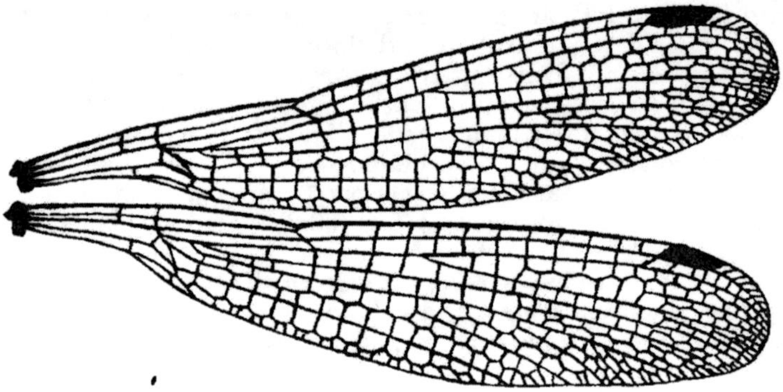

Fig. 10.—Wings of *Lestes præmorsa* Selys, male.

Distribution.—Cosmopolitan. Five genera known from within Indian limits : *Lestes* Leach, *Platylestes* Selys, *Indolestes* Fraser, *Ceylonolestes* Kennedy, and *Sympycna* Charp.

Key to the Genera of the Lestinæ.

1.	Pterostigma nearly as broad as long, nearly squared	[p. 58. PLATYLESTES Selys,
	Pterostigma at least twice as long as broad, usually more	2.
2.	Discoidal cell of hind-wing longer and much narrower than that of fore-wing .	3.
	Discoidal cell of hind-wing shaped similarly to that of fore-wing	LESTES Leach, p. 29.
3.	Posterior lobe of prothorax trilobed, the middle lobe projecting far beyond the laterals ; *ac* almost at the level of the basal antenodal nervure ; palæarctic...	[p. 84. SYMPYCNA Charp.,
	Posterior lobe of prothorax a single simple lobe unifomly arched ; *ac* almost at the level of the distal antenodal nervure ; tropical and montane.......	4.
4.	Ground-colour of body pale brown marked with black ; wings tinted with pale yellow or palest brown..............	[p. 77. INDOLESTES Fras.,
	Ground-colour of body azure blue marked with bronzed black ; wings not enfumed or tinted	[Kennedy, p. 62. CEYLONOLESTES

Genus LESTES Leach (1815).

Lestes Leach, Edinb. Encycl. vol. ix, p. 137 (1815); Steph., Ill. Brit. Ent., Mand. vi, p. 76 (1836); Selys, Mon. Lib. Europ. p. 135 (1840); Ramb., Ins. Névr. p. 243 (1842); Selys, Rev. Odon. p. 146 (1850); id., Bull. Acad. Belg. (2) vol. xiii, p. 295 (1862); Kirby, Cat. Odon. p. 160 (1890); Lucas, British Dragonflies, p. 58 (1900); Tillyard, Biology of Dragonflies, pp. 276, 277 (1917); Laid., Rec. Ind. Mus. vol. xix, pp. 145, 146 (1920); Tillyard, Ins. Aust. & New Zeal. p. 78 (1926); Fras., J. Bombay Nat. Hist. Soc. vol. xxxiii, pp. 840, 841 (1929).
Puella Brulle, Expl. Morée, vol. ii (i), p. 104 (1832).
Anapates Charp., Lib. Europ. p. 18 (1840).

Characters of the subfamily; always resting with the wings nearly fully open; petiolation beginning a little before the level of *ac*, which latter is situated about midway between the level of the two antenodal nervures; pterostigma at least twice as long as broad; discoidal cell similar in fore- and hind-wings, its base two-thirds the length of costal side; *IRiii* only slightly zigzagged; body with or without metallic markings.

Distribution.—Cosmopolitan. Represented in INDIA, BURMA, and CEYLON by several species and races.

Genotype, *Lestes sponsa* Hans.

Key to the Species and Subspecies of Lestes *and* Platylestes.

1.	Thorax without metallic markings	2.
	Thorax with green metallic antehumeral stripes	5.
2.	Pterostigma unicolorous	3.
	Pterostigma bicolorous	4.
3.	Vertex of head pale brown	*L. umbrina* Selys, p. 53.
	Vertex of head mat black	*L. thoracica* Laid., p. 56.
4.	Thorax with about 10 small black spots on each side; pterostigma white at both ends	*P. platystyla* (Ramb.), [p. 59.
	Thorax immaculate; pterostigma with its costal third palest brown, posterior two-thirds dark brown	*L. nodalis* Selys, p. 51.
5.	Thorax with a uniform narrow antehumeral stripe on each side	6.
	Thorax with narrow antehumeral stripes expanded outwardly at the upper ends.	8.
	Thorax with antehumeral stripes deeply scalloped or crenulate on the outer sides	9.
	Thorax with the whole of dorsum and forepart of sides brilliant metallic green	11.
6.	Vertex of head metallic green; thorax largely black; only 11 postnodal nervures to fore-wings	*L. angularis* Fras., [p. 43.
	Vertex of head non-metallic; 12 to 14 postnodal nervures to fore-wings ...	7.

7.	Vertex of head pale brown; dorsum of thorax pale brown, with very prominent brilliant metallic green antehumeral stripes; 12 postnodal nervures to fore-wings	*L. viridula* Ramb., [p. 45.
	Vertex of head mat black; dorsum of thorax black with an obscure narrow metallic green stripe on each side; 14 postnodal nervures to fore-wings..	*L. patricia* Fras., p. 47.
8.	Sides of thorax pale brown with a few black points	*L. elata* Hagen, p. 37.
	Sides of thorax with two thick black bands enclosing a medial narrow brown stripe	*L. malabarica* Fras., [p. 40.
9.	Small species, with abdomen 30–32 mm. and hind-wing 20–21 mm.	10.
	Larger species, with abdomen 36–39 mm. and hind-wing 24–26 mm.	*L. dorothea* Fras., p. 35.
10.	Mesothorax with a long irregular post-humeral stripe and a large lower spot of metallic green	*L. præmorsa sikkima* [Fras., p. 34.
	Mesothorax without metallic or other markings	*L. præmorsa præmorsa* [Selys, p. 30.
11.	Small species, with abdomen about 30 mm. and hind-wing 21 mm.; pterostigma bicolorous; palæarctic..	*L. barbara* Fabr., p. 49.
	Large species with abdomen 52 mm. and hind-wing 38 mm.; pterostigma unicolorous; oriental	*L. orientalis* Hagen, [p. 48.

4. Lestes præmorsa præmorsa Selys. (Figs. 10, 11, & 12.)

Lestes præmorsa Selys, Bull. Acad. Belg. (2) vol. xiii, p. 320 (1862); id., Mitt. Mus. Dresd. vol. iii, p. 317 (1878); id., Ann. Soc. Españ. Hist. Nat. vol. xi, p. 49 (1882); id., Ann. Mus. Civ. Genova, (2) vol. x (xxx), pp. 495, 496 (1891); Kruger, Stett. Ent. Zeit. p. 130 (1898); Laid., Proc. Zool. Soc. Lond. p. 382 (1902); id., Rec. Ind. Mus. vol. xix, pp. 147, 154, & 155 (1920); Fras., Rec. Ind. Mus. vol. xxvi, p. 484 (1924); id., J. Bombay Nat. Hist. Soc. vol. xxxiii, pp. 841–3, text-fig. 3 (1929); id., Rec. Ind. Mus. vol. xxxiii, p. 448 (1931).

Lestes præmorsus Kirby, Cat. Odon. p. 162 (1890).

Lestes decipiens Kirby, J. Linn. Soc. Lond., Zool. vol. xxiv, pp. 565, 566 (1893); Laid., Rec. Ind. Mus. vol. xix, p. 154 (1920).

Lestes præmorsa subsp. *andamanensis* Fras., Rec. Ind. Mus. vol. xxvi, p. 410 (1924).

Male.—Abdomen 32–35 mm. Hind-wing 21–22 mm.

Head: labium yellowish; labrum turquoise-blue; bases of mandibles glossy brown; anteclypeus, cheeks, and a narrow basal streak to postclypeus turquoise-blue; rest of head mat black, pruinosed over lower and back part of eyes; eyes deep sapphire-blue. *Prothorax* black, with obscure yellow markings largely hidden by pruinescence. *Thorax* black on dorsum, marked by a pair of dark green metallic antehumeral stripes, straight on their inner border, deeply crenulate or lobed on their outer (this stripe subject to some variation, but usually with a somewhat quadrate lobe projecting from the upper end of stripe, a medial hooked, with or without a serrate hinder border, and an elongate clubbed lobe formed by an expansion of the lower part of the stripe); laterally blue or pale greenish-yellow, or in others (*decipiens*) black, heavily pruinosed so as to appear violaceous; in non-pruinosed examples several irregular spots present on the sides, one on the upper part of the humeral suture,

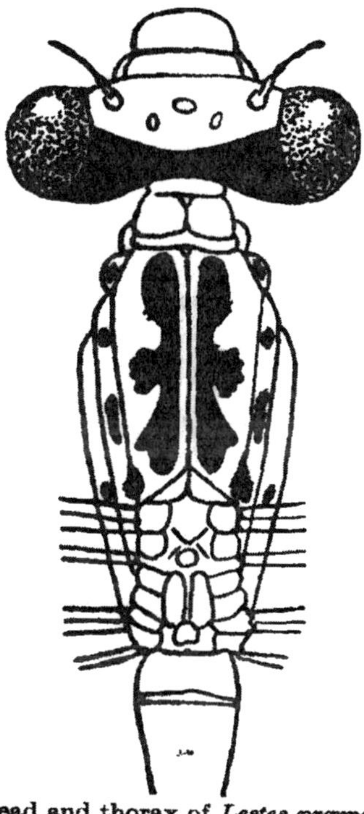

Fig. 11.—Head and thorax of *Lestes præmorsa* Selys, showing markings.

two spots below this on the mesepimeron, one a short distance behind the humeral suture and the second or lower just touching the suture; two spots on middle of mesepimeron, one at upper end of postero-lateral suture, and another pair, rather wide apart, situated on the ventral border; beneath thorax black heavily pruinosed; in melanotic examples the metallic antehumeral stripes and lateral spots are almost or entirely obscured, and the whole overlaid by pruinescence. *Legs* black with a pale bluish-green longitudinal stripe on the outer side of femora and extensor surface of tibiæ. *Wings* hyaline or palely enfumed in adults; pterostigma dark reddish-brown or blackish-brown, covering two cells; 10 to 12 postnodal nervures in all wings; *ac* slightly nearer the basal antenodal in the fore-wing, but about midway between the two antenodals in the hind; *IRiii* situated about 5 to 6 cells proximal to the pterostigma in fore-wing, 8 to 9 in the hind. *Abdomen* pale blue, marked broadly on the dorsum with bronzed green or coppery metallic, changing to mat black on the hinder segments; segment 1 with a broad quadrate spot of bronzed green on dorsum, not extending to apical border in subadults; segment 2 with a broad stripe shaped like an elongate thistle-head, falling short of both ends so as to leave a narrow basal bluish annule and a similar, broadly

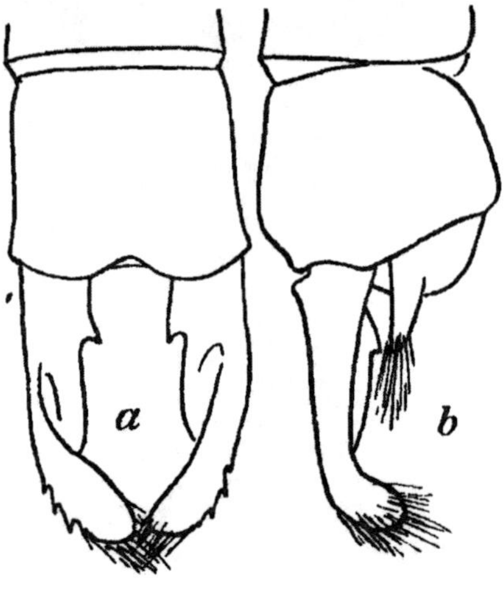

Fig. 12.—Anal appendages of *Lestes præmorsa* Selys, male. *a*. Dorsal view. *b*. Right lateral view.

interrupted one at the apical border; segments 3 to 7 with similar bands, the apical ends abruptly expanded and then contracted before joining narrow apical black rings, and with narrow basal and apical interrupted rings blue; segment 8 with only a vestige of the basal ring; segment 9 with a very large lateral spot of blue; segment 10 with a small ventro-lateral spot of the same colour; segment 8 often with a large apico-lateral blue spot, obscured in some. *Anal appendages*: superiors as long as segment 9, pale yellow or ochreous, with the base and the apex black, forcipate, finely spined on the outer border near apex, which is blunt and curved in strongly; on the inner side near the base a robust spine followed by a thin shell-like expansion inwards which ends in a minute spine ; inferior appendages much shorter, apposed at base, where they are very broad and rounded, then abruptly contracted to form a short, cylindrical, ungulate process tipped with stiff whitish hairs.

Female.—Abdomen 30–32 mm. Hind-wing 20–21 mm.

Very similar to the male; differs as follows :—Labrum and cheeks more olivaceous or greenish-yellow than blue ; antehumeral stripes more distinct and more liable to variation ; sides of *thorax* yellowish or pale greenish-blue, heavily pruinosed beneath and on the lower parts of sides, which are often chalky white ; *legs* ochreous, with the two anterior pairs of femora striped longitudinally with dark brown on the outer side ; sides of *abdomen* greenish-yellow ; segment 1 with better defined markings, a baso-dorsal spot confluent with a small subdorsal spot on either side, the apical area pale and unmarked ; borders and apical ends of segments 9 and 10 narrowly blue. *Anal appendages* creamy white, shortly conical, pointed. Vulvar scale pale, robust, extending to end of abdomen.

Distribution.—The distribution may be said to run from Western India across to Assam and Burma, and so throughout Southern Asia as far as the Philippines.

The *type*, a female from Manila, Philippines, is now in the Selysian collection. The allotype, a male, also in the same collection, was taken at Pwepoli, Burma, in June. I have specimens from many parts of India, Burma, and Ceylon which do not differ markedly from one another nor from the types. After examining a great number, I have come to the conclusion that the species cannot be split up, even into subspecies, beyond what I have already done by separating *dorothea* from it, and by giving subspecific rank to examples from Sikkim, which I call *sikkima*. Owing to the gradual development of melanosis, a large number of varying forms are found ranging from pale brown with distinct markings in the teneral condition to melanotic forms which are quite black and exhibit no markings. Kirby's form, known as *L. decipiens*, is merely one such as the last, very heavily pruinosed. The race *andamanensis* Fras., from Port Blair, is, I now think, not more than a variety, and differs only by segment 10 being without markings, and by the pronounced character of the metallic stripes of the thorax, which have the hinder border of the middle lobe serrate, a character since observed in other examples. These stripes differ rather widely according to whether the body of the stripe is thick or very fine, the lobes being much better developed when the latter happens. As regards the size, which, Selys mentions, differs widely, I think this is due to error in confusing *L. dorothea* with *L. præmorsa*.

5. Lestes præmorsa sikkima Fraser.

Lestes præmorsa sikkima Fras., J. Bombay Nat. Hist. Soc. vol. xxxiii, pp. 843, 844 (1929).

Male.—Abdomen 34 mm. Hind-wing 24 mm.

Very similar to *præmorsa præmorsa*, differing principally by the metallic colouring of the vertex of the head and the metallic posthumeral stripe. *Head*: labium yellowish; labrum, clypeus, and cheeks olivaceous (possibly blue in the living state); back of eyes yellow, with a large black spot below; vertex dark metallic emerald-green except for a small oval spot of warm brown on the outer side of each hinder ocellus. *Prothorax* pale brown, with a triangular emerald-green metallic dorsal spot and a linear subdorsal spot similarly coloured on each side; the whole base of posterior lobe metallic emerald-green. *Thorax* with a green metallic antehumeral narrow stripe with large lateral lobes, the medial one finely crenulate along its border; mesothorax with a posthumeral metallic green stripe made up of an upper spot finely confluent with a long wedge-shaped spot, and a small inferior lower posthumeral spot (some specimens have the upper spot discrete as well as the lower, but this is unusual); beneath thorax pale yellow or white, and without the large triangular black spot behind the legs seen in *L. præmorsa præmorsa*. *Wings* with 11 to 12 postnodal nervures; pterostigma blackish-brown, covering two cells, otherwise details as for *præmorsa præmorsa*. *Abdomen* similar to *præmorsa præmorsa*, but segments 8 to 10 very broadly yellow at the sides (possibly blue during life). *Anal appendages* not markedly different, but the inner expansion of the superiors ending in a more robust spine, and the inferiors with a longer apical process.

Female.—Abdomen 34–36 mm. Hind-wing 26 mm.

Similar in most respects to the male, but a much more robust insect. Anterior femora black except for the extreme base, which is yellow, other femora black on the flexor surface only, and stippled with a chain of small yellow spots. *Wings* with 12 to 13 postnodal nervures; pterostigma black, over two cells. *Abdomen*: segment 1 pale brown, unmarked; segment 2 with a small pale lateral spot; segment 9 with a very large lateral yellowish spot; whilst segment 10 is almost entirely yellow. *Anal appendages* shortly conical, yellow at apices, basal half black. Vulvar scale black broadly marked with yellow.

Distribution.—A number of specimens collected by Mr. Oscar Lindgren at Kurseong, SIKKIM, all agreeing by having the metallic posthumeral stripe, which at once serves to distinguish it from typical *L. præmorsa præmorsa*.

Type in the author's collection; paratypes in the Morton collection.

6. Lestes dorothea Fraser. (Fig. 13.)

Lestes dorothea Fras., Rec. Ind. Mus. vol. xxvi, pp. 484–6 (1924); id., J. Bombay Nat. Hist. Soc. vol. xxxiii, pp. 844, 845 (1929).
Lestes præmorsa dorothea Fras., Rec. Ind. Mus. vol. xxxiii, p. 448 (1931).

Male.—Abdomen 40 mm. Hind-wing 26 mm.

Head: labium whitish-yellow; labrum, cheeks, bases of mandibles and frons turquoise-blue, rest of head mat black; eyes deep sapphire-blue above, turquoise-blue beneath. *Prothorax* mat black. Thorax black, pruinosed white laterally, citron-yellow beneath; marked on the dorsum with metallic green antehumeral stripes shaped like those seen in *L. præmorsa præmorsa*, the area between these and on the

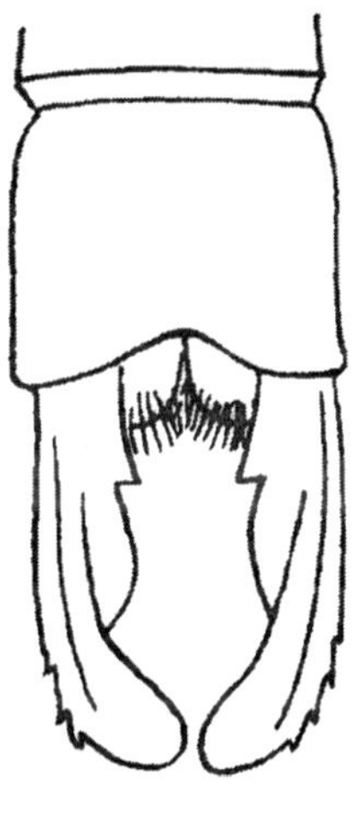

Fig. 13.—Dorsal view of anal appendages of *Lestes dorothea* Fras., male.

outer side paler, followed by a diffuse black stripe on the humeral suture; a large diffuse black spot just in front of the upper part of the postero-lateral suture, another smaller spot at the middle of the antero-lateral suture, and a third spot over the spiracle; tergum and beneath thorax heavily pruinosed, obscuring all markings. In melanotic specimens, especially those from Assam, all the above spots are entirely obscured, and even the metallic stripes are made out with great difficulty. *Legs* greenish-yellow, striped laterally with black; tibial spines long, femoral short, 14 on the hind femora. *Wings* hyaline, sometimes evenly enfumed; pterostigma black, less than two and a half times as long as broad, covers 2 to 3 cells; 15–16 postnodal nervures in fore-wings, 14–15 in the hind. *Abdomen* blue or greenish-blue marked with black, as follows:—Segment 1 broadly black on dorsum, apical border narrowly blue; segment 2 with a broad thistle-shaped mark on dorsum extending from base to apex; segments 3 to 6 with a broad dorsal stripe which expands abruptly near the apical end of the segments, and then contracts as abruptly to become confluent with a narrow apical black ring, whilst the basal end tapers to a fine point which just meets the apical ring of the segment before, leaving a small blue spot on either side; the apical expanded portion forms a complete broad ring round each segment; segment 7 with a similar stripe which falls short of the base of the segment; segment 8 similar, with a fine basal blue ring; segments 9 and 10 entirely black, the latter pruinosed white on dorsum. *Anal appendages* bluish during life, pale yellow after death, broadly black at base and apex, running straight back, the apices turned down and inwards at an angle of about 45°; basal half expanded on the inner side and with a very robust inner basal spine; inferiors much shorter, not visible from above, widely separated, thick at base, with a short ungulate process at apex tipped with whitish hairs.

Female.—Abdomen 36 mm. Hind-wing 27 mm.

Differs rather markedly from the male in its ground-colour and by its broader markings. Eyes olivaccous above and below, bluish behind; *thorax* olivaceous green, pale greenish-yellow laterally; segment 9 with a large medio-lateral blue spot on each side; segment 10 entirely blue save for the mid-dorsal carina, which is finely black. *Anal appendages* small, conical, pointed, black.

Distribution.—I had thought this species to be strictly confined to Coorg, but I now think that the large melanotic specimens from Assam, which I had thought might be a distinct species, are actually a near race of *L. dorothea*, characterized by their slightly smaller dimensions, greater melanosis, and bright yellow thoracic sides. Some of these latter specimens are so black that I am unable to discern any thoracic markings whatever, even the metallic ones being obscured. Younger ones, however, show the markings of *L. dorothea.* It is probably these specimens that Selys mixed up with true *L. præmorsa.* In Coorg it is found on every tank, the pulping-tanks of coffee estates being especially favoured, and here it may be taken usually in company with true *L. præmorsa.* Its much larger size and the absence of markings on segments 8 and 9 in the male, and its higher postnodal index, will serve to distinguish it from *L. præmorsa.*

Type in the British Museum; paratypes in the Indian and Pusa Museums and in many private collections, including the author's.

7. **Lestes elata** Hagen. (Fig. 14.)

Lestes elata Hagen, Verh. Zool.-bot. Ges. Wien, vol. viii, p. 478 (1858); Selys, Bull. Acad. Belg. (2) vol. xiii, p. 319 (1862); Laid., Rec. Ind. Mus. vol. xix, pp. 146, 153, 154, fig. 2 (1920); Fras., ibid. vol. xxvi, p. 484 (1924); Laid., Spolia Zeylan. vol. xii, pp. 47–8, 357 (1924); Fras., Rec. Ind. Mus. vol. xxxiii, p. 448 (1931).

Lestes elatus Kirby, Cat. Odon. p. 162 (1890); id., Proc. Zool. Soc. Lond. p. 203 (1891); id., J. Linn. Soc. Lond., Zool. vol. xxiv, p. 565 (1893).

Lestes elata elata Fras., J. Bombay Nat. Hist. Soc. vol. xxxiii, pp. 845–7, text-fig. 4 *b* (1929).

Male.—Abdomen 34–36 mm. Hind-wing 23–24 mm.

Head: labium white; labrum, bases of mandibles, cheeks, and anteclypeus turquoise-blue; postclypeus olivaceous,

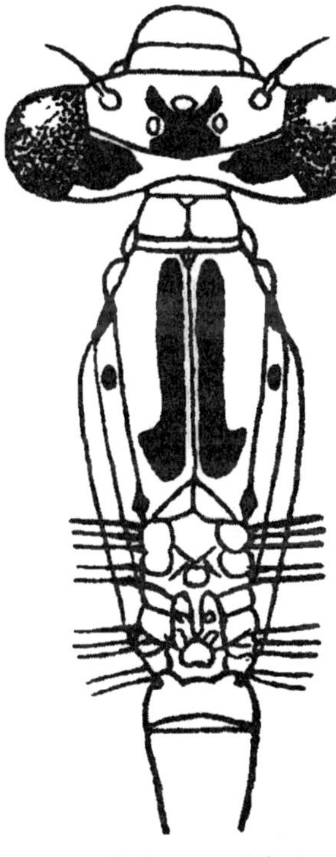

Fig. 14.—Head and thorax of *Lestes elata* Hagen, male, showing markings.

marked at the base on each side with a small triangle of reddish-brown narrowly bordered with black; frons olivaceous green; vertex between the ocelli and bases of antennæ black, bordered outwardly by a broad band of light brown which extends obliquely forwards and outwards as far as the blue cheeks; a large triangular spot of coppery bronze lying to the inner side of each eye above and narrowly edged with black; behind eyes and the upper surface of occiput olivaceous yellow; eyes turquoise-blue. Fully adult specimens have the whole of the upper surface of the head mat black and the bronzed spots obscured, whilst behind the eyes is black, heavily pruinosed white below. *Prothorax* with mid-dorsum olivaceous, with a longitudinal diffuse, narrow, warm brown subdorsal stripe on each side, in the middle of which is an irregular black spot; posterior lobe with a large black spot on each side of the middle line. Adults have the dorsum black and the sides pruinosed white, the markings obscured. *Thorax* in tenerals warm brown on dorsum, changing to pale olivaceous and finally yellow on the sides, or almost white

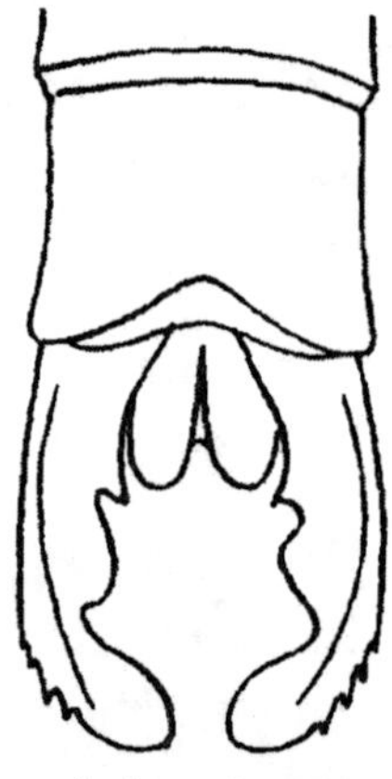

Fig. 15.—Dorsal view of anal appendages of *Lestes elata* Hagen, male.

beneath; in adults dark reddish-brown to almost black, heavily pruinosed at the sides and quite chalky white below and beneath; dorsum marked with a pair of narrow antehumeral metallic green stripes expanded outwardly abruptly above; these stripes very brilliant and prominent in young examples, but becoming gradually obscured by melanosis in adults, although rarely, if ever, quite obscured; on the sides, in tenerals, an upper tiny humeral spot and two or three equally small ones in a line on the lower part of mes- and metepimeron, these spots not sharing in the general pruinescence of adult age, and standing out prominently black against the white pruinescence; in addition, in pruinosed adults a narrow glossy stripe of the ground-colour on the mesepimeron, which does not partake of the surrounding pruinescence. *Legs* palest yellow in tenerals, with a black stripe on the outer side of the femora, one on the flexor surface of the tibiæ, and the whole of the tarsi black. Adults similar, but more pale greenish-yellow. *Wings* hyaline; pterostigma black in adults, pale brown in tenerals, or dark reddish-brown by transmitted light, covering two cells, about twice or two and a half times as long as broad; 11 to 12 postnodals in fore-wings, 9 to 10 in the hind; petiolation begins at *ac*, which is situated about midway between the levels of the two antenodals. *Abdomen* pale yellow at the sides in tenerals, pale bluish-green in adults, marked broadly on dorsum with metallic green or bronze, changing to dull black on the terminal segments; segment 1 unmarked in tenerals, obscurely brownish on dorsum in adults; segments 2 to 6 with a broad metallic stripe on dorsum, which expands abruptly near the apical ends of segments, completely encircling them, and then contracts to become confluent with a narrow black apical ring, whilst at the base it is rounded, and leaves a narrow ring of the ground-colour, which is pale blue in adults; segments 7 to 9 broadly dark brown on dorsum in tenerals, black in adults, and leaving on the latter segment a broad apical ring, yellowish in tenerals, bluish in adults. In very adult specimens all the terminal segments black, unmarked, the apical half of segment 9 and the whole of 10 white with pruinescence and very conspicuous when flying. *Anal appendages* creamy yellow, broadly tipped with black in tenerals, entirely black in adults, the apices tipped with yellow hairs; a robust inner basal tooth, followed by a wide dilatation which extends for half the length of appendage and is finely denticulate; apical half narrow and curved strongly inward, so that apices usually meet; finely spined outwardly; inferiors very short, about one-third the length of superiors, rounded at base, conical.

Female.—Abdomen 34 mm. Hind-wing 24 mm.

Somewhat similar to the male, but always less melanotic, the markings always well defined and the ground-colour pale brown. *Head* similar to the male, but the metallic markings always in evidence and bronzed or coppery in adults. *Thorax* with the following lateral spots (undeveloped for the greater part in tenerals):—A spot on the upper part of each part of each lateral suture, five spots on the mesepimeron, three of which are posthumeral, one elongate, and one situated on the spiracle; two spots on the ventral border of metepimeron, the two pairs showing up very prominently when the thorax is viewed from beneath. *Abdomen* similar to the male, dorsal bands brown, non-metallic in tenerals, dark reddish-brown with coppery reflex in adults; segment 9 with the band arrested short of the apical border and deeply bifid on dorsum; segment 10 and the apical part of 9 bluish, unmarked (yellowish in tenerals). *Anal appendages* creamy white in tenerals, black in adults, shortly conical. Vulvar scale dark brown, very robust, extending to end of abdomen.

Distribution.—Confined to PENINSULAR INDIA and CEYLON. It is the commonest species of the genus found within Indian limits, and may be taken throughout the year in southern districts, abounding around ponds and tanks during the monsoon months, hiding in scrub-jungle during the drier seasons. Very inconspicuous in the latter situations on account of its dull melanotic colouring, but a conspicuous and pretty insect when on the wing over tanks and paddy. I have seen it in hundreds over swamps in Mysore and equally so over paddy lands in Bangalore. The flight is short, like most *Lestes*, and, when settled, it has a peculiar habit of swaying its abdomen, the reason for which is not at all clear, but may be sexual, as it is confined to the males.

I have been unable to detect any racial forms, although varieties due to age-differences are common. Females are comparatively rare over water unless *in cop.* and ovipositing, so that this sex must be sought for and beaten up in scrub adjoining water, where it will be found in great numbers. Distinguished from other species by its metallic thoracic stripes, with only an upper dilatation instead of two as in the *præmorsa* group.

The above description has been written from living specimens, and may, therefore, be noted as differing considerably from the original description of its author; it is, however, carefully compiled, and made from insects undoubtedly conspecific with the type of *L. elata* in the Selysian collection. The differences arise from the fact that the original description was made from a dried specimen which had lost its colours from decomposition. The living insect is a much more beautiful insect than the drab creature depicted by Selys.

The wet-season brood has a remarkably short larval existence of not more than ten weeks, as I have a note of imagos emerging in the second week of September from eggs which had been laid in the last week of the previous June.

8. **Lestes malabarica** Fraser. (Fig. 16.)

Lestes malabarica Fras., J. Bombay Nat. Hist. Soc. vol. xxxiii, pp. 847, 848 (1929); id., Rec. Ind. Mus. vol. xxxiii, pp. 448, 463 (1931).

Male.—Abdomen 24–25 mm. Hind-wing 21 mm.

Head: labium white; labrum, bases of mandibles, cheeks, and anteclypeus turquoise-blue; postclypeus and vertex mat black ; occiput and behind eyes ferruginous in tenerals, dark brown to blackish in full adults ; a large triangular spot on the inner side of each eye dark metallic green, extending forwards as far as cheeks as a dark cupreous band ; eyes turquoise-blue, paler beneath. *Prothorax* blackish-brown in adults, thinly pruinosed laterally, pale yellow in tenerals, changing to warm reddish-brown on dorsum and dark brown on the posterior lobe. *Thorax* with dorsum bright metallic green in tenerals and early adults, the mid-dorsal carina narrowly and an antehumeral stripe bright reddish-brown, the metallic stripe but slightly or not at all expanded in its upper part ; the anterior half of mesepimeron in tenerals blackish-brown, with the rest of the sides and beneath thorax creamy-white, adults wholly black save for the lower part

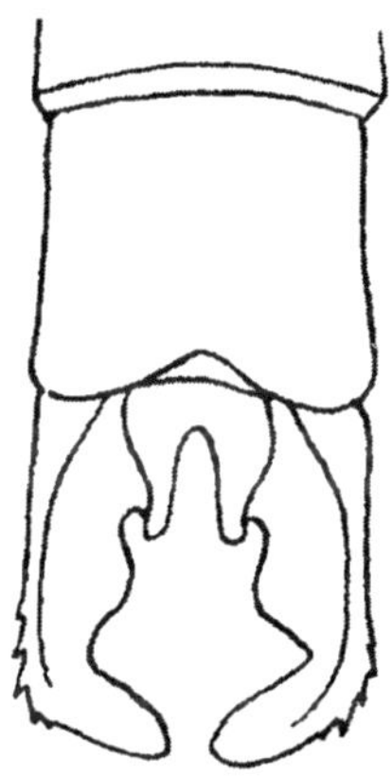

Fig. 16.—Dorsal view of anal appendages of *Lestes malabarica* Fras., male.

of metepimeron and a medio-lateral stripe interrupted by the spiracle, which are light brown. The black portions densely pruinosed ; beneath black, overlaid by pruinescence save for an anterior ventro-lateral black spot on each side (the posterior corresponding spots found in *L. elata* absent in this species). *Legs* yellow, with a parallel pair of black stripes on outer sides of femora ; flexor surfaces of tibiæ black and, in adults, the inner sides of the femora. *Wings* palely and evenly enfumed in adults, hyaline in tenerals ; pterostigma decidedly longer and narrower than in *L. elata*, covering two cells, black in adults, pale brown in tenerals ; 11 to 13 postnodal nervures in fore-wings, 10 to 12 in the hind ; petiolation ceases at *ac*, which is situated rather nearer the distal antenodal nervure. *Abdomen* with the sides azure blue, the dorsum broadly dull metallic; segment 1 with apical half of dorsum metallic green in tenerals, dull bronzed in adults, the sides broadly azure blue, with an oblique black stripe running upwards and basalwards from the apico-lateral corner of the segment, but not meeting the base; segment 2 with a broad thistle-shaped stripe on dorsum extending from apical border to base of segment, the sides blue; segments 3 to 6 with a narrow basal ring of blue and a subapical interrupted ring of pale yellow, the dorsum metallic green in tenerals, bronzed black in adults, this band tapered abruptly at base of segments and expanded as abruptly subapically so as to encircle segments, finally contracted again and confluent with a narrow apical black ring; segments 7 to 10 black, the last two in tenerals with large subdorsal pale blue to pale brown spots; adults with the dorsum of segments 9 and 10 pruinosed white. *Anal appendages*: superiors differing widely from those of *L. elata* in that they are bent sharply inwards at the junction of apical and middle thirds at nearly a right angle; apices bent strongly downwards as seen in profile; a sub-basal robust spine, longer than that seen in *L. elata*, on the inner side of appendage, followed by a broad thin expansion minutely denticulate on the inner side, after which the appendage is abruptly angulated inwards; outer border near apex coarsely spined, but no tuft of hairs at apex as in *L. elata*. Inferior appendages also differing from those of *L. elata*; seen from the side about half the length of superiors, broad and rounded at base, then narrowly constricted as a short ungulate process resembling that seen in *L. præmorsa*, but broader, and without the apical pencil of hairs; superiors white, tipped with black at apex and broadly black at base in tenerals, wholly black in adults, inferiors white at base, apical process glossy black.

Female.—Abdomen 32–33 mm. Hind-wing 22–23·5 mm.

Very similar to the male, but the ground-colour more greenish than blue and decidedly yellowish in tenerals. The metallic antehumeral stripes dilated above as in *L. elata*; sides of thorax presenting only five black spots or points, the medial posthumeral one being metallic green and very small. *Wings* hyaline, but evenly enfumed in adults, 11 to 13 postnodal nervures in fore-wings, 10 to 11 in the hind; pterostigma dark reddish-brown to nearly black, according to age of specimen. *Abdomen* with dorsum bright metallic green to cupreous in tenerals, but dull cupreous in adults as far as segment 6, then dull mat black; segment 10 entirely, segment 9 laterally yellow, and segment 8 yellow for its apical half and sides. In adults these markings change to blue, and finally mat black. *Anal appendages* shortly conical, yellow in tenerals, bluish in adults. Adults have a dense-

pruinosity beneath thorax and prothorax, but rarely exhibit any on the sides or terminal segments of the abdomen.

Distribution.—S. INDIA: South Malabar and Cochin. An annual migration is seen northwards to North Malabar just previous to the S.W. monsoon, but there is no evidence that the species ever establishes itself, as there are no suitable breeding-places after the paddy lands have dried up. South Malabar and Cochin, on the other hand, abound with suitable tanks and ponds, in which the species breeds in great abundance and almost completely crowds out *L. elata* and *L. præmorsa*.

From the latter of these two species it is to be distinguished by the shape of the superior and inferior anal appendages, and also by the absence of the conspicuous hinder ventro-lateral black spot on the thorax; from the former the shape of the metallic humeral stripes will at once serve to distinguish it.

Type in the Fraser collection; paratypes in the Laidlaw and Morton collections. Paratypes will be placed in the collections of the Indian, Pusa, and British Museums.

9. **Lestes angularis** Fraser. (Fig. 17.)

Lestes angularis Fras., J. Bombay Nat. Hist. Soc. vol. xxxiii, pp. 848 849 (1929).

Male.—Abdomen 34 mm. Hind-wing 22 mm.

Head: labium pale ochreous; labrum olivaceous; base of mandibles and cheeks bluish: epistome, frons, and vertex mat black, marked with a small tawny triangular spot on the outer side of each hinder ocellus; a large triangular area to the inner side of each eye dark metallic green: behind the eyes black. (Probably the labrum is blue during life.) *Prothorax* mat black, pruinosed on the sides. *Thorax* with the mid-dorsal carina narrowly dark ochreous, the alar sinus and dorsum as far out as the humeral suture mat black, marked by a narrow antehumeral metallic green stripe of even width on each side; mesepimeron black, marked by a U-shaped spot anteriorly and an interrupted stripe posteriorly ochreous; metepimeron black save for the posterior border, which is narrowly ochreous; beneath black. *Legs* yellow, femora striped longitudinally with black on inner and outer sides: tibiæ blackish on flexor surface. *Wings* hyaline; pterostigma dark reddish-brown, broad, about twice as long as broad, inner and outer ends oblique; 11 postnodal nervures to fore-wings, 9 to 11 to the hind; *ac* meeting *ab* at hinder border of wing: *ac* situated about midway between the two antenodal nervures. *Abdomen* yellow or bluish at the sides and beneath, metallic green on dorsum from segments 1 to 7, mat black from segment 8 to the end; dark metallic bands on segments 3 to 7 dilated subapically so as to encircle the segments completely, then narrowed to become confluent, with fine terminal black rings, whilst basally the bands fall short of the base to leave narrow blue basal annules; segment 10 deeply notched. *Anal appendages*: superiors tawny, tipped broadly with black, the base with a robust inner spine, which is immediately followed by a dilatation which extends for a little beyond the middle of the appendage, at which point the latter is bent abruptly inward at a right angle; apices blunt, coarsely spined on the outer side; inferior appendages very short conical processes, barely extending as far as the basal spine of superiors, dark reddish-brown in colour.

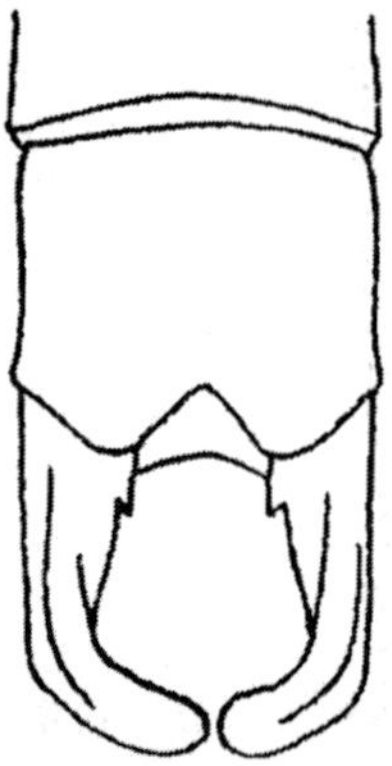

Fig. 17.—Dorsal view of anal appendages of *Lestes angularis* Fras., male.

Female.—Abdomen 32 mm. Hind-wing 22 mm.

Ground-colour of head and body generally paler, that of head dark reddish-brown on frons, and vertex with smaller black areas, the triangular metallic spots much more conspicuous, the tawny spots on outer sides of ocelli larger, and finally the occiput and upper part of back of eyes pale ochreous. *Prothorax* pale ochreous, with an irregular black dorsal spot shaped like a bull's head with diverging horns. *Thorax* bright ochreous, the mid-dorsal carina and the shape of the metallic antehumeral stripes as for the male; two small black spots on anterior part of mesepimeron. *Legs* marked as for the male, but brighter ochreous. *Wings* similar to the male, 11 postnodal nervures in fore-wings, 10 in the hind; pterostigma dark yellow between black nervures, more than twice as long as broad. *Abdomen* with segment 1 pale ochreous, unmarked; segment 10 and apical half of segment 9 pale blue, this colour extending broadly on the sides of the latter segment. *Anal appendages* ochreous, shortly conical.

Distribution.—LOWER BURMA.

Type, a male from Minbu, now in the Pusa collection, collected 8. viii. 14.

Closely resembles *L. elata*, from which it differs by the angulation of the superior appendages, and the antehumeral metallic stripes of even width throughout. In the latter respect it resembles *L. viridula*, as, indeed, also by the angulation of the appendages, but the latter species has no black markings and has no metallic markings on the head.

10. **Lestes viridula** Rambur. (Fig. 18.)

Lestes viridula Ramb., Hist. Nat. Ins. Névrop. pp. 252–3 (1842); Selys, Syn. Agrion. p. 38 (1862); Laid., Rec. Ind. Mus. vol. xix, p. 153 (1923); Fras., Rec. Ind. Mus. vol. xxvi, p. 487 (1924); id., J. Bombay Nat. Hist. Soc. vol. xxxiii, pp. 849, 850 (1929); id., Rec. Ind. Mus. vol. xxxiii, p. 448 (1931).
Lestes viridulus Kirby, Cat. Odon. p. 163 (1890).

Male.—Abdomen 34–35 mm. Hind-wing 23–24 mm.

Head: labium palest yellow; labrum, clypeus, frons, vertex, and occiput pale brown, deepening to chocolate brown

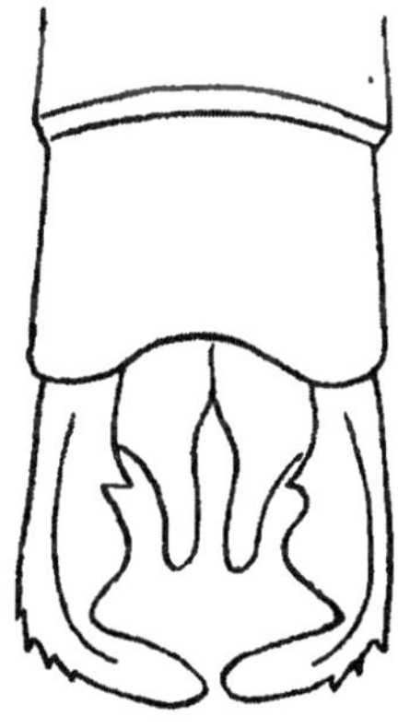

Fig. 18.—Dorsal view of anal appendages of *Lestes viridula* Ramb., male.

on vertex and occiput; bases of mandibles and cheeks palest brown tinted with golden yellow; behind the eyes pale yellow; eyes brown above, golden yellow below. *Prothorax* uniform pale khaki brown, unmarked. *Thorax* pale khaki brown or fawn, paling to creamy white low down on the sides and beneath, which latter is pruinosed white; dorsum of thorax with two very narrow conspicuous metallic green stripes running closely parallel to the mid-dorsal carina, the area between them warm reddish-brown; no spots on the thorax. *Legs* pale yellow with black spines. *Wings* hyaline: pterostigma khaki brown, very narrow, four times as long as broad, covering two cells; *ac* meeting *ab* at hinder border of wing at a level rather nearer the distal antenodal nervure, at which level petiolation begins; 11 to 12 postnodal nervures to fore-wings, 9 to 12 in the hind. *Abdomen* pale yellow at the sides and almost white beneath; warm reddish-brown on dorsum, with a cupreous reflex, paler on segments 1, 9, and 10, the latter very pale brown or yellow; all segmental joints ringed narrowly with blackish-brown; segments 3 to 6 with diffuse, rather obscure dark brown subapical rings; segment 9 with a short dark brown carinal stripe on its basal half. *Anal appendages* yellow, tipped with black; superiors rather longer than segment 10, broad at base, where there is a very robust sub-basal spine, after which the usual expansion occurs, broadening to about the middle of the appendage and finely denticulate along its inner border; at the termination of this dilatation the appendage bent abruptly in at an obtuse angle, and, as viewed from the side, rather strongly downwards; coarsely spined on outer side of angulation. Inferiors half the length of superiors, broad and rounded at base, then truncate and blunt at apex, directed straight back somewhat like the same appendage in *L. præmorsa*, but without the stiff pencil of hairs at the apex.

Female.—Abdomen 34 mm. Hind-wing 24 mm.

Coloured exactly as in the male except segment 8, which has a narrow blackish-brown mid-dorsal spot on the carina extending nearly to apex of segment, and a broader and more conspicuous mid-dorsal mark on segment 9 on the basal three-fourths of segment; segment 10 very pale brown. *Wings* very similar to the male, but pterostigma distinctly longer and narrower; 12 to 13 postnodal nervures to fore-wings, 11 to 12 in the hind. *Legs* with a longitudinal dark brown stripe on all femora not quite extending to base of limb, and more conspicuous on the two anterior pairs. *Anal appendages* palest brown, shortly conical, pointed; vulvar scale very robust, extending nearly to end of the abdomen, darkish brown.

Distribution.—Confined to Peninsular India, especially to the Deccan and Western India. It is commonest during the dry seasons, and is to be sought for among long dry grass, the colour of which it simulates closely. Where found usually quite common; in a colony found in South Coorg I took a score of couples in as many minutes.

The *type* comes from Bombay, and is in the Rambur collection or Paris Museum. Easily distinguished from all others by its uniform pale brown colour combined with a pair of dorsal thoracic metallic green stripes of uniform width.

11. **Lestes patricia** Fraser. (Fig. 19.)

Lestes patricia Fras., Rec. Ind. Mus. vol. xxvi, p. 486 (1924); id., J. Bombay Nat. Hist. Soc. vol. xxxiv, p. 965 (1931); id., Rec. Ind. Mus. vol. xxxiii, p. 448 (1931).

Male.—Abdomen 37 mm. Hind-wing 24 mm.

Head: labium greyish; labrum, genæ, epistome, and frons azure blue; vertex and occiput mat black; eyes deep azure blue above, turquoise-blue beneath (fading to olivaceous brown after death). *Prothorax* greenish-blue laterally, mat black above. *Thorax* bluish-green on the dorsum and laterally, pale greenish-white beneath; dorsum marked with a broad stripe of mat black extending out to about half-way between

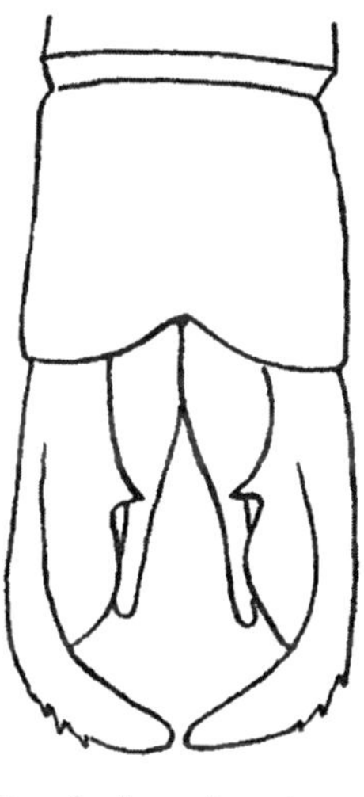

Fig. 19.—Dorsal view of anal appendages of *Lestes patricia* Fras., male.

the dorsal carina and humeral suture, and with straight borders; laterally the upper portion of the area between the humeral and postero-lateral sutures greyish-green, whilst beneath are two pairs of black spots partly obscured by pruinescence. *Legs* bluish outwardly, reddish on flexor surfaces, black on the extensor; femoral spines short, tibial long; about 11 spines on the femora. *Wings* hyaline, palely enfumed, this especially noticeable with the wings superposed; pterostigma dark brown, about four times as long as broad; 14 postnodal nervures in fore-wings, 10 in the hind; discoidal cells equal; *ac* situated midway between the antenodal nervures. *Abdomen* bluish-green, marked with mat black on segments 1 and 2 and from segments 6 to 10;

segments 3 to 5 warm brown dorsally, non-metallic; all segments with a broad dorsal stripe which dilates apically but does not extend quite to base of segments; segment 1 with its basal half only black, 2 with a mid-dorsal stripe slightly dilated apically, 7 with the dorsal stripe gradually widening apically, whilst segments 8 to 10 are entirely black, the last two being somewhat pruinosed dorsally. *Anal appendages*: superiors black, with the inner dilatation white, this latter rather broad, and with a robust black tooth at its base; apical ends turned in abruptly at nearly a right angle, and with a few small spines on the outer border; inferiors very short, digitate, extending nearly to the end of expanded part of superiors, apices naked, obtuse, black.

Female unknown.

Distribution.—COORG only.

The *type* is the only specimen known, and is in the author's collection. It was found lurking in bushes besides a small pond near Virajpet, on the Sidapur road. It bears a superficial appearance to *Ceylonolestes pulcherrima*; it is easily distinguished from other Indian species of the genus *Lestes* by the single mid-dorsal black band with straight borders.

12. **Lestes orientalis** Hagen.

Lestes orientalis Hagen, Verh. Zool.-bot. Ges. Wien, vol. ix, p. 206 (1859); Selys, Bull. Acad. Belg. (2) vol. xiii, p. 322 (1862); Kirby, Cat. Odon. p. 163 (1890); id., J. Linn. Soc. Lond., Zool. vol. xxiv, p. 566 (1893); Laid., Rec. Ind. Mus. vol. xix, p. 155 (1920); id., Spolia Zeylan. vol. xii, p. 357 (1924); Fras., J. Bombay Nat. Hist. Soc. vol. xxxiv, p. 87 (1930).

Male.—Abdomen 52 mm. Hind-wing 38 mm.

Head: labium yellow; labrum, cheeks, and bases of mandibles pale yellow, rest of head coppery-bronze or metallic green; eyes brown; behind head pale yellow. *Prothorax* and *thorax* metallic green-bronze on dorsum, with antehumeral stripes pale yellow on the latter; laterally and beneath pale yellow, with two brown spots on each side of the chest. *Legs* black, bases of femora paler, especially on flexor surface. *Wings* hyaline, fore-wings with 18 to 20 postnodal nervures; *IRiii* not zigzagged; pterostigma dark yellow, framed in black nervures, covering from 2 to 2½ cells. *Abdomen* metallic green on dorsum and sides. *Anal appendages* black, superiors forcipate, apices curving towards each other and terminating in a point, furnished near the base with a blunt spine, and at about the middle a small tubercle below; inferiors rudimentary, conical, very short.

Female.—Abdomen 49 mm. Hind-wing 40 mm.

Closely similar to the male, differing only in sexual characters; segment 10 brown, notched at its apical border. *Anal appendages* conical, pointed at apex; vulvar scale yellow, extending to end of abdomen, with serrate border.

Distribution.—Ramboda Pass, CEYLON.

Type in the Hagen collection.

This species, which greatly resembles a *Megalestes*, must be either extremely local or very rare, as it has never been taken since the type was procured in 1858, seventy years ago. It differs from *Megalestes major* by the presence of intercalated sectors between *IRiii* and *Riv+v*, but is possibly closely allied to it. It is to be hoped that some one of the entomologists in Ceylon may rediscover this beautiful and interesting insect.

13. **Lestes barbara** Fabricius. (Fig. 20.)

Agrion barbara Fabr., Suppl. Ent. Syst. p. 286 (1798); Fonscolomb, Ann. Soc. Ent. France, vol. vii, p. 554, t. 33, fig. 2 (1838).

Agrion barbarum Charp., Hors. Ent. p. 9 (1825); id., Libell. Eur. p. 142, t. 35, figs. 3, 4 (1840).

Agrion nympha Hansem, Wieden. Zool. Mag. vol. ii (1), p. 161 (1823).

Lestes barbara (♀) Lind., Mon. Lib. Eur. p. 36 (1825); Selys (♂ et ♀), Mon. Lib. Eur. p. 142 (1840); Ramb., Ins. Névrop., p. 251 (1841); Selys, Rev. Odon. p. 159 (1850); id., Bull. Acad. Belg. (2) vol. xiii, p. 318 (1862); Laid., Rec. Ind. Mus. vol. xix, p. 155 (1920); Fras., J. Bombay Nat. Hist. Soc. vol. xxxiv, pp. 87, 88 (1930).

Lestes barbarus Kirby, Cat. Odon. p. 162 (1890); Calv., Proc. Acad. Nat. Sci. Philad. p. 147 (1898).

Male.—Abdomen 26–34 mm. Hind-wing 21–25 mm.

Head: labium pale brownish-white; labrum, cheeks, and bases of mandibles yellowish or pale olivaceous; penultimate joint of antennæ and behind eyes yellow; rest of head dark bronzed green with a coppery or golden reflex; eyes brown. *Prothorax* metallic green on dorsum, yellow at the sides and finely along borders of posterior lobe, which is rounded, very small and narrow. A black spot low down on each side. *Thorax* brilliant metallic green on dorsum as far lateral as the middle of mesepimeron; old specimens with a golden or coppery reflex; dorsum marked with a narrow humeral yellow stripe narrowly bordered with black; in Kashmir examples this black border is more extensive, so that in some specimens the yellow line is greatly narrowed at its middle or entirely obliterated by a confluence of the black borders; mid-dorsal carina finely yellow in European examples, dull black in Kashmir ones; laterally and beneath bright citron-yellow from beyond the middle of mesepimeron; postero-lateral suture finely mapped out in black. *Legs* yellow, femora and tibiæ with a narrow black line on the outer side, flexor surface of tibiæ black, tarsi black. *Wings* hyaline, palely tinted with yellow; pterostigma bicolorous, rather more than the proximal half blackish-brown, outer part white or creamy, braced, covering two cells, framed in thick black nervures, 1·5 mm. in length; 10 to 14 postnodals to fore-wings, 10 to 12 in the hind; *Riii* arising 2½ to 3 cells beyond the node in both wings; *IRii* arising 7 cells after the node in fore-wing, 5 in the hind. *Abdomen* yellow at the sides, metallic green to coppery on the dorsum, and with narrow apical rings to most segments; segment 2 with the mid-dorsal carina finely yellow; segments 3 to 6 with narrow basal yellow rings often interrupted at the middle line; segment 10 mat black on dorsum, non-metallic, yellow at the sides, pulverulent white on the dorsum in old specimens. *Anal appendages*: superiors yellow tipped with black, Kashmir examples being yellow at the base and outwardly only, the whole of the inner side, including the basal tooth and the apex, being black; rather

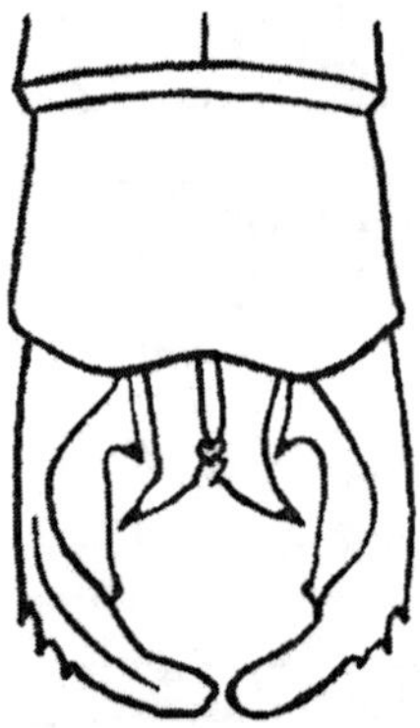

Fig. 20.—Dorsal view of anal appendages of *Lestes barbara* Fabr., male.

longer than segment 10, forcipate, the apices curling in, with rounded overlapping ends, a robust inner basal spine followed by a shell-like dilatation minutely denticulated on its inner free border; inferior appendages yellow, finely tipped with black, thick, tumid, and apposed at the bases, then strongly divaricate, markedly tapered, turned up to end in a fine point.

Female.—Hind-wing 22–27 mm. Abdomen 29–33 mm.

Similar to the male except for sexual differences. A small oval yellow spot on the outer side of each posterior ocellus; occiput bordered narrowly with yellow; humeral stripe not bordered with black, the Kashmir examples not differing from the type in this respect; *legs* rather more broadly striped with black; *wings* similar to male, 11 to 14 postnodal nervures to all wings; *abdomen* coppery-bronze on dorsum, segment 10 yellow, with a rather narrow well defined blackish-brown stripe on dorsum, which in some examples tapers to the apical border of segment. *Anal appendages* yellow, tipped with black in Kashmir specimens, conical, pointed, slightly shorter than segment 10; vulvar scale very robust, yellow or yellow bordered with black in Kashmir specimens.

Distribution.—Restricted to KASHMIR and N.W. INDIA, within Indian limits, but widely distributed throughout S. Europe, Asia Minor, and Algeria. As a rule, Kashmir examples exhibit some melanism, and are decidedly larger than European specimens. Mr. T. Bainbrigge Fletcher took fair numbers at Yusimarg, 7,500 ft. (Kashmir), during August, and I have specimens from France taken during the same month.

14. **Lestes nodalis** Selys. (Fig. 21.)

Lestes nodalis Selys. Ann. Mus. Civ. Genova, (2) vol. x (xxx), pp. 496, 497 (1891); Laid., Rec. Ind. Mus. vol. xix, p. 151 (1920); Fras., J. Bombay Nat. Hist. Soc. vol. xxxiv, pp. 88, 89, pl. i, fig. 2 (1930).

Male.—Abdomen 30–35 mm. Hind-wing 19–21 mm.

Head: labium brownish-white; labrum, bases of mandibles, and cheeks pale blue, but piceous in the dried state; rest of head blackish-brown, but paler behind the eyes; eyes brown. *Prothorax* olivaceous brown on dorsum, bluish laterally, two black horizontal stripes on the anterior trochanters. *Thorax* dark brown on dorsum as far out as the humeral suture, and slightly overlapping this above, laterally pale blue or palest olivaceous in the dried state, a small black spot on the upper part of the postero-lateral suture and a large, more conspicuous spot near the ventral border on the fore part of metepimeron; beneath olivaceous, with a triangular black area just behind the hind pair of limbs. *Wings* palely and evenly tinted with yellow, with dark, conspicuous neuration; pterostigma bicolorous longitudinally, blackish-brown at centre, yellow along costal and inner borders, elongate, tapering distally, oblique at both ends, non-braced, covering two to three cells; node thickened and showing as a conspicuous black point at costa; 13 to 15 postnodals to forewings, 9 to 13 in the hind. *Legs* reddish-yellow, the anterior pair of femora and the distal ends of the others on the outer side brownish-black. *Abdomen* light olivaceous brown, peppered with black and marked with blackish-brown on the dorsum of segments 1 and 2 and the apical ends of segments 3 to 6, where this colour forms a broad apical ring, enclosing on the dorsum a pale spot shaped like a barbed arrow-head with its point directed basally; on the dorsum of segment 2 this spot is reproduced on a much larger scale; segments 7 and 8 dark brown; 9 and 10, especially the latter, appear to be bluish during life, but are pale olive in the dried state. *Anal appendages* reddish-brown, superiors about equal in length to segment 10, broad and robust, the apices abruptly right-angled inwards to meet in the middle line, the apex broad, but tapering rapidly to a fine point and bordered outwardly by blackish spines; the usual inner scale-like lamina broad, occupying about the middle third of appendage and with thickened inner and outer borders, the latter projecting slightly as a fine spine; inferior appendages shortly conical, coated with short, coarse hairs.

Fig. 21.—Dorsal view of anal appendages of *Lestes nodalis* Selys, male.

Female.—Abdomen 28–32 mm. Hind-wing 20–21 mm.

Very similar to the male. In the single female I posses, the labrum, bases of mandibles, and cheeks are dark reddish-brown, as also the rest of the *head*; the *prothorax* and *thorax* are similar to the male; the *wings* are more deeply tinted a pale golden brown; the pterostigma is similar, but the nervures forming its inner angle are creamy-white; postnodal nervures to fore-wings 13 to 16, 9 to 12 in the hind; the small spines along the costa are conspicuously blackish-brown against the yellow costa, and the space after the pterostigma is enfumed brown; the *abdomen* differs by the marking on segment 2 being very obscure, the apical rings on 3 to 7 present as paired spots on the subdorsum, whilst segment 9 has on each side a very large black spot extending for about the basal two-thirds; segment 10 appears to have been bluish during life, as also the sides of segment 2. *Legs* similar to the male. *Anal appendages* conical, pointed at apex, as long as segment 10, carneous. Vulvar scale round, robust, dark brown.

Distribution.—Assam (Margherita), sparingly during May and June; Burma (Palon), in September; Yunnan. Probably widely and sparingly distributed throughout N.E. India and Burma ; its dull colouring may render it very inconspicuous, so that it may have often been overlooked. The longitudinally bicolorous pterostigma, the small black nodal point on the costa, and the characteristically broad right-angled anal appendages easily serve to distinguish it from other species.

15. **Lestes umbrina** Selys. (Fig. 22.)

Lestes umbrina Selys, Ann. Mus. Civ. Genova, (2) vol. x (xxx), pp. 497, 498 (1891); Laid., Rec. Ind. Mus. vol. xix, pp. 150–1 (1920); Fras., J. Bombay Nat. Hist. Soc. vol. xxxiv, pp. 89, 90 (1930).

Male.—Abdomen 32 mm. Hind-wing 20 mm.

Head: labium dirty yellow or pale brown ; labrum, bases of mandibles, and cheeks pale yellowish-brown ; rest of head

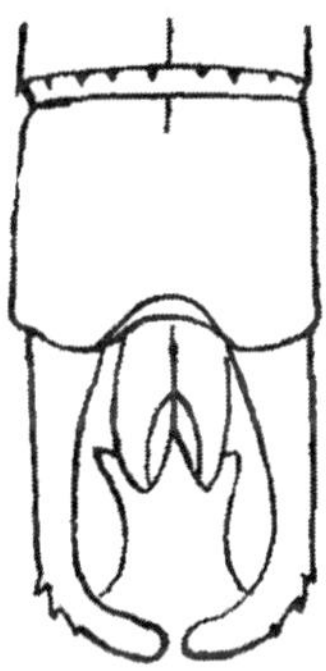

Fig. 22.—Dorsal view of anal appendages of *Lestes umbrina* Selys, male.

pale reddish-brown ; antennæ dark brown, except the two basal segments, which are yellowish ; eyes brown above, yellow beneath. *Prothorax* and *thorax* dark reddish-brown on dorsum, pale yellowish-brown laterally, without markings. *Legs* reddish-yellow ; the anterior limbs black on the outer surface ; all femora and tibiæ black on the flexor surface, and the mid-pair of femora with an ill-defined interrupted stripe on the outer surface. *Wings* palely enfumed, hyaline ; pterostigma elongate, pale brown, the outer and inner borders creamy-white or palest brown, covering about 1½ cells, braced ; 9 to 11 postnodal nervures in fore-wings, 9 in the hind. *Abdomen* reddish-brown, the dorsum darker brown, but paler in some specimens ; the intersegmental sutures dark brown. *Anal appendages* brownish-yellow, darkening at apices ; superiors slightly longer than segment 10, coarsely spined along the outer border, forcipate, the apices curling in to meet each other and rounded at the ends, furnished on the inner border, at junction of basal and middle thirds, with a small spine, the middle third occupied with the usual scale-like expansion, which terminates distally in a small inconspicuous spine; inferior appendages conical and apposed, but the apices slightly divergent, about half the length only of the superiors.

Female.—Abdomen 29–30 mm. Hind-wing 21 mm.

Almost exactly similar to the male, differing mainly in sexual characters. *Wings* as for male, with 9 to 10 postnodal nervures to fore-wings and 9 in the hind; *head, thorax,* and *legs* as for male; *abdomen* darker on the dorsum from segments 3 to 7, with a pair of subdorsal, subapical, comma-like spots placed transversely; segments 8 to 10 and the apical half of 7 broadly blackish-brown on dorsum, this colour tapering apically on 9 and 10; segments 8 and 9 also with a large blackish-brown spot on each side of the ventral border. *Anal appendages* shortly conical, pale yellow, and with a short, conical protuberance, notched at its apex, placed between them at the apical border of segment 10. Vulvar scale short but robust, brownish-yellow, minutely serrate along its under border.

Distribution.—The *type* comes from BURMA (Bhamo), August, and is now in the Selysian collection. In this same collection is a pair from Pamizah, BENGAL ?, and in the MacLachlan collection a female from Yunnan. I possess a female from Baghwonie, Duars, BENGAL, taken in June. The Indian Museum possesses specimens from Cutch and Panch Mahals (BOMBAY PRES.), and Allahabad (UNITED PROV.). A specimen in the same collection is labelled as from Waltair, but during two years' residence in that district I never once came across it, so that the label may be wrong. The species, however, appears to be widely scattered, although sparsely so. Laidlaw is of opinion that it is synonymous with *L. concinna* Selys, from the Philippines and Java; after carefully comparing the description of this species with *L. umbrina*, I feel inclined to share his opinion. If, as he says, Selys had compared the two insects, he would probably have hesitated before describing *L. umbrina* as a separate species; the shape of the male anal appendages and the markings of segments 8 to 10 in the female do not differ from *L. concinna*. Superficially, *L. umbrina* resembles *Ceriagrion olivaceum*, found in similar places, among long dried grasses. It is distinguished from other Indian species by its uniform colouring and its pterostigma, with outer and inner borders paler than the body of the organ.

It is to be noted that artefacts in the colouring of the thorax, usually in the form of stripes on the dorsum and sides, are quite common in the dried state of these small insects, which may account for a number of errors which have crept into the Selysian descriptions.

16. **Lestes nigriceps** Fraser.

Lestes nigriceps Fras., Mem. Dept. Agric. India (Ent.), no. 8, vol. vii (1924); Fras., J. Bombay Nat. Hist. Soc. vol. xxxiv, pp. 90, 91 (1930).

Male.—Abdomen 32 mm. Hind-wing 21 mm.

Head: labium, labrum, cheeks, and bases of mandibles pale brown; rest of head mat black; behind eyes and head pruinosed white. *Prothorax* black, largely obscured by white pruinescence. *Thorax* black, marked with a narrow reddish-brown (possibly blue during life) humeral stripe and a narrow, irregular, pale yellowish-green stripe on the mesepimeron; the sides near the ventral border also pale greenish-yellow. On the dorsum, internal to the humeral stripe, an obscure metallic-green stripe of even width similar to that seen in *L. viridula*. *Legs* yellow, femora and tibiæ striped longitudinally with black. *Wings* hyaline; pterostigma long and narrow, about four times as long as broad, pale brown; 11 to 12 postnodal nervures in forewing, 9 to 10 in the hind. *Abdomen* brownish-yellow, broadly marked on dorsum with bronzed black or metallic green as follows:—Segments 2 to 6 and the basal half of 7 metallic green or bronzed black, remaining segments black; segment 2 with the mid-dorsal carina finely black for its basal three-fourths, but widely so thereafter; on all segments the dorsal black expanding apically and confluent with a narrow black apical ring. *Anal appendages* yellowish-brown, superiors black at apex and base and furnished with a robust inner basal spine followed by the usual expansion, which occupies the middle third of appendage, the apical third sharply angulated inward and downward and spined along its outer border; inferior appendages short, angulate, about half the length of superiors.

Female.—Abdomen 30–32 mm. Hind-wing 21–22 mm.

Somewhat similar to the male, but without the extensive melanism, especially on the thorax. The black marking of the *head* is more restricted and has a bronzed or metallic reflex; the back of the head is reddish-brown, without pruinescence; the *prothorax* is pale brown, with two fine dark brown lines on the mid-dorsum and without overlying pruinescence. *Thorax* pale brown, without black markings, but with the narrow metallic-green humeral stripes very conspicuous. *Abdomen* paler, segment 1, the sides and apical end of 9, and the whole of 10 pale brown. *Anal appendages* brown, conical, short, pointed at apex; vulvar scale brown, robust, finely serrate below.

Distribution.—The *type* is a male in the British Museum from Pusa, BIHAR, and, apart from two females from the same locality, is the only specimen known. It is related to *L. viridula* by the green thoracic stripe and by the close similarity of the anal appendages, but it differs by the extreme melanism, unknown in *L. viridula*, and by the longer pterostigma. I have hesitated to include this species, as I have not had the opportunity of comparing it with *L. malabarica*, which it closely resembles; should the two species be synonymous, *L. nigriceps* will have priority. The ground-colour of *L. malabarica* is blue, but the type of *L. nigriceps* may have undergone colour-changes from decomposition.

17. **Lestes thoracica** Laidlaw. (Fig. 23.)

Lestes thoracica Laid., Rec. Ind. Mus. vol. xix, pp. 152–3 (1920); Fras., J. Bombay Nat. Hist. Soc. vol. xxxiv, pp. 91, 92, pl. i, fig. 3 (1930).

Male.—Abdomen 30–32 mm. Hind-wing 20 mm.

Head: labium white; labrum, bases of mandibles, and cheeks turquoise-blue; rest of head mat black, except

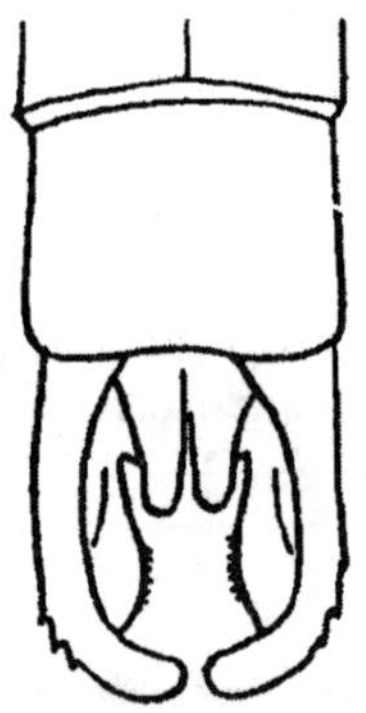

Fig. 23.—Dorsal view of anal appendages of *Lestes thoracica* Laid., male.

occiput and behind head, which are creamy-white; eyes dark blue above, pale beneath. *Prothorax* black on dorsum, creamy-white laterally, marked with a large black spot on the anterior trochanter. *Thorax* pale olive-green, with a pinkish tinting towards the mid-dorsum; the mid-dorsal carina finely black, as also the upper parts of the humeral and lateral sutures. *Legs* yellow, striped in their length with black on the outer side; tarsi black; flexor surface of femora also black. *Wings* hyaline; 10 to 11 postnodal nervures to both fore- and hind-wings; pterostigma pale brown, framed in blackish nervures and with the outer distal end pale; braced, covering 1 to 2 cells. *Abdomen* black on dorsum, pale azure blue at the sides, the dorsal black expanded subapically on segments 2 to 7, completely ringing the latter segment; segments 8 to 10 entirely black on dorsum, sides, and beneath, except for a small latero-apical spot, bluish-white on both sides of the latter two segments. *Anal appendages* creamy white tipped with black; superiors one and a half times the length of segment 10, forcipate, apices rounded at the ends, curling in to meet each other, coarsely spined along the outer border, furnished at the junction of basal and middle thirds with a robust inner spine which is followed by the usual scale-like expansion, which occupies rather more than the middle third of the appendage and is finely denticulate along its inner border; inferior appendages not quite half the length of superiors, thick at base, ungulate thereafter, rounded and furnished at apex with coarse hairs.

Female.—Abdomen 28–31 mm. Hind-wing 20–22 mm.

Very similar to the male.

Head: labrum, cheeks, and bases of mandibles pale yellowish-green, rest of head olivaceous, with a small dark brownish mark on each side of the clypeus and some small spots in the ocellar space. *Prothorax* and *thorax* olivaceous green on dorsum, paler laterally and beneath, without any markings. *Legs* as for male, but the lateral stripes restricted to the distal ends of the femora. *Wings* palely enfumed; pterostigma as for male; 10 to 12 postnodal nervures to fore-wings, 9 to 10 in the hind. *Abdomen* pale olivaceous, greenish-yellow at the sides, dark markings of dorsum poorly defined; segments 2 to 7 with a pair of small spots transversely placed subapically on dorsum; segments 8 and 10 and the apical two-thirds of 7, with the dorsum, black only, broad on 7 and 8, abruptly narrowed on 9 and 10; ventral borders of 8 to 10 blackish-brown. *Anal appendages* very short, conical, pointed, creamy-white. Vulvar scale pale, robust, extending to end of abdomen, finely serrate along lower border.

Distribution.—The species appears to be confined to ORISSA, BENGAL, BIHAR, and the UNITED PROVINCES. The *type* is a male taken at Agra, now in the Indian Museum, and there are other specimens in the same Museum from the Chilka Lake. I have a pair from Pusa, taken in July. It is easily recognized by the jet-black head (spotted with black in the female). The female resembles that of *L. umbrina* rather closely, but is distinguished by the head marked with black and by the pterostigma bordered outwardly only with pale brown.

Genus PLATYLESTES Selys (1862). (Fig. 24.)

Platylestes **Selys, Bull. Acad. Belg. (2) vol. xiii, p. 337 (1862).**

Small dragonflies of the size of *Lestes* and considerably smaller than *Megalestes*, resting with wings expanded, rather dull in colouring, non-metallic; wings long and narrow, hyaline; pterostigma subquadrate, about twice as long as broad, braced; postnodal nervures moderately numerous; *ac* (postcostal nervure) situate about midway between the two antenodal nervures; *ab* meeting *ac* at border of wing; discoidal cells narrow, acute distally, differing but slightly in shape and of the same length in fore- and hind-wings, that of fore-wing with the hinder border more than twice the length

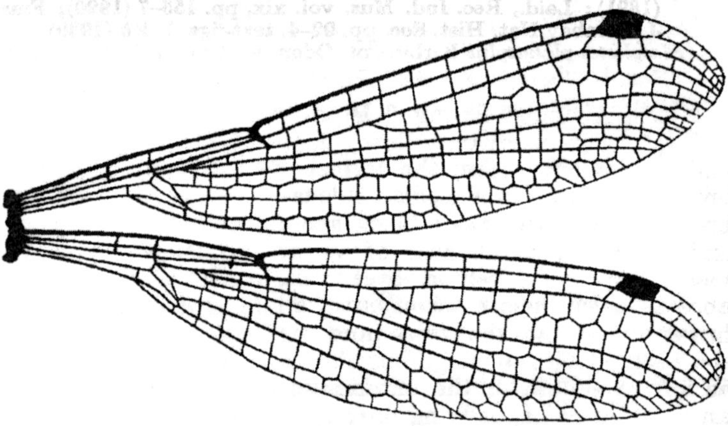

Fig. 24.—Wings of *Platylestes platystyla* (Ramb.), male.

of the costal and at least three times the length of the basal, that of hind-wing only twice the length of the costal, but four times the length of the basal; *IRiii* and *Riv+v* arising much closer to arc than to node, and running parallel for some distance; *Riii* arising from *Rii* about 3 to 4 cells after the node; an oblique nervure always present between *Riii* and *IRiii*; intercalated sectors at apical end of wing similar to genus *Lestes*. Abdomen slender, longer than wings; superior anal appendages forcipate, long and attenuated, with two spines on the inner border. Female with anal appendages markedly flattened and spatulate.

Genotype, *Platylestes platystyla* (Ramb.).

Venationally *Platylestes* scarcely differs from *Lestes*, although its author emphasized that the nervure *IRiii* is zigzagged, in contradistinction to *Lestes*, where it is not so; this character is not, however, at all evident in any of the three specimens mentioned below under *P. platystyla*, so that it may have been an aberration in the wing of the type female. The slight but very evident difference between the discoidal cells of the fore- and hind-wings, together with the elongate attenuated anal appendages of the male, point to a close relation to *Ceylonolestes*, or at least to a passage from *Lestes* to that genus; *Platylestes* may therefore be considered as a link between the two.

18. **Platylestes platystyla** (Rambur). (Figs. 24, 25, & 26.)

Lestes platystyla Ramb., Ins., Névrop. p. 254 (1842).
Platylestes platystyla Selys, Bull. Acad. Belg. (2) vol. xiii, p. 338 (1862); id., Ann. Mus. Civ. Genova, (2) vol. x (xxx), p. 499 (1891); Laid., Rec. Ind. Mus. vol. xix, pp. 155–7 (1920); Fras., J. Bombay Nat. Hist. Soc. pp. 92–4, text-figs. 1, 2 *b* (1930).
Platylestes platystylus Kirby, Cat. Odon. p. 164 (1890); id., J. Linn. Soc. Lond., Zool. vol. xxiv, p. 566 (1893).

Male.—Abdomen 33 mm. Hind-wing 21 mm.

Head: labium dirty yellow; labrum, bases of mandibles, and cheeks olivaceous, the former with a medio-basal dark brown spot lying in a median sulcus; rest of head violaceous brown, with three small black spots at base of postclypeus and similar spots at base of antennæ and anterior to the median ocellus; behind eyes yellowish; eyes olivaceous above, paler below. *Prothorax* and *thorax* palest khaki brown, paler at the sides and pruinosed white beneath, the thorax with a large number of black spots, disposed as follows:—One at the anterior end of the antealar sinus, an antehumeral spot lying near the middle of the mid-dorsal carina, three humeral spots, a superior, a medial, and an inferior, four spots on the mesepimeron, one at its upper third, one over the spiracle, and two at its lower part, a spot at the upper part of the postero-lateral suture, two spots on the ventro-lateral border, and two small points, closely apposed, beneath thorax behind the hind limbs. *Legs* pale whitish-yellow with black spines, the anterior pair of tibiæ with a black outer stripe, and a similar line along the extensor surface of femora. *Wings* palely enfumed, petiolated as far as *ac*, rather pointed at apices; pterostigma short and broad, not quite twice as long as broad, distal end nearly straight, proximal oblique and in line with the brace, costal border shorter than hinder, covering about two cells, dark brown, with white or pale inner and outer ends; 10 to 11 postnodal nervures in the fore-wings, 9 in the hind, more rarely 8. *Abdomen* olivaceous, changing to warm reddish-brown on the hinder segments; segments 1 to 6 with small paired subdorsal, lanceolate, subapical dark brown spots, these segments paler at both ends, especially at the base, where the ground-colour tapers to a point; segments 8 to 10 with similar subapical spots, but yellow and conspicuous against the dark background; segments 2 to 7 with black apical rings. *Anal appendages* whitish, the superiors black at base,

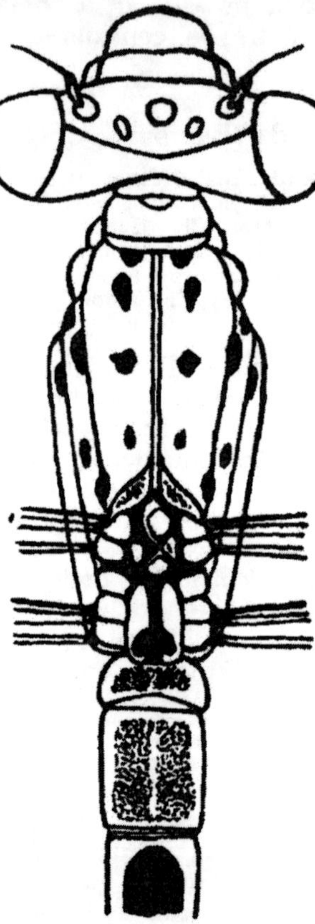

Fig. 25.—Head and thorax of *Platylestes platystyla* (Ramb.), showing markings.

curling in at apices to meet each other in a gradual and regular curve, but the extreme apices curling a little outwards again, the outer border coarsely spined, the inner border with a thin shell-like plaque beginning from near the base as an obtuse spine and terminating about the junction of the middle and outer thirds of the appendage in an acute spine; apex furnished with a tuft of stiff hairs; the whole appendage longer and narrower than in *Lestes*, and the inner expansion much narrower and resembling rather closely the type of *Ceylonolestes*; inferior appendages about half the length, thick at base, apposed at extreme apices, then divergent and abruptly narrowed and ungulate, finally convergent at apices so as to enclose a small oval space.

Female.—Abdomen 33 mm. Hind-wing 23 mm.

Closely resembles the male in most respects, differing mainly in sexual characters. The black spots on the postclypeus and *head* are absent; labrum and bases of mandibles pale brown or yellowish; eyes brown; *thorax*, *legs*, and *wings* similar to the male; fore-wings with 10 postnodal nervures, hind with only 8; pterostigma about twice as long as broad; *abdomen* dark brown on dorsum, segments 8 and 9 dark

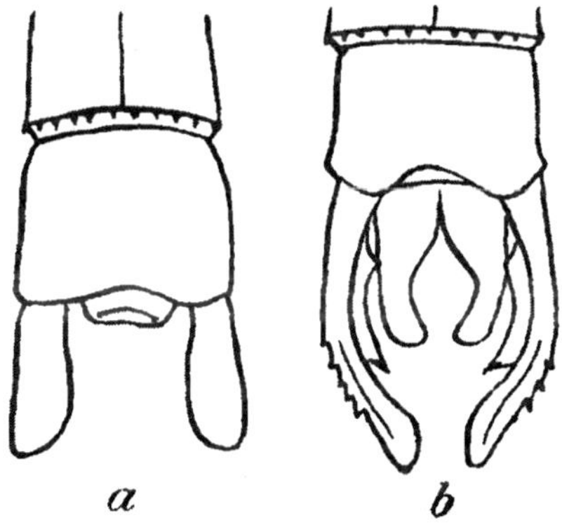

Fig. 26.—Anal appendages of *Platylestes platystyla* (Ramb.).
a. Dorsum of female. *b*. Dorsal view of male.

ochreous, 10 golden brown, clouded with dark brown near the base and along the mid-dorsal carina, otherwise marked as for the male. *Anal appendages* yellow, blackish-brown at the base, as long as segment 10, flattened, lanceolate, but inner border rather sinuous, apex obtusely pointed. Vulvar scale robust, not extending to the end of abdomen, yellow, clouded with brown.

Distribution.—BENGAL and BURMA.

Very few specimens of this rare and interesting species are known; the *type* is a female in the Selysian collection, the original Rambur type. Its description differs considerably from the above, but the differences are probably due to its poor condition and, perhaps, decomposition. The above description is made from a fine specimen from Hasimara, Duars, Bengal,

taken by Mr. H. V. O'Donel, 7. viii. 23, a female with one anal appendage missing, and now in my collection. The description of the male is from a specimen in the Indian Museum taken in Calcutta in November. There is another male in the Pusa collection from Burma, and these four specimens are the only ones known. The quadrate pterostigma will at once distinguish the species from any other *Lestes*. It is possible that more than one species exists among these four, but I doubt it. It is to be hoped that more material will come to hand to settle this point.

Genus **CEYLONOLESTES** Kennedy (1920). (Fig. 27.)

Ceylonolestes Kennedy, Ohio Journ. Sci. vol. xxi, no. 2, p. 84 (1920); Fras., J. Bombay Nat. Hist. Soc. vol. xxxiv, p. 94 (1930).
Ceylanicolestes Fras., Rec. Ind. Mus. vol. xxvi, p. 487 (1924).
Lestes (Section 2, 1st Group, A) Selys, Syn. Lestes, p. 42 (1862).

Dragonflies of small stature, resting with wings closed over dorsum; body non-metallic, ground-colour bright azure blue; wings hyaline, untinted, petioled as far as *ac*,

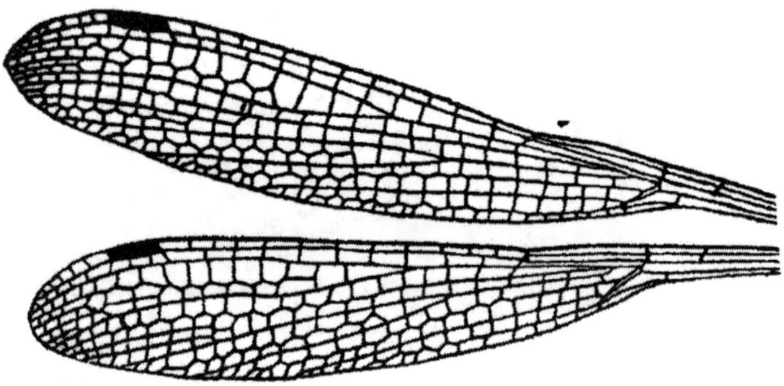

Fig. 27.—Wings of *Ceylonolestes cyanea* (Selys), male.

ab meeting ac at hinder border of wing, ac situated at a level more or less distal to the mid-point between the two antenodals, discoidal cells differing in shape and size in fore- and hind-wings, narrow, long, and acutely pointed at the distal angle, discoidal cell of fore-wing with posterior border more than twice the length of the costal and nearly four times the length of basal, that of the hind wing with posterior border not quite twice the length of costal and five times the length of basal, *Riii* arising about 4 cells distal to the node in fore-wing, 3 cells in the hind; *IRiii* and *Riv+v* arising much closer to arc

than to node, *IRiii* zigzagged towards its apical end, *MA* zigzagged from level of node, oblique vein present between *Riii* and *IRiii*; pterostigma unicolorous, narrow, more than twice as long as broad, braced, outer end not oblique; intercalated sectors similar to genus *Lestes*. Posterior lobe of prothorax simple, rounded, not lobed. Superior anal appendages forcipate, long and attenuated, furnished along the middle third of the inner border with a narrow expansion which begins and ends in a sharp spine; inferior appendages variable, anal appendages of female conical, short.

Distribution.—CEYLON, S. INDIA, BURMA, BENGAL, PUNJAB, and Japan. Four species are confined to South India and Ceylon, two to Burma, one to N. India, and the remaining one (*C. peregrinus* Ris) to Japan.

Genotype, *Ceylonolestes gracilis* (Selys).

Key to the Species of Ceylonolestes.

1.	Thorax with a single mid-dorsal dark band only, no enclosed antehumeral stripes	2.
	Thorax with humeral and lateral as well as dorsal dark markings, and more or less enclosed antehumeral stripes	4.
2.	Wings enfumed pale brown; pterostigma very elongate, longer in the hind-wing than in the fore, bright rusty red in colour	*inflata*, sp. n., p. 67.
	Wings not enfumed; pterostigma of the same size in fore- and hind-wings, black or blackish-brown in colour	3.
3.	Dorsal thoracic marking and dorsal markings of segments 1 and 2 dark metallic green; pterostigma comparatively short and stout, black	*divisa* (Hagen-Selys), [p. 65
	Dorsal thoracic marking dark reddish cupreous brown; dorsum of segments 1 and 2 black; pterostigma elongate and narrow, brownish-black	*pulcherrima* Fras., [p. 72.
4.	Antehumeral stripes partially enclosed on the outer side by a row of 2 or 3 black spots	*gracilis* (Hagen), p. 64.
	Antehumeral stripes entirely enclosed on the outer side by a complete black band.	5.
5.	Abdominal segments 2 to 7 with paired dorsal apical spots or stripes, the mid-dorsum blue between them	*cyanea* (Selys), p. 74.
	Abdominal segments 2 to 7 broadly black on dorsum	6.
6.	Burmese species	*birmana* (Selys), p. 71.
	South Indian species	*davenporti* Fras., p. 69.

19. Ceylonolestes gracilis (Hagen). (Fig. 28.)

Lestes gracilis Hagen, Verh. Zool.-bot. Ges. Wien, vol. viii, p. 478 (1858); Selys, Bull. Acad. Belg. (2) vol. xiii, p. 327 (1862); Kirby, Cat. Odon. p. 163 (1890); id., Proc. Zool. Soc. Lond. p. 206 (1891); id., J. Linn. Soc. Lond., Zoöl. vol. xxiv, p. 566 (1893); Laid., Rec. Ind. Mus. vol. xix, p. 158 (1920).

Lestes gracilis gracilis Ris, Suppl. Ent. no. v, pp. 13–15, pl. i, fig. 4 (1916); Laid., Spolia Zeylan. vol. xii, p. 358 (1920).

Ceylonolestes gracilis Fras., J. Bombay Nat. Hist. Soc. vol. xxxiv, pp. 94, 95, pl. i, fig. 8 (1930).

Male.—Abdomen 30–33 mm. Hind-wing 20–22 mm.

Head: labium yellowish-white; labrum, cheeks, and bases of mandibles pale azure blue; rest of head mat black, with a cupreous reflex; eyes deep blue above, paler beneath. *Prothorax* blue at the sides, with two thick brownish-black bands on the dorsum. *Thorax* azure blue, paler on sides and beneath, dorsum with a thick cupreous or bronzed black

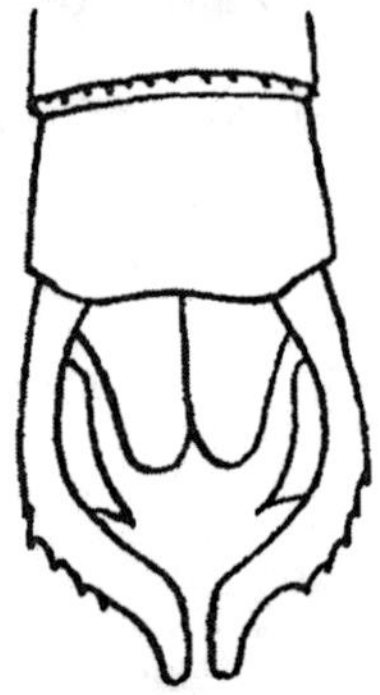

Fig. 28.—Dorsal view of anal appendages of *Ceylonolestes gracilis* (Hagen), male.

band extending outwards nearly as far as the humeral suture; upper part of humeral suture black, and slightly posterior to it a chain of three bronzed black spots; a spot on the lower part of mesepimeron just behind the trochanter and another on the upper part of the postero-lateral suture. Beneath, two small triangular blackish areas on the paired sclerites. *Legs* ochreous on the extensor surfaces, black on the flexor, as also the tarsi; spines moderately long and numerous. *Wings* hyaline; 10 to 12 postnodal nervures to fore-wings, 10 to 11 in the hind; pterostigma rather more than twice as long as broad, outer end not oblique, covering two cells, braced, dark reddish-brown. *Abdomen* azure blue, marked on dorsum with bronzed black; segment 1 with a small baso-dorsal quadrate black spot, 2 with a broad dorsal band falling short of the apical border, sometimes split up into two broad parallel stripes by the dorsal crest, which is bluish, but in adults the stripes confluent for rather more than the basal half, after which is seen a small round blue spot on the dorsal carina continuous with a fine blue line which runs into an apical blue ring; segments 3 to 8 with narrow blue basal rings and the mid-dorsal carina finely yellow, rest of dorsum bronzed black; segment 9 with the dorsal black ending well before apical border as two widely divergent points, the apical portion blue, as also the whole of segment 10, except for a small basal spot of black on each side. *Anal appendages* black, very long, sinuous, and forcipate, curling gradually inwards until the apices meet, after which they are directed straight back in a bayonet-like angulation to end in a moderately acute point; the inner border at its middle third furnished with a scale-like dilatation which begins rather abruptly as a rounded angle and ends in a long fine spine; beneath this dilatation, at its middle, a small pointed tubercle visible only in profile, outer border near apex coarsely spined; inferior appendages tumid, apposed, rounded at apex, about half the length of superiors, which are about half as long again as segment 10.

Female.—Abdomen 29–30 mm. Hind-wing 23 mm.

Marked exactly as in the male, but the ground-colour greenish-blue. In some specimens the dorsal markings of segments 2 to 5 or 6 are metallic green and the mid-dorsal carina is finely ochreous as far as segment 7. *Wings* with 9 to 10 postnodal nervures in the fore and 9 in the hind; pterostigma similar to the male. *Anal appendages* brown, paler at the base, black at apices, pointed, conical, about as long as segment 10; vulvar scale yellow or pale brown, broadly black along lower border, which is not serrate.

Distribution.—CEYLON: hill tracts up to 6,000 ft. The *type*, a male in the Hagen collection, is from Ramboda. I have a large series collected by Col. F. Wall, I.M.S., at Bandarawela, 5,000 ft., during October.

The species is distinguished from others by the bifid dorsal marking on segment 9 and by the posthumeral chain of three spots. As pointed out by Dr. Ris, these spots are not mentioned in the original description.

20. **Ceylonolestes divisa** (Hagen-Selys). (Fig. 29.)

Lestes divisa Hagen-Selys, Bull. Acad. Belg. (2) vol. xiii, p. 328 (1862); Laid., Rec. Ind. Mus. vol. xix, p. 158 (1920); Laid., Spolia Zeylan. vol. xii, pp. 358–60 (1924).

Lestes divisus Kirby, Cat. Odon. p. 163 (1890); id., J. Linn. Soc. Lond., Zool. vol. xxiv, p. 566 (1893); Ris, Ent. Suppl. no. 5, p. 13 (1916).

Ceylonolestes divisa Fras., J. Bombay Nat. Hist. Soc. vol. xxxiv, pp. 95, 96 (1930).

Male.—Abdomen 33 mm. Hind-wing 19–21 mm.

Head: labium pale yellow; labrum, genæ, bases of mandibles, and anteclypeus turquoise-blue, as also a small spot at the centre of postclypeus; rest of head, including beneath, black; eyes dark olivaceous. *Prothorax* pale blue at sides, dorsum blackish-brown, this area split by a narrow reddish-brown line on the mid-dorsum; posterior lobe broadly rounded, shallow. *Thorax* with a broad dorsal metallic-green band extending nearly up to the humeral suture, narrowing somewhat below; laterally pale blue, with a small black spot at the upper end of humeral suture, a second at the upper end of postero-lateral suture, and a third at the base of the hind legs; two elongate oval metallic-green spots lying just posterior to the humeral suture. *Legs* ochreous, flexor surfaces dark blackish-brown. *Wings* hyaline; pterostigma rather short, only twice as long as broad, outer end squared, black, covering two cells; 10 to 12

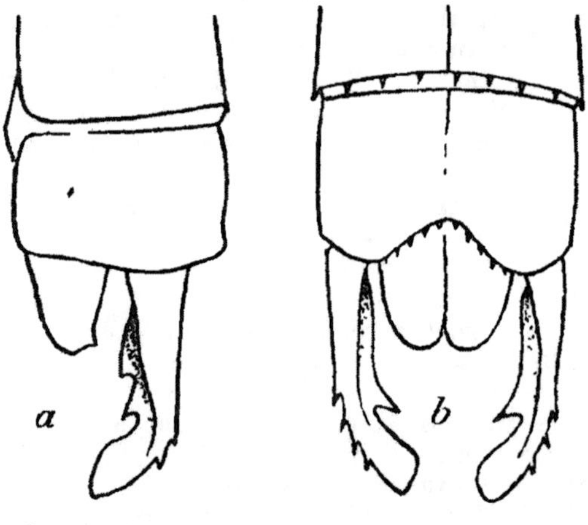

Fig. 29.—Anal appendages of *Ceylonolestes divisa* (Hagen-Selys), male. *a.* Left lateral view. *b.* Dorsal view.

postnodal nervures in fore-wings, 10 in the hind. *Abdomen*: segment 1 with a broad black basal spot, the sides blue; segment 2 with a narrow dorsal thistle-shaped marking extending the whole length of segment, black for its apical third, dark metallic green for the basal two-thirds; segments 3 to 8 broadly black on dorsum, this band expanding somewhat subapically, narrowing basally to leave narrow blue basal annules; segment 9 black for its basal half, the black tapering apically, apical half and the whole of segment 10 azure blue. *Anal appendages* ochreous, black at apical ends; superiors longer than segment 10, forcipate, narrow, with some spines along the outer apical border, furnished within by an expansion which begins gradually and is bevelled off into the appendage, whilst at its apical end it terminates in a long fine spine; apex glossy black, subrotundate, slightly bevelled within. Inferiors only half the length of superiors, very tumid, conical, broad at base, bluntly obtuse at apex.

Female.—Abdomen 31 mm. Hind-wing 21 mm.

Very similar to the male, but the ground-colour greenish or pale olivaceous, with markings similar to the male. *Anal appendages* shortly conical, brown; vulvar scales short, dark brown, minutely serrate along lower border.

Distribution.—Confined to the hilly tracts of CEYLON. *Type*, a male, from the Ramboda Pass, in the Hagen collection; co-type, a female in the same collection. Laidlaw reports it from Kandy, 2,000 ft., during the months of May to July. Paratypes in the Laidlaw and author's collections.

Closely allied to and similar to *C. gracilis*, from which it differs by the short black pterostigma, metallic-green dorsum to thorax, and marking on dorsum of segment 2; also by the inner dilatation of the superior anal appendages beginning gradually instead of abruptly.

21. **Ceylonolestes inflata**, sp. n. (Fig. 30.)

Male.—Abdomen 35 mm. Hind-wing 19 mm.

Head: labium palest yellow; labrum, bases of mandibles, and genæ turquoise-blue; anteclypeus yellowish-brown; postclypeus with four quadrate black spots on a tawny ground-colour as in *C. pulcherrima*; frons, vertex, and occiput dark reddish-brown, with a cupreous reflex, but the back of latter and beneath the head pale bluish-green; eyes olivaceous, darker brown above. *Prothorax* bluish-green on the sides and subdorsum, with a broad dark reddish-brown stripe on mid-dorsum extending from anterior to posterior lobe and narrowly divided by a thin medial reddish stripe; posterior lobe broadly rounded, its border slightly crenate. *Thorax* pale blue on sides, bluish-green on the humeral region, the mid-dorsum with a moderately narrow black stripe with a cupreous reflex. The mid-dorsal carina very finely reddish; a small black spot at the upper end of each lateral suture; beneath immaculate, pale yellow. *Legs* bluish, darker at distal ends of femora, the tibiæ bright ochreous on the extensor surface, spines black. *Wings* enfumed brown, reticulation reddish-brown; pterostigma bright ferruginous framed in reddish nervures, very elongate, especially in the hind-wings, where it is nearly half as long again as in the fore, ends slightly oblique; 13 postnodal nervures in fore-wings, 11 in the hind. *Abdomen* pale azure blue at sides, marked broadly with black on dorsum as follows:—Segment 1 with a broad quadrate spot extending from end to end of segment; segment 2 with a narrow thistle-shaped marking connected by its stalk to a narrow apical ring; segments 3 to 5 with a broad dorsal stripe expanding apically and confluent here with a narrow apical ring, tapering to a point at base; segments 6 to 9 and the apical third of 5 dark reddish-brown, but the sides narrowly blue; segment 9 with the apical half bluish; segment 10 almost entirely blue, marked with two broad basal spots confluent at their bases. *Anal appendages* black; superiors slightly longer than segment 10, but the apical half curled strongly down and in, so as to give a foreshortened appearance to the appendage; a broad inner shelf at the middle of each appendage ending distally in a very robust spine; some

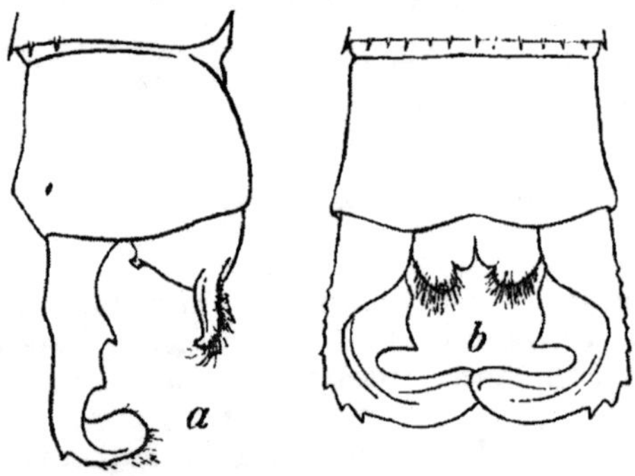

Fig 0.—Anal appendages of *Ceylonolestes inflata*, sp. n., male. *a*. Seen from the right side. *b*. Seen from above.

robust spines along the outer border; apex rounded, obtuse, very glossy; inferiors half as long, very tumid and rounded or globate, with a short foliate process perched on top of each, directed straight back.

Female unknown.

Distribution.—A single adult male from Maymyo, North Shan States, BURMA, collected by Col. F. Wall, 5. vi. 25, now in the author's collection.

Bears a strong resemblance to *C. pulcherrima*, from which it is distinguished by the strongly angulated and much more robust appendages, and also by the enfumed wings with bright rusty elongate pterostigma. The same characters will serve to separate it from the rest of the genus. The only other species of the genus from Burma has enclosed blue antehumeral stripes.

22. Ceylonolestes davenporti Fraser. (Fig. 31.)

Lestes gracilis birmanus Ris, nec Selys, Suppl. Ent. no. 5, pp. 13, 14 (1916); Laid., Rec. Ind. Mus. vol. xix, p. 158 (1920).
Ceylonolestes davenporti Fras., J. Bombay Nat. Hist. Soc. vol. xxxiv, pp. 96, 97, pl. I, fig. 7 (1930); id., Rec. Ind. Mus. vol. xxxiii, pp. 448, 464 (1931).

Male.—Abdomen 30–33 mm. Hind-wing 19–23 mm.

Head: labium pale yellow; labrum, bases of mandibles, cheeks, and clypeus turquoise-blue, postclypeus with its outer ends broadly black, only the central part blue; head above bronzed or cupreous black; occiput yellow at the median parts of its hinder border. *Prothorax* black on dorsum, broadly blue laterally, and with a median dorsal stripe on the middle lobe tapering anteriorly; anterior lobe blue, with a small black spot posteriorly on its mid-dorsum;

Fig. 31.—Dorsal view of anal appendages of *Ceylonolestes davenporti* Fras.

posterior lobe with a fine mid-dorsal blue line which broadens anteriorly to become confluent with the mid-dorsal blue of middle lobe. *Thorax* black on dorsum, with a cupreous reflex as far back as nearly to middle of mesepimeron, the border of the black being broadly angulate or crenulate; a narrow antehumeral blue stripe bordering the humeral suture, incomplete above, where it slightly overlaps the suture; laterally azure blue, with a small linear spot on mesepimeron near its upper part, and a second black spot on upper part of postero-lateral suture. Beneath pale blue or yellow clouded with brown, two oblique spots on the paired sclerites just behind the hinder pair of limbs and a pair of tiny pyriform black spots at centre. *Legs* bluish-green on whole of extensor surfaces from femora to tarsi, black on flexor. *Wings* hyaline; pterostigma blackish or dark reddish-brown, the terminations of the bordering nervures posteriorly white, covering nearly two cells, braced, nearly three times as long as broad; 10 to 13 postnodal nervures in fore-wings, 10 to 11 in the hind. *Abdomen* azure blue laterally, broadly black on dorsum, with a metallic green reflex as far as segment 5, then with cupreous reflex to the end of segment 8; the greater part of 9 and the whole of segment 10 azure blue save for a narrow medio-basal black line and a lateral black spot on each side of segment 10; segment 1 black on dorsum from base to apex; segment 2 marked as for *C. divisa*, the metallic band extending from base to apex, but tapering to a point at apex of segment to become confluent with a narrow apical black ring, mid-dorsal carina finely ochreous, expanding into a small blue spot just apical to the middle of segment; segments 3 to 5 with the mid-dorsal carina finely ochreous, with narrow blue basal rings and with the dorsal black expanded subapically on each segment; segments 7 and 8 entirely black save for the ventral borders of 8 narrowly; segment 9 with a broad triangular black spot on each side narrowly confluent over dorsum at base, tapering apically, but not nearly reaching end of segment. *Anal appendages* blue, apices and inner border black, or entirely black in old specimens, the apices thickly coated with white hairs; superiors forcipate, long, attenuated, and curved gently towards each other to meet at apices, which are thickened, with the point directed straight back, coarsely spined along outer border near apex, and with the usual expansion on the inner border at the middle third beginning as an abrupt rounded angle and ending in a long acute spine; beneath the appendages, near the middle of the expansion, a small tubercle seen in profile; inferior appendages half the length of superiors, which are about half as long as segment 10; broad, rounded, apposed at bases; apices rounded, black.

Female.—Abdomen 31 mm. Hind-wing 24 mm.

Closely similar to the male, the markings only differing on segments 8 and 9, where the sides are broadly blue; segment 9 has the whole of dorsum cupreous black, whilst 10 has two large triangular black spots confluent or not over dorsum as on segment 9 in the male. *Wings* hyaline, palely enfumed in very old specimens; 11 postnodal nervures in the fore-wings, 9 in the hind; pterostigma reddish-brown. *Anal appendages* greenish-blue, about as long as segment 10, conical, pointed; vulvar scale yellow, brownish-black along the border, which is very minutely serrate.

Distribution.—Western Ghats south of the Palghat Gap, from 4,000 ft. to 6,000 ft. I have specimens from the Palni Plateau and from the Mudis and Anaimalai Hills, the latter being decidedly smaller than the Palni specimens, which are taken at a greater altitude. Recently I found it moderately common at Manar, Travancore, during July.

Mr. T. Bainbrigge Fletcher and Col. Frere took this species plentifully in the Palnis from May to July, but I found it by no means common in the Mudis, where it required much searching for. Here it frequented only one locality amongst scrub, near a river, resting almost vertically and with closed wings. I was much struck with how blue it appeared as compared with dried specimens, or even with *C. pulcherrima* in the living state.

It is distinguished easily from *C. gracilis* and *C. divisa* by the blue antehumeral stripe. From *C. birmana* Selys, with which Dr. Ris thinks the species to be synonymous, the differentiation is far more difficult, but as the Selysian description is very brief, I have no doubt differences will easily be found. All species of the genus are so local in their distribution that I am unable to conceive *C. birmana* as occurring in Continental India and in so far removed a district as the Western Ghats. It is thus, for geographical reasons only at present, that I consider them to be two distinct species.

23. **Ceylonolestes birmana** (Selys).

Lestes birmana Selys, Ann. Mus. Civ. Genova, (2) vol. x (xxx), p. 495 (1891).

Lestes gracilis birmanus Ris, nec Selys, Suppl. Ent. no. 5, pp. 13–14 (1916); Laid., Rec. Ind. Mus. vol. xix, p. 158 (1920).

Ceylonolestes birmanus Fras., J. Bombay Nat. Hist. Soc. vol. xxxiv, pp. 97, 98 (1930).

Male.—Abdomen 30 mm. Hind-wing 21 mm.

The Selysian description of a single male taken at Pwepoli, BURMA, in June, is as follows :—" I am not certain that this single male is *divisa* Hagen, which I do not possess and which I know only from a short description which I published in 1862 in 'Syn. Lestes,' no. 40. In the Puepoli specimen there is, immediately after the humeral suture on the sides of the thorax, a moderately broad black band of which the outer border has three angles *en échelon* somewhat like that seen in *L. colensonis* of New Zealand. This black band is not mentioned in the description of *divisa* and if not a simple omission, the species of Fea will be new and I propose for it the name of *Lestes birmana*, distinguished from *L. divisa*, Ceylon, by the front of thorax black with a blue antehumeral band on each side."

Thus the description, if such a brief one may be so called, fits *C. davenporti* exactly; but, even so, such is the extremely local character of all species of *Ceylonolestes* that the vast distance separating Burma from the Western Ghats of India seems to me an insuperable difficulty in regarding the two insects as conspecific. Further material from Burma is needed to settle this point.

24. Ceylonolestes pulcherrima (Fraser). (Figs. 32 & 33.)

Ceylanicolestes pulcherrima Fras., Rec. Ind. Mus. vol. xxvi, pp. 487–9, pl. xxvi, fig. 6 (1924).
Ceylonolestes pulcherrima Fras., J. Bombay Nat. Hist. Soc. vol. xxxiv, pp. 98, 99, pl. i, fig. 10, and text-fig. 2 *a* (1930); id., Rec. Ind. Mus. vol. xxxiii, p. 448 (1931).

Male.—Abdomen 34 mm. Hind-wing 21 mm.

Head: labium white; labrum, bases of mandibles, cheeks, and anteclypeus turquoise-blue; postclypeus darker blue,

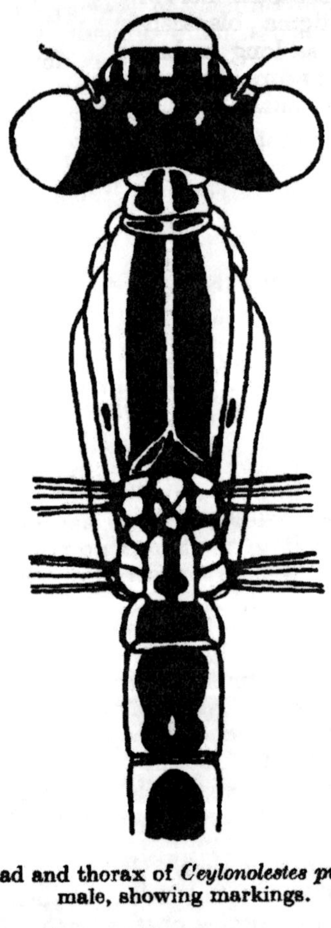

Fig. 32.—Head and thorax of *Ceylonolestes pulcherrima* Fras., male, showing markings.

marked with two small black points, rest of head black, with a dull green metallic reflex; eyes deep blue above, turquoise-blue in front, greenish-yellow beneath. *Prothorax* blue laterally, brownish above, marked with two dark metallic green parallel stripes. *Thorax* turquoise-blue laterally, the dorsum narrowly dark metallic green, the mid-dorsal carina and margins of antealar sinus finely blue; laterally marked with a large black spot behind the upper part of the first lateral suture and a smaller one behind its middle; the upper ends of sutures black, beneath dirty white, with two large black spots converging at the anterior part of the unpaired sclerites, and a larger blackish-brown spot on either side posterior to them. *Legs* brown, flexor surfaces black. *Wings* hyaline, postnodal nervures to fore-wings 12, in the hind 11; pterostigma blackish-brown, with paler borders, about three times as long as broad, inner side oblique, outer straight. *Abdomen* azure blue marked with black, as follows:—Segment 1 with a small basal black spot; segment 2 with a narrow apical ring and a broad dorsal band shaped like the

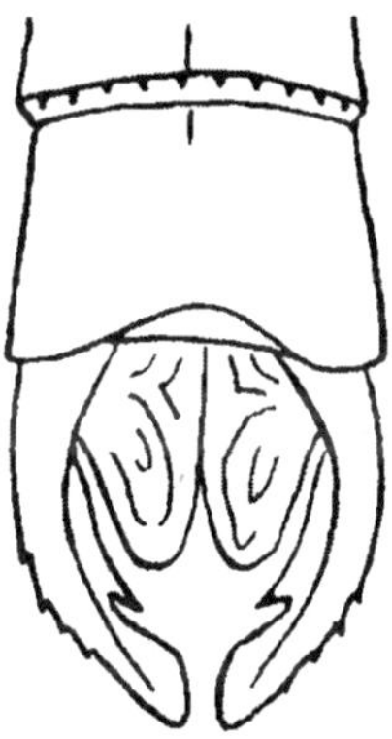

Fig. 33.—Dorsal view of anal appendages of *Ceylonolestes pulcherrima* Fras.

head of a thistle extending from base to apex; segments 3 to 7 with dorsal bands not quite extending to base of segments, where they leave a small blue ring, apically expanding, and then contracting again to become confluent with narrow apical black rings; segment 8 all black, except the apical suture, which is blue; segment 9 with the basal third black, the remainder blue, as is the whole of segment 10. *Anal appendages*: superiors blue at base, turning to white and finally brown at apices, which are tipped with white hairs, narrow, forcipate, apices curling gradually in until they meet, outer border near apices coarsely spined, furnished on the inner border with a very narrow expansion, which begins to emerge from the appendage gradually, not abruptly as in other species (except *C. divisa*), and ends in a long fine spine; inferior appendages about half the length of superiors, which are about half as long again as segment 10, very tumid, apposed except at extreme apices, broadly conical.

Female.—Abdomen 33 mm. Hind-wing 24 mm.

Head: eyes olivaceous green above, pale greenish-yellow beneath; labrum very pale blue, rest of head as in the male. *Prothorax* and *thorax* pale olivaceous brown, with a somewhat violaceous tint, the mid-dorsal carina and margins of antealar sinus finely reddish-brown, black markings exactly similar to the male. *Legs* pale yellowish-brown, flexor surfaces black. *Wings* similar, but in old adults evenly tinted with pale brown; 11 to 12 postnodal nervures to fore-wings, 10 in the hind; pterostigma as for the male. *Abdomen* violaceous brown, changing to reddish-brown on the distal segments, marked with black, as follows:—Segment 1 with a basal black dorsal spot; segment 2 with a broad dorsal band split by the reddish-brown dorsal carina for its apical two-thirds, the band unevenly expanded towards the basal border of segment, abruptly expanded towards the apical; segments 3 to 5 as for the male; segments 5 and 7 with blue basal rings; segments 6 and 7 with the dorsal band rather ill defined towards the base of segments, whilst segments 8 to 10 are uniform dark brown. *Anal appendages* as long as segment 10, conical, pointed, pale blue; vulvar scale robust, dark brown, very minutely serrate along free border.

Distribution.—Found only in COORG, but in many swampy localities at about 2,500 ft. to 3,500 ft. in that area. Usually found in jungly retreats; a belt of dense scrub or trees surrounding a small pond is the favourite situation. In such a place they may be found perched on the tips of twigs at any height up to about 15 feet from the ground. The species is essentially arboreal, males being found in the above-described swampy retreats, females in the surrounding jungle not far from the breeding-places.

25. **Ceylonolestes cyanea** (Selys). (Figs. 27 & 34.)

Lestes cyanea Selys, Bull. Acad. Belg. (2) vol. xiii, p. 335 (1862); Laid., Rec. Ind. Mus. vol. xix, pp. 158–61 (1920).
Lestes cyaneus Kirby, Cat. Odon. p. 163 (1890).
Indolestes helena Fras., Mem. Dept. Agric. India (Ent.), vol. vii, pp. 60–1, pl. vii, fig. 1 (1922).
Indolestes veronica Fras., ibid. vol. viii, no. 8, p. 85 (1924).
Ceylonolestes cyanea Fras., J. Bombay Nat. Hist. Soc. vol. xxxiv, pp. 99–101, pl. i, fig. 9, & text-fig. 3 (1930).

Male.—Abdomen 32–36 mm. Hind-wing 21–24 mm.

Head: labium palest brown or dirty white; labrum, cheeks, and bases of mandibles turquoise blue; rest of head mat black, with a coppery reflex behind the bases of antennæ; eyes dark blue above, paler below. *Prothorax* black on dorsum, with a blue mid-dorsal stripe, narrow on the posterior lobe, broadening abruptly on the median lobe; laterally pale blue, traversed by a thick longitudinal black stripe. *Thorax* with the dorsum and sides as far posterior as the middle of mesepimeron black, with a violet or coppery reflex, the borders of the antealar sinus and the mid-dorsal carina finely mapped out in pale greenish-blue, a narrow antehumeral stripe gradually broadening below, limited posteriorly by the humeral suture and incomplete above, but continued posterior to the suture by a narrow dagger-shaped stripe, the upper portion of which may be cut off to form an isolated small spot; laterally blue, the margin where it meets the black anteriorly being irregularly zigzagged or angulated, the black extending back above and gradually tapering away on the metepimeron and descending for a short

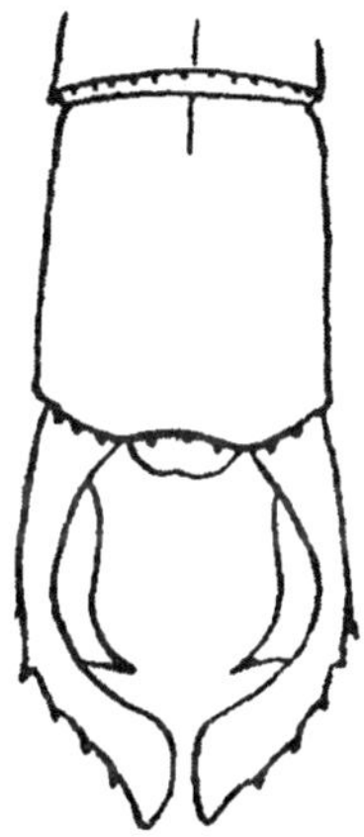

Fig. 34.—Dorsal view of anal appendages of *Ceylonolestes cyanea* (Selys).

distance along the postero-lateral suture; beneath pale brown, with two triangular areas dark brown. *Legs* with the flexor surfaces of femora and tibiæ and the tarsi black, the extensor surfaces of femora and tibiæ bright ochreous. *Wings* hyaline, palely enfumed in very old adults only; 13 to 14 postnodal nervures to fore-wings, 11 to 12 in the hind; pterostigma rather squared at both ends, especially distally, braced, but the brace occasionally proximal to the pterostigma, about 3½ times as long as broad, black or blackish-brown, covering three or more cells; *Riii* arising 3 to 4 cells distal to the node; posterior border of discoidal cell in the fore-wing about 4 times as long as the basal, and about 5 times as long in the hind; *ac* slightly nearer the level of the distal antenodal nervure. *Abdomen* azure blue marked with black, as follows:—Segment 1 with the base black, the outer borders of this marking prolonged apically as a narrow black line as far as the apical border; segment 2 with a narrow subdorsal black stripe on each side, which becomes confluent at a point over the dorsum at the apical border of the segment and sends up a subapical pointed prolongation over the dorsum to nearly enclose a small subapical spot of the ground-colour; segments 3 to 6 with apical paired wedge-shaped spots confluent finely over the dorsum at apical border but tapering basally, the mid-dorsal carina finely blue between them, these spots gradually lengthening from segments 3 to 6, on the latter of which they extend for about half-way to base; segment 7 nearly entirely black, only a narrow basal ring and the whole length of the mid-dorsal carina finely blue; segments 8 and 9 all black except the mid-dorsal carina, which is finely blue; segment 10 entirely blue, but some black spines along its apical border. *Anal appendages* blue, changing to ferruginous and finally blackish-brown at extreme apex, about half as long again as segment 10, forcipate, curving gradually in to meet at apices, which are directed straight back (the two appendages resembling the arms and hands of a man in the act of diving), furnished near the base with a narrow expansion occupying the middle third of appendage, arising insensibly from the body of appendage and not as a blunt spine, and terminating in a long, thin, acutely pointed spine directed towards apex of appendage; apex obtusely pointed and directed a little downwards and slightly expanded as seen in profile, tipped with fine white hairs, outer border rather coarsely spined; inferior appendages rudimentary, tumid, rounded at apices, closely apposed, about one-fourth the length of superiors.

Female.—Abdomen 33–36 mm. Hind-wing 24–28 mm.

Head, thorax, and *legs* marked exactly as in the male, but the ground-colour pale greenish-blue; *wings* in elderly adult specimens more deeply enfumed, especially towards the apices; 12 to 14 postnodal nervures to fore-wings, 11 to 12 in the hind; pterostigma dark reddish-brown, the inner and outer borders finely bright ochreous. *Abdomen* with the black markings more extensive, as follows:—Segment 1 with whole of dorsum black save for a small mid-dorsal spot of blue; segment 2 with the subdorsal black stripe thicker and confluent at base as well as apical border of segment, the included ground-colour narrow at apex and base but broadening at the middle three-fifths; segments 3 to 6 with the apical spots prolonged basally as narrow black stripes almost as far as the base, where they leave a broad basal blue ring confluent with a broad mid-dorsal blue stripe, which extends as far as apex of segments; segment 7 similar to the male, but the basal blue ring broader; segment 8 with the dorsum all black save for the basal half of mid-dorsal carina, which is finely blue; segment 9 entirely black and segment 10 all blue save for a minute mid-dorsal basal triangle of black. *Anal appendages* pale ochreous or carneous, about as long as segment 10, conical, pointed; vulvar scale blackish-brown, yellow along attached border, lower margin not serrate.

Distribution.—BENGAL and hills of the NORTHERN PUNJAB from 3,500 to 9,000 ft. during May and June. I possess specimens from Simla and Darjeeling, and at the latter place I found larvæ breeding in swift cold mountain-streams at Ghum. A thousand feet lower down I found the imago on the wing at the same date.

This species is one of the most beautiful Lestids known, and is easily distinguished from all others by its abdominal markings. It is much larger than all other species of the genus *Ceylonolestes*, and differs from them also by the short spines on its legs. It appears to be more closely allied to the genus *Austrolestes*, and should, perhaps, be placed in a genus of its own. It rests, however, with the wings closed and has narrow differing discoidal cells like *Ceylonolestes*.

After due consideration I have come to the conclusion that my species *Indolestes helena* and *I. veronica* are merely varieties of this species, the difference in ground-colour and markings being due probably to their age. I have come to this conclusion after perusing Dr. Laidlaw's discussion on some teneral specimens of *C. cyanea* which greatly puzzled him, as did the specimens of my *I. helena* and *I. veronica*, and which Dr. Laidlaw grouped in his second category, "B." The anal appendages of *I. helena* are similar to those of *C. cyanea*, and the difference in the markings may well be explained through decomposition. The specimens of *I. veronica* are the largest known specimens of *C. cyanea*, if indeed they are synonymous with that species. The type of *C. cyanea* is in the Selysian collection, of *I. veronica* and *I. helena* in the British Museum collection.

Genus **INDOLESTES** Fraser (1922).

Indolestes Fras., Mem. Dept. Agric. India (Ent.), vol. vii, pp. 57, 58 (1922); id., J. Bombay Nat. Hist. Soc. vol. xxxiv, pp. 101, 102 (1930).

Dragonflies of small stature resting with wings closed over dorsum; body non-metallic or this restricted to some small spots on the basal abdominal segments, ground-colour pale brown; wings hyaline, always more or less tinted evenly pale brown, very narrow and with pointed apices, petiolated as far as *ac* ; *ac* meets *ab* at hinder border of wing ; postnodal nervures numerous ; discoidal cells differing in shape and size in fore- and hind-wings, very narrow and elongate and very acutely pointed at distal angle, discoidal cell of fore-wing with posterior border 3 times as long as basal and twice the length of costal, that of hind-wing with posterior border not quite double the length of costal and 6 times as long as basal ; *Riii* arising 3½ cells distal to node in fore-wing, 2½ cells in the hind ; *IRiii* and *Riv+v* arising much nearer arc than node ; *IRiii* slightly angulated towards its apical end, *MA* zigzagged from level of node ; oblique vein present between *Riii* and *IRiii* ; pterostigma bicolorous, narrow, 3 times as long as broad, outer end not oblique ; intercalated sectors similar to genus *Lestes*. Posterior lobe of prothorax simple, rounded, not lobed. Superior anal appendages narrow, elongate, forcipate, furnished at the middle third of inner border with a narrow expansion, which begins with an obtuse spine and ends with a long acute one ; inferior appendages short, tumid ; anal appendages of female conical, short.

Distribution.—N.E. INDIA and BURMA.

This genus is very similar to *Ceylonolestes* and so nearly related that I have been unable to find any very satisfactory characters by which to differentiate them. The points which have influenced me in bestowing separate generic rank are the bright blue ground-colour of all species of *Ceylonolestes* contrasting so strikingly with the dull pale browns of *Indolestes*. The enfumed tinted wings and the bicolorous pterostigma of the latter are other striking points ; the discoidal cell, especially that of the hind-wing, is narrower in *Indolestes* and strongly suggests that there is an evolutionary tendency for that structure to disappear entirely by a fusion of the costal and hinder borders. The straight distal border of the pterostigma and the simple posterior lobe of the prothorax separate it from *Sympycna*, which it otherwise greatly resembles.

26. **Indolestes indica** Fraser. (Figs. 35 & 36.)

Indolestes indica Fras., Mem. Dept. Agric. India (Ent.), vol. vii, pp. 58–59 (1922) ; id., J. Bombay Nat. Hist. Soc. vol. xxxiv, pp. 102, 103, pl. i, fig. 11, text-fig. 4 *a* (1930).
Lestes sp. Laid., Rec. Ind. Mus. vol. xix, pp. 161–2 (1920).
Indolestes buddha Laid., *nov. nom.* Fras., l. c. pp. 57, 59 (1922).

Male.—Abdomen 30 mm. Hind-wing 19 mm.

Head : labium palest brown ; labrum brown, paler at the sides ; base of mandibles, cheeks, the basal three segments of antennæ except the apex of the third, and two small triangular spots just posterior to and to the outer side of the hinder ocelli pale yellowish-brown ; rest of head black with a cupreous sheen, or dull blackish-brown ; middle of occipital border yellowish ; eye bluish-grey, with five dark brown bars running from above out and forwards. *Prothorax* dark reddish-brown on dorsum, with the mid-dorsum of middle lobe paler brown and the sides pale brown.

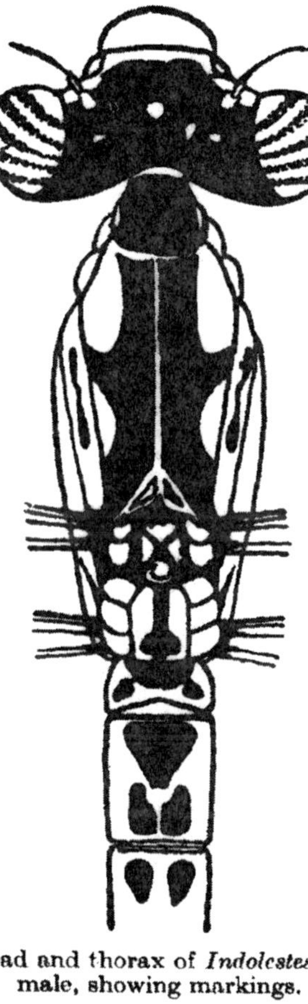

Fig. 35.—Head and thorax of *Indolestes indica* Fras., male, showing markings.

Thorax black on dorsum with a cupreous reflex, this colour forming a band extending from the mid-dorsal carina outwards to about half-way to the humeral suture, the outer border sending out processes somewhat like those seen in *L. præmorsa*, one above extending to the humeral suture, another at the middle of band which falls far short of the suture, the border of the band thus exhibiting two deep concavities, the upper prolongation enclosing a tiny point of the ground-colour; laterally very pale brown marked with linear black spots at the upper ends of the lateral sutures and with two or three small black or metallic green spots posterior to the humeral suture; beneath whitish-brown marked with two oblique streaks of blackish-brown. *Legs* yellow or ochreous on the extensor surface, black on the flexor; spines moderately numerous and short, black. *Wings* hyaline, neuration dark brown, membrane evenly enfumed with pale brown; pterostigma pale brown with the distal end bright yellow, proximal end oblique, distal nearly straight; 11 to 13 post-nodal nervures to fore-wings, 11 to 12 in the hind. *Abdomen* pale reddish-brown or ochreous marked with metallic green and black as follows: segment 1 with a basal lunar spot,

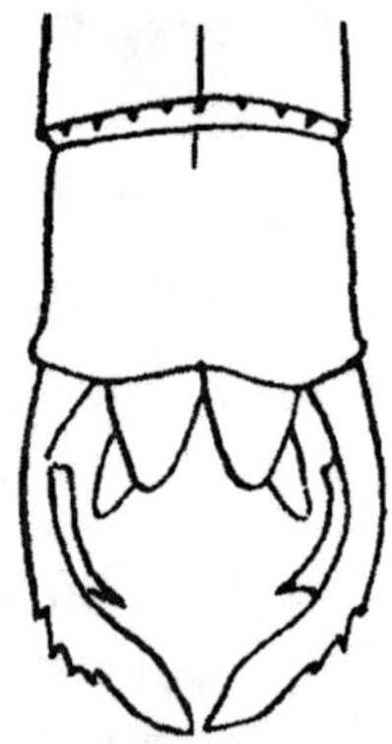

Fig. 36.—Dorsal view of anal appendages of *Indolestes indica* Fras., male.

followed immediately by, or actually confluent with, a small triangular subdorsal metallic-green spot on each side; segment 2 with a broad basal spot on dorsum, shaped like the ace of clubs, but with the middle lobe tapered apically for rather more than half the length of the segment, immediately following this a broad subdorsal comma-like spot on each side metallic green or cupreous black (in some specimens the narrow apical end of the basal spot is squared and each corner becomes confluent with the apical spots to enclose a bright yellow oval dorsal spot); segments 3 to 6 with paired apical and sub-basal spots, the former pyriform, with the thick end towards base of segment and separated narrowly by the ochreous mid-dorsal carina, the apical spots triangular and confluent over dorsum, where they are narrowly separated from fine black apical rings; segments 7 and 8 wholly black on dorsum, but on 8 the black not quite reaching base of segment and bisected finely by the ochreous mid-dorsal carina for nearly its whole length; segment 10 palest brown, unmarked. *Anal appendages* almost white; superiors very long, about half as long again as segment 10, forcipate, curved evenly and gradually towards each other until the apices meet, dilatation at inner third barely perceptible, but beginning as an angle and ending as a long sharp spine, apex sinuous and tapered to a point, its outer border finely spined; inferior appendages not half the length of superiors, bases formed by two broad cones, from the outer side of the apex of which springs a short, thick, tapered, ungulate process coated with hairs.

Female.—Abdomen 23·5–28·5 mm. Hind-wing 19–21 mm.

Closely similar to the male, the ground-colour and markings not differing except on the abdomen. *Wings* more deeply enfumed; 13 postnodal nervures to fore-wings, 12 in the hind, pterostigma as in the male; segment 1 with the sub-dorsal spots quadrate and always confluent with the basal; segment 2 with the spots confluent and forming a dumb-bell-shaped marking not quite extending to apex of segment, the mid-dorsal carina in the apical half of the marking ochreous, expanding into a small fusiform spot near the middle of segment (the combined comma-like spots are together broader than the basal spot, although shown in the reverse order in Laidlaw's figure, but they may be subject to some variation); segments 3 to 5 similar to the male, but on 5 the basal pair of spots are very small and may be quite absent; segment 6 with only a diffuse, poorly defined brown apical ring; segment 7 with the apical pair of spots continued basally as a diffuse broad brown stripe and with the mid-dorsal carina pale brown; segment 8 with two broad subdorsal stripes separated by the ochreous mid-dorsal carina; segment 9 with a broad subdorsal reddish-brown stripe on each side. *Anal appendages* shorter than segment 10, pale yellow, flattened, slightly but distinctly notched at apex; vulvar scale palest brown or yellow, not extending to end of abdomen, minutely serrate along lower border.

Distribution.—Known only from ASSAM.

The type of Laidlaw's *Lestes* sp. is a female from Cherrapunji, Assam, 4,400 ft., 8. x. 14, in the Indian Museum collection, no. 8204/20. The author's type is a male in the British Museum from Shillong, Assam, 6,000 ft., June 20. Mr. T. Bainbrigge Fletcher took this species in June and again throughout October, and found them resting with closed wings on bushes beside a stream, but others were found far from water in a pine tope at the top of a hill.

Since the author first described the species more material has come to hand, and from the variability in the size and markings of the insects it is clear that it is conspecific with Dr. Laidlaw's *Lestes* sp. The latter was described from a sub-adult specimen in which the markings were not fully developed and of which the size is decidedly smaller than the Shillong specimens. Distinguished from *I. bilineata* and *I. assamica* by the abdominal markings and presence of posthumeral thoracic markings.

27. **Indolestes bilineata** (Selys).

Lestes bilineata Selys, Ann. Mus. Civ. Genova, (2) vol. x (xxx), p. 498 (1891); Laid., Rec. Ind. Mus. vol. xix, p. 163 (1920).
Indolestes bilineata Fras., Mem. Dept. Agric. India (Ent.), vol. vii, p. 58 (1922); id., J. Bombay Nat. Hist. Soc. vol. xxxiv, pp. 103, 104 (1930).

Male unknown.

Female.—Abdomen 30 mm. Hind-wing 20 mm.

Head: labium palest brown; labrum, cheeks, bases of mandibles, clypeus, and frons as far as origin of antennæ olivaceous; rest of head and behind eyes bronzed black, except the space between the ocelli and occiput, which is pale brown; eyes brown. *Prothorax* pale brown, with a small darkish spot on each side of the middle lobe. *Thorax* pale reddish-brown on dorsum, passing to whitish-brown laterally and beneath; the mid-dorsum narrowly black. bisected by the ochreous mid-dorsal carina; laterally a short dark spot at the upper ends of the lateral sutures. *Legs* pale brown, the anterior pair of femora with a black stripe on the outer side; spines long, black. *Wings* hyaline, palely enfumed brown, with brown reticulation; pterostigma greyish-brown between black nervures, short, thick, only twice as long as broad, distal end straight, proximal oblique, covering 1½ to 2 cells; 10 to 12 postnodal nervures to fore-wings. *Abdomen* pinkish-brown marked with bronzed green as follows: segment 1 with a mid-dorsal spot; segment 2 with a broad dorsal stripe finely bisected by the ochreous mid-dorsal carina; segments 3 to 6 with a poorly defined dorsal bronzed fascia ending in a darker subapical ring; segment 7 with the dorsum bronzed metallic except at both ends, which are pale; segments 8 to 10 brown; beneath black. *Anal appendages* as long as segment 10, palest brown, sublanceolate, pointed; vulvar scale brown, not quite extending to end of abdomen, finely serrate beneath.

Distribution.—Burma only.

The *type* is a female in the Selysian collection from Palon, taken in September. Only three specimens are known, the third being also a female, in the British Museum. It is evidently closely allied to *I. indica*, from which it is distinguished by the totally different markings of the abdomen. From *I. assamica* it is distinguished by the relatively short pterostigma, by the different character of the anal appendages, and by the markings of the anterior pair of femora, etc.

28. **Indolestes assamica** Fraser.

Indolestes assamica Fras., J. Bombay Nat. Hist. Soc. vol. xxxiv, p. 104 (1930).

Male unknown.

Female.—Abdomen 32 mm. Hind-wing 21 mm.

Head: labium palest brown, almost white; labrum, bases of mandibles, cheeks, clypeus, and anterior border of frons pale yellow, the postclypeus with a black point at each end and two larger submedian black spots; the three basal joints of antennæ yellow, as also a tiny point behind each posterior ocellus and the whole of the occiput from eye to eye; rest of head mat black; eyes olivaceous brown. *Prothorax* pale brown, with a pair of parallel black subdorsal stripes expanding on the posterior lobe, narrow on the middle lobe, where they are very narrowly separated. *Thorax* pale brown, the mid-dorsum mat black as far as midway from mid-dorsal carina to humeral suture, the outer border of the band quite straight, the carina finely ochreous; laterally a small black spot on the upper part of the humeral suture and a similar but smaller one on the upper part of the postero-lateral suture; beneath unmarked. *Legs* pale straw-coloured with black spines, quite unmarked. *Wings* evenly enfumed pale brown, reticulation darker brown; pterostigma long and narrow, inner and outer ends oblique, reddish-brown between dark brown nervures, covering 2 to 3 cells; discoidal cells as for genus; 12 to 13 postnodal nervures to fore-wings, 10 in the hind. *Abdomen* pale brown or yellowish-brown marked with dark brown, black, and reddish-brown as follows: segment 1 with a quadrate dorsal spot extending from end to end of segment; segment 2 with a narrow black band, its ends slightly expanded, extending from base to apex of segment, the mid-dorsal carina finely dark ochreous; segments 3 to 7 similar to 2, but the bands tapered abruptly at each end of the segment; segments 8 to 10 reddish-brown on dorsum. *Anal appendages* yellow, conical, pointed, as long as segment 10; vulvar scale robust, extending to end of abdomen, yellow.

Distribution.—Confined to ASSAM so far as known.

The *type* is a female in the author's collection, and is the only specimen known; the specimen has been labelled "*Lestes bilineata* Selys?" since 1918, but after a careful re-examination and comparison with the description of that species the following insuperable differences were noted :—The pterostigma is nearly four times as long as broad instead of "nearly quadrate" as in *I. bilineata*, as noted by Selys; it is also oblique at both ends, whereas *I. bilineata* has the outer margin straight; the legs have no markings; and the vulvar scale extends to the end of the abdomen.

From *I. indica* it is easily distinguished by the outer border of the mid-dorsal marking of the thorax being quite straight, whereas it is deeply indented in *I. indica*; the totally different character of the abdominal markings, and the shape and size of the pterostigma are additional distinctions.

Genus **SYMPYCNA** Charpentier (1840).

Sympycna Charp., Lib. Eur. p. 19 (1840); Selys, Rev. Odon. p. 161 (1850); id., Bull. Acad. Belg. (2) vol. xiii, p. 336 (1862); Kirby, Cat. Odon. p. 163 (1890); Laid., Rec. Ind. Mus. vol. xix, pp. 145–6 (1920); Fras., J. Bombay Nat. Hist. Soc. vol. xxxiv, pp. 104, 105 (1930).
Sympecma Selys, Mon. Lib. Eur. p. 144 (1840).

Dragonflies of small stature very similar to *Indolestes*, and resting, like them, with wings closed over dorsum; ground-colour pale brown with metallic markings on head, thorax, and abdomen; wings hyaline tinted with pale brown, very narrow and with pointed apices, petiolated as far as *ac*, *ac* meeting *ab* at hinder border of wing; postnodal nervures fairly numerous; discoidal cells differing in shape and size in fore- and hind-wings, very narrow and elongate and very acute at distal angle, discoidal cell of fore-wing with posterior border at least 3 times as long as basal and twice the length of costal, that of hind-wing with posterior border 5 or 6 times the length of basal and slightly less than twice the length of costal; *Riii* arising 5 to 4 cells after the node in fore-wing and hind-wing respectively; *IRiii* and *Riv+v* arising much nearer arc than to node; *IRiii* not or scarcely angulated; *MA* zigzagged from level of node; oblique nervure present between *Riii* and *IRiii*; pterostigma uni- or bi-colorous, three or four times as long as broad, inner and outer ends oblique; intercalated sectors as for *Lestes*. Posterior lobe of prothorax trilobed, the middle lobe being much more prominent than the lateral. Superior anal appendages narrow, elongate, forcipate, furnished at the middle third of the inner border with a narrow dilatation which begins with a very robust tooth or spine and ends in an obtuse angulation; inferior appendages short, tumid; anal appendages of female sublanceolate, pointed, longer than segment 10; vulvar scale robust, extending only to middle of segment 10.

Distribution.—Europe, Asia Minor, North Africa, Central Asia (Mesopotamia, Persia, Kashmir).

The genus is closely allied to *Indolestes*, but is easily recognized by the long pterostigma, oblique at both ends, and by the trilobate character of the posterior lobe of the prothorax. Only one species or subspecies found within Indian limits (from Kashmir and Quetta).

Genotype, *S. fusca* Lind.

29. **Sympycna pædisca annulata** Selys. (Figs. 37 & 38.)

Sympycna pædisca annulata Selys, Ann. Soc. Ent. Belg. vol. xxxi, p. 43 (1887); Morton (nec *fusca*), Trans. Ent. Soc. Lond. p. 308 (1907); Bartenef, Annu. Mus. Zool. Acad. Imp. Sci. St. Pet. t. xvii (1912); Morton, Ent. Month. Mag. ser. 3, vol. v, pp. 145–6 (1919); Fras., J. Bombay Nat. Hist. Soc. vol. xxxiv, pp. 105, 107, pl. i, fig. 12, text-fig. 4 *b* (1930).

Sympycna pædisca pædisca Laid., Rec. Ind. Mus. vol. xix, p. 163 (1920).

Male.—Abdomen 30 mm. Hind-wing 21 mm.

Head: labium whitish; labrum, bases of mandibles, and cheeks pale olivaceous brown; rest of head slightly darker warmer brown, marked with blackish-brown and metallic green as follows: postclypeus with a small spot at each end and a large submedian quadrate blackish-brown spot at each side of middle line; frons with a sinuous linear stripe of metallic green curling outwards on each side and narrowly interrupted by the middle line, an hour-glass-shaped spot on the outer side of each posterior ocellus, the whole of the ocellar space, the occiput, and a foliate spot behind each eye which is confluent with it, all brilliant metallic green; eyes brown above, yellow beneath. *Prothorax* pale brown, its middle lobe with a large dorsal spot on each side, its posterior lobe metallic emerald-green and border finely yellowish. *Thorax* palest brown, its dorsum broadly metallic green nearly as far out as the humeral suture, the outer border of this marking irregular and usually with a hook-like prolongation about its middle; antealar sinus metallic green within its borders, which latter, as well as the mid-dorsal carina, are finely ochreous. Laterally a narrow, very irregular, and somewhat zigzagged bright metallic-green stripe running from the antero-lateral suture above obliquely down to lower end of humeral suture. *Wings* hyaline, enfumed pale brown, venation brownish-yellow; node thickened; pterostigma yellow to dark brown with bordering nervures yellow, oblique at both ends, covering two cells; 14 postnodal nervures to fore-wings, 12 in the hind. *Legs* yellow, femora with a chain of small dark brown spots; numerous and moderately long dark spines. *Abdomen* pale brown or coloured like dry grass, marked with metallic green as follows: segment 1 with a pair of triangular apical spots barely confluent over dorsum; segment 2 with a pair of parallel stripes narrowly separated by the pale yellow mid-dorsal carina, each shaped like a note of exclamation, the basal half pyriform, the apical rounded apposed surfaces flat; segments 3 to 7 similar, the stripes being longer and narrower and separated by the

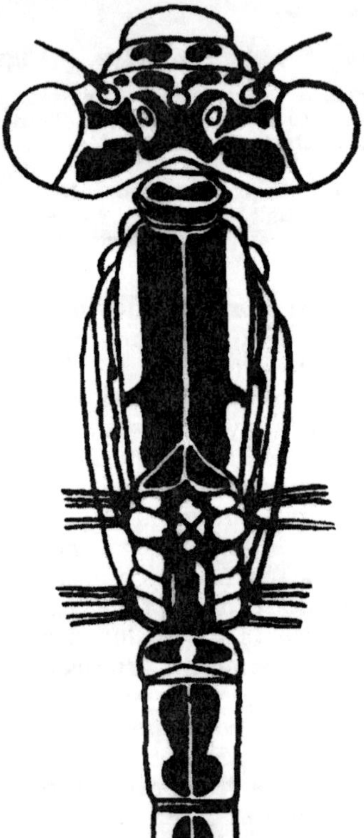

Fig. 37.—Head and thorax of *Sympycna pædisca annulata* Selys, male, showing markings.

yellow mid-dorsal carina and falling a little short of the base; segments 8 and 9 with stripes approximating to that on segment 2 but broader and clumsier in build and with a cupreous reflex; segment 10 with the dorsum narrowly dark cupreous brown, deeply notched at apex and bordered with black spines. *Anal appendages* yellowish-white, bordered near apices with black spines; superiors narrow, forcipate, apices curling gradually in to meet each other, furnished at the middle third of inner border with a dilatation which begins with a very large and robust tooth and ends in an angulation tipped with a minute black spine ; apices rounded, tipped with white hairs, appendages about one and a half times as long as segment 10 ; inferiors rudimentary, about one-third the length of superiors, conical, closely apposed.

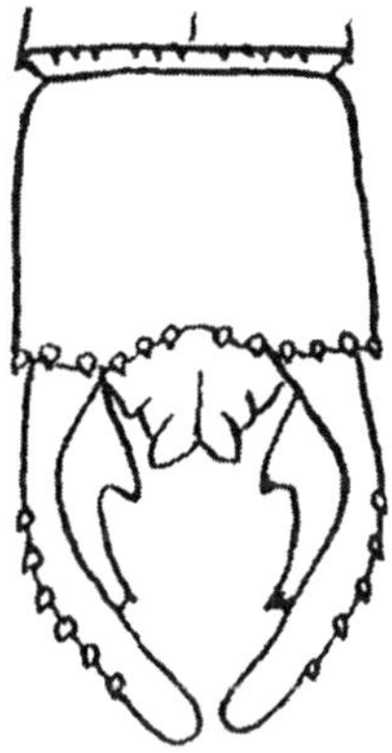

Fig. 38.—Dorsal view of anal appendages of *Sympycna pædisca annulata* Selys, male.

Female.—Abdomen 30 mm. Hind-wing 21 mm.

Exactly similar to the male, but some variation of the mid-dorsal stripe of *thorax*, which may have the outer border quite straight or with a markedly conspicuous medial prolongation ; the wavy posthumeral stripe often interrupted once or twice. Pterostigma palest brown, covering 1½ to 2 cells. Femora of first two pairs of legs with a black outer stripe or chain of spots. Segment 10 with a dorsal stripe tapering apically but not reaching the apex. *Anal appendages* very large and robust, cylindrical, pointed, longer than segment 10, pale buff in colour ; vulvar scale short, extending only to middle of segment 10, lower border minutely serrated with black teeth.

Distribution.—Kashmir and N. W. India within our limits. The specimen described by Morton from Quetta as *S. fusca* is really this subspecies of *pædisca*. Laidlaw mentions specimens from the Jhelum Valley, Kashmir, 5,000 ft., and I have seen a number of specimens collected by Mr. T. Bainbrigge Fletcher at Srinagar, Kashmir, in July. So far as known, it extends from Asia Minor to Kashmir through Mesopotamia and Persia and the N.W. Frontier Provinces.

Mesopotamian specimens have the wavy posthumeral stripe split up into three widely separated spots ; the prothorax has a median spot on the anterior lobe, a pair of angulated spots on the middle lobe, and the whole middle section of the posterior lobe dark cupreous metallic; pterostigma bright ochreous. The shape of the pterostigma and the posterior lobe of the prothorax will serve to distinguish this species from any of *Indolestes*.

Subfamily MEGAPODAGRIINÆ.

Podagrion Selys, *nom. prœoc.*, Syn. Agrion. Bull. Acad. Belg. (2) vol. xiv, p. 12 (1862); id., Mém. Cour. vol. xxxviii, p. 40 (1886).
Megapodagrioninæ Fras., J. Bombay Nat. Hist. Soc. vol. xxxiv, p. 966 (1931).

Dragonflies of medium size, usually of more robust build than the average Cœnagriine, body-colouring non-metallic, resting with wings widely or partly spread. Wings with a long petiole, narrow, hyaline, discoidal cell elongate, moderately narrow, variably acutely angulated distally; sectors of arc arising from lower half of arc; *Riii* arising far distal to node; origins of *IRiii* and *Riv+v* variable; *Cuii* arched slightly costalwards at its origin; no oblique vein present between *Riii* and *IRiii*; pterostigma variable, usually slightly more than twice as long as broad; intercalated sectors present in most, especially at apical area of wings, these sectors giving a pectinated effect to *IA* and *MA* in some species; anal bridge vestigial or complete. Abdomen long and slender, or moderately long and robust, especially in the females. Superior anal appendages of male usually more or less forcipate; inferiors vestigial.

Genitalia: lobe usually depressed, hamules large and prominent, penis with the end curled over its stem and ending in two long, more or less curled branches which embrace the stem, lobe tumid, more or less pyriform. Vulvar scales very robust and extending well beyond end of abdomen. Larvæ unknown.

Distribution.—Only three genera are known from within Indian limits, beyond which the subfamily extends throughout the tropics of both hemispheres, Australasia, and the Philippines.

The three Indian genera stand somewhat isolated in the fauna, each being represented by a single species within our limits and all confined to N.E. India and Burma. *Burmargiolestes melanothorax* Selys, which was formerly included under the genus *Argiolestes*, has since been placed in a genus of its own, so that the distribution of this species is no longer anomalous. A Rhinagrion which I had incorrectly placed as *M. tibetana* turns out to be a large race of *R. mima*

Karsch, the distribution of this species now extending from by Borneo to Burma. Lastly, *Mesopodagrion*, as represented *M. tibetana* MacL., still remains a monotypic genus, confined, so far as known, to Tibet and S.E. China.

Genus **BURMARGIOLESTES** Kennedy (1925). (Fig. 39.)

Burmargiolestes Kennedy, Bull. Mus. Comp. Zoöl. Harvard, vol. lxvii, no. 7, p. 296, pl., fig. 6 (1925); Fras., J. Bombay Nat. Hist. Soc. vol. xxxiv, p. 967 (1931).

Wings long and narrow, petiolated nearly to level of distal end of discoidal cell and far distal to *ac*; node situated at less than one-third the distance from base of wing to pterostigma; *ac* situated almost opposite the basal antenodal nervure; *ab* vestigial, only a trace of its distal end present; discoidal cell acute at distal end, its costal side about two-thirds the length of postcostal, these two sides diverging slightly and the outer side oblique; *IRii* arising about 3 cells distal

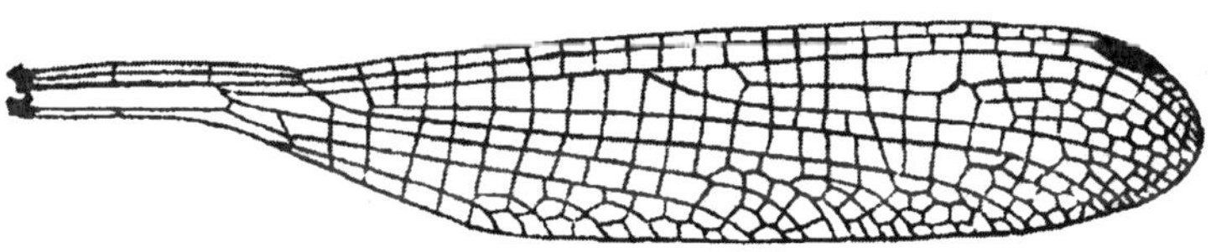

Fig. 39.—Fore-wing of *Burmargiolestes melanothorax* (Selys), male.

to the origin of *Riii*; *Riii* arising slightly nearer the node than pterostigma; *IRiii* arising 1 cell distal to the subnode; *Riv+v* arising at the subnode; *MA* and *IA* pectinate (this pectination being an expression of intercalated sectors between *Riv+v* and *MA* and *Cuii* and *IA*); intercalated sectors also present between *IRii* and *Riii* and between *IRiii* and *Riv+v* at apical ends of wings; only a single row of cells between *IA* and hinder border of wings; pterostigma elongate, at least twice as long as broad, slightly dilated at the middle, both inner and outer ends oblique, unbraced, covering less than two cells. Head narrow; epistome rather prominent. Prothorax with posterior lobe sub-bilobate; thorax robust; legs rather short, tibiæ not dilated, claw-hooks situated near end of claws. Abdomen long and slender, slightly dilated at base and anal ends. Superior anal appendages nearly twice the length of segment 10, forcipate; inferior appendages vestigial. Genitalia (*cf.* fig. 46, *e*, *f*): hamules broad, subquadrate plates; penis broad at apex and slightly notched, curling up over the stem, its branches embracing that structure and spiral-like at the ends; lobe of penis flask-shaped, tumid. Vulvar scale robust, extending slightly beyond end of abdomen.

Genotype, *B. melanothorax* (Selys).

Distribution.—South Asia and the eastern Himalayan tracts.

30. **Burmargiolestes melanothorax** (Selys). (Figs. 39 & 40.)

Argiolestes melanothorax Selys, Bull. Acad. Belg. (2) vol. xiv, p. 38 (1862); id., Mém. Cour. vol. xxxviii, p. 80 (1880); Kirby, Cat. Odon. p. 125 (1890); Selys, Ann. Mus. Civ. Genova, (2) vol. x (xxx), p. 500 (1891); Laid., Rec. Ind. Mus. vol. xiii, pp. 323–5 (1917).

Burmargiolestes melanothorax Kennedy, Bull. Mus. Comp. Zoöl. Harvard, vol. lxvii, no. 7, p. 296 pl., fig. 6 (1925); Fras., J. Bombay Nat. Hist. Soc. vol. xxxiv, pp. 967, 968, text-figs. 1 & 2 (1931).

Male.—Abdomen 40 mm. Hind-wing 28 mm.

Head: labium brown; labrum pale blue; bases of mandibles and genæ black; epistome, frons, and basal joints of antennæ pale blue; vertex and occiput black; eyes oliva-

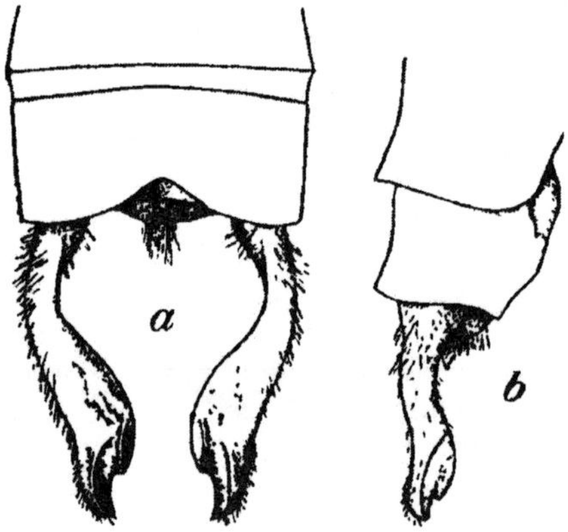

Fig. 40.—Anal appendages of *Burmargiolestes melanothorax* (Selys). *a.* Dorsal view. *b.* Right lateral view.

ceous brown. *Prothorax* black, with a broad subdorsal pale brown fascia on each side. *Thorax* glossy black, the upper ends of humeral and lateral sutures pale yellowish-brown. *Wings* hyaline; pterostigma dark reddish-brown to almost black; 23 postnodal nervures to fore-wings, 21 in the hind. *Legs* yellow, the tarsi and spines ferruginous, the distal ends of femora with a diffuse spot of dark brown on the extensor surface. *Abdomen* glossy black or blackish-brown in subteneral specimens, segments 3 to 6 with an obscure yellowish spot or annule near the base. *Anal appendages* blackish-brown; superiors twice the length of segment 10, forcipate, slightly tumid at base, then subcylindrical finally and compressed at the apex, where they are dilated and broadly notched and hollowed out on the inner surface, and present a flattened facet just proximal to the notch; outer borders finely spined, broadly sinuous; inner border of base minutely tuberculate and spined; inferiors not visible from above, rounded, vestigial plates.

Female.—Abdomen 34 mm. Hind-wing 29 mm.

Very similar to the male, but of more robust and shorter build. Differs as follows: labium brownish-black; black of vertex encroaches on frons, which has two punctate black points; eyes darker brown. *Wings* palely suffused with brown in adults, especially at apices, which may be quite dark brown along extreme borders; pterostigma shorter and broader, often imperfectly braced, covering 2 to 2½ cells; 29 postnodal nervures in fore-wings, 21 in the hind; *ac*, especially in the fore-wings, often slightly proximal to the level of the basal antenodal nervure. *Abdomen* with sides paler and the dorsum of segments 8 and 9 yellowish-brown; segment 10 pruinosed white in adults; *anal appendages* conical, pointed, as long as segment 10; vulvar scales brown, robust, minutely spined below.

Distribution.—Upper Burma, Sikkim, and Assam. The species from Siam and Tonkin appears to be quite distinct. The discoidal cell in this species is much shorter than in typical *Argiolestes*, the proportion of its costal and subcostal sides is quite different, the nervure *ac* lies much more basally, whilst the origins of *IRiii* and *Riv+v* differ entirely from what is found in that genus; important differences are also found in the genitalia.

Type in the Selysian collection; paratypes in most national museum collections.

Genus **RHINAGRION** Calvert (1913). (Fig. 41.)

Amphilestes Selys, Bull. Acad. Belg. (2) vol. xiv, p. 42 (1867); id., Mém. Cour. vol. xxxviii, p. 92 (1886), *nom præoc.*; Kirby, Cat. Odon. p. 126 (1890).

Rhinagrion Calv., *nov. nom.* (*Amphilestes*), Proc. Acad. Phil. vol. lxv, p. 258 (1913); Fras., J. Bombay Nat. Hist. Soc. vol. xxxiv, p. 968 (1931).

Wings long and narrow, petiolated to the level of *ac*; node situated at slightly less than one-third the distance from base to pterostigma; *ac* situated nearly opposite the level of the distal antenodal nervure, slightly proximal to that structure; *ab* complete, meeting *ac* at border of wing; discoidal cell subacute at apex, its costal side two-thirds to three-quarters the length of subcostal, these sides diverging distally, outer side rather oblique; *IRii* arising 4 to 5 cells distal to the origin of *Riii* or half-way between the origin of the latter and the pterostigma ; *Riii* arising in the fore-wing 5 to 6 cells distal to the node and much nearer the latter than pterostigma ; *IRiii* arising at the subnode ; *Riv+v* arising well before level of subnode ; *MA* and *Cuii* simple ; *IA* zigzagged at its distal half only ; no intercalated sectors except short ones at the apices of wings ; pterostigma elongate, more than twice as long as broad, slightly dilated at its middle, inner and outer ends oblique, only occasionally braced, covering two cells.

Head moderately narrow ; epistome not prominent ; occiput deeply concave. Prothorax with a simple rounded and arched posterior lobe ; thorax robust, furnished in the female with prominent hooks at the anterior ends of humeral sutures. Legs rather short, robust, tibiæ slim, claw-hooks near ends of claws, femoral and tibial spines short, robust, not very numerous. Abdomen long, cylindrical, moderately robust, especially in the female, dilated at base and gradually so at anal end. Anal appendages forcipate, nearly twice the length

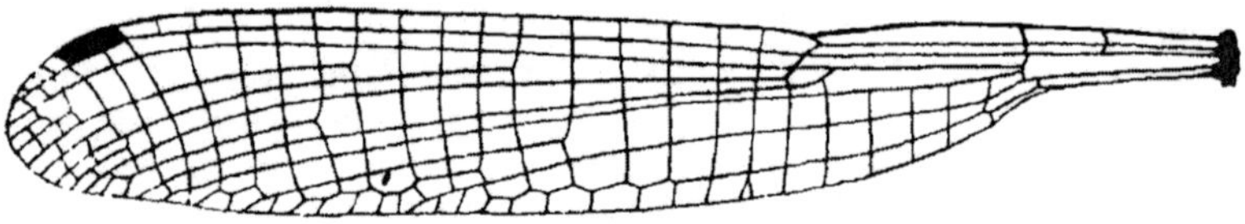

Fig. 41.—Hind-wing of *Rhinagrion mima* (Karsch), male.

of segment 10, inferiors vestigial. Genitalia (*cf.* fig. 46 *a, b*) : hamules moderately broad quadrate plates, pointed posteriorly ; penis narrow at apex, curling over stem of organ, its branches strongly curled to embrace the stem and shaped like the horns of an ox as seen from the dorsum ; lobe small, inconspicuous, pyriform. Vulvar scale robust, projecting well beyond end of abdomen, stylets of great length.

Genotype, *R. macrocephalum* (Selys).

Distribution.—Burma, Borneo, Malaysia, and Sumatra.

31. **Rhinagrion mima** (Karsch). (Figs. 41, 42, & 43.)

Amphilestes mima Karsch, Ent. Nachr. vol. xvii, pp. 242, 243 (1891) ; Kruger, Stett. Ent. Zeit. p. 100 (1898) ; Laid., Proc. Zool. Soc. Lond. p. 382 (1902).
Rhinagrion tibetana Fras. (=*mima*), Rec. Ind. Mus. vol. xxvi (1926).
Rhinagrion mima Fras., J. Bombay Nat. Hist. Soc. vol. xxxiv, pp. 969, 970, text-figs. 3, 4, & 5 (1931).

Male.—Abdomen 36 mm. Hind-wing 27 mm.

Head black, marked with bright ochre or citron-yellow ; labium pale straw-yellow ; labrum bright ochreous broadly bordered with glossy black; epistome similar; bases of mandibles and genæ bright yellow; frons and vertex bright ochreous, with a broad median uneven stripe of mat black; occiput, antennæ, and the area round base of the latter, as well as a small point in front of them, mat black; eyes olivaceous, bordered behind by a narrow black band and this latter by another ochreous band equally wide. *Prothorax* with an anterior collar, the whole of posterior lobe and a large oval spot on each side bright ochreous, middle lobe and dorsum black. *Thorax* velvety black, marked bright yellow or

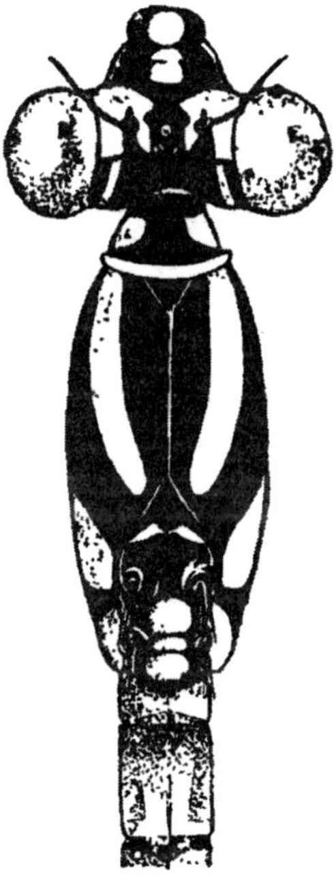

Fig. 42.—Head and thorax of *Rhinagrion mima* (Karsch), male, showing markings.

ochreous as follows: a broad antehumeral stripe not extending to upper and lower part of dorsum, gently curved, with the concave sides of the stripes facing one another, also the upper halves of the antealar sinus; laterally bright yellow, with a moderately broad median black stripe. *Legs* bright ochreous, the flexor surfaces paler yellow, the spines, tarsi, flexor surfaces of tibiæ, and the extensor surfaces of the two anterior pairs of femora black. *Wings* hyaline, palely enfumed in old adults; pterostigma black, covering 1½ to 2 cells; 13 to 14 postnodal nervures in fore-wings, 12 in the hind. *Abdomen* brick-red, ringed subapically on segments 2 to 6 with bright ochreous, followed by moderately broad black apical rings, segment 1 paler yellow or greenish-yellow, with a narrow apical bordering of black laterally and a finely delineated dorsal marking shaped like the handle of a dagger; segment 10 with a broad dorsal marking shaped like a cross; ventral surfaces and ventro-lateral borders of segments 8 to 10 black. *Anal appendages* black, the superiors with the apical third bright ferruginous except the extreme apex which is black; superiors forcipate, curving medialwards to meet at apices, apical halves compressed, the inner surface shallowly concave, apices bevelled within and ending in an acute point; inferiors rounded, very short, vestigial, not visible from above.

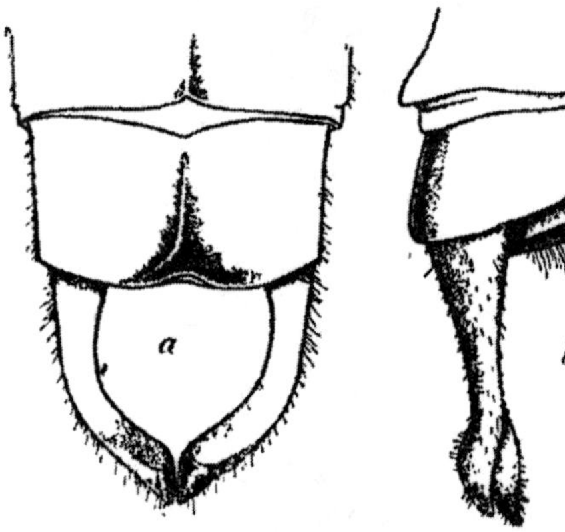

Fig. 43.—Anal appendages of *Rhinagrion mima* (Karsch), male. *a.* Dorsal view. *b.* Right lateral view.

Female.—Abdomen 33 mm. Hind-wing 29 mm.

Very similar to the male but more robust, differs as follows:—Bright markings of *head*, *prothorax* and *thorax* citron-yellow, those on the former more restricted, a large irregular spot on the vertex being isolated by the black; occiput with a narrow dark ferruginous border behind. *Thorax* with the black on anterior border of metepimeron reduced to an irregular dark blackish-brown stripe, incomplete above and below. The anterior part of mid-dorsal carina bifurcating to enclose a small triangular space, on either side of which is a large laterally compressed spine ending in a finely pointed hook, the outer part of the spine continued as a shell-like plaque overhanging the posterior lobe of *prothorax* or sometimes hidden partly by the latter. *Abdomen* with segment 1 greenish-yellow, the subapical annules on segments 2 to 6 brighter yellow and very conspicuous between the apical black annules and the ferruginous colour of the dorsum; on segments 3 to 7 the apical black is continued basalwards laterally, only slightly so on 3, but as far as the base on segments 6 and 7, on which segments the black also passes basalwards on the mid-dorsum so as to enclose a spot of the ground-colour; segment 8 black, the mid-dorsum finely, a subapical subdorsal spot and the ventral border broadly ochreous; segments 9 and 10 entirely black. *Anal appendages* black, short, conical. Vulvar scales and stylets black. *Legs* with the extensor surfaces of all femora ochreous or ferruginous. *Wings* with 15 postnodal nervures in the fore, 14 in the hind.

Distribution.—Maymyo, UPPER BURMA; Sumatra and Malay Peninsula.

A number of both sexes taken by Col. F. Wall, I.M.S., along the borders of a small stream. The species, which appears to be locally common, is most closely allied to *R. borneense* (Selys). Col. Wall's specimens, which I incorrectly placed as *M. tibetanum*, have since been determined as *R. mima* by a comparison which I have been able to make with specimens in the British Museum. They are considerably larger than the type, and the markings are more yellow or ochreous than greenish-yellow.

Genus **MESOPODAGRION** MacLachlan (1896). (Fig. 44.)

Mesopodagrion MacL., Ann. Mag. Nat. Hist. ser. 6, vol. xvii (1896); Fras., J. Bombay Nat. Hist. Soc. vol. xxxiv, p. 971 (1931).

Wings long and narrow, but moderately dilated about the middle area, petioled to a little proximal to *ac*; node situated at a little more than one-third the distance from the

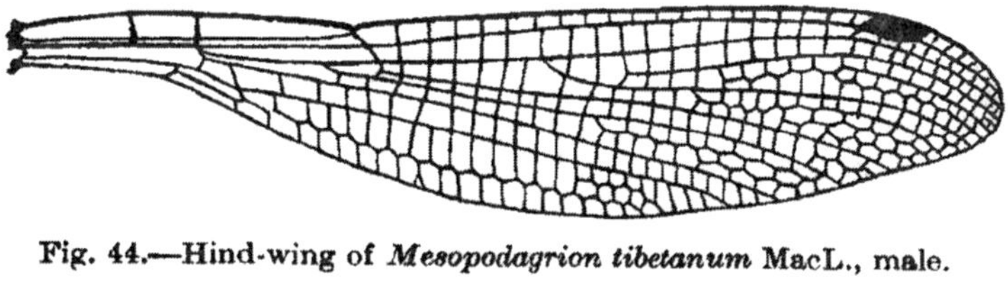

Fig. 44.—Hind-wing of *Mesopodagrion tibetanum* MacL., male.

base of wing to pterostigma; *ac* situated at a level slightly nearer the distal antenodal nervure; *ab* complete, extending a little proximal to *ac*; discoidal cell more acute than in *Rhinagrion*, its costal border rather less than two-thirds the length of subcostal, outer border very oblique, costal and subcostal sides diverging slightly; *IRii* arising about three cells distal to the origin of *Riii* and about midway between the node and pterostigma ; *Riii* arising six cells distal to the node in fore-wing and much nearer node than pterostigma ; *IRiii* arising at the subnode ; *Riv+v* rising well before subnode ; *MA* and *Cuii* simple ; *IA* zigzagged from its origin ; two rows of cells between *Cuii* and *IA* near their terminations ; intercalated sectors between *IRii* and *Riii*, between *Riii* and *IRiii*, and between *Riv+v* and *MA* ; pterostigma elongate, dilated at middle, oblique at outer and inner ends, well braced, covering about three cells.

Head moderately narrow ; labium deeply cleft, lobes distant, subacute at apex, 2nd joint of antennæ rather longer than the first ; epistome not prominent ; occiput concave. Prothorax with posterior lobe simple, rounded, arched ; thorax robust. Legs robust, rather short, spines of moderate length, not numerous, tibiæ slim, claw-hooks situated at extreme ends of claws. Abdomen robust, cylindrical, moderately short as compared with species of *Rhinagrion*, dilated somewhat at base and anal end. Superior anal appendages not twice the length of segment 10, forcipate, tapering, acute at apices ; inferiors rounded, vestigial. Genitalia (*cf.* fig. 46, *c*, *d*) : lamina arched ; hamules quadrate, plates pointed posteriorly ; penis narrow at apex, but squared and curling strongly over body of organ, its branches embracing the stem and making a complete curl like a ram's horn ; lobe small, pyriform, grooved longitudinally. Vulvar scales robust, strongly curved, projecting well beyond end of abdomen. Anal appendages of female conical, acuminate, as long as segment 10.

Genotype, *Mesopodogrion tibetanum* MacL.

Distribution.—Tibet and S.W. China.

32. **Mesopodagrion tibetanum** MacLachlan. (Figs. 44, 45, & 46.)

Mesopodagrion tibetana MacL., Ann. Mag. Nat. Hist. ser. 6, vol. xvii (1896) ; Fras., J. Bombay Nat. Hist. Soc. vol. xxxiv, p. 972 (1931).

Rhinagrion tibetana Fras., Rec. Ind. Mus. = *Rhinagrion mima* Karsch.

Male.—Abdomen 33 mm. Hind-wing 30 mm.

Head: labium brownish-yellow ; labrum, bases of mandibles, and genæ greenish-yellow ; vertex and occiput mat black, with a short oblique line on the outer side of each ocellus, a short line on the posterior border of occiput, and a large spot behind each eye yellow. *Prothorax* black, with a broad greenish-yellow stripe on each side continuous with a similarly coloured antehumeral stripe on thorax. *Thorax* black on dorsum, yellow on the sides ; dorsum marked with a pair of slightly curved antehumeral greenish-yellow stripes, the concavity of the stripe facing inwards; laterally a moderately broad oblique black stripe broadening anteriorly. *Legs* black, coxæ and trochanters with yellow spots. *Wings* hyaline, palely enfumed and faintly tinted at the bases, pterostigma dark brown to blackish framed in black; 21 postnodal nervures in fore-wings, 20 in the hind. *Abdomen* glossy black, basal segments clothed with long black hairs, segment 1 broadly yellow at the sides, segment 2 with a pair of parallel yellow stripes along the sides, segments 3 and 4 and, in teneral specimens, segments 5 to 7 with a large baso-lateral spot which is confluent with a rather broad lateral greenish-yellow or yellow stripe extending from base nearly to apex, but falling short of the latter in adult specimens on segments 4 to 7; segments 9 and 10 black, the latter

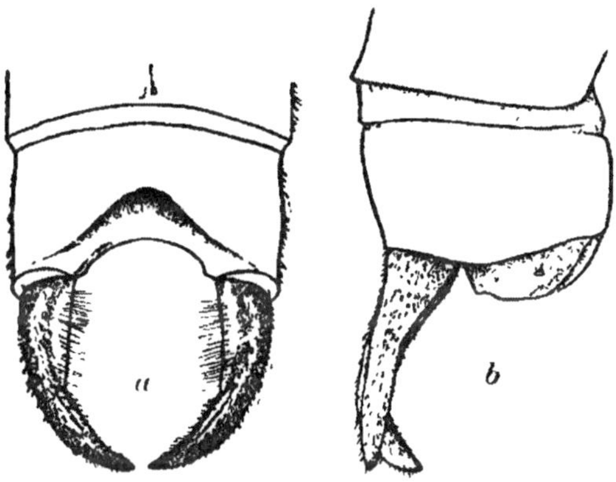

Fig. 45.—Anal appendages of *Mesopodagrion tibetanum* MacL., male. *a*. Dorsal view. *b*. Right lateral view.

with a large lateral yellow spot. *Anal appendages* black; superiors forcipate, rather longer than segment 10, subcylindrical at base but compressed nearer apex, which is acuminate, whilst on the inner side near the middle of the appendages is a false joint or sulcus, at which point the appendage is slightly but distinctly angulated inwards; inferiors rounded, vestigial, not visible from above. Genitalia as for genus.

Female.—Abdomen 35 mm. Hind-wing 34 mm.

Closely similar to the male, but the markings more greenish-yellow and the spot behind the eyes absent or obsolete. *Wings* faintly enfumed; pterostigma as for male (yellowish-white in the allotype, which, however, is slightly teneral); 21 postnodal nervures in fore-wings, 19 in the hind. The markings on the *abdomen* similar to those on teneral males, the stripes being continued on to the 3rd and 4th segments and the basal spots on to the 7th; segment 9 with a very large lateral yellow spot, whilst the 10th is largely yellow on the sides. *Anal appendages* as long as segment 10, slightly curved laterally, acuminate, pointed, yellowish at base, and separated by a large conical structure as viewed from above. Vulvar scales black, strongly curved.

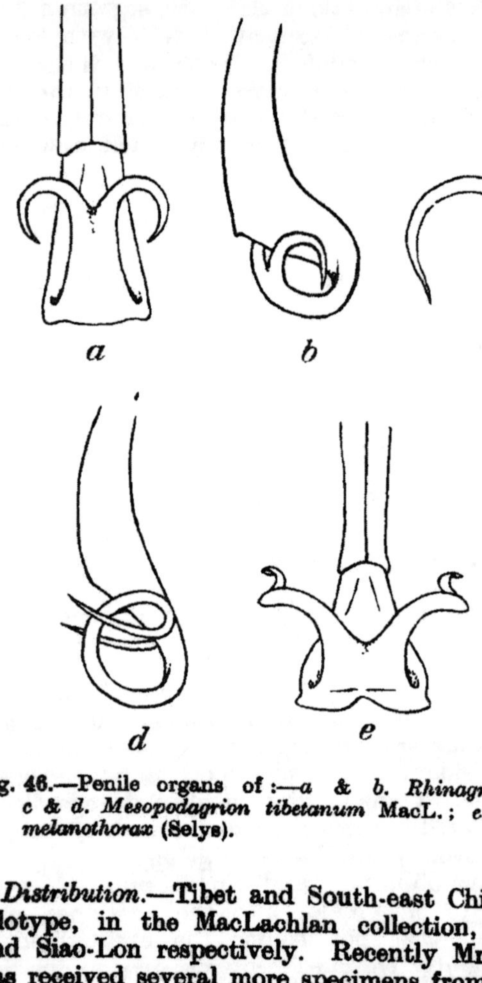

Fig. 46.—Penile organs of:—*a* & *b*. *Rhinagrion mima* (Karsch); *c* & *d*. *Mesopodagrion tibetanum* MacL.; *e* & *f*. *Burmargiolestes melanothorax* (Selys).

Distribution.—Tibet and South-east China. The *type* and allotype, in the MacLachlan collection, are from Moupin and Siao-Lon respectively. Recently Mr. Kenneth Morton has received several more specimens from the same district, which I have been able to compare with the types. The markings in the latter are definitely apple-green but, so far as the male is concerned, may be blue in the living insect.

Subfamily PLATYSTICTINÆ Laidlaw (1924).

Genus *Platysticta* Selys, Bull. Acad. Belg. (2) vol. x, p. 436 (1860); id., ibid. pp. 144, 145 (1886).
Legion *Platysticta* Laid., Rec. Ind. Mus. vol. xiii, p. 323 (1917).
Subfamily Platystictinæ Laid., Spolia Zeylan. vol. xii, pp. 300, 301 (1924); Fras., J. Bombay Nat. Hist. Soc. vol. xxxv, pp. 66–8 (1931).

Dragonflies of small size, coloured black or brown with white markings, or, more rarely, reddish marked with blue, rarely, if ever, metallic; resting with wings folded over dorsum; wings very narrow, falcate at apex, hyaline or tipped with black at apices; the nervure *ac* situated midway between the two antenodal nervures; an accessory basal postcostal nervure always present and situated well proximal to *ac*; petiolation beginning well distal to the nervure *ac*; the nervure *ab* reduced or absent, when present extending from *ac* or the posterior border of wing to the posterior side of the discoidal cell; *IA* absent; *Cuii* markedly reduced, falling well short of middle of wings; discoidal cell elongate, about four times as long as broad, the ends squared; sectors of arc arising from the lower end of arc, separate or confluent for a short distance at origin; individual cells mostly four-sided, more rarely five-sided; *Riii* arising well distal to node, nearer node than pterostigma; origin of *Riv+v* variable, either slightly proximal to, at, or slightly distal to the oblique nervure descending from the subnode; pterostigma short and broad, its costal side usually shorter than the posterior, its inner or proximal side oblique, its distal straight or slightly rounded; no intercalated sectors present except *IRii*.

Abdomen of great length and very slender, in some genera twice or more than twice the length of hind-wing, the relative lengths of segments 8, 9, and 10 variable in individual species; anal appendages of male complex, variable in the species, the superiors usually forcipate and often chelate at the apices or spatulate; inferior appendages more simplified, usually tapering to a point, but spatulate in others. Vulvar scales very robust, extending beyond end of abdomen in most species, and with a robust dorsal spine-like protuberance before the apex. Penile organ variable, with or without a frill or ridge on the dorsum, the end curling strongly over the stem, which its terminal branches embrace.

Larvæ slender, cylindrical, furnished with three triquetral gills, mask flat, subquadrate, Gomphine-like, without setæ, middle lobe cleft, lateral lobe robust, furnished with a long movable hook, legs long and slim. Breeding in torrential montane streams or seepages on the sides of hills in dense virgin jungle.

Distribution.—Tropical and neotropical. Within our limits from CEYLON, WESTERN GHATS, N.W. INDIA, and BURMA, in submontane and montane tracts, rarely at sea-level.

Species are found haunting the banks of mountain streams of small size, often a mere trickle over rocks or a chain of pools below a spring on a steep jungly hillside in dense shade. Many species inhabit small caves among boulders and rocks beside streams, where they exist in a perpetual twilight. In flight they are much given to hovering with the long attenuated abdomen held out stiffly and horizontally, whilst advancing or retiring in a series of short jerky movements, and, if approached head on, will retire backwards instead of reversing and heading off, this backward flying seeming to be accomplished as easily as the forward movement. Owing to their dull colouring, small size, and dark surroundings, they are remarkably inconspicuous during flight, and would be invisible were it not for the chain of white and blue spots on the abdomen seen to be moving stealthily about the dark recesses. The vivid blue identification marks on the terminal segments of the abdomen show up with remarkable conspicuousness even in the darkest retreats when the insect is at rest.

Ceylon is remarkably rich in Platystictas, whilst Continental India and Burma are equally rich in Protostictas and Drepanostictas. The former group is sharply separated from the Indian forms by the penile organ, which possesses a ridge on its dorsum absent in the two latter groups, and I have employed this character as a means of classifying an otherwise difficult group of insects. Selys classed the whole of the Ceylon species as *Platysticta*, and divided these into two subgenera by venational characters. To the first subgenus a south Indian species has to be added, whilst the second subgenus has been incorporated by Dr. Laidlaw in his genus *Drepanosticta*. The penile organ of the Ceylon forms of this subgenus differs from that of the Indian forms, which furnish the genotype; and it is, therefore, clear that the two groups are not congeneric, although venationally they agree. This being the case, I have here retained Laidlaw's genus *Drepanosticta* for the Indian species and have created a new genus, *Ceylonosticta*, for the Ceylon forms which possess *Drepanosticta* venation. The chain of evolution is probably *Protosticta*, *Drepanosticta*, *Ceylonosticta* and, finally, *Platysticta*; or the two latter genera may have evolved independently from *Protosticta* and on parallel lines to *Drepanosticta*.

The enormous length of the abdomen of these insects and the complex nature of the anal appendages, admirably adapted as they are for obtaining an enormous grip of the female prothorax, are probably coordinated with the function of oviposition, and I surmise that they enable the male to lower his mate into the swiftly flowing current of the streams they frequent without fear of being swept away.

Key to the Indian Genera of the Platystictinæ.

1.	Ceylon forms : penis with ridge on dorsum simulating a dorsal spine as seen in profile	2.
	Indian and Burmese forms : penis without ridge on dorsum	3.
2.	Sectors of arc separate and divergent from origin ; the nervure *IRiii* zigzagged almost from origin	PLATYSTICTA, p. 119.
	Sectors confluent at and for some distance from origin ; the nervure *IRiii* not zigzagged	CEYLONOSTICTA, p. 127.
3.	The nervure *ab* entirely absent..........	PROTOSTICTA, p. 101.
	The nervure *ab* present	DREPANOSTICTA, p. 141.

Genus **PROTOSTICTA** Selys (1885). (Fig. 47.)

Protosticta Selys, C. R. Soc. Ent. Belg. vol. xxix, p. cxlv (1885); id., Mém. Cour. vol. xxxviii, p. 157 (1886); Laid., Rec. Ind. Mus. vol. xi, p. 391 (1915); id., ibid. vol. xiii, pp. 339, 342 (1917); Fras., ibid. vol. xxiv, pp. 4, 5 (1922); id., ibid. vol. xxvi, p. 499 (1924); id., J. Bombay Nat. Hist. Soc. vol. xxxv, pp. 68, 69 (1931).

Zygopterous dragonflies of small size and slender build with characters of the subfamily, coloured steely black marked with white and turquoise blue; wings hyaline, immaculate,

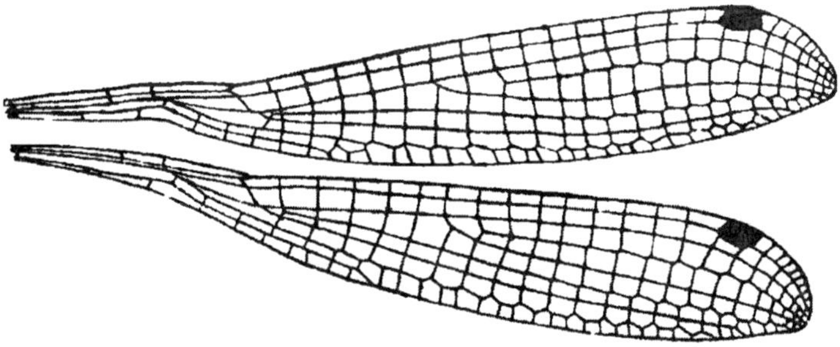

Fig. 47.—Wings of *Protosticta himalaica* Laid., male.

long, narrow, falcate, cells mostly four-sided; sectors of arc fused for a short distance from origin, and thus arising from a common stalk; *IRiii* straight, never zigzagged; *Riv+v* arising at or a little proximal or distal to the oblique nervure descending from the subnode ; *ab* entirely absent. Abdomen of great length, especially in the male, double or more than double the length of hind-wing (excepting *P. hearseyi*), segments 3 to 7 very slim and very long, segments 8 to 10 varying in length in the species and sexes, 9 sometimes as long as 8, 10 very short. Anal appendages about twice the length of segment 10, angulated downwards and inwards at their middle and chelate at apices (except in *P. himalaica* and *P. uncata*, which are spatulate), inferiors simple, broad at base, tapering thereafter to a fine point, incurved or not. Vulvar scale as for subfamily. Larvæ as for subfamily (but only those of *P. gravelyi* and *P. mortoni* are known).

Genotype, *Protosticta simplicinervis* Selys.

Distribution.—The WESTERN GHATS of India from sea-level to 4,000 ft., and N.E. INDIA and BURMA from 3,000 to 6,000 ft. Habits those of the subfamily : see also under species.

Species of this genus are remarkably local in distribution, and in some cases have been found confined to one spot for years, streams close by never rendering a single specimen ; this may be explained by their weak flight, although some at least (*P. mortoni* and *P. gravelyi*) do wander far from their habitats. Col., F. Wall has taken one species at light, and I have on several occasions found specimens hovering about the windscreen of my car or actually coming inside. On one occasion I took a male in my bungalow, but this may have come to light the night before. As a rule, species form definite and lasting colonies, restricted to small districts or even streams.

Key to the Species of Protosticta.

1.	Small species, with ground-colouring cupreous and abdomen of both sexes considerably less than 40 mm. in length ; male and female of the same length ...	*hearseyi* Fras., p. 113.
	Larger species, with ground-colouring steely black and abdomen more than 40 mm. in length ; male always considerably longer than the female	2.
2.	Pterostigma in both sexes blood-red	*sanguinostigma* Fras., [p. 109.
	Pterostigma in both sexes black	3.
3.	Superior anal appendages chelate at apex.	4.
	Superior anal appendages spatulate at apex	7.
4.	Inferior anal appendages bifid at apex ; posterior lobe of prothorax with a pair of long horn-like processes	*antelopoides* Fras., [p. 107.
	Inferior anal appendages not bifid at apex, and the posterior lobe of prothorax without any horn-like processes	5.

5.	Basal half of segment 8 pale blue, with the mid-dorsal carina finely black; superior anal appendages with the claw-like arms narrow and of subequal length...	*gravelyi* Laid., p. 103.
	Basal half of segment 8 pale blue, the mid-dorsal carina not marked with black; superior anal appendages varying as shown below..................	6.
6.	Outer fork of superior anal appendages thickened, obtuse at apex, and but slightly longer than the inner	*davenporti* Fras., [p. 105.
	Outer fork of superior anal appendages very narrow, and nearly four times the length of inner....................	*mortoni* Fras., p. 111.
7.	Inferior anal appendages with a long finger-like medial spine directed inwards	*himalaica* Laid., p. 115.
	Inferior anal appendages without an inner spine	*uncata* Fras., p. 117.

33. Protosticta gravelyi Laidlaw. (Fig. 48.)

Protosticta gravelyi Laid., Rec. Ind. Mus. vol. xi, pp. 389, 390, text-fig. 2 (1915); id., ibid. vol. xiii, p. 342 (1917); Fras., ibid. vol. xvi, p. 499 (1924); id., J. Bombay Nat. Hist. Soc. vol. xxxv, pp. 69, 70, pl. i, figs. 7, 8 (1931); id., Rec. Ind. Mus. vol. xxxiii, pp. 449, 466 (1931).

Protosticta stevensi Fras., Rec. Ind. Mus. vol. xxiv, pp. 7, 8, pl. i, figs. 1, 2, 7 (1923); id., ibid. vol. xxvi, p. 499 (1924).

Male.—Abdomen 46–49 mm. Hind-wing 20–22 mm.

Head: labium brownish-black; labrum turquoise-blue, rather broadly bordered with black along its free margin; clypeus pale turquoise-blue; frons, vertex, and occiput glossy black; eyes dark bottle-green paling to pale greenish beneath. *Prothorax* creamy white, marked dorsally with a broad black triangle, with its base occupying the greater part of posterior lobe and its apex tapering to the middle of the mid-lobe. *Thorax* steely glossy black, marked with a broad, oblique, creamy-white stripe on each side extending to the middle pair of legs, and a similar stripe on the posterior part of metepimeron extending on to the hinder pair of legs; beneath marked with broad black stripes on each side, which converge and fuse anteriorly. *Legs* creamy white, the knees darker; hind pair of femora occasionally marked on the extensor surface with a brown stripe. *Wings* hyaline; 13 to 14 postnodal nervures to fore-wings, 12 to 13 in the hind; pterostigma black, nearly half as long again as broad, the costal side shorter than the posterior, outer border nearly straight, inner oblique; *Riv+v* arising slightly or well distal to the level of subnode. *Abdomen* black, marked with white and turquoise-blue as follows:—Segments 1 and 2 white laterally, segment 3 with a narrow basal annule finely divided with black on the dorsum, segments 4 to 7 with broad basal annules gradually broadening as far as 7 and partially divided on the mid-dorsal carina by the ground-colour, segment 8 with nearly its basal half turquoise-blue, this extending apically on the sides and ventrally, but divided along the mid-dorsal carina by a fine black line, segments 9 and 10 unmarked ; segments 9 double the length of 10, and 8 double the length of 9. *Anal appendages* black ; superiors forcipate, chelate, half as long again as segment 10, broad at base, where they are furnished with an inner and dorsal blunt rounded tooth, then tapering, but again expanded at the apical half into a "finger-and-thumb"-like structure, the finger process being a little longer than the thumb ; inferior appendages about three-fourths the length of superiors,

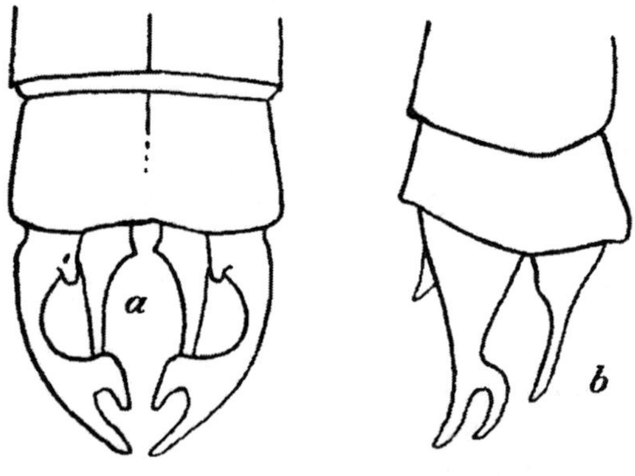

Fig. 48.—Anal appendages of *Protosticta gravelyi* Laid., male. *a.* Dorsal view. *b.* Right lateral view.

broad and conical at base, then tapering rather rapidly to apex, furnished with an inner stout spine at base.

Female.—Abdomen 33–35 mm. Hind-wing 19–23 mm.

Very similar to the male, but shorter and more robustly built; differs as follows:—Black bordering of labrum broader ; eyes in the living state with a diffuse white spot on the outer sides ; *wings* all with 13 to 14 postnodal nervures. *Abdomen* shorter and stouter, especially from segment 7 to the anal end, the former segment nearly three times as long as 8, which is itself actually shorter than segment 9, the latter segment being three times the length of segment 10. The basal annule on segment 7 occupying only about one-fifth the length of segment and often interrupted on the dorsum, whilst 8 is unmarked save for a large white spot situated at the base on each side ; segments 9 and 10 immaculate, but the former sometimes with a large lateral white spot.

Anal appendages black, barely as long as segment 10, broad, conical, pointed at apex.

Distribution.—S. INDIA. This species is the most widely distributed of all the Protostictas, and occurs in many ravines from 2,000 to 4,000 ft., in many parts of the Western Ghats. It occurs in the Nilgiri Hills on both sides of the plateau, at Kallar, Burliyar, and Gudalur, from May to June. In Malabar and Cochin it occurs in the Vayittiri, Nilambur, and Kavalai Ghats from May to June and again in September and October. It may be found lurking amongst rocks and ferns or in the dark tunnels formed by trees and shrubs overhanging torrential beds. In flight it holds its abdomen rigidly horizontal and advances in short jerks on its objective or reverses and flies backwards with equal ease, and is visible in the prevailing gloom only by the chain of white dots marking the bases of the segments.

Type in the Indian Museum; paratypes in the British Museum, etc.

Some confusion has arisen over this species, probably due to the fact that the type is immature and in very poor condition, with its abdomen crumpled up and appendages distorted. The marking on segment 8 has been given as that for 7, and the length of the abdomen has been given as that for the body. It was due to these excusable errors that *P. stevensi* was described by myself as a separate species. Since then I have had an opportunity of examining the type and female co-type and making new measurements. From this I find that *P. gravelyi* and *P. stevensi* are conspecific, Dr. Laidlaw's name having priority. The female, I find, is really that of *P. mortoni*. It is to be remembered that when Dr. Laidlaw described these sexes, *P. gravelyi* represented the only Indian *Protosticta* then known, so that it was but natural he should consider the insects as the two sexes of one species, although taken in different localities.

34. **Protosticta davenporti** Fraser. (Fig. 49.)

Protosticta davenporti Fras., J. Bombay Nat. Hist Soc. vol. xxxv, pp. 70, 71, pl. i, figs. 9, 10 (1931); id., (♀) Rec. Ind. Mus. vol. xxxiii, pp. 449, 467 (1931).

Male.—Abdomen 43–45 mm. Hind-wing 23–24 mm.

Very similar to *P. gravelyi*, from which it differs, however, by its much more robust build, this character being very striking when the insects are confronted. It is then seen that, although the abdomen of *P. gravelyi* is longer, the head and thorax are of much slighter build and the wings are correspondingly shorter and narrower; the anal appendages, although built on the same plan as those of *P. gravelyi*, have very marked differences. *Head, prothorax,* and *thorax* similar to *P. gravelyi* but the markings more definitely bluish and the black dorsal marking of the prothorax restricted to the middle portion of the posterior lobe and ending as two points on the dorsum of the mid-lobe. *Wings* broader and longer ; pterostigma black, half as long again as broad, 14 to 15 postnodal nervures to fore-wings, 13 to 14 in the hind (one specimen is remarkable in possessing four postcostal nervures in one of its wings, instead of the usual nervure *ac* and one additional postcostal) ; *Riv+v* arising opposite the level of subnode. *Abdomen* black, with the sides of segments 1 and 2 and the base of 3 bluish-white ; segments 4 to 7 with narrow white basal annules, which broaden apically on the sides ; segment 8 with its basal third or half pale turquoise-blue, this colour extending nearly to apex of segment laterally, but not divided

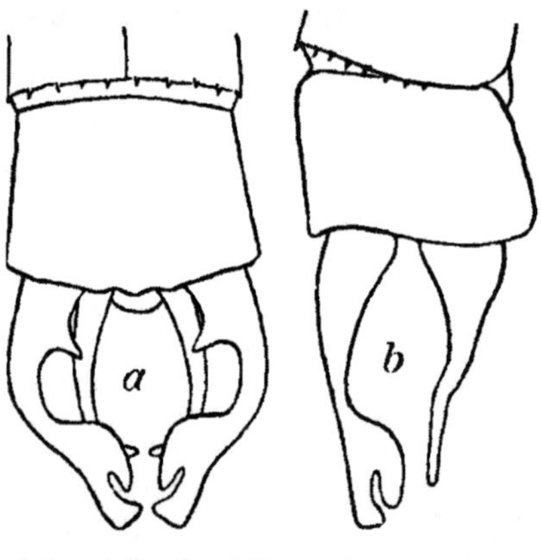

Fig. 49.—Anal appendages of *Protosticta davenporti* Fras., male.
a. Dorsal view. *b*. Right lateral view.

on the mid-dorsum as in *P. gravelyi* ; the blue annule on this segment separated from the extreme base by a narrow triangle of black ; segments 9 and 10 immaculate and of the same relative lengths as in *P. gravelyi*. *Legs* differing from the latter species by the femora being black, the two hinder pairs white on the extensor surfaces and the middle pair with a large white distal spot at the distal ends on the outer side ; tibiæ dirty white, tarsus black. *Anal appendages* black ; superiors forcipate, chelate, half as long again as segment 10, broad at base, where they present a large inner sub-basal pointed tooth or spine, then narrowing abruptly and expanding again in the distal half to form a "finger-and-thumb"-like structure which differs from that of *P. gravelyi* by the finger process being shorter and broader and with its apex broadened and flexed inwards almost to meet the apex of the thumb ; outer border of this appendage very sinuous, not gradually curved as in the last species ; inferior appendages broad at base, then tapered gradually to apex, slightly sinuous as seen in profile, but with a distinct angulation about the middle as seen from above, and with the apical fourth curled rather abruptly inwards ; nearly as long as superior appendages.

Female.—Abdomen 35–37 mm. Hind-wing 23–24 mm.

Very similar to the male; basal half of labrum and anteclypeus turquoise-blue, rest of head glossy black. The middle three-fifths of posterior lobe and posterior half of dorsum of middle lobe of *prothorax* black, remainder creamy-white ; *thorax* steely blue-black on dorsum, laterally pale blue, marked with a thick oblique black stripe and a short linear streak of the same colour on hinder angle of metepimeron; beneath creamy white marked with a black spur-shaped bifid streak. *Legs* as for male. *Wings* all with 14 postnodal nervures, otherwise as for the male. *Abdomen* marked similarly to the male except for segments 7 to 10, the former segment with a broad basal pale blue ring occupying rather less than a quarter of the segment and indented strongly at two points by the black, so that the hinder border of the ring is serrate ; segment 8 with a large basal lateral pale blue spot ; segments 9 and 10 unmarked. Vulvar scale black, very robust.

Distribution.—S. INDIA : Anaimalai, Mudis, and Travancore Hills at elevations of about 3,000 to 4,000 ft., in situations similar to the last.

Type in the British Museum ; co-type, female, and paratypes in the author's collection.

This species is far rarer than most others of the genus and less than a dozen examples are known, of which four males and two females were taken on the Western Ghat road running from Munnar, Travancore, June 3rd, 1931. The shape of the anal appendages will serve to distinguish this fine insect from all other Protostictas.

35. **Protosticta antelopoides** Fraser. (Fig. 50.)

Protosticta antelopoides Fras., Rec. Ind. Mus. vol. xxxiii, pp. 449, 467, 468 (1931).

Male.—Abdomen 53 mm. Hind-wing 30 mm.

Head : labium pale yellow ; labrum and anteclypeus turquoise-blue, the former narrowly bordered with black ; rest of head steely black ; eyes bottle-green. *Prothorax* pale at sides, greyish-black on dorsum ; posterior lobe furnished with a short robust spine at each outer end, and internal to these a pair of very long divaricate horns which extend nearly half way up the thoracic dorsum, on which they lie. *Thorax* steely blue-black on dorsum and anterior part of sides, beyond which the latter are palest blue marked with the usual oblique black stripe on hinder border of mesepimeron; beneath unmarked. *Legs* dirty white, with a broad pale blue ring followed by a narrow black ring towards the distal ends of femora; tibiæ dark on flexor surface. *Wings* hyaline; pterostigma black, with a fine yellow lining to the enclosing nervures, of the usual Protostictine shape; 15 to 16 postnodal nervures to all wings. *Abdomen* black, pale yellow beneath and on lower parts of sides, this colour extending on to dorsum at the base of segments 3 to 6 to form narrow basal rings; remaining segments unmarked. *Anal appendages*: superiors more than twice the length of

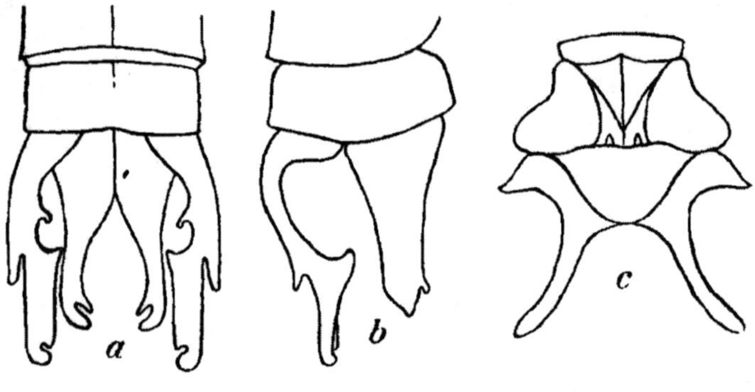

Fig. 50.—Anal appendages and prothorax of *Protosticta antelopoides* Fras.

a. Dorsal view of anal appendages. *b*. Right lateral view of same. *c*. Prothorax, showing horn-like processes on the posterior lobe.

segment 10, black, broad at base and strongly convex outwards, subcylindrical in the basal two-fifths, then dilated to form the usual "finger-and-thumb"-process, but the two digits here fused so that only the final joints, as it were, are left free; on the inner side of the convexity a small ventral spine, and at the end of the convexity a robust dorsal spine; the inner basal part of the dilated terminal portion strongly angulated so as to form a third spine; inferior appendages white for the basal half, black for the apical, very robust as compared with the superior appendages, and of equal length, very broad at base and closely apposed for the basal half, thinner and cylindrical at the apical black half, the apex curled in and split into two chelate short processes.

Female.—Abdomen 40 mm. Hind-wing 29 mm.

Exactly similar to the male in markings and wings, differing only in sexual characters and in the posterior lobe of prothorax, which lacks the long medial horns but has the lateral spines.

Distribution.—A single pair taken at the top of the Munnar, Western Ghats, TRAVANCORE, May 30, 1931.

The male is remarkable for several features unknown in other species of the genus—the armament of the prothorax, which is quite unique, the total absence of any identification mark on the end segments of the abdomen, the unmarked under surface of the thorax, and lastly, the large size of the insect, far greater than any other known species of the genus. As for the female, its large size and the absence of identification mark on the abdomen will easily serve to distinguish it from other females.

Type and paratype in the author's collection.

36. **Protosticta sanguinostigma** Fraser. (Fig. 51.)

Protosticta sanguinostigma Fras., Rec. Ind. Mus. vol. xxiv, pp. 6, 7, pl. i, figs. 5, 6 (1922); id., ibid. vol. xxvi, p. 499 (1924); id., J. Bombay Nat. Hist. Soc. vol. xxxv, pp. 71, 72, pl. i, figs. 5, 6; (1931); id., Rec. Ind. Mus. vol. xxxiii, p. 449 (1931).

Protosticta cerinostigma (var.) Fras., Rec. Ind. Mus. vol. xxvi, pp. 499, 509 (1924).

Male.—Abdomen 42–45 mm. Hind-wing 22–26 mm.

Head: labium yellowish; labrum turquoise-blue, narrowly bordered with black along the free margin; clypeus turquoise-blue; frons, vertex, and occiput glossy black, the latter marked with a transverse coppery-brown fascia; eyes bottle-green, pale greenish-blue below, these two colours separated by an equatorial band of blackish-brown which crosses the head from above and behind somewhat obliquely. *Prothorax* black on dorsum, whitish laterally, the mid-lobe marked with a large oval blue spot on each side of the middle line. *Thorax* glossy steely or bronzed black, pale blue on the sides, where a narrow oblique stripe of black bordered with brown traverses the anterior border of the metepimeron; beneath immaculate, thus contrasting strongly with the two former species. *Legs* pale yellow, femora lined with black on extensor surface and with a broad brown annule at the distal ends; tibiæ clouded or stippled with brown; tarsi dark brown. *Wings* hyaline; pterostigma blood-red, covering one cell, of similar shape to the former species; 15 to 18 postnodal nervures to fore-wings, 14 to 15 in the hind; $Riv+v$ arising slightly or well distal to the oblique nervure descending from the subnode. *Abdomen* blackish-brown on dorsum, paler on the sides, marked with white and pale turquoise-blue; segments 1 and 2 white laterally, segments 3 to 6 with broad distal black annules, segment 3 with a very narrow obscure white annule at the base, segments 4 to 7 with broader annules gradually increasing in depth from 4 to 7, segment 8 turquoise-blue, its apical border black, this colour continued basally as a gradually narrowing tongue of black along the mid-dorsal carina as far as the base of segment, segments 9 and 10 entirely black; segment 7 very long and broadening apically, segment 8 only slightly longer than 9, which is nearly three times the length of 10. *Anal appendages* black; superiors forcipate, chelate, seen from above appearing only slightly longer than segment 10, but in profile, owing to the ends being bent strongly downwards, seen to be nearly twice the length of the segment, very broad at base, then tapering and inclined strongly down and inward to end in the usual "finger-and-thumb"-process, the finger portion much broader and longer than the thumb and notched at its apex, which

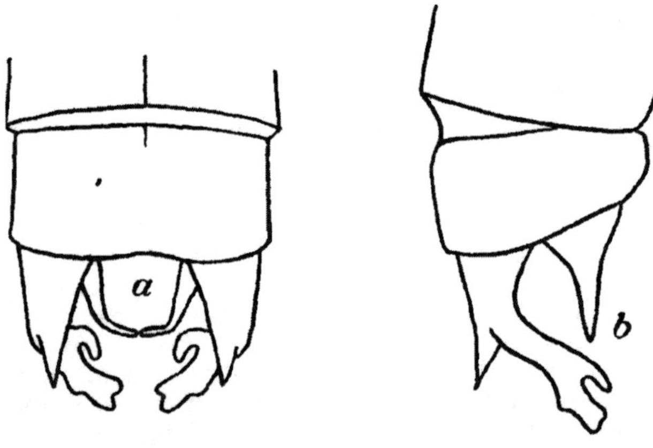

Fig. 51.—Anal appendages of *Protosticta sanguinostigma* Fras., male. *a*. Dorsal view. *b*. Right lateral view.

is broad and blunt; the thumb curled in nearly to meet the opposing process; dorsally the basal portion continued horizontally back to end in a robust pointed spine; inferior appendages about three-fourths the length of superiors, broad at base, then rapidly tapering to a fine point, the apical fourth angulated strongly inwards to meet or overlap its fellow on the opposite side.

Female.—Abdomen 39 mm. Hind-wing 26 mm.

Very similar to the male, but of much shorter and stouter build; differs as follows: the equatorial belt of black encircling the eye much better defined. *Prothorax* a dirty white laterally and without the mid-dorsal blue spots. *Wings* similar to the male, pterostigma blood-red; 16 to 17 postnodal nervures to fore-wings, 14 to 15 in the hind; *Riv+v* arising at the same level or nearer to the subnode. *Abdomen* black marked with white and turquoise-blue as follows: segment 1 with a blue spot on each side, 2 with a lateral blue spot at the base prolonged lateralwards, apically, and ventrally, segment 3 with its middle two-thirds laterally and beneath brownish-white, 4 similar and with a narrow whitish basal annule, 5 and 6 with well-marked white basal annules, the ventro-lateral markings more obscure, segment 7 with its basal third turquoise-blue, the black not continued basally along the mid-dorsal carina as in the male, remaining segments entirely black. *Anal appendages* black, barely as long as segment 10, small, conical, triangular processes. Vulvar scale robust, extending beyond end of abdomen.

Distribution.—S. INDIA; Nilgiri Hills, Coorg, and Malabar Wynaad. Specimens from the two latter districts are much smaller than those from the Nilgiris. In the latter district it is moderately common in the Burliyar ravine, Mettupalayam Ghat, at 1,500 ft., but very rare on the opposite side of the plateau, where the variety *cerinostigma* was found. (Structurally this latter insect does not differ from the type, so that I consider it to be merely a variety.) In Coorg it is confined to Hallery, whilst in the Wynaad it appears to be restricted to the Vayittiri and Nilambur Ghats from May to October.

The blood-red pterostigma and the shape of the anal appendages, the relative lengths of the terminal abdominal segments, etc., will serve to differentiate this beautiful species from all other Protostictas.

Type in the British Museum; paratypes in Morton, Fraser, and Laidlaw collections, etc.

37. **Protosticta mortoni** Fraser. (Fig. 52.)

Protosticta mortoni Fras., Rec. Ind. Mus. vol. xxvi, pp. 500, 501 (1924); id., J. Bombay Nat. Hist. Soc. vol. xxxv, pp. 72, 73, pl. i, figs. 3, 4 (1931); id., Rec. Ind. Mus. vol. xxxiii, pp. 449, 466 (1931).

Protosticta gravelyi Laid. (*mortoni*, ♀, nec *gravelyi*), Rec. Ind. Mus. vol. xiii, p. 342 (1917).

Male.—Abdomen 42–43 mm. Hind-wing 20 mm.

Head: labium blackish-brown; labrum turquoise-blue, moderately broadly bordered with black along its free margin; clypeus turquoise-blue; vertex and occiput glossy steely or bronzed black; frons black; eyes ultramine blue capped with black above, paler blue beneath. *Prothorax* pale blue except the posterior lobe, which is black. *Thorax* steely black on dorsum, pale blue laterally, traversed along the postero-lateral suture by a broad black stripe; beneath thorax yellowish marked by two thick black stripes, which converge and become confluent anteriorly as in *P. gravelyi*.

Legs white, the knees and femora sparsely stippled with black. *Wings* hyaline; pterostigma black, framed narrowly in pale brown, and this again framed within thick black nervures; 13 postnodal nervures to fore-wings, 12 in the hind; *Riv+v* arising distal to the oblique nervure descending from the subnode. *Abdomen* black marked with turquoise-blue and white as follows: segment 1 blue laterally, segment 2 with the basal two-thirds of the sides white, 3 with a narrow white basal annule prolonged for a short distance along the sides and narrowly interrupted on the dorsum, segments 4 to 7 with broad white basal annules occupying about one-sixth the length of the segments and prolonged along the sides and venter, segment 8 with the basal half turquoise-blue, but this colour prolonged along the sides nearly as far as the apical end of segment, and separated from the base by a narrow

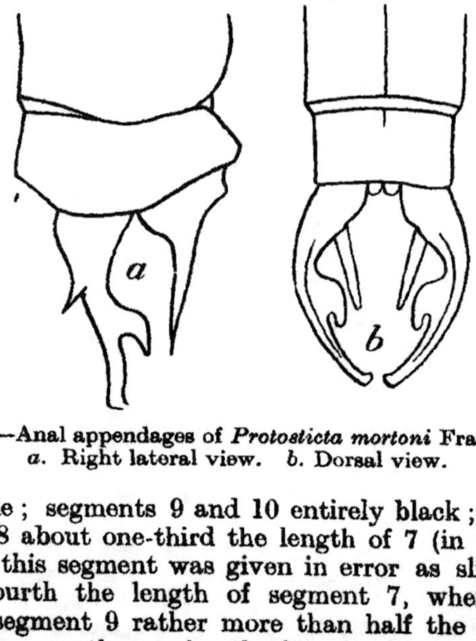

Fig. 52.—Anal appendages of *Protosticta mortoni* Fras., male. *a*. Right lateral view. *b*. Dorsal view.

black annule; segments 9 and 10 entirely black; segment 7 very long, 8 about one-third the length of 7 (in the original description this segment was given in error as slightly more than one-fourth the length of segment 7, whereas 9 was intended), segment 9 rather more than half the length of 8 and slightly more than twice the length of 10. *Anal appendages* black; superiors forcipate, chelate, about twice the length of segment 10, broad at base, then constricted and curved downwards and again dilated in the first part of the apical half to form the usual "finger-and-thumb"-process, the finger thrice the length of the thumb and with its extreme end angulated inward, the thumb portion short and stout; dorsally the basal portion ending in a robust pointed tooth or spine very much as in *P. sanguinostigma*; inferior appendages very broad at base, then abruptly narrowed, curled gently in and tapered to a fine point, about three-fourths the length of superiors.

Female.—Abdomen 32–33 mm. Hind-wing 19–20 mm.

Very similar to the male, but much shorter and more robust. Markings differ only on the *abdomen*, where segment 7 has a blue basal annule occupying from one-third to one-fourth the length of the segment, and segment 8 has a large baso-lateral whitish spot on each side, 9 and 10 entirely black; segment 7 is rather more than four times the length of segment 8, which is itself slightly shorter than segment 9, the latter being nearly four times the length of 10. *Anal appendages* very short, barely the length of segment 10, conical, pointed, black. Vulvar scale very robust, extending beyond end of abdomen.

Distribution.—S.W. INDIA: very local, usually found in isolated colonies. The original colony in the Sampaji Ghat, Coorg, was restricted to one part of a tributary of the Sampaji River. Since then I have found other colonies along the foot of the Western Ghats in Malabar, and a single female has been taken on Salsette Island by Mr. Prater. The species thus differs from all others by being found at or near sea-level.

The colouring of the prothorax forms a ready means of identification, whilst the appendages will serve to separate it from all other species. As mentioned above, the female described by Dr. Laidlaw as that of *P. gravelyi*, which was taken at Talewadi, Castle Rock, Kanara (a locality far removed from that where the type of *gravelyi* was found), is really the female of *P. mortoni*.

Type in the British Museum.

38. **Protosticta hearseyi** Fraser. (Fig. 53.)

> *Protosticta hearseyi* Fras., Rec. Ind. Mus. vol. xxiv, pp. 5, 6, pl. i, figs. 3, 4 (1922); id., ibid. vol. xxvi, p. 499 (1924); id., J. Bombay Nat. Hist. Soc. vol. xxxv, pp. 73, 74, pl., figs. 1, 2 (1931); id., Rec. Ind. Mus. vol. xxxiii, pp. 449, 466 (1931).

Male.—Abdomen 30–35 mm. Hind-wing 18–21 mm.

Head: labium ashy white; labrum turquoise-blue narrowly bordered with black along the free margin; the two basal joints of antennæ blue; clypeus turquoise-blue; frons, vertex, and occiput black; eyes olivaceous above, paling and changing to pale blue beneath. *Prothorax* pale blue, immaculate, fading to pale brown in the dry state. *Thorax* cupreous on the dorsum, pale blue laterally, the mid-dorsal carina finely blue, whilst laterally a broad oblique black stripe traverses the length of the postero-lateral suture and anterior part of metepimeron; beneath whitish marked with a round spot of black between the legs and a pair of short stripes which converge and become nearly confluent with the anterior spot. *Legs* bluish-white, the two hind emora with narrow blackish transverse lines on the extensor surface; the knees black. *Wings* hyaline, pterostigma black, shaped similarly to the rest of the genus, covering one cell; 10 to 14 postnodal nervures to fore-wings, 10 to 12 in the hind; *Riv+v* arising opposite or distal to the oblique nervure descending from the subnode. *Abdomen* enfumed brown or cupreous marked with pale blue, this colour more in evidence on the final segments; segments 1 and 2 with the sides bluish-white and with a diffuse longitudinal stripe on the mid-dorsum, incomplete on the apical half of 2; segments 3 to 7 with narrow pale basal annules extending more broadly along the sides and with broad black apical annules; segment 8 turquoise-blue, with a narrow black

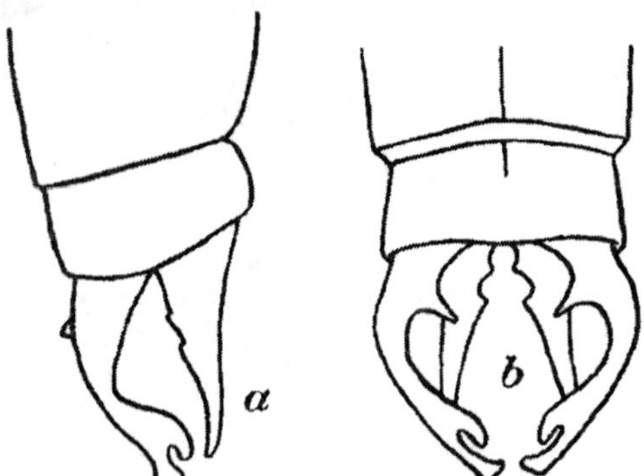

Fig. 53.—Anal appendages of *Protosticta hearseyi* Fras., male. *a.* Right lateral view. *b.* Dorsal view.

apical annule; 9 and 10 black, the former with a bluish crescentic spot on each side; segment 7 twice the length of segment 8; 8 half as long again as 9, which is more than twice the length of 10. *Anal appendages* black; superiors about twice the length of segment 10, forcipate, chelate, broad at the base and furnished here with a robust inwardly directed sub-basal spine, then constricted and subcylindrical, but again slightly expanded at the apical third to form a "finger-and-thumb"-process, the finger portion considerably longer than the thumb and squared at its apex; the whole appendage curving gently in and downwards; inferior appendages about four-fifths the length of superiors, truncate, broad at base and tapering to a slightly and gently upturned point, the base furnished with a blunt inner tooth and one or two minute teeth thereafter.

Female.—Abdomen 32–33 mm. Hind-wing 22 mm.

Very similar to the male and approximately of the same length, although more robust in build. Differs as follows: labrum more broadly bordered with black; eyes olivaceous brown above, changing to pale green below; mid-dorsal carina of thorax obscurely white at its upper part only. *Wings* with 12 to 13 postnodal nervures to fore-wings, 12 in the hind; *Riv+v* arising at or but slightly distal to the subnode. Segment 7 with a broader basal annule; segment 8 brownish-black, marked only with a baso-lateral dirty white quadrate spot; segment 9 paler brown, marked with a broad white spot on each side confluent with a smaller subdorsal spot of the same colour. *Anal appendages* black, very small, about equal in length to segment 10, conical, pointed. Vulvar scale robust, extending beyond end of the abdomen.

Distribution.—S. INDIA: Nilgiris, Anaimalais, and Travancore. The *type* was taken at a small brook running through a coffee tote on the Gudalur Ghat, in June; out of 17 specimens only one was a male. A single male was taken later in the neighbouring Ochterlony valley. Recently I found the species in the Mudis, Anaimalai Hills, 3,000 ft., in May, where about 18 males, but no females, were taken. The locality was a rocky hillside in virgin jungle where a seepage not amounting to a brook found its way down through fern and moss to the river below, and the whole of the specimens were found in an area not covering more than a quarter of an acre; search in neighbouring jungle revealed none. These specimens were in company with *Heliogomphus promelas*, which were equally restricted to the spot. Returning a few days later via the Nilgiris, I again found *P. hearseyi* in the same spot on the Gudalur Ghat and, by a curious coincidence, took a number of females and only one male. This curious incidence of the sexes is very difficult to explain.

It is the commonest *Protosticta* in Travancore, May to July. The small size of the species and its cupreous colouring will serve to identify it from all others of the genus. It is the only species in which the two sexes are approximately of the same length.

39. **Protosticta himalaica** Laidlaw. (Figs. 47 & 54.)

Protosticta himalaica Laid., Rec. Ind. Mus. vol. xiii, pp. 342, 343 (1917); Fras., J. Bombay Nat. Hist. Soc. vol. xxxv, pp. 74, 75, pl., figs. 11, 12 (1931).

Protosticta lindgreni Fras., J. Bombay Nat. Hist. Soc. vol. xxvii, p. 150 (1920); id., ibid. vol. xxix, p. 741 (1923).

Male.—Abdomen 40–43 mm. Hind-wing 25–26 mm.

Head: labium pale brown or brownish-yellow; labrum turquoise-blue, narrowly bordered with black along its free margin; clypeus turquoise-blue; frons, vertex, and occiput steely bronzed black; eyes black above, changing to dark olivaceous and, finally, pale bluish-green beneath. *Prothorax* black on dorsum and upper part of sides, marked with a broad subdorsal, longitudinal, creamy-yellow stripe on each side extending from the posterior lobe to the anterior end, yellow low down on the sides. *Thorax* bronzed black on dorsum, pale blue at the sides, which are marked with a moderately broad black oblique stripe extending along the postero-lateral suture; pale blue beneath, unmarked with black. *Legs* pale sandy yellow, extensor surfaces of all femora black. *Wings* hyaline, pterostigma dark reddish-brown, almost black, with a fine frame of light yellow bordering the enclosing nervures within, about one-third longer than broad, costal

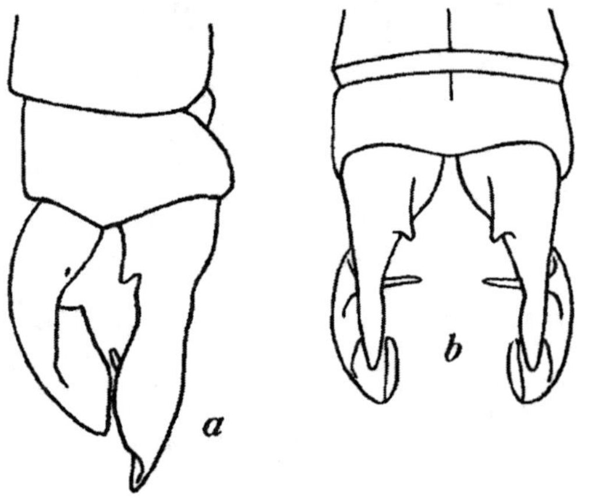

Fig. 54.—Anal appendages of *Protosticta himalaica* Laid., male. *a.* Right lateral view. *b.* Dorsal view.

side only slightly longer than the posterior, and the outer side but slightly longer than the inner; *Riv+v* in the fore-wings arising very slightly distal to the oblique nervure descending from the subnode, but opposite that nervure in the hind; 14 to 16 postnodal nervures in fore-wings, 13 to 15 in the hind. *Abdomen* blackish-brown, marked with white and pale blue as follows:—Segment 1 bluish at the sides and with a white apical ring, 2 bluish laterally and marked on the dorsum with a bluish longitudinal stripe not extending to the apical border, segment 3 with a small white dorsal basal triangle and a broad apical black annule, segments 4 to 7 with narrow white basal annules, 4 to 6 with broad black apical annules, segment 7 with its apical half, and the whole of 8 and 9 pale blue, segment 10 blue at its extreme base, black apically; segments 7 to 10 gradually decreasing in length towards the last segment. *Anal appendages* about twice the length of segment 10, dark brown; superiors broad at base, compressed, with an angulation on the inner side near the middle not amounting to a spine, after which the appendages are curved downward and abruptly broadened into a scoop-like expansion hollowed out on the inner side; inferiors slightly longer than the superiors, broad at base, then constricted and again broadened, and with the edges strongly curled like a drying leaf; a long inwardly directed narrow spine springing from the upper side of this expansion about the middle of the appendage.

Female.—Abdomen 37 mm. Hind-wing 26 mm.

Very similar to the male, differing only in its shorter and more robust abdomen and in some of the markings of this structure. The ventro-lateral aspect of segment 3, as well as 1 and 2, are definitely blue; segments 4 to 6 have broader basal annules, especially the latter segment, 8 has a large bluespot on each side, whilst 9 and 10 are reddish-brown (possibly blue in the living state). *Anal appendages* small, rather shorter than segment 10, conical, pointed, blackish-brown. Vulvar scale robust, extending a little beyond the end of abdomen.

Distribution.—N. Bengal, Assam, and Sikkim, at altitudes of 3,000 to 6,000 ft. I have examined specimens from Kalimpong and Pashok and have taken the insect at Mangpu, Darjeeling District, in May and June.

Although so widely spread, it does not appear to be nearly as common as some of the South Indian species. Its habits are entirely similar and it is found in similar localities. The abdominal markings and the shape of the anal appendages, which are entirely different from those of the genotype, will serve easily to distinguish it from all others. A fresh comparison of the types of *P. himalaica* and *P. lindgreni* has convinced me that they are conspecific, the former name having priority.

40. **Protosticta uncata** Fraser. (Fig. 55.)

Protosticta uncatus Fras., J. Bombay Nat. Hist. Soc. vol. xxxv, pp. 75, 76 (1931).

Male.—Abdomen 42 mm. Hind-wing 21 mm.

Head: labium pale brown; labrum turquoise-blue, narrowly bordered with black along the free margin; anteclypeus turquoise-blue; postclypeus, frons, vertex, and occiput bronzed black; eyes dark olivaceous green, paler below. *Prothorax* pale yellow, marked with a pair of ill-defined, longitudinal, subdorsal dark brown stripes made up of a row of three spots, the largest of which is situated on the posterior lobe. *Thorax* bronzed black on dorsum, pale blue laterally, changing to pale yellow beneath, which is unmarked; a narrow oblique black stripe on each side mapping out the postero-lateral suture, and a small upper posthumeral blue spot. *Legs* pale yellow, knees darker, hind femora with a black stripe along the extensor surface. *Wings* hyaline; pterostigma as broad as long, costal side only slightly longer than posterior, the inner side very oblique, blackish-brown, framed in black nervures which are lined inwardly with pale yellow; 15 to 16 postnodal nervures to fore-wings, 15 in the hind; *Riv+v* well distal to the oblique nervure descending from the subnode. *Abdomen* shaded with brown and ringed with blue, white, and black as follows: segment 1 with the sides blue,

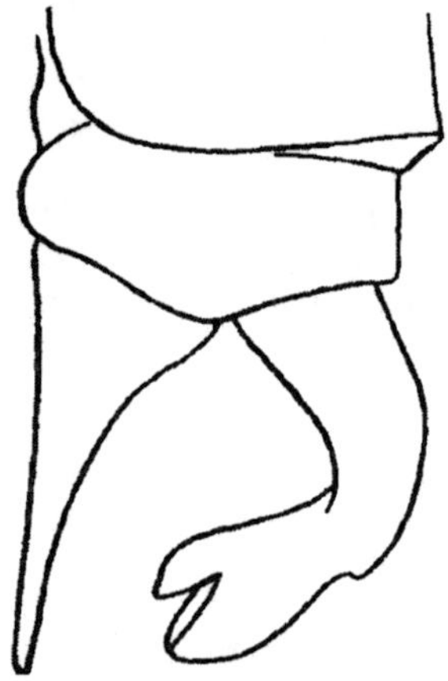

Fig. 55.—Left lateral view of anal appendages of *Protosticta uncata* Fras.

dorsum blackish-brown, 2 with the apical half blackish-brown, basal half on dorsum warm brown, the sides blue for the basal three-fourths, but this colour invaded by the brown of dorsum at its middle, segments 3 to 6 with broad basal pale bluish-white annules and equally broad apical black annules, the intervening part warm brown, 7 with only the basal annule, the rest brown, gradually deepening to black at apex, 8 black, with a small baso-lateral spot on each side, 9 entirely turquoise-blue, 10 entirely black; segment 7 three times as long as 8, which is half as long again as 9, the latter nearly four times as long as 10. *Anal appendages* black; superiors as long as segment 10, broad at base, then tapering as far as their middle, at which point they are bent down rather abruptly and terminate in an expansion shaped like a parrot's beak;

inferior appendages slightly longer, broad at base, spatulate, compressed and furnished at the apex with a tuft of about 10 to 12 stout bristles.

Female.—Abdomen 35 mm. Hind-wing 21 mm.

Closely similar to the male save for its sexual characters and the slightly shorter and more robust abdomen. Markings of *head, thorax*, and *abdomen* not differing markedly from the male, but segment 9 has a narrow basal black annule and the sides low down dark brown. *Wings* similar, 15 to 16 post-nodal nervures in fore-wings, 15 in the hind. *Anal appendages* very short, not quite as long as segment 10, conical, pointed, blackish-brown. Vulvar scale robust, extending beyond the end of abdomen.

Distribution.—BURMA: Maymyo, Mandalay Dist., and Gokteik, N. Shan States. One male and two females collected by Col. F. Wall, during the early part of June.

The species is remarkable for its anal appendages which, like those of *P. himalaica*, differ markedly from those of the genotype. The situation of the "recognition" mark on segment 9 instead of 8 is also quite at variance with other species of the genus. These specimens were taken at light, but should the habitat of the insect be discovered it will probably prove to be similar to that of other species, and the species itself equally common.

Genus PLATYSTICTA Selys (1860). (Fig. 56.)

Platysticta Selys, Bull. Acad. Belg. (2) vol. x, p. 436 (1860); id., Syn. Agr., sep. pp. 5, 6, 8, 9 (1860); id. (subgenus *Platysticta*, 1st group), Mém. Cour. vol. xxxviii, p. 150 (1886); Kirby, Cat. Odon. p. 132 (1890); Laid., Rec. Ind. Mus. vol. xiii, pp. 339, 340 (1917); id., Spolia Zeylan. vol. xii, p. 360 (1924); Fraser, Rec. Ind. Mus. vol. xxvi, p. 498 (1924); Laid., J. Malay Branch Roy. As. Soc. vol. vii, pp. 303–4 (1924); Fras., J. Bombay Nat. Hist. Soc. vol. xxxv, pp. 325, 326 (1931).

Dragonflies of slender build with the characteristics of the subfamily, of rather larger stature than the Protostictas, but resembling them closely otherwise. Body-colouring non-metallic black, brown or bright brick-red marked with azure blue. Wings closed when at rest, long, narrow, falcate at apex, with a long petiole, hyaline or opaque at apices; discoidal cell elongate, costal border about four times as long as the proximal and slightly shorter than the posterior; sectors of arc arising at lower part of arc, *but diverging from origin*; an accessory basal-postcostal nervure always present near base of wing, in addition to the nervure *ac*, which is situated about midway between the two antenodal nervures; *ab* always present but incomplete, meeting *ac* at or well away from the posterior margin of wing and extending to the posterior side of discoidal cell, which it meets at a point near its middle; *IA* absent; *Cuii* of about 6 to 9 cells in length; *MA* and *IRiii* zigzagged almost from origins; *Riv+v* arising a short distance proximal to the oblique nervure descending from the subnode, *IRiii* slightly proximal to the level of the subnode: pterostigma subquadrate, costal side shorter than posterior, which is somewhat convex, inner border oblique, unbraced, outer straight; cells of wings mainly pentagonal.

Head, thorax, and abdomen similar to *Protosticta*, but *the latter less than double the length of wings*; anal appendages variable, superiors forcipate, spatulate at apices; inferiors usually shorter than superiors, broad at base, tapering to the apex, which is acute or obtuse, directed straight back or curled inwards towards its fellow. Genitalia: hamules

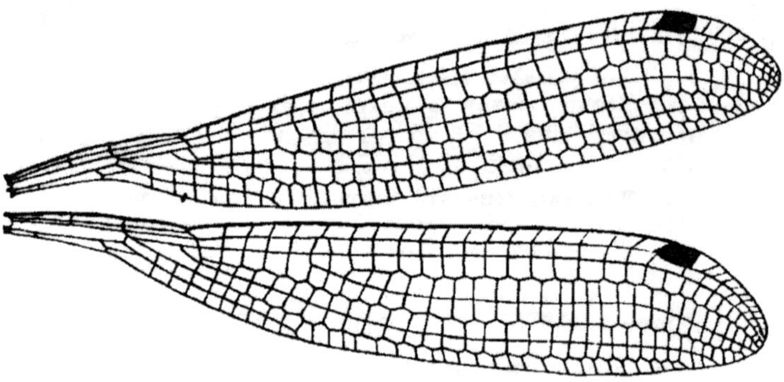

Fig. 56.—Wings of *Platysticta deccanensis* Laid., male.

broad quadrate plates, penis with apex curled strongly over dorsum and its branches long, curled, and embracing the stem of the organ; a crest or ruff on the dorsum, which simulates a spine as viewed in profile; lobe flask-shaped. Vulvar scales very robust, minutely spined below, usually extending beyond end of abdomen. Larvæ unknown.

Genotype, *Platysticta maculata* Selys.

Distribution.—INDIA and CEYLON only. Three species only are included in this genus from within our limits, and it is doubtful if other species outside of this area can be strictly classed as congeneric with *P. maculata*. Two species are from Ceylon, the third from Cochin in South India. The genus as defined here includes only those species which fall within the Selysian subgenus *Platysticta*, group 1. Like *Protosticta*, species of the genus are to be found along the banks of montane streams clinging to mosses and ferns overhanging shady spots.

Key to the Species of Platysticta.

1.	Wings of male tipped with black	*apicalis* Kirby, p. 123.
	Wings hyaline throughout.............	2. [p. 124.
2.	Thorax uniform brick-red	*deccanensis* Laid.,
	Thorax black marked with blue	*maculata* Selys. p. 121.

The author of the species *P. maculata* and *C. hilaris* has been shown to be Selys, since Hagen's papers published in the Verh. zool.-bot. Ges. Wien, vols. viii and ix of 1858 and 1859 respectively, gave only a list of names without descriptions. The descriptions of these two species were published later by Selys, who adopted Hagen's names. It is even possible that the descriptions were communicated to Selys by Hagen, as in the case of many others.

41. **Platysticta maculata** Selys. (Fig. 57.)

Disparoneura maculata Hagen, Verh. zool.-bot. Ges. Wien, vol. ix, p. 206 (1859). (Name only.)
Platysticta maculata Selys, Bull. Acad. Belg. (2) vol. x, p. 437 (1860); id., Rev. Syn. Agr., sep. p. 151 (1886); Kirby, Cat. Odon. p. 132 (1890); id., J. Linn. Soc. Lond., Zool. vol. xxiv, p. 561 (1893); Laid., Rec. Ind. Mus. vol. xi, p. 387 (1915); id., ibid. vol. xiii, p. 341 (1917); id., Spolia Zeylan. vol. xii, p. 361 (1924); Fras., J. Bombay Nat. Hist. Soc. vol. xxxv, pp. 326, 327, pl. ii, figs. 1, 2 (1931).
Platysticta greeni Kirby, Proc. Zool. Soc. Lond. p. 204, pl. xx, figs. 3, 3 *a* (1891).

Male.—Abdomen 47–48 mm. Hind-wing 31–33 mm.

Head velvety black; labium brown, paling to dirty white at base; labrum glossy black, pale azure blue for its basal half; anteclypeus pale azure blue, as also bases of mandibles adjacent to it; postclypeus and rest of head unmarked; eyes black above, dark olivaceous brown below. *Prothorax* black, its sides low down palest blue. *Thorax* velvety black on dorsum as far lateral as the anterior border of metepimeron, and traversed on each side midway between the humeral and postero-lateral sutures by a narrow oblique pale blue stripe; the hinder three-fourths of metepimeron palest blue, changing to creamy-white beneath thorax. *Legs* brown, trochanters and coxæ creamy-white, the base and flexor surface of femora pale, spines pale golden-yellow. *Wings* hyaline, palely tinted with yellow; pterostigma subquadrate, but slightly longer than broad, inner border very oblique, outer slightly convex, costal border four-fifths the length of posterior, dark reddish-brown framed finely in yellow and enclosed in thick black nervures; 20 to 25 postnodal nervures in fore-wings, 19 in the hind; *ac* and *ab* meeting the posterior border of wing at a common point, *ab* meeting discoidal cell nearer its base than the outer end. *Abdomen* black above, yellow along the ventral border and below, the sides of segments 1 and 2 and base of 3 pale yellow, this colour forming basal annules interrupted on the dorsum on segments 3 to 7; segments 8 to 10 azure blue on dorsum, this colour broad at apical end of segment 8 and tapering to a fine point on the mid-dorsum just short of the base of the segment; segment 10 black at sides. *Anal appendages* black, paler at apices; superiors nearly twice the length of segment 10, which is very short; seen from above, these appendages are forcipated, curling gently inwards at apices, broad at base, then tapered at the middle, again abruptly dilated on the inner side, from which point, as far as the apex, they are of even width and spatulate (the whole appendage viewed from above resembling a Kropp

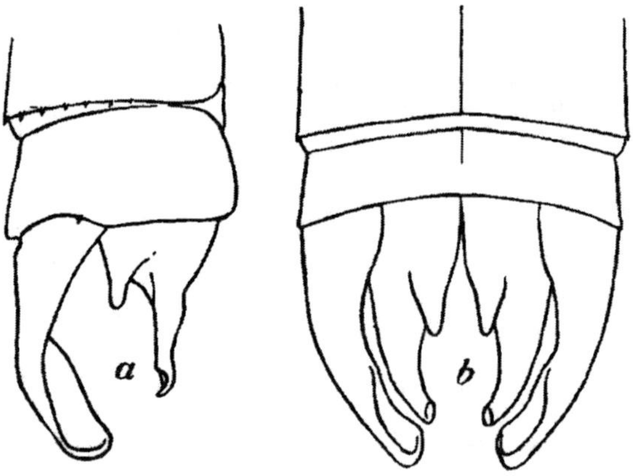

Fig. 57.—Anal appendages of *Platysticta maculata* Selys, male.
a. Dorsal view. *b*. Right lateral view.

razor with its blade nearly fully opened); inferior appendages rather more than half the length of the superiors, very broad at base, tapering rapidly to an obtuse point, the apical fourth angulated inwards; a small sub-basal tubercle on the inner side.

Female.—Abdomen 36 mm. Hind-wing 29 mm.

Closely similar to the male save for sexual characters. *Prothorax* with two curious bosses on the dorsum of middle lobe; segments 8 and 9 blackish-brown on dorsum, unmarked with blue; segment 10 with a large spot of blue on the mid-dorsum, its apical border ridged. Vulvar scales robust, extending beyond end of abdomen, minutely spined below; *anal appendages* short, pointed at apex, conical, as long as segment 10.

Distribution.—Confined to the hilly tracts of CEYLON at elevations of 3,000 to 4,000 ft.: Ramboda, Punduloya, Urugala during April and May.

In the British Museum collection there are no less than three species included under the head of *P. maculata*, but I do not think that there is much doubt as to which of these is the actual species under discussion. The description given here is from specimens collected by Col. F. Wall at Urugala, and compares so closely with the original description that there can be no doubt as to the correct identification of the specimens.

42. **Platysticta apicalis** Kirby. (Fig. 58.)

Platysticta apicalis Kirby, J. Linn. Soc. Lond., Zool. vol. xxiv, pp. 561, 562, pl. xlii, fig. 1 (1893); Laid., Rec. Ind. Mus. vol. xi, p. 387 (1915); id., ibid. vol. xiii, p. 341 (1917); id., Spolia Zeylan. vol. xii, p. 361 (1924); Fras., J. Bombay Nat. Hist. Soc. vol. xxxv, pp. 327, 328, pl. i, figs. 5, 6 (1931).

Male.—Abdomen 48 mm. Hind-wing 34 mm.

Head velvety black with a steely blue reflex; labium pale brown; labrum, bases of mandibles, ante- and postclypeus azure blue, rest of head unmarked. Eyes black above, dark

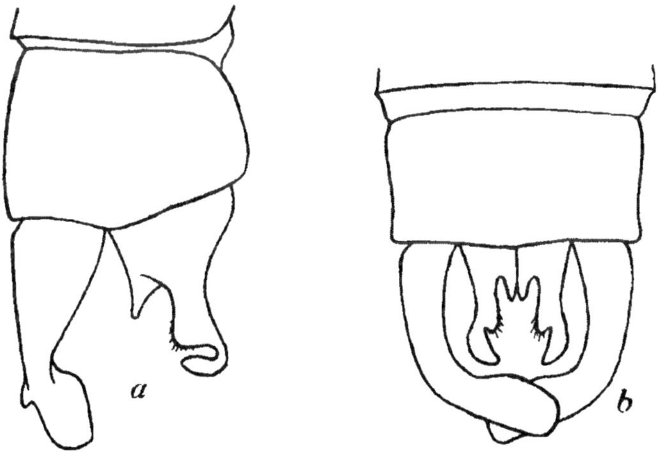

Fig. 58.—Anal appendages of *Platysticta apicalis* Kirby, male. *a*. Right lateral view. *b*. Dorsal view.

brown below. *Prothorax* azure blue, posterior lobe black. *Thorax* steely black with a blue reflex, marked on each side with a narrow oblique azure blue stripe on the mesepimeron; beneath and along lower border of metepimeron pale yellow. *Legs* black; coxæ, trochanters, and bases of all femora pale yellow. *Wings* hyaline, with the apices of all blackish-brown as far proximal as the inner end of pterostigma, which structure is twice as long as broad, inner end oblique, outer end angulated or convex, covering one or two cells, dark blackish-brown; 19 postnodal nervures in fore-wings, 18 in the hind; *ac* very oblique and meeting *ab* very near the posterior margin of wings; other features of venation as for the genus. *Abdomen* black on dorsum, yellow low down on the sides and beneath except for the last few segments; segments 8 to 10 azure blue on dorsum, but the base of the former segment narrowly black, which projects into the blue along the mid-dorsal carina as a tapering point. *Anal appendages* black; superiors one and a half times as long as segment 10, which is very short, broad at base, then tapered as far as the middle, where it again dilates as a compressed spatulate apex somewhat squared at the end; seen in profile, the appendage is convex dorsally and rather strongly curved at its middle; inferiors about four-fifths the length of superiors, very broad at base, then tapered as far as apex, which ends in an acute, abruptly turned up point; a very robust spine projecting back from the dorsum near its base.

Female.—Abdomen 42 mm. Hind-wing 36 mm.

Closely similar to the male except for sexual characters, differing only by its wings, which are tipped with yellow instead of blackish-brown, and by the terminal segments, the 8th being entirely black, 9 with a small greenish-blue or yellow spot on each side, and 10 with its dorsum greenish-blue or yellow. *Anal appendages* conical, pointed at apex, as long as segment 10. Vulvar scales robust, extending to end of abdomen.

Distribution.—CEYLON only, confined to the montane tracts: Belihuloya during June, and Madulsima, where Mr. Bainbrigge Fletcher took a few specimens, during August. It is probably on the wing from June to August or September. The species is probably very local, and has only been taken in the years 1892 and 1908; it should not be difficult for Ceylon entomologists to rediscover this beautiful insect and give us some account of its habits.

43. **Platysticta deccanensis** Laidlaw. (Figs. 56 & 59.)

Platysticta deccanensis Laid., Rec. Ind. Mus. vol. xi, p. 388, text-fig. 1 (1915); id., ibid. vol. xiii, pp. 340, 341, pl. xv, fig. 4 (1917); Fras., ibid. vol. xxvi, pp. 498, 499 (1924); id., J. Bombay Nat. Hist. Soc. vol. xxxv, pp. 328, 329, pl. i, figs. 3, 7 (1931); id., Rec. Ind. Mus. vol. xxxiii, pp. 444, 449 (1931).

Male.—Abdomen 45–47 mm. Hind-wing 31–32 mm.

Head: labium yellowish; labrum and anteclypeus azure blue, the former narrowly bordered with black; postclypeus azure blue, bordered above with black; rest of head mat black;

3rd joint of antennæ brownish at its distal end ; eyes black above, olivaceous-brown below. *Prothorax* dark reddish-brown above, paling to light red at the side and yellowish beneath. *Thorax* bright brick-red, changing to golden yellow low down on the sides and pale yellow beneath ; mid-dorsal carina and antealar sinus narrowly marked out in black. *Legs* reddish-brown, femora at proximal ends and coxæ and trochanters golden yellow. *Wings* palely enfumed, hyaline ; pterostigma dark red framed finely in light yellow and heavy black nervures, not quite twice as long as broad, inner end pointed, outer border straight, costal border shorter than posterior, covering 1 to 2 cells ; *ac* meeting *ab* at the posterior margin of wing, or a short stalk intervening, *ab* short and often sinuous ; 17 to 21 postnodal nervures to fore-wings,

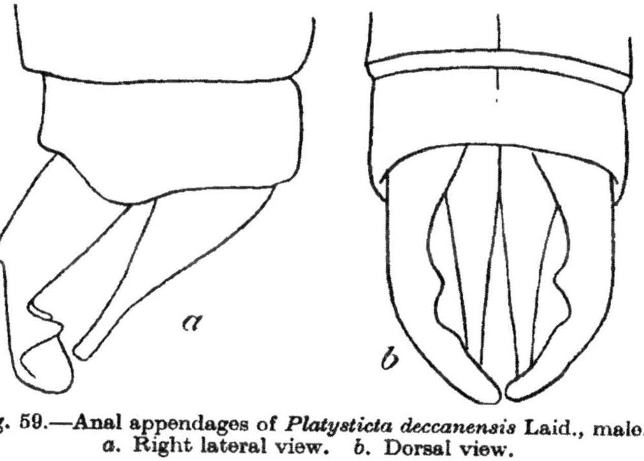

Fig. 59.—Anal appendages of *Platysticta deccanensis* Laid., male. *a*. Right lateral view. *b*. Dorsal view.

17 to 19 in the hind. *Abdomen* dark reddish-brown, paling to golden yellow beneath at bases of segments 2 to 6 ; segments 8 to 9 azure blue, but segment 8 has a narrow black basal annule, and both it and 9 are bordered below with black ; segment 10 blackish-brown, very narrow. *Anal appendages* black, subforcipate as seen from above ; superiors broad at base, constricted at about the middle, where a robust obtuse spine is seen on the inner side, then broadened and somewhat spatulate for the apical half, which is curled down and in and hollowed out ; inferiors nearly as long as superiors, broad at base, then tapered as far as apex, which ends in an obtuse point turned slightly inward.

Female.—Abdomen 36–40 mm. Hind-wing 28–29 mm.

Differs rather widely from the male, far more so than is usual in the subfamily. *Head* similar to male ; *prothorax* coloured similarly, the posterior lobe very broad, rounded. *Thorax* brick-red, the mid-dorsal carina and antealar sinus mapped out in black, the dorsum and anterior half of mesepimeron black, but the ground-colour showing as a fine line against the mid-dorsal carina and upper humeral region, the humeral suture being finely delineated in reddish and the upper and lower part of the mesepimeron more broadly so; traversing the black of the dorsum is a narrow antehumeral pale blue stripe, convex inwards and not extending to the upper part of the thorax; bordering the black area of the mesepimeron another similar blue stripe (a beautiful ensemble of colours); beneath as for male. *Legs* and *abdomen* similar to the male, but segment 1 bright brick-red at the sides, segment 8 unmarked with blue, segment 9 with a large oval pale blue lateral spot; segments 7 to 10 are blacker than in the male, segment 8 only about one-fourth the length of 7, segment 9 nearly twice the length of 8, and segment 10 very short, as in the male. *Anal appendages* short, not longer than segment 10, reddish-brown, pointed, conical; vulvar scale very long and robust, extending well beyond end of abdomen, and with a robust spine above the end. *Wings* usually more deeply enfumed than in the male, pterostigma of similar shape, but more blackish than reddish-brown; 20 postnodal nervures in the fore-wings, 17 in the hind.

Distribution.—Confined, so far as known to, COCHIN STATE, S. INDIA.

The *type* is in the Indian Museum, and was taken at Kavalai, Cochin State, on 24–27. ix. 14. Wishing to obtain more of these interesting and beautiful insects, and to study their habits, I revisited Kavalai in September 1928, and took 17 males and 4 females at the same stream as the types came from. Most were found resting on maidenhair fern in dark shady spots, a number being seen in a culvert running beneath the forest railway. The red thorax and the blue identification marks, especially the latter, were very conspicuous. The flight of the insect is short and weak. Although very local, it was by no means uncommon, and I suspect this to be the case with its near relation *P. apicalis*. The colouring of the female, with its blue striped thorax, is nothing less than remarkable, and without parallel in an order where, if the females differ from the male, it is usually on the side of duller and less conspicuous colouring. It would suggest a more primitive insect in which both sexes were black with blue antehumeral and lateral stripes, as in *Indoneura*, the male having subsequently lost these markings in favour of heliochromatic tints.

CEYLONOSTICTA, gen. nov.

Platysticta, 2me group, Selys, Syn. Agr. Dernière Légion, sep. p. 9 (1860).

Ceylonosticta Fras., J. Bombay Nat. Hist. Soc. vol. xxxv, pp. 329, 330 (1931).

Dragonflies of smaller size than the last genus, but with similar habits and characters. Body-colouring black and brown with blue markings. Wings closed when at rest, long, narrow, falcate at apex, with a long petiole, hyaline, never opaque; discoidal cell shaped as in *Platysticta*; sectors of arc arising from lower part of arc, *but fused for a short distance* from origin; an accessory basal postcostal nervure present near the base of the wings in addition to the nervure *ac*, which is situated rather nearer the distal antenodal nervure; *ab* always present, but incomplete, meeting *ac* at or a little away from the posterior margin of the wing, and extending to the posterior side of the discoidal cell, which it meets at a point near its proximal end; *IA* absent; *Cuii* of about 6 to 8 cells in length; *MA* and *IRiii not zigzagged at origins*; *Riv+v* arising a short distance proximal to, at, or slightly distal to the oblique nervure descending from the subnode, *IRiii* at or slightly distal to the level of subnode; pterostigma similar to the genus *Platysticta*; cells of wings mainly quadrangular.

Head, thorax, and abdomen similar to *Protosticta*, but the abdomen less than double the length of wings; anal appendages variable, superiors usually somewhat forcipate and spatulate at apices; inferiors tapering from base to apex or more widely variable in the species. Genitalia of both sexes very similar to those of *Platysticta*.

Larvæ unknown.

Genotype, *Ceylonosticta hilaris* (Selys).

Distribution.—Confined to CEYLON in montane and submontane tracts.

Habits similar to those of *Platysticta* so far as known. The original descriptions are poor, so that without the actual types before me it has been difficult or impossible accurately to determine the available material. Four species were originally described by Selys, but a fifth has been described since by Laidlaw under the heading of *Drepanosticta* sp., which he surmises, however, and, I think, with good reason, may be conspecific with *P. hilaris* Selys. Another pair, also described as *Drepanosticta* sp., but presumably differing from the above, is said by the same author to be too damaged to identify. I have also found some difficulty in determining some material before me collected by Col. F. Wall, several species of which do not correspond exactly with the Selysian descriptions. Thus it is certain that more species remain to be described, and three of such have been named below. Further material is greatly to be desired, and it is hoped that Ceylon entomologists will assist us in elucidating this problem.

Key to the Species of Ceylonosticta.

1.	Moderately large species, with abdomen more than 40 mm. in length	2.
	Smaller species, with abdomen less than 40 mm. in length	4.
2.	Anterior lobe of prothorax prolonged forward as two stalked processes	*tropica* (Selys), p. 132.
	Anterior lobe of prothorax simple	3.
3.	Inferior anal appendages acutely pointed; labrum bordered with black	*hilaris* (Selys), p. 128.
	Inferior anal appendages obtuse at apex; labrum not bordered with black	*montana* (Selys), p. 130.
4.	Inferior anal appendages, viewed dorsally, squared at apex and with a very robust subapical spine directed inwards to meet its fellow across the middle line...	*digna* (Selys), p. 134.
	Inferior anal appendages simple, without subapical spine	5.
5.	Very small species, with abdomen less than 30 mm. in length; labrum finely bordered with black; inferior anal appendages acuminate at apex; posterior lobe of prothorax simple........	*lankanensis* Fras., [p. 138.
	Very small species with abdomen less than 30 mm. in length; labrum broadly bordered with black; posterior lobe of prothorax prolonged into a long tapering point	*walli* Fras., p. 139.
	Larger species, with abdomen more than 30 mm. in length; inferior anal appendages trilobate at apex..............	*nietneri* Fras., p. 136.

44. **Ceylonosticta hilaris** (Selys). (Fig. 60.)

Agrion hilare Hagen, Verh. zool.-bot. Ges. Wien, vol. viii, p. 479 (1858). (Name only given.)

Disparoneura hilaris Hagen, Verh. zool.-bot. Ges. Wien, vol. ix, p. 206 (1859). (Name only given.)

Platysticta hilaris Selys, Bull. Acad. Belg. (2) vol. x, p. 438 (1860); id., Syn. Agr., sep. pp. 11 & 12 (1860); id., Rev. Syn. Agr., sep. p. 151 (1886); id., Mém. Cour. vol. xxxviii, p. 152 (1886); Kirby, Cat. Odon. p. 132 (1890); id., J. Linn. Soc. Lond., Zool. vol. xxiv, p. 562 (1893); Laid., Rec. Ind. Mus. vol. xi, p. 387 (1915).

Drepanosticta hilaris Laid., Spolia Zeylan. vol. xii, pp. 362, 364 (1924).

Drepanosticta sp. Laid., Spolia Zeylan. vol. xii, pp. 362–4 (1924).

Ceylonosticta hilaris Fras., J. Bombay Nat. Hist. Soc. vol. xxxv, pp. 330, 331, pl. ii, figs. 5, 6 (1931).

Male.—Abdomen 42 mm. Hind-wing 25 mm.

Head: labium brown; labrum pale turquoise-blue narrowly bordered with glossy black; bases of mandibles glossy black, the extreme base only blue; anteclypeus turquoise-blue; postclypeus glossy black, rest of head bronzed black, with a large patch of obscure reddish-brown partially encircling the ocelli behind and on each side; eyes black above, olivaceous brown beneath; occiput with a projecting scale-like ledge, pointed at either end, overlapping anterior lobe of prothorax. *Prothorax* bronzed brown on dorsum, paler brown laterally; anterior lobe sloping forward, encircling the neck like a stiff collar; middle lobe with a deep central pit situated between two prominent rounded bosses; posterior lobe simple, rounded, the posterior border straight or very shallowly concave. *Thorax* bronzed black or deep coppery brown on dorsum, changing to reddish-brown and then ochreous on the sides and beneath; a narrow oblique azure blue stripe on each

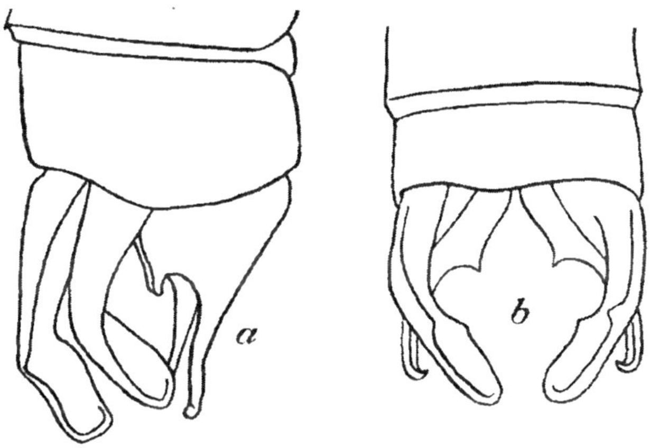

Fig. 60.—Anal appendages of *Ceylonosticta hilaris* (Selys) male. *a*. Right lateral view. *b*. Dorsal view.

side traversing the centre of mesepimeron, after which is a thin black line on the postero-lateral suture (in the original description the blue stripe is omitted, probably because decomposition had obscured it). *Wings* hyaline, enfumed and tinted palely with yellow; pterostigma reddish-brown finely framed in yellow and thick brown nervures, rather longer than broad, inner side a little oblique, outer nearly straight; 15 to 16 postnodal nervures in fore-wings, 14 to 15 in the hind; *Riv+v* arising slightly distal to or in continuation of the oblique nervure descending from the subnode; *ab* meeting *ac* near the hinder border of wings. *Legs* yellow or olivaceous, outer surfaces of femora and articulations blackish. *Abdomen* dark reddish-brown, deepening to bronzed black at ends of segments; sides of segments 1 and 2 paler; segments 3 to 7 with moderately broad basal annules; segment 8 with its apical border broadly, and the whole of segments 9 and 10 azure blue, reddish-brown below and along ventral borders. *Anal appendages* blackish-brown; superiors more than twice the length of segment 10, broad at base, then tapering and curving down in the apical half, which is broadened out, excavated on its inner side, and squared at the apex; a small dorsal spine situated at the point where the appendage begins to broaden out; inferior appendages slightly shorter than superiors, very broad at base, where is situated a robust dorsal spine, then slender and tapering to an acute point at apex.

Female.—Abdomen 36–37 mm. Hind-wing 26–27 mm.

Closely similar to the male in colouring and markings, differing as follows:—Border of labrum less broadly bordered with black and the black border itself bordered with reddish-brown; legs yellow, articulations only blackish. *Wings* more deeply enfumed; 15 to 17 postnodal nervures to fore-wings, 14 to 15 in the hind; *Riv+v* arising proximal to the oblique nervure descending from the subnode. *Abdomen* similar, but the basal annules pale blue in old specimens, yellow in others and tenerals; segment 8 unmarked with blue, 9 with only a small subdorsal oval spot near the apical border of each side of segment, 10 with a large dorsal spot. *Anal appendages* reddish-brown, conical, pointed at apex, barely as long as segment 10; vulvar scale robust, extending well beyond end of abdomen.

Distribution.—CEYLON only: Ramboda Ghat, Kandy, and other montane areas from May to September.

The figure given of the anal appendages by Dr. Laidlaw for his *Drepanosticta* sp. is sufficient proof that this species is *C. hilaris*, as he surmised. Post-mortem decomposition accounts for the differences to be noted between the above and the Selysian description. From *C. digna* this species is easily distinguished by the shape of its inferior appendages; from *C. tropica* by the simple collar-like anterior lobe of prothorax; from *C. montana* by the labrum bordered with black and by the inferior appendages ending in an acute point: from the remaining species by its much larger size and more numerous postnodal nervures.

45. **Ceylonosticta montana** (Selys). (Fig. 61.)

Platysticta montana Selys, Bull. Acad. Belg. (2) vol. x, p. 438 (1860); id., Syn. Agr., sep. pp. 10–11 (1860); id., Mém. Cour. Rev. Syn. Agr. p. 151 (1886); Kirby. Cat. Odon. p. 132 (1890); id., J. Linn. Soc. Lond., Zool. vol. xxiv, p. 363 (1893); Laid., Rec. Ind. Mus. vol. xi, p. 387 (1915).

Drepanosticta montana Laid., Spolia Zeylan. vol. xii, p. 362 (1924).

Ceylonosticta montana Fras., J. Bombay Nat. Hist. Soc. vol. xxxv, pp. 331, 332, pl. i, fig. xi (1901).

Male.—Abdomen 43 mm. Hind-wing 28 mm.

Head: labium dark brown; labrum turquoise-blue narrowly

bordered with reddish-brown; bases of mandibles and anteclypeus turquoise-blue; postclypeus glossy black, rest of head bronzed black with a broad patch of reddish-brown encircling the ocelli on either side and behind; eyes black above, olivaceous brown beneath. *Prothorax* bronzed brown on dorsum, paler reddish-brown laterally; anterior lobe collar-like; posterior lobe simple, rounded. *Thorax* bronzed black on dorsum, changing to reddish-brown on the sides and yellow beneath (probably an oblique blue stripe on the middle of mesepimeron, which is obscured by decomposition in the type). *Legs* yellowish, the outer surface of femora and articulations blackish. *Wings* hyaline, palely enfumed; pterostigma slightly longer than broad, similar in shape to that of *C. hilaris*; 16 to 18 postnodal nervures in fore-wings, 15 to 16 in the hind; *Riv*+*v* arising a little distal to the oblique

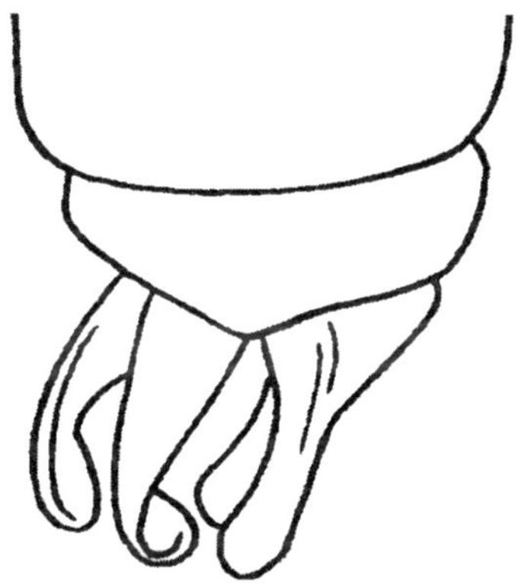

Fig. 61.—Anal appendages of *Ceylonosticta montana* (Selys), seen from the right side.

nervure descending from the subnode. *Abdomen* bronzed brown on dorsum deepening to black at the distal ends of segments 3 to 7, with moderately broad basal annules; segment 8 unmarked (so far as can be seen in the type, but probably its apical border bluish during life); segments 9 and 10 azure blue on the dorsum (in the type segment 9 brown from decomposition). *Anal appendages* blackish-brown, superiors more than twice the length of segment 10, broad at base, then tapered as far as the middle, at which point they are dilated internally and curved rather strongly downward and furnished with a small obtuse spine at the point of the angulation on the upper inner border of appendages; the dilated part of even width, excavate on the inner side, and ending in a squared apex directed somewhat inward and downward; inferiors slightly shorter than superiors, broad at base, then tapered and slim as far as apex, which is again broadened and curved inwards towards its fellow ; a robust tooth on the upper inner border near the base.

Female.—Abdomen 37 mm. Hind-wing 26 mm.

Closely similar to the male, differing only in sexual characters and by the markings on the terminal segments of *abdomen.* A small subdorsal subapical spot of blue on each side of segment 9, and a broad dorsal spot of the same colour on segment 10. *Anal appendages* as long as segment 10, which is very short and without an apical notch, conical, pointed at apex, brown ; vulvar scale very robust, extending well beyond end of abdomen.

Distribution.—CEYLON only, in montane tracts: Ramboda Ghat and Kandy.

Differs from *C. tropica* by the simple shape of the lobes of prothorax ; from *C. hilaris,* to which it is very closely related, by the tumid end of inferior appendages and by the labrum unbordered with black ; from *C. digna* by the absence of a middle inner spine on inferior appendages ; from the remaining species by its much larger size.

46. **Ceylonosticta tropica** (Selys). (Fig. 62.)

Platysticta tropica Selys, Bull. Acad. Belg. (2) vol. x, p. 438 (1860) ; id., Syn. Agr., sep. p. 10 (1860) ; id., Mém. Cour. Rev. Syn. Agr. p. 151 (1886) ; Kirby, Cat. Odon. p. 132 (1890) ; id., J. Linn. Soc. Lond., Zool vol. xxiv, p. 562 (1898) ; Laid., Rec. Ind. Mus. vol. xi, p. 387 (1915) ; id., ibid. vol. xiii, p. 341 (1917).
Drepanosticta tropica Laid., Spolia Zeylan. vol. xii, p. 362 (1914).
Ceylonosticta tropica Fras., J. Bombay Nat. Hist. Soc. vol. xxxv, pp. 332, 333, pl. i, figs. 1, 2 (1931).

Male.—Abdomen 41 mm. Hind-wing 26 mm.

Head: labium pale brown ; labrum, bases of mandibles, and anteclypeus pale turquoise-blue, the former narrowly bordered with brown ; postclypeus bronzed brown above, rest of head bronzed black, with a small obscure yellow oval spot on the outer side of each outer ocellus ; eyes black above, olivaceous brown beneath ; 2nd and 3rd segments of antennæ pale yellowish. *Prothorax* pale olivaceous brown, the middle lobe possibly bluish during life ; posterior lobe bronzed black, arched, rounded ; anterior lobe with a deep and wide cleft at its middle, the lobe on each side of this prolonged into a curious stalked scale resembling those seen on the posterior lobe of some of the *Caconeura* group ; between these two scales the lobe prolonged forward as a short blunt process. *Thorax* dark bronzed brown or cupreous black on dorsum, the lower part of sides paling to reddish-brown and then yellow beneath thorax ; a moderately broad oblique azure-blue stripe on each side traversing the centre of mesepimeron from above down.

Legs olivaceous, the external and extensor surfaces of femora and the articulations mottled with black. *Wings* hyaline, palely enfumed: pterostigma blackish-brown finely framed in creamy white, and again by thick black nervures, half as long again as broad, costal border shorter than posterior, distal border straight, proximal very oblique, braced; *Riv+v* arising slightly before the oblique nervure descending from the subnode; 16 to 17 postnodal nervures to fore-wings, 14 to 16 in the hind; *ab* meeting *ac* at a short distance from posterior margin of wing. *Abdomen* bronzed brown, deepening to black at apical ends of segments, and with narrow basal annules on segments 3 to 7; segment 1 yellow laterally; segments 2 to 4 yellow along the ventral border; segments 9 and 10 azure blue on the dorsum, the latter narrowly black at the apex. *Anal appendages* blackish-brown; superiors

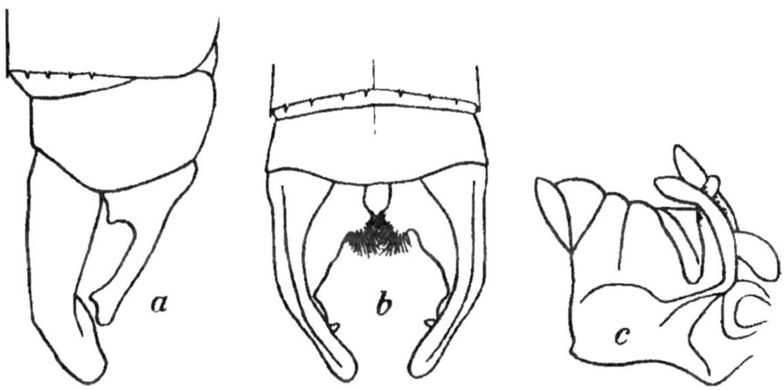

Fig. 62.—Anal appendages and prothorax of *Ceylonosticta tropica* (Selys).
a. Right lateral view of anal appendages. *b*. Dorsal view of same. *c*. Prothorax, the posterior lobe to the left.

more than twice as long as segment 10, seen from above curving gently inward towards one another, broad at base, then tapered to apex, which is obtuse, the apical half presenting a spatulate dilatation on the inner border, which is hollowed out on its inner aspect; inferior appendages rather more than half as long as superiors, very broad at base, where is seen an inner tubercle which, viewed from above, is broad and coated with numerous stiff bristles; the apical three-fourths slim, sloping up and back and ending in an obtuse shallowly notched apex as viewed in profile, or as an acutely pointed, inwardly directed spine as viewed from above.

Female.—Abdomen 40 mm. Hind-wing 29 mm.

Closely similar to the male in colour and markings, but rather more robustly built. Labium dark brown; labrum entirely blue; *prothorax* with similar scale-like processes on anterior lobe; *wings* more deeply enfumed; 16 to 17 postnodal nervures to fore-wings, 16 in the hind; other details of venation similar to male. No trace of blue to be seen on the terminal abdominal segments, but the markings may be obscured by post-mortem decomposition, and during life are probably present as subdorsal spots on the sides fo segment 9 and a large dorsal spot on segment 10. *Anal appendages* short, barely as long as segment 10, conical, pointed, brown; segment 9 about one-third longer than 8 and more than twice the length of 10; vulvar scales moderately robust, blackish-brown, *not extending to end of abdomen.*

Distribution.—Confined to the montane tracts of CEYLON: Passara, Haycock Hill, May to August; Hakgala, March and April.

The description of the female is from a specimen in the author's collection from Hakgala. There is no possibility of confusing this species with any other, as the curious prothoracic processes are unique in the genus, and are sufficient to determine the species at a glance. It is of interest to note that these accessory sexual appendages are found in both sexes.

47. **Ceylonosticta digna** (Selys). (Fig. 63.)

Agrion digna Hagen, Verh. zool.-bot. Ges. Wien, vol. viii, p. 479 (1858.) (Name only.)

Disparoneura digna Hagen, Verh. zool.-bot. Ges. Wien, vol. ix, p. 207 (1859). (Name only.)

Platysticta digna Selys, Bull. Acad. Belg. (2) vol. x, p. 440 (1860); id., Syn. Agr., sep. p. 12 (1860); id., Mém. Cour. Rev. Syn. Agr. vol. xxxviii, p. 151 (1886); Kirby, Cat. Odon. p. 132 (1890); id., J. Linn. Soc. Lond., Zool. vol. xxiv, p. 362 (1893); Laid., Rec. Ind. Mus. vol. xi, p. 387 (1915).

Drepanosticta digna Laid., Spolia Zeylan. vol. xii, pp. 361, 362, text-fig. 9 (1924).

Ceylonosticta digna Fras., J. Bombay Nat. Hist. Soc. vol. xxxv, pp. 333, 334, pl. i, figs. 8, 9 (1931).

Male.—Abdomen 35 mm. Hind-wing 23 mm.

Head: labium pale brown; labrum, bases of mandibles, and anteclypeus pale turquoise-blue; postclypeus glossy black, rest of head blackish-brown; eyes black above, olivaceous brown beneath; 3rd joint of antennæ palest brown; occiput with the usual scale-like ledge posteriorly. *Prothorax* ochreous, possibly blue on the dorsum during life; anterior lobe arched, sloping forward, collar-like; posterior lobe simple, rounded, dark bronzed brown. *Thorax* dark bronzed brown on dorsum, paling to reddish-brown laterally and yellowish beneath; mid-dorsal carina finely blue, and an oblique stripe of the same colour traversing the length of the mesepimeron midway between the two lateral sutures. *Legs* ochreous, outer surface of femora and all articulations blackish-brown. *Wings* hyaline, palely enfumed in adults; pterostigma blackish-brown finely framed in yellow, and an outer frame of thick black nervures, shaped similarly to that of *C. hilaris*; 15 to 16 postnodal nervures to fore-wings, 14 in the hind; *Riv+v* arising in prolongation of the oblique nervure descending from the subnode; *ac* and *ab* meeting at a common point on the posterior margin of wing or a short stalk intervening at the junction. *Abdomen* blackish-brown, segments 2 and 3 at the sides and the bases of 2 to 7 with basal annules yellow; segment 8 black; segments 9 and 10 azure blue above, black along the ventral borders. *Anal appendages* blackish-brown; superiors more than twice the length of segment 10, broad at base, then tapering somewhat to apex,

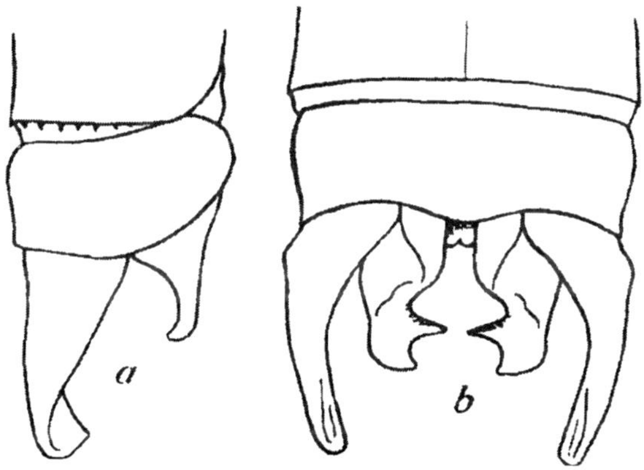

Fig. 63.—Anal appendages of *Ceylonosticta digna* (Selys).
a. Right lateral view. *b.* Dorsal view.

which is obtuse, slightly curved in and down as viewed from above, and the apical half, which is a little expanded on the inner side, markedly excavate; inferior appendages about two-thirds the length of superiors, of unusual and irregular shape, very broad at base, outer border sinuous as seen from above, inner border deeply serrate, presenting an obtuse basal tubercle followed by a sub-basal, which is followed successively by a deep notch, a median robust spine, and then a final shallow notch; the median spine is directed straight inwards and nearly meets its fellow of the opposite side; the apex squared and slightly bevelled; seen from the side the appendage tapers rapidly to an obtuse point.

Female unknown.

Distribution.—CEYLON, only in montane tracts: Ramboda Ghat, Urugala, and Haragama from May to August.

The above revised description was made from a specimen in the British Museum. The species is easily determined from all others by the curiously shaped inferior appendages.

Type in the Selysian collection.

48. Ceylonosticta nietneri Fraser. (Fig. 64.)

Ceylonosticta nietneri Fras., J. Bombay Nat. Hist. Soc. vol. xxxv, pp. 334, 335, pl. ii, figs. 7, 8 (1931).

Male.—Abdomen 31–32 mm. Hind-wing 20–21 mm.

Head: labium pale brown; labrum, bases of mandibles, and anteclypeus pale turquoise-blue; postclypeus glossy black, rest of head bronzed black; 3rd joint of antennæ pale brown; eyes black above, olivaceous brown beneath; occiput with a scale like that seen in *C. hilaris*. *Prothorax* pale ochreous; posterior lobe, except its hinder border and the whole of the dorsum of middle lobe, pale blue; anterior

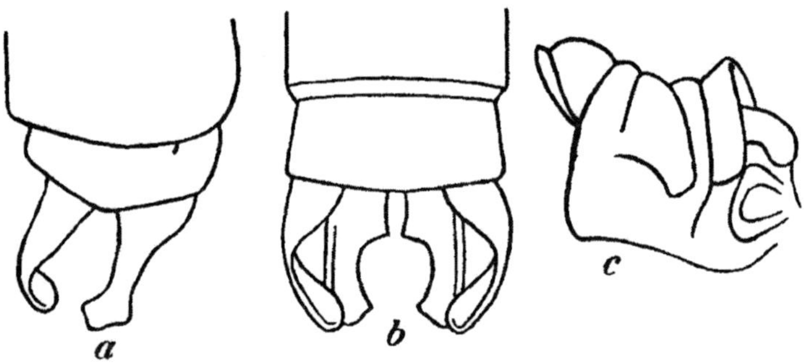

Fig. 64.—Anal appendages and prothorax of *Ceylonosticta nietneri* Fras., male. *a*. Right lateral view. *b*. Dorsal view. *c*. Prothorax, the posterior lobe to the left.

lobe collar-like, posterior rounded, simple. *Thorax* bright reddish-ochreous, with a cupreous reflex on dorsum and a broad oblique pale blue stripe on each side situated between the two lateral sutures; pale ochreous beneath. *Legs* pale olivaceous, the articulations speckled with black. *Wings* hyaline, very palely enfumed; pterostigma blackish-brown, finely framed in creamy white and thick black nervures, nearly square, but slightly longer than broad, distal side convex, proximal oblique, braced; 14 to 16 postnodal nervures in fore-wings, 13 to 14 in the hind; *ab* meeting *ac* at a short distance from the posterior margin of wing, the former ending very near the proximal end of discoidal cell; origin of *Riv+v* at or slightly distal to the oblique nervure descending from the subnode. *Abdomen* yellow beneath and on the sides of segment 1, dark enfumed olivaceous on dorsum and sides, deepening to black at apical ends of segments and on the whole of segment 7 and most of 8 ; bases of segments 3 to 4 or 5 with incomplete blue basal annules which extend apically along the ventral borders of segments ; segment 8 blackish-brown, marked dorsally with an apical triangle of azure blue which extends variably towards base of segment, covering from one-fourth to half the segment ; segments 9 and 10 entirely blue on dorsum, the ventral borders black. *Anal appendages* blackish-brown, apices of inferiors paler brown ; superiors twice the length of segment 10, simple, arched up and then down, broad and depressed at base, then at the middle twisted on the long axis of appendage, the outer half broadened, compressed, and hollowed out within, *not bearing any spines* ; inferior appendages of the same length, broad at base, directed straight back or a little obliquely upward, of even thickness throughout as far as apex, which is expanded into three angles like the webbed foot of a duck ; *without a basal spine.*

Female.—Abdomen 30–31 mm. Hind-wing 22 mm.

Closely similar to the male, but of stouter build, the abdomen more especially robust. Differs as follows : *prothorax* more prominently marked with blue on middle and posterior lobes, the latter very simple, rounded, the scale-like border very narrow and of even width. *Wings* with 14 postnodal nervures to fore-wings, 13 in the hind, otherwise exactly similar to those of male, but *Riv+v* always arising distal to the oblique nervure descending from the subnode. *Abdomen* dark reddish-brown, segments 3 to 7 with broad basal azure-blue annules interrupted on the mid-dorsum on the latter segment ; segment 8 dark reddish-brown, the apical articulation only blue ; segment 9 and 10 broadly azure blue on dorsum, reddish-brown on lower part of sides. *Anal appendages* shorter than segment 10, small, pointed, conical processes, brown ; vulvar scale brown, robust, extending beyond end of abdomen.

Distribution.—CEYLON only, Kandy, 2,000 ft. : three males and a single female taken at the latter place by Col. F. Wall, I.M.S., 14. x. 24.

This species and the two following, *C. lankanensis* and *C. walli*, are the three smallest species known in the subfamily, not excepting *P. hearseyi*, and may be distinguished from all other species at a glance by reason of this feature. The curious shape of the apex of the inferior anal appendages, like a duck's foot, will serve to distinguish this from the other two species mentioned as being of the same size. The female is to be distinguished by the arched round shape of its prothorax, this being produced and strongly angulated in *C. walli*, whilst the female of *C. lankanensis* is unknown.

49. **Ceylonosticta lankanensis** Fraser. (Fig. 65.)

Platysticta montana Kirby, J. Linn. Soc. Lond., Zool. vol. xxiv, p. 562 (1893).
Ceylonosticta lankanensis Fras., J. Bombay Nat. Hist. Soc. vol. xxxv, p. 335, pl. ii, figs. 11, 12 (1931).

Male.—Abdomen 29 mm. Hind-wing 20 mm.

Head: labium whitish-brown; labrum, bases of mandibles, and anteclypeus pale turquoise-blue, the former very narrowly bordered with black, rest of head blackish-brown; eyes dark brown above, olivaceous brown beneath. *Thorax* dark reddish-ochreous on dorsum and sides, but paling to ochreous on metepimeron and yellowish beneath; a broad, oblique, pale blue stripe on each side lying parallel with and midway between the lateral sutures, contrasting strikingly with the

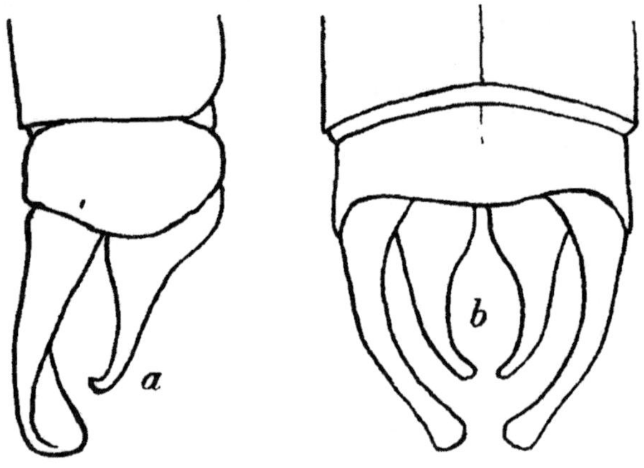

Fig. 65.—Anal appendages of *Ceylonosticta lankanensis* Fras., male. *a.* Right lateral view. *b.* Dorsal view.

adjacent reddish-brown. *Legs* olivaceous, marked with dark brown on the outer sides of femora and all articulations. *Wings* hyaline, palely enfumed; pterostigma blackish-brown, finely framed in white and thick black nervures, similar in shape to that of *C. nietneri*; 12 to 13 postnodal nervures in fore-wings, 11 to 12 in the hind; *Riv+v* arising in continuation with the oblique nervure descending from the subnode; *ab* meeting *ac* near the posterior margins of wings. *Abdomen* reddish-brown, deepening to black at apical ends of segments; segment 1 yellow laterally; segments 3 to 7 with narrow basal yellow annules; segment 8 with a small apical triangular bordering of blue; segments 9 to 10 azure blue dorsally, black along the ventral borders. *Anal appendages* blackish-brown, superiors more than twice the length of segment 10, broad at base, then tapering as far as the middle, at which point they are dilated, compressed and rotated on their long axis, the apex broad and squared; seen from above, these appendages are curved gently towards one another, of almost even breadth throughout and slightly clubbed at apex; inferiors of about two-thirds the length of superiors, very broad at base, then abruptly slim and cylindrical, curved gently in, the extreme apices more abruptly so and ending in a fine point.

Female unknown.

Distribution.—CEYLON only, confined probably to montane districts.

The *type*, in the British Museum, is labelled "Kottawa, 19. iv. 92", and was incorrectly determined by the late Mr. Kirby as *P. montana*. This specimen is undoubtedly Col. Yerbury's, referred to by Kirby, "with some doubt", as *P. montana*, in the J. Linn. Soc. (*l. c.*). Its small size—the smallest species of the genus—is sufficient to distinguish it from *P. montana*, and several other differential points may be made out with ease. The acutely pointed inferior appendages are sufficient to distinguish it from *C. nietneri*, in which these are broad and clubbed at the apices and as long as the superiors. It is possible that the next species described below is the female of this, but the curious shape of the hinder margin of the prothorax and the extensive black bordering of the labrum appear to place it as an entirely distinct species.

50. **Ceylonosticta walli** Fraser. (Fig. 66.)

Ceylonosticta walli Fras., J. Bombay Nat. Hist. Soc. vol. xxxv, pp. 335, 336 (1931).

Male unknown.

Female.—Abdomen 31 mm. Hind-wing 22 mm.

Head: labium blackish-brown; labrum pale turquoise-blue broadly bordered with glossy black; bases of mandibles black with a spot of blue at the extreme base: anteclypeus pale turquoise-blue, postclypeus and rest of head bronzed black; eyes black above, olivaceous brown beneath; distal end of 3rd segment of antennæ pale brown. *Prothorax* with middle lobe pale blue, anterior and posterior lobes reddish-brown, the former collar-shaped, the latter with the medial portion of its thinned-out border prolonged into a long tapering point. *Thorax* reddish-ochreous, deepening to dark bronzed brown on dorsum, paling on lower part of metepimeron and beneath; a moderately broad azure blue stripe traversing the whole length of mesepimeron midway between the lateral sutures. *Legs* olivaceous, articulations blackish, outer surfaces of femora and whole of tibiæ enfumed brown. *Wings* hyaline, palely enfumed; pterostigma nearly quadrate, especially in the fore-wings, inner end slightly oblique, outer slightly convex, dark reddish-brown framed finely in white and thick black nervures; 15 postnodal nervures in fore-wings, 14 in the hind; *Riv+v* opposite or a little distal to the oblique nervure descending from the subnode; *ac* and *ab* arising from different points on the posterior border of wings (in the type the nervure *ab* is absent in the right pair of wings). *Abdomen* reddish-brown, changing to blackish-brown at distal ends of segments; segments 2 to 7 with azure blue basal annules, very narrow on segment 2, broad on 4 to 6; segment 8 unmarked; segments 9 and 10 with the dorsum azure blue, the ventral

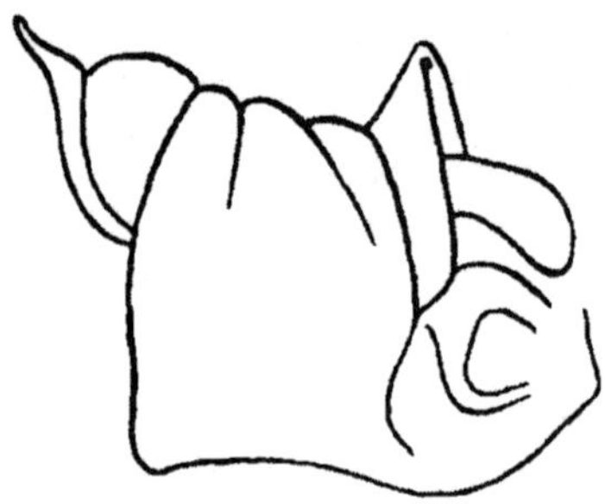

Fig. 66.—Prothorax of *Ceylonosticta walli* Fras., female. Posterior lobe to the left.

borders dark blackish-brown. *Anal appendages* reddish-brown, slightly longer than segment 10, conical, acute at apex; vulvar scale robust, extending beyond end of abdomen.

Distribution.—Ceylon only, confined to montane tracts.

Type, a female in the author's collection, taken at Kandy, 2,000 ft., 13. ix. 24, by Col. F. Wall, I.M.S. Differs from the female of *C. nietneri* by the posterior lobe of prothorax, which is prolonged into a long tongue-like point, and also by the labrum broadly bordered with black, by the higher nodal index, etc. From the male of *C. lankanensis* it also differs by the shape of the posterior lobe of prothorax, by the labrum broadly bordered with black, by the broader basal blue annules of the abdomen, and by the different origin of *ab*, etc.

Genus DREPANOSTICTA Laidlaw (1917). (Fig. 67.)

Platysticta Selys, *pars*, Bull. Acad. Belg. (2) vol. x, p. 436 (1860); id., Mém. Cour. vol. xxxviii, p. 150 (1886); Kirby, Cat. Odon. p. 132 (1890); Laid., Rec. Ind. Mus. vol. xi, pp. 378, 379 (1915).
Drepanosticta Laid., Rec. Ind. Mus. vol. xiii, pp. 339, 341 (1917); id., J. Malay Branch Roy. As. Soc. vol. ii, pp. 304–6 (1924); Fras., J. Bombay Nat. Hist. Soc. vol. xxxv, pp. 336, 337 (1931).

Dragonflies of the same small size as the last genus and with similar habits and characteristics. Body-colouring usually black, but in one species at least reddish-brown marked with blue. Wings closed when at rest, long, narrow, falcate at apex, with a long petiole, hyaline; discoidal cell shaped as in *Platysticta*; sectors of arc arising from lower part of arc, *but fused for a short distance from origin*; an accessory basal postcostal nervure present near base of all wings in addition to the nervure *ac*, which is situated rather nearer the distal antenodal nervure; *ab* usually present, but sometimes absent, and, when present, always incomplete, meeting *ac* at or near the hinder margin of wing, or arising from that margin at

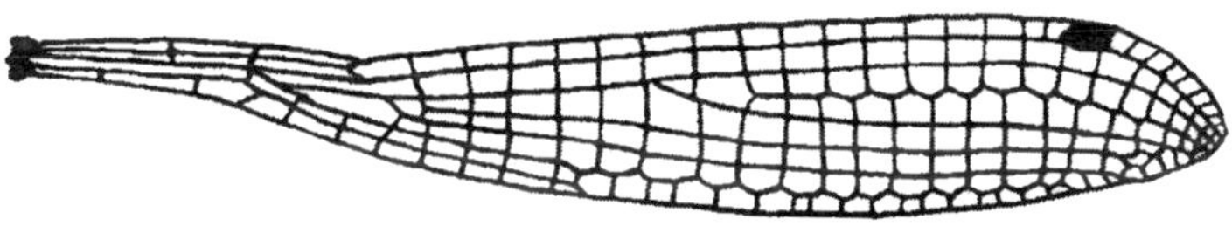

Fig. 67.—Wing of *Drepanosticta carmichaeli* Laid., male.

some point more or less distal to *ac* and extending to the posterior border of discoidal cell near its proximal end; *IA* absent; *Cuii* of variable length, usually 6 to 8 cells in length; *MA* and *IRiii* not zigzagged at origins; *Riv+v* arising slightly before, at, or slightly distal to the oblique nervure descending from the subnode; *IRiii* at or a little distal to that nervure; pterostigma subquadrate, costal side slightly shorter than the posterior, inner end oblique, unbraced, outer slightly convex. Cells of wings mainly quadrangular.

Head, thorax, and abdomen similar to *Protosticta*, but the abdomen less than twice the length of wings except in *D. viridis*; anal appendages variable, superiors subforcipate, spatulate in the apical half, which is curved down and usually carries a robust dorsal spine at the middle; inferiors variable, with or without a long inner narrow spine; genitalia similar to those of *Protosticta*, *penis without a dorsal ruff or spine*. Larvæ unknown.

Genotype, *Drepanosticta carmichaeli* (Laid.).

Distribution.—N.E. INDIA, BURMA, Indo-China, Malaya to New Guinea.

The habits of species of the genus are unknown save for *D. carmichaeli*, in which they resemble those of *Protosticta*. Dark shaded retreats are the localities in which to seek these insects, where, clinging to ferns, etc., they may be beaten up and driven into the open. Four or five species are known from within Indian limits, one of which is from Bengal, two or possibly three from Burma, and a fifth from the Andamans.

Key to the Species of Drepanosticta.

1.	Length of abdomen twice that of hind-wing.	*viridis* Fras., p. 145.
	Length of abdomen considerably less than twice that of hind-wing	2.
2.	Species coloured black, marked with white and pale blue	3.
	Species coloured reddish-brown, marked with blue	4.
3.	Anal appendages complex, furnished with spines	*quadrata* Selys, p. 147.
	Anal appendages simple, without any spines	*annandalei* Fras., [p. 148.
4.	Pterostigma slightly longer than broad; thorax with ill-defined blue stripes, blackish-brown beneath	*carmichaeli* Laid., [p. 142.
	Pterostigma squared; thorax with well-defined blue stripes, beneath pale yellow	*polychromatica*, sp. n., [p. 144.

51. **Drepanosticta carmichaeli** Laidlaw. (Figs. 67 & 68.)

Protosticta carmichaeli Laid., Rec. Ind. Mus. vol. xi, p. 390, fig. 1 (1915).

Drepanosticta carmichaeli Laid., Rec. Ind. Mus. vol. xiii, pp. 341, 342, pl. xv, fig. 5 (1917); id., J. Malay Branch Roy. As. Soc. vol. ii, pp. 304–6 (1924); Fras., J. Bombay Nat. Hist. Soc. vol. xxxv, pp. 337, 338, pl. ii, figs. 3, 4 (1931).

Male.—Abdomen 36–37 mm. Hind-wing 23–24 mm.

Head: labium brown; labrum turquoise-blue, narrowly bordered with dark brown; genæ, bases of mandibles, and anteclypeus turquoise-blue; postclypeus, frons, and vertex as far back as the level of posterior ocelli blackish-brown, behind which an irregular broad band of pale blue extending from eye to eye; occiput behind black, presenting a vestigial ridge behind as seen in *Platysticta*; 2nd joint of antennæ pale blue; eyes blue above, paler below, and marked with a darker equatorial band of grey. *Prothorax* with anterior lobe projecting forward, collar-like; posterior lobe, simple, rounded; mid-dorsum olivaceous, the rounded bosses on each side of middle lobe azure blue; posterior lobe reddish-ochreous, blue at either side; middle lobe dark chocolate-brown on each side and beneath. *Thorax* olivaceous on dorsum, changing to golden brown laterally, and then dark chocolate-brown ventrally and beneath; an ill-defined pale blue antehumeral stripe on each side of dorsum confluent with the blue on posterior lobe of prothorax; laterally a broad, slightly oblique pale blue stripe traversing the whole length of mesepimeron. *Legs* pale ochreous, hind femora with a small distal point of black on extensor surface. *Wings* palely enfumed, hyaline; pterostigma rather longer than broad, costal side slightly shorter than posterior, proximal side very oblique, distal side slightly convex, covering one cell, unbraced, brown, framed finely in creamy white and thick black nervures; 15 to 16 postnodal nervures in fore-wings, 14 in the hind; *Riv+v* arising distal to the oblique nervure descending from the subnode; *ac* midway between the two antenodal nervures or nearer the distal one; *ab* arising from

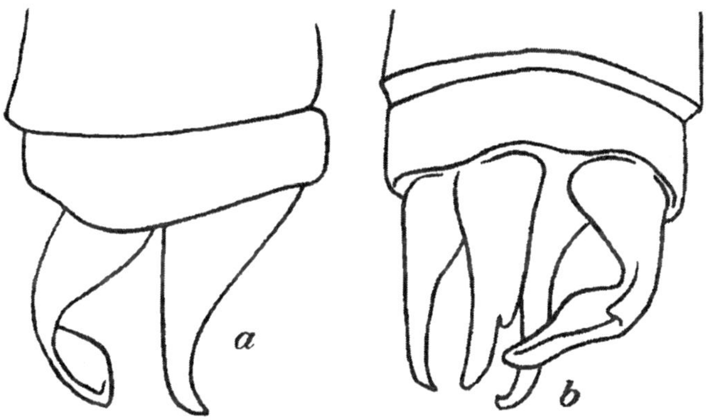

Fig. 68.—Anal appendages of *Drepanosticta carmichaeli* Laid. *a.* Right lateral view. *b.* Dorsal view.

the posterior border of wing well distal to *ac*, very oblique, meeting the discoidal cell near its proximal end, hardly longer than *ac*; *Cuii* 8 to 9 cells long in fore-wings, 12 in the hind. *Abdomen* dark golden brown, deepening to blackish-brown at apical ends of segments and on the whole of segments 7 to 10; segment 1 with an apical blue annule; 2 with the mid-dorsum narrowly turquoise-blue from base nearly to apex; segments 3 to 7 with bright blue basal cordate spots; 8, 9, and 10 pale azure blue on dorsum, the lower parts of sides, especially of 10, black. *Anal appendages* blackish-brown, apices of superiors paler brown. Superiors broad at base, then tapering strongly till about the middle, the apical half expanded and curved strongly downward; apex falcate; a very obtuse spine on dorsum at the site of angulation; equal to more than twice the length of segment 10; inferiors equal in length to superiors, very broad at base, then tapered, cylindrical and ending in a fine point which is curved inward; a large robust acutely pointed spine situated at the base, directed inward and upward.

Female.—Abdomen 34 mm. Hind-wing 23 mm.

Almost exactly similar to the male, but the *abdomen* shorter and stouter. Segments 8 and 10 unmarked with blue; basal blue spots on segments 4 to 6 larger and more conspicuous; *wings* similar to those of male, but *Riv+v* arising rather more distal to the subnode; 16 postnodal nervures to fore-wings, 15 in the hind. *Anal appendages* shortly conical, pointed, brown; vulvar scales robust, but not extending beyond end of abdomen.

Distribution.—Sikkim: Kalimpong, 4,000 ft.; Pashoke, 3,000 ft., from April to July, in similar situations to those frequented by *Protosticta*.

The beautiful colour-complex of this insect will serve to distinguish it from all others of the genus save its near relative *D. polychromatica*, from which it is distinguished by its larger size, different character of the thoracic markings, different shape of pterostigma, distal origin of *Riv+v*, and longer *Cuii*.

Type in the Indian Museum; paratypes in the author's collection.

52. **Drepanosticta polychromatica** Fraser.

Drepanosticta polychromatica Fras., J. Bombay Nat. Hist. Soc. vol. xxxv, p. 338 (1931).

Male.—Abdomen 32 mm. Hind-wing 21 mm.

Head: labium pale brown; labrum, genæ, bases of mandibles, and anteclypeus palest blue, the former narrowly bordered with brown; rest of head blackish-brown as far back as the level of the posterior ocelli, at which point is a moderately broad blue transverse band which falls well short of the eyes save for a minute point anteriorly. *Prothorax* with dorsum chocolate-brown, laterally blackish-brown, marked on the centre of the middle lobe with a narrow yellow line, and on the sides by a broad band of blue extending from the anterior to the posterior lobe, the latter ochreous on the mid-dorsum and shaped as in *D. carmichaeli*. *Thorax* black on dorsum, dark chocolate-brown laterally, the former marked with a well defined narrow blue antehumeral stripe, broad anteriorly where it is confluent with the blue on prothorax, narrowing and tapering to a fine point near the antealar sinus posteriorly; laterally a sharply defined oblique moderately broad blue stripe lying just anterior to the postero-lateral suture, followed by an equally broad stripe of the ground-colour; lower part of metepimeron and beneath thorax pale yellow. *Legs* pale ochreous, unmarked. *Wings* hyaline, palely enfumed ; pterostigma almost quadrate, costal side only a shade shorter than posterior, the proximal side but slightly oblique, dark brown finely framed in creamy white and thick black nervures ; 15 postnodal nervures to fore-wings, 14 in the hind ; *Riv+v* arising in continuation of the oblique nervure descending from the subnode ; *Cuii* only 7 cells long in fore-wings, 10 in the hind ; *ac* lying midway between the two antenodal nervures ; *ab* arising from *ac* or, more usually, from the posterior margin of wing a little distal to *ac*, and longer and more oblique than in *D. carmichaeli*. *Abdomen* yellow at the sides and beneath, enfumed on dorsum, and deepening to blackish-brown at the apical ends of segments 3 to 6 to form rather sharply defined apical annules ; segment 2 with a narrow mid-dorsal streak of blue not quite extending to apical border ; segment 3 with a pale blue dorsal basal spot ; segments 6 to 7 with narrow pale blue complete basal annules ; dorsum and sides of segments 8 and 9 broadly pale azure blue ; segment 10 black, unmarked. *Anal appendages* black, paler at apices, shaped similarly to those of *D. carmichaeli*, but the inferiors stouter and the basal spine less pronounced.

Female unknown.

Distribution.—BENGAL : Gopaldhara, Darjeeling District, May to July.

Type in the author's collection. Distinguished from all other species of the genus by its colouring, and from *D. carmichaeli* by the points enumerated under the description of that species. The sharply defined blue and dark brown stripes on the sides form the best medium for separating these two closely allied species, but in addition it is to be noted that whilst the thorax of *D. carmichaeli* is lighter than that of *D. polychromatica*, the abdomen is strikingly darker ; the 10th abdominal segment is blue in the former, unmarked in the latter.

53. **Drepanosticta viridis** Fraser. (Fig. 69.)

Drepanosticta viridis Fras., Mem. Dept. Agric. India (Ent.), vol. vii, no. 7, pp. 39, 40 (1922) ; id., J. Bombay Nat. Hist. Soc. vol. xxxv, p. 339, pl. ii, figs. 9, 10 (1931).

Male.—Abdomen 50 mm. Hind-wing 25 mm.

Head : labium palest brown ; labrum, bases of mandibles, and anteclypeus palest blue, the former bordered narrowly with glossy black ; postclypeus glossy black, rest of head steely blue-black ; eyes pale yellow, capped above with puce ; occiput ridged behind as in *Platysticta*. *Prothorax* dirty yellow ; posterior lobe dark bronzed brown, broadly arched, simple ; anterior lobe collar-like. *Thorax* dark metallic green or blue on dorsum, changing to reddish-brown at humeral region and pale brownish-yellow laterally ; lateral sutures mapped out obscurely in pale brown, the anterior suture bordered broadly and diffusely with black ; beneath yellow. *Legs* yellow with black spines ; extensor surfaces of femora blackish-brown, tibiæ olivaceous. *Wings* hyaline ; pterostigma reddish-brown, very small, almost quadrate, costal border but slightly shorter than posterior, distal border convex, the proximal but slightly oblique ; 14 to 15 postnodal nervures to fore-wings, 14 in the hind ; *ac* lying much nearer the level of distal antenodal nervure ; *ab* arising from posterior margin of wing well distal to *ac*, very short, hardly longer than *ac* ; *Riv+v* arising well distal to the oblique nervure

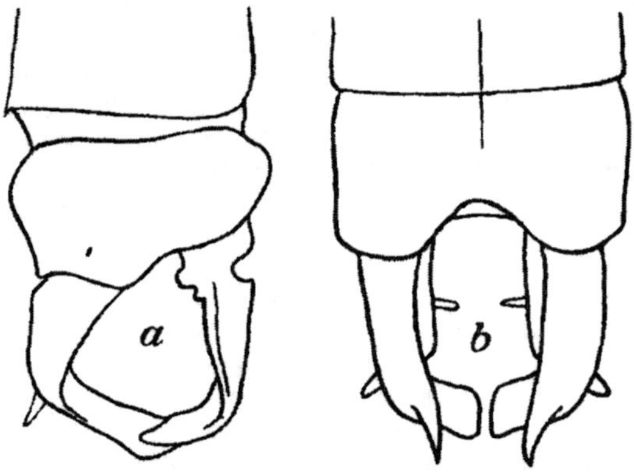

Fig. 69.—Anal appendages of *Drepanosticta viridis* Fras., male. *a*. Right lateral view. *b*. Dorsal view.

descending from the subnode. *Abdomen* of great length, very slim, resembling species of *Protosticta* more closely than any other species of the genus ; segments 1 to 7 brownish-black, this colour deepening at apices of segments 2 to 7, but falling short of bases of segments, where are seen narrow white basal annules ; segment 8 with the apical half azure blue on dorsum, black laterally and mid-dorsally, so that the blue of apex presents two points directed basalwards subdorsally ; segment 9 entirely pale azure blue on dorsum ; segment 10 black, emarginate, and with a deep depression on dorsum extending nearly as far as its base. *Anal appendages* black ; superiors twice the length of segment 10, bent at a right angle at their middle, downwards, and a little inwards, broad at base, presenting a needle-like dorsal spine at the point of angulation ; apex expanded, bevelled at end ; inferiors nearly as long as superiors, broad at base, then flattened, twisted spirally on their long axis at the middle of appendages, apical half angulated sharply upwards to meet and overlap apices of superiors; a small needle-like spine on the inner border, directed inwards near the middle of appendages.

Female.—Abdomen 37 mm. Hind-wing 25 mm.

Closely similar to the male in colour and markings, but shorter and more robustly built. Segments 8 and 9 without blue markings; segments 1 to 6 with well defined basal white annules, 7 without, otherwise all details, including venational ones, similar to those of the male.

Distribution.—King Island, Mergui, LOWER BURMA. Found beside marshes and small irrigation channels and streams in rubber plantations.

Distinguished from all other species of the genus by the extreme length and attenuation of its abdomen, which is exactly double the length of the wings, as in species of *Protosticta*. The armature of the anal appendages recalls that of *P. himalaica*.

Type in the British Museum; paratype in the author's collection.

54. **Drepanosticta quadrata** (Selys).

Platysticta quadrata Selys, Bull. Acad. Belg. (2) vol. x, p. 441 (1860); id., Syn. Agr., Dernière Légion, sep. p. 13 (1860); Kirby, Cat. Odon. p. 133 (1890); Selys, Ann. Mus. Civ. Genova, (2) vol. x (xxx), pp. 507, 508 (1891); Laid., Fasc. Malay., Zool. pt. iv, p. 11, fig. 1 (1907).

Drepanosticta quadrata Laid., Rec. Ind. Mus. vol. xiii, p. 343 (1917); id., J. Malay Branch Roy. As. Soc. vol. ii, pp. 304–6 (1924); Fras., J. Bombay Nat. Hist. Soc. vol. xxxv, pp. 339, 340 (1931).

Male.—Abdomen 35 mm. Hind-wing 21 mm.

Head: labrum, bases of mandibles, and anteclypeus palest blue, the former broadly bordered with glossy black; post-clypeus and rest of head black. *Prothorax* creamy white; posterior lobe black, simple. *Thorax* bronzed black on dorsum, pale bluish laterally, pale yellow beneath; laterally an oblique black stripe traversing the length of mesepimeron. *Legs* pale, the extensor surfaces of femora and articulations black. *Wings* hyaline; pterostigma subquadrate, black, finely framed in white and thick black nervures; *Riv+v* arising slightly distal to the oblique nervure descending from the subnode; 11 to 12 postnodal nervures in fore-wings, 10 to 11 in the hind. *Abdomen* blackish-brown, segments 2 to 7 with a narrow yellow basal annule, the black expanding at apical ends of each segment to form broad apical annules; segment 8 with the apical half and 9 with the whole of dorsum pale blue; segment 10 black. *Anal appendages* black, superiors twice as long as segment 10, slightly curved towards each other and downwards, a robust spine on the upper border of middle of appendage, apex slightly flattened and dilated; inferior appendages of similar length, pale at base, which is broad and presents an inner upper obtuse spine or tubercle, then tapering, subcylindrical, apex curved in, notched, and preceded by a long inner spine.

Female.—Abdomen 31 mm. Hind-wing 18 mm.

Head: labrum dirty white; anteclypeus pale yellow; postclypeus and front of frons traversed by an obscure band; rest of head black. *Prothorax* pale yellow; posterior lobe emarginate at middle, yellow, with obscure greyish fascia. *Thorax* bronzed black on dorsum as far lateral as the antero-lateral suture, the mid-dorsum greyish-yellow. Laterally and beneath pale yellow, with a broad oblique brown stripe traversing the mesepimeron. *Legs* pale yellow, articulations obscure. *Wings* similar to the male, but *Riv*+*v* arising in continuation with the oblique nervure descending from the subnode. *Abdomen* brownish-black, segments 2 to 7 marked similarly to the male; segment 9 yellow on dorsum, 10 black. *Anal appendages* short, conical, black; vulvar scale robust.

Distribution.—The type comes from Singapore, but it is by no means certain that the allotype female from Burma (teneral and in poor condition) belongs to it, the probability being otherwise. The female, taken on Mt. Karen in May, has the paler yellow markings probably pale blue in the adult stage.

Type in the Selysian collection.

55. **Drepanosticta annandalei** Fraser. (Fig. 70.)

Drepanosticta annandalei Fras., Rec. Ind. Mus. vol. xxvi, pp. 412, 413 (1924); id., J. Bombay Nat. Hist. Soc. vol. xxxv, pp. 340, 341, pl. i, fig. 4 (1931).

Male.—Abdomen 32 mm. Hind-wing 22 mm.

Head: labium brown; labrum, bases of mandibles, and adjacent part of anteclypeus palest blue, the former narrowly bordered with glossy black; postclypeus and rest of head mat black; basal and 2nd segments of antennæ white; eyes olivaceous brown above, paler beneath. *Prothorax* brownish-white, with a large black spot on each side; posterior lobe simple, rounded. *Thorax* blackish-brown on dorsum, with the mid-dorsal carina and upper part of humeral suture paler; laterally white, with a diffuse brown fascia traversing the sides obliquely as far as metepimeron. *Legs* creamy white; femora on extensor surfaces stippled with black; tibio-femoral joints black. *Wings* hyaline; pterostigma blackish-brown, shaped similarly to that of *D. carmichaeli*; 16 postnodal nervures to fore-wings, 14 in the hind; *Riv+v* arising in continuation with the oblique nervure descending from the subnode; *ab* very oblique, arising from posterior border of wing somewhat distal to *ac*. *Abdomen* creamy white laterally and beneath, black on dorsum, the black not extending to extreme base of segments, but expanding apically to form broad annules; segment 8 white or pale blue for its basal two-thirds, but the black of the apical third extending finely along the mid-dorsal carina as far as base of segment; segments 9 and 10 black, unmarked with blue. *Anal appendages* black, slightly longer than segment 10, the inferiors slightly shorter than superiors; superiors broad at base, rotated on their long axis at the middle, at which point they

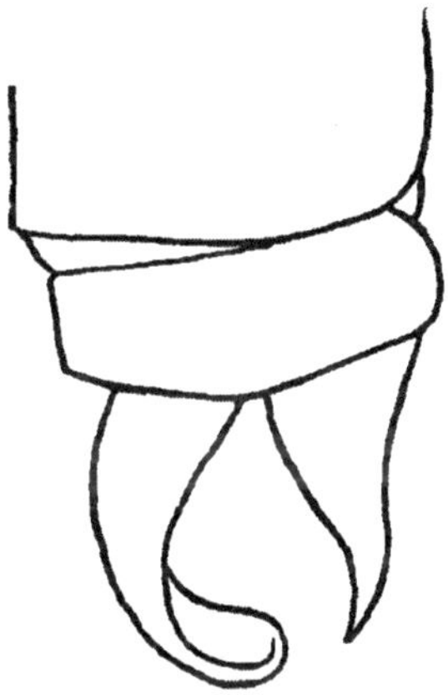

Fig. 70.—Right lateral view of anal appendages of *Drepanosticta annandalei* Fras., male.

curve somewhat inwards and strongly downwards; apices obtuse and slightly bevelled; inferiors broad at base, then tapering, cylindrical, ending in an acute point.

Female.—Abdomen 25 mm. Hind-wing 20 mm.

Very similar to the male, but abdominal segment 8 black for its apical half; segment 9 with the basal two-thirds white (or blue), apical third black. *Anal appendages* shortly conical, pointed, pale; vulvar scale robust, not extending beyond end of abdomen.

Distribution.—ANDAMANS only.

Type in the Indian Museum, from Mt. Harriet, 800 ft., South Andamans, taken in deep jungle at the edge of a stream in December. Its simple black colour, with white and blue markings, serves to distinguish it from *D. carmichaeli* and *D. polychromatica*, whilst the comparatively short abdomen will separate it from *D. viridis*. Its extremely simple anal appendages will distinguish it from *D. quadrata* as well as from other Malayan species of the genus. In the original description I compared this species with *D. hilaris*, from Burma, whereas *D. quadrata* was of course intended.

Subfamily PLATYCNEMINÆ.

Platycnemis, 4me Légion, Selys, Bull. Acad. Belg. (2) vol. xvi, p. 150 (1863).
Platycnemis, 3me Légion, Selys, Rev. Syn. Agr. p. 95 (1886).
Platycnemis, Legion 2, Laid., Rec. Ind. Mus. vol. xiii, p. 325 (1917).
Platycneminæ Fras., J. Bombay Nat. Hist. Soc. vol. xxxiii, p. 835 (1929); Laid., J. Fed. Malay States Mus. vol. xvi, p. 192 (1930); Fras., J. Bombay Nat. Hist. Soc. vol. xxxv, p. 645 (1932).

Dragonflies of small size, coloured black, marked with blue, red or yellow, rarely, if ever, metallic; resting with wings closed over dorsum; wings moderately narrow or narrow, rounded at apex, hyaline; the nervure *ac* situated midway between the two antenodal nervures; no accessory basal postcostal nervures ever present; petiolation beginning at or slightly proximal to *ac*; the nervure *ab* always present and complete and continued on as the nervure *IA*, which is always present; *Cuii* extending distally beyond the middle of wings; discoidal cell elongate, three or four times as long as broad, dilating distally, the distal end more or less subacute; sectors of arc arising from lower end of arc, always separate from their origins; individual cells mostly four-sided; *Riii* arising well distal to node, midway between the node and pterostigma or nearer the node; *Riv+v* arising a little proximal to the level of the oblique nervure descending from the subnode, *IRiii* from that level, but these origins variable in some genera; pterostigma slightly longer than broad, costal and posterior sides usually equal, both distal and proximal sides oblique; no intercalated sectors save *IRii*.

Abdomen of moderate length, never twice the length of hind-wing, usually much shorter than this; segments 8, 9, and 10 decreasing gradually in length from 8 to 10. Anal appendages of male not markedly complex, the superiors usually shorter than inferiors, but very variable in the genera and species; vulvar scales robust, extending to the end of abdomen, but rarely beyond that level, and always without the robust dorsal spine-like protuberance seen in the PLATYSTICTINÆ. Penile organ very variable in the genera and species, but the end always curled over and embracing the stem of the organ, branched or unbranched at apex.

Larvæ, so far as known, slender, cylindrical, furnished with three caudal gills which are broadly dilated into paddle-like processes at the extremities; mask kite-shaped, flat, furnished with setæ, mid-lobe entire, lateral lobe robust and furnished with long movable hooks; legs long and slim. Breeding in small montane streams, where they cling to twigs or roots, resting with the abdomen curled well over the dorsum and waving freely in the current.

Distribution.—Old World only. Palæarctic, Ethiopian, Oriental, and Australian. Within our limits from CEYLON, INDIA, and BURMA, but with the exception of the genus *Copera*, all genera are confined to N.E. India and Burma, and usually to submontane and montane tracts.

Species of the subfamily are found haunting the banks of small montane streams or rivulets, or hiding in the undergrowth of dense jungle near their parent streams. The teneral forms of some species are quite white and without markings, and form conspicuous objects as they move furtively through the dense undergrowth. A few species haunt small streams flowing through marsh-lands, but most are found in wooded country.

Five genera are found within Indian limits, of which *Platycnemis* is Palæarctic and confined solely to the far north, *Copera* is Ethiopian and Oriental in distribution and found throughout India, Burma and Ceylon, whilst the other three —*Calicnemis*, *Cœliccia*, and *Indocnemis*—are purely Oriental and confined to N.E. India and Burma.

Key to the Genera of the Platycneminæ.

1.	Costal side of discoidal cell in fore-wing one-fifth or more shorter than the posterior	2.
	Costal and posterior sides of discoidal cell equal or nearly so	4.
2.	Wings petiolated to the level of *ac*, i. e., *ab* ends at the level of *ac*	CŒLICCIA Kirby, [p. 152.
	Wings petiolated short of *ac*, i. e., *ab* ends some distance proximal to *ac*	3.
3.	Wings long and narrow; 4 cells between the discoidal cell and the oblique nervure descending from the subnode	INDOCNEMIS Laid., [p. 185.
	Wings broader and shorter; only 3 cells between the discoidal cell and nervure descending from the subnode	CALICNEMIS Selys, [p. 171.
4.	Second segment of antennæ as long as the third; Oriental and Ethiopian species	COPERA Selys, p. 191.
	Third segment of antennæ equal in length to the first two taken together; solely Palæarctic species	PLATYCNEMIS Charp., [p. 187.

Genus CŒLICCIA Kirby (1890). (Fig. 71.)

Trichocnemis Selys, in Sagra, Hist. Cuba, Ins. p. 464 (1857); id., Bull. Acad. Belg. (2) vol. xvi, p. 155 (1863); id., Mém. Cour. vol. xxxviii, p. 114 (1886); Laid., Fasc. Malay. (Odonata), pt. ii, pp. 2–7 (1907).

Cœliccia Kirby, Cat. Odon. p. 128 (1890); Ris, Suppl. Ent. no. 1, pp. 60–2 (1912); Laid., Rec. Ind. Mus. vol. xiii, pp. 322, 325, 331–3 (1916); id., ibid. vol. xxxiv, pp. 7–12 (1932).

Zygopterous dragonflies of medium size and slender build, with abdomen less than twice the length of hind-wings; coloured black, marked with blue, yellow, or bright chrome-yellow. Wings hyaline, moderately narrow, rounded at apices, never falcate, with a moderately long petiole; discoidal cell elongate, costal side about four-fifths or slightly less than four-fifths the length of the posterior side, distal side at least half as long again as basal and very oblique, so that the cell is acutely pointed at its lower distal angle; sectors of arc slightly separate at origin, arising from lower end of arc, divergent from origin; no accessory basal postcostal nervure

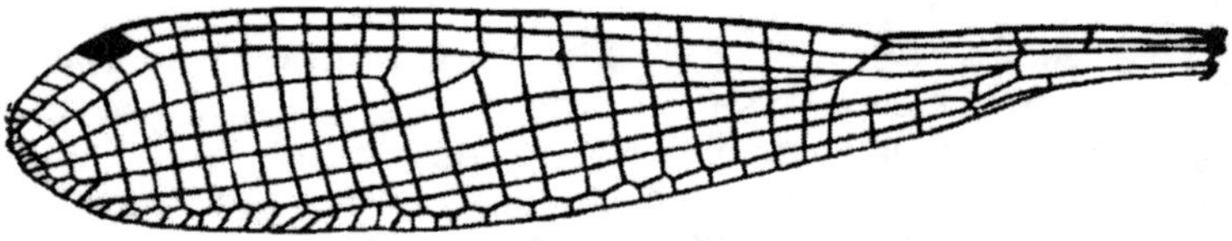

Fig. 71.—Wing of *Cœliccia membranipes* (Ramb.), male.

present; the nervure *ac* situated at a level slightly nearer the distal antenodal nervure; *ab* always present and always complete, continued on as *IA*, which latter extends for nearly three-fourths the length of wings; petiolation beginning slightly proximal to *ac*; *Cuii* extending a short distance beyond the end of *IA*, 20 cells or more in length; *Riv+v* arising a little proximal to level of node, *IRiii* a little distal, but the position of these two latter very variable in the species; pterostigma small, its sides nearly equal, oblique and diamond-shaped, poorly braced; cells of wings mainly quadrangular; 2 to 3 cells between the discoidal cell and the nervure descending from the node.

Head narrow; eyes smaller than in species of *Platysticta*; occiput simple; prothorax with the posterior lobe rounded and simple in the male, notched or armed with elongate spines in the female; abdomen slim, cylindrical, broadened at extreme base and gradually broadening from segment 7 to the end, about one-third longer than the wings; anal appendages variable in the species, but inclined to be rather homogeneous, superiors usually broad at base, acute at apex, and furnished with an inner subapical spine, inferiors longer than superiors, broad at base, abruptly tapered and subcylindrical, slightly sinuous, subacute or acute at apex.

Genitalia: lamina narrowly but deeply cleft; hamules broad, quadrate plates meeting broadly across the genital pit; penis with apical end curling strongly over stem of organ, and with the end variable, deeply cleft, bifid or branched; lobe pyriform, short, with a shallow longitudinal groove running its whole length; vulvar scales robust, extending slightly beyond end of abdomen.

Genotype, *Cœliccia membranipes* (Ramb.).

Distribution.—N.E. INDIA, BURMA, Malaysia, Indo-China, Formosa, Philippines, Sondaic Archipelago, and Borneo. Only nine species found within Indian limits, of which five occur in Assam and the others in Burma.

Laidlaw has divided the genus into three groups of species, according to the positions of *Riv+v* and *IRiii* relatively to the subnode. In the 1st group *IRiii* arises at the subnode, *Riv+v* proximal to it; in the 2nd group *IRiii* arises distal and *Riv+v* proximal to it; whilst in the 3rd, *IRiii* arises distal to the subnode and *Riv+v* from it. Most Indian species belong to the first and second groups.

Key to the Indian Species of Cœliccia.

1.	Dorsum of thorax wholly blue	*poungyi* Fras., p. 165.
	Dorsum of thorax wholly golden yellow...	*chromothorax* (Selys), [p. 167.
	Dorsum of thorax with a single pair of oval pale blue spots	*bimaculata* Laid., [p. 154.
	Dorsum of thorax pale blue for the lower half	*renifera* (Selys), p. 156.
	Dorsum of thorax with two pairs of pale blue spots	2.
	Dorsum of thorax with a pair of antehumeral stripes	3.
2.	Lower pair of spots simple, pyriform; segment 10 and the basal two-thirds of segment 9 pale blue	*didyma* (Selys), p. 159.
	Lower pair of spots broad and branching into two above; segment 10 blue, marked broadly with black at apical border; segment 9 wholly blue; segment 8 with its apical half blue........	*loringæ* Laid., p. 161.
3.	Antehumeral stripes lying nearer the humeral suture than mid-dorsal carina.	4.
	Antehumeral stripes lying closely apposed to the mid-dorsal carina	*fraseri* Laid., p. 169.
4.	Posterior lobe of prothorax of female simple	*loogali* Laid., p. 163.
	Posterior lobe of prothorax of female prolonged and deeply and broadly notched, the corners of the notch prolonged into long recurved horns. (Male unknown.)	*vacca* Laid., p. 155.

56. Cœliccia bimaculata Laidlaw. (Fig. 72.)

Cœliccia bimaculata Laid., Rec. Ind. Mus. vol. viii, p. 341, pl. xvi, fig. 1 (1914); id., ibid. vol. xiii, pp. 335, 336 (1917); id., ibid. vol. xxxiv, pp. 11–14 (1932).

Male.—Abdomen 36 mm. Hind-wing 22 mm.

Head: labium pale yellow; labrum black; anteclypeus, genæ, and the bases of mandibles partly blue; postclypeus, frons, and vertex black, the latter traversed at the level of ocelli by a broken pale blue stripe; occiput with pale yellowish pyriform postocular spots. *Prothorax* black, the sides palest blue or creamy white. *Thorax* black on dorsum, palest blue laterally, the former with a pair of oval pale blue spots lying in close apposition to the anterior half of the mid-dorsal carina; a narrow oblique black stripe

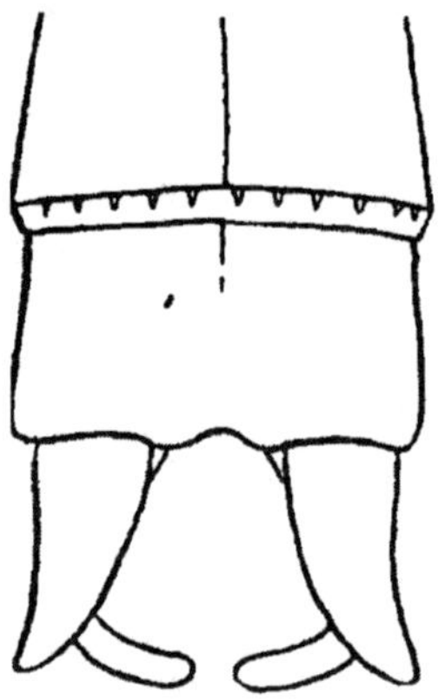

Fig. 72.—Dorsal view of anal appendages of *Cœliccia bimaculata* Laid., male.

on each side bordering the mesepimeron anteriorly; beneath creamy white. *Legs* palest yellow, femora with two black stripes, tibia black on flexor surface only, joints and cilia black. *Wings* hyaline; pterostigma blackish-brown, framed palely in brown, its proximal border very oblique, covering 1 cell; *Riv+v* arising proximal to the subnode, *IRiii* at the subnode; 18 to 20 postnodal nervures in fore-wings, 17 to 18 in the hind. *Abdomen* dark reddish-brown on dorsum, deepening to black at apical ends of segments and on dorsum of segment 2 and segments 7 to 10: all segments pale yellow laterally, this colour extending dorsalwards on 3 to 6 so as to form paired spots. *Anal appendages* creamy white; superiors triangular as seen from above, deeply excavated within, the inner border ending in a minute black point;

inferiors slightly longer than the superiors, which are themselves slightly longer than segment 10, slender, curling in at apices, forcipate, and also ending in a minute black point at apex. Penis with a prominent terminal lobe, the apex bifid, and ending in a pair of long whip-like processes.

Female.—Abdomen 35 mm. Hind-wing 23 mm.

Very similar to the male, but the blue replaced by creamy white except the facial markings and lower half of eyes, which are blue. *Prothorax* with black anterior and posterior lobes, these two areas confluent by a fine mid-dorsal black line on the middle lobe; posterior lobe with the middle portion of hinder border prolonged into a small tongue-like process, on either side of which is a minute lobe. Dorsal thoracic markings present as narrow stripes lying closely apposed to the mid-dorsal carina and prolonged and narrowing as far as the antealar sinus very much as in *C. fraseri*. Segment 8 of *abdomen* with its apical two-thirds palest yellow or, in tenerals, segments 8 to 10 creamy white. *Anal appendages* and vulvar scales pale yellow.

Distribution.—Several males and females from Tura, Garo Hills, ASSAM.

Type, an incomplete male in the Indian Museum, taken at an altitude of 1500 ft. in June–July. Distinguished from other species by the two oval spots on thoracic dorsum and by the pale under surface of thorax.

57. **Cœliccia vacca** Laidlaw.

Cœliccia vacca Laid., Rec. Ind. Mus. vol. xxxiv, pp. 11, 15, 16 (1932).

Male unknown.

Female.—Abdomen 36 mm. Hind-wing 25 mm.

Head: labium pale yellow, labrum black; anteclypeus, genæ, and bases of mandibles pale blue, rest of head black, with a chain of four small spots crossing the ocellar space and a pyriform postocular spot on each side of occiput; eyes black above, greenish below. *Prothorax* with anterior and posterior lobes black, this dark area connected by a broad median black stripe on the dorsum of middle lobe, the sides pale blue or yellowish, posterior lobe prolonged back as a quadrate lobe, deeply and broadly notched, the angles of the notch prolonged as two horns which are angulated strongly forwards, and in profile are seen to be shaped like the horns of a cow. *Thorax* black on dorsum, pale blue laterally and beneath, the dorsum marked with a narrow pale blue antehumeral stripe on each side which lies nearer the humeral suture than the mid-dorsal carina; on each side a narrow oblique black stripe incomplete below and bordering the mesepimeron. *Legs* creamy white, extensor surfaces of femora, the lower or distal half of anterior femora on flexor surface, and the flexor surface of tibiæ black. *Wings* hyaline; pterostigma black, bordered finely along costal border with whitish, covering one cell, braced; 19 postnodal nervures in fore-wings, 17 in the hind; *Riv+v* arising at subnode, *IRiii* well distal to it. *Abdomen* blackish-brown on dorsum, yellowish laterally, the sides of segment 8 broadly yellow, this colour extending up on to dorsum to meet above and form a narrow apical annule; basal half of segment 9 also yellow. *Anal appendages* black, conical, short; vulvar scales black, robust.

Distribution.—Tura, Garo Hills, Assam.

Only a single specimen of this curious insect is known—the *type*, a female in the Indian Museum, Calcutta. Distinguished from all other species of the genus by the unique development of the posterior lobe of prorothax.

58. **Cœliccia renifera** (Selys). (Fig. 73.)

Trichocnemis renifera Selys, Mém. Cour. vol. xxxviii, p. 119 (1886); Laid., Fasc. Malay. (Odonata) pt. ii, p. 2 (1907).
Trichocnemis didyma Selys (*C. renifera*, nec *T. didyma*), Mém. Cour. vol. xxxviii, pp. 118–19 (1886).
Cœliccia renifera Kirby, Cat. Odon. p. 128 (1890); Laid., Rec. Ind. Mus. vol. xiii, pp. 322, 332, 333 (1917); Fras., J. Bombay Nat. Hist. Soc. vol. xxix, p. 744 (1923); Laid., Rec. Ind. Mus. vol. xxxiv, pp. 12, 18, 19 (1932)

Male.—Abdomen 33–37 mm. Hind-wing 22–28 mm.

Head: labium pale brown tipped with black: labrum and bases of mandibles steely blue-black; anteclypeus and genæ pale sky-blue; postclypeus glossy black; vertex, occiput, and beneath the head velvety black, the former marked with small obscure triangular spots, one between each outer ocellus and the bases of antennæ; a well defined pale yellow, transversely oval spot on each side of back of occiput, and a small yellow spot beneath head against the eyes; eyes olive-green below, black above. *Prothorax* velvety black, unmarked; posterior lobe rounded, simple. *Thorax* velvety black on dorsum, marked with pale blue or pale greenish-yellow as follows: a broad oval spot on each side of the mid-dorsal carina extending about half-way up the dorsum, narrowly separated from each other by the black carinal ridge; laterally broadly coloured pale blue or yellowish in tenerals for the anterior two-thirds of the mesepimeron, except the part just beneath the wings, and the whole of the metepimeron, the whole area narrowly divided by a fine black stripe on the postero-lateral suture. *Beneath entirely black*. *Legs* black on extensor surfaces of femora and flexor surfaces of tibiæ, opposite surfaces of these pale carneous or creamy white; coxæ and trochanters yellow anteriorly, black behind. *Wings* hyaline; pterostigma black, finely framed in pale brown and thick black nervures, braced, covering one cell; 18 to 20 postnodal nervures in fore-wings, 16 to 18 in the hind; *Riv+v* arising a short distance proximal to the subnode (or in Simla specimens at the node or a shade proximal to it, or sometimes in conjunction with *IRiii*), the latter arising at the subnode or a shade proximal to it. *Abdomen* black, yellow beneath; segment 1 with a small lateral pale yellow spot; segment 2 with its ventral border broadly pale blue or yellow (according to age); segments 3 to 6 with elongate subapical lateral creamy white or pale blue spots; remaining segments unmarked. *Anal appendages* creamy white, superiors beneath and inferiors at base brownish-black; superiors about as long as segment 10, conical and rather

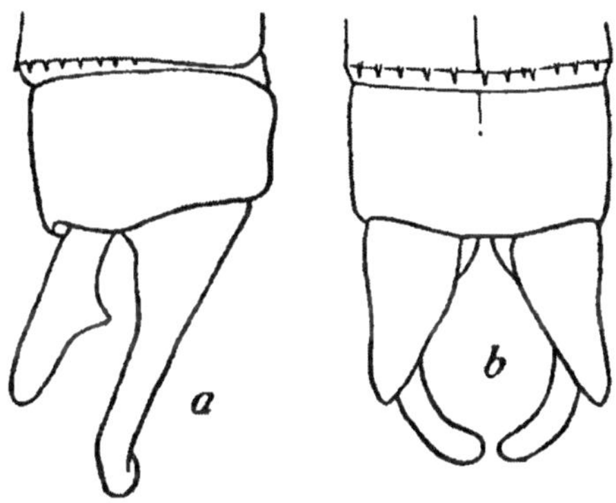

Fig. 73.—Anal appendages of *Cœliccia renifera* (Selys), male.
a. Right lateral view. *b*. Dorsal view.

obtuse at apex as seen from above, cigar-shaped as seen from the side, and with a small median black ventral spine; inferiors one-third as long again as superiors, broad at base, then cylindrical and curled gently in, their apices nearly meeting, sloping strongly up as seen in profile, the apex obtusely pointed. Penis much broadened anteriorly and shallowly concave at anterior border, the apical portion curled strongly over, presenting a small spine on the dorsum within the arch of the curve, and a small lappet-like lobe on either side just before the apex bifurcates into short branches, which turn directly out at right angles to the body of the organ and end obtusely.

Female.—Abdomen 42–46 mm. Hind-wing 29–33 mm.

Differs in several respects from the male, the markings being blue only in very adult specimens, and generally creamy yellow. *Head* similar to the male in most respects, differing as follows: the markings of face pale greenish-blue; the small triangular spot on vertex replaced by a chain of six spots, one in the same situation as in the male, and connected narrowly with a second which extends out to the eye; lastly, a third rounded spot lying between the outer and medial ocelli on each side; the spot on the under side of head against the eyes very much larger and broadly triangular. *Prothorax* citron-yellow, with the whole of the anterior lobe, the mid-dorsal ridge rather broadly, and the hinder border black; the posterior lobe vestigial, only a lateral rudimentary process on each side. *Thorax* black on dorsum nearly as far lateral as the antero-lateral sinus, and marked with a pair of antehumeral narrow, complete, bluish-green stripes; laterally pale blue or yellow in sub-adults and tenerals, marked with a very fine line of black on the postero-lateral suture; beneath pale blue or yellow, unmarked. *Legs* yellow, all tibiæ black on flexor surface; anterior pair of femora with a black stripe on front and back, confluent at distal end of limb; hinder two pairs with this marking much restricted, especially on the hinder pair. *Wings* hyaline; pterostigma with the pale brown frame much more in evidence; 20 to 21 postnodal nervures in fore-wings, 18 to 19 in the hind; *Riv+v* arising more proximally in regard to the subnode, and *IRiii* always rising distal to that level; the wings palely enfumed in old adults. *Abdomen* blackish-brown, black towards the end segments, marked beneath and laterally broadly with greenish-blue, changing to yellow on the end segments, and laterally and subapically expanding into largish spots; segment 8 with a broad apical yellow or bluish-green annule covering nearly half the segment; segment 9 nearly entirely pale blue or bluish-green, its base narrowly and the sides broadly black, the black of base prolonged apically as two submedian points. Occasionally the yellow annule on 8 narrowly interrupted on the mid-dorsal carina; segment 10 brownish-black, unmarked. *Anal appendages* shorter than segment 10, conical, pointed, black; vulvar scales robust, bright yellow, projecting well beyond end of abdomen.

Distribution.—UPPER ASSAM and BENGAL, SIKKIM and SIMLA HILL STATES. I have specimens from Darjeeling, Bengal, collected in July; from Pashoke, British Sikkim, collected in May; and from Gopaldhara, Bengal, collected in September.

Some specimens from Barogh, Simla Hills, 5,000 ft., collected in August, are very small, and evidently similar to the specimen described by Selys from Darjeeling, and determined by him in error as *C. didyma*. The single thoracic dorsal marking and the black under side of the thorax, which is unmarked, are in themselves sufficient to distinguish the insect from *C. didyma*, which has two spots on the dorsum and a pale blue under side to thorax.

These small Simla specimens, except for size, do not differ in the slightest from the larger forms found elsewhere. The single blue spot restricted to the lower half of the dorsum of the thorax will serve to determine this insect.

59. **Cœliccia didyma** (Selys). (Fig. 74.)

Trichocnemis didyma Selys, Bull. Acad. Belg. (2) vol. xvi, p. 155 (1863); id. (*C. renifera*, nec *didyma*), Mém. Cour. vol. xxxviii, p. 118 (1886).

Cœliccia didyma Kirby, Cat. Odon. p. 128 (1890); Laid., Rec. Ind. Mus. vol. xiii, pp. 322, 332, 335 (1917); Fras., J. Bombay Nat. Hist. Soc. vol. xxix, p. 745 (1923); Laid., Rec. Ind. Mus. vol. xxxiv, pp. 12, 16, 17 (1932).

Male.—Abdomen 40 mm. Hind-wing 26 mm.

Head: labium yellow; labrum and postclypeus steely blue-black; bases of mandibles, genæ, and anteclypeus pale azure blue, the former narrowly black at base; vertex and occiput velvety black, the former with a small triangular blue spot on each side lying between the ocelli and antennæ, the latter with a pyriform elongate pale yellow spot on each side; beneath broadly pale blue; eyes olive-green below, black above. *Prothorax* black, with a large pale blue spot on each side of the middle lobe and the lower border narrowly pale blue; posterior lobe simple, rounded. *Thorax* velvety black on dorsum, marked with two pyriform elongate antehumeral spots on each side of mid-dorsal carina, a lower large spot extending nearly half-way up the dorsum, pointed above, and an upper, smaller spot extending from the antealar sinus for about one-fourth the length of the dorsum and pointed below; laterally pale blue, the postero-lateral suture broadly mapped out in black; beneath palest blue, unmarked. *Wings* hyaline; pterostigma blackish, framed in pale brown and thick black nervures, braced, covering 1½ cells; 18 postnodal nervures in fore-wings, 17 in the hind; *Riv+v* arising a short distance proximal to the subnode, *IRiii* a similar distance distal to that structure; *ab*, especially in the fore-wings, arising a short distance distal to the level of *ac*. *Legs* yellow, femora with a black stripe running the whole length of flexor and extensor surfaces, more extensive on the latter, where it is continued on to the trochanters and coxæ; tibiæ black on flexor surface. *Abdomen* black on dorsum and sides, pale yellow beneath, the sides of segment 1 broadly yellow, the lower part of sides of segment 2 narrowly so; a linear streak of yellow on dorsum of the latter segment, not quite extending to apical end; segments 3 to 6 with small paired pale blue baso-dorsal spots; segments 7 and 8 entirely black save for a narrow blue border on the ventral surface of the latter; segments 9 and 10 pale blue, but the basal third of former with a broad black annule. *Anal appendages* livid or pale yellow, the base of superiors and the apices of the inferiors black; superiors slightly longer than segment 10, conical and hollowed out within as seen from the dorsum, narrow at base, with a broad ventral tooth tipped with black, and the obtuse apex with a shallow dorsal notch as viewed from the side; in semi-profile view a fine black spine seen near the base within; inferiors slightly longer than the superiors, very sinuous as seen from the side, the apex obtuse, rounded, and curled slightly down and very markedly in.

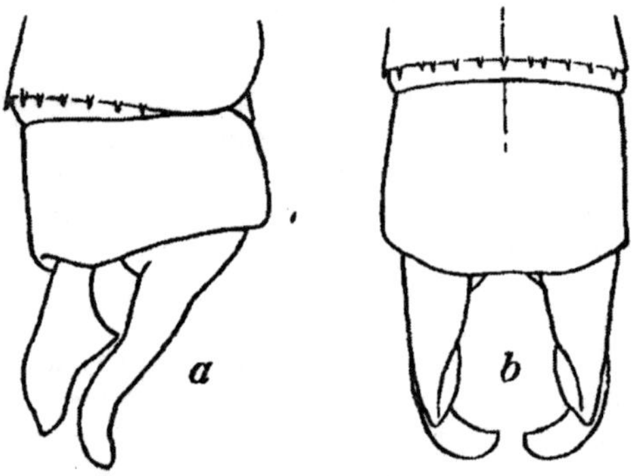

Fig. 74.—Anal appendages of *Cœliccia didyma* (Selys), male. *a*. Right lateral view. *b*. Dorsal view.

Female unknown.

Distribution.—Described from a specimen from Kedah Peak, Simla Hill States, PUNJAB, 3,000 ft., taken by Capt. Cardew, 29.iii.28. The *type*, in the Selysian collection, is an incomplete male with the end of the abdomen missing; locality, Tibet. Subsequently Selys described a small form of *C. renifera*, from Darjeeling, as a male of this species, giving the supposed and completed description of *C. didyma*.

The species is easily distinguished from all others by the two pyriform dorsal thoracic spots, pale blue in colour. *C. loringæ*, which is very closely allied to it, differs by the lower of the two dorsal spots being broader and deeply bifid along the upper border, and also by the blue markings of the abdomen extending well on to segment 8, instead of being restricted to the apical half of segments 9 and 10.

·60. **Cœliccia loringæ Laidlaw.**

Cœliccia loringæ Laid., Rec. Ind. Mus. vol. xxxiv, pp. 12, 17, 18 (1932).

Male.—Abdomen 42 mm. Hind-wing 27 mm.

Head: labium white; labrum and postclypeus glossy black; bases of mandibles, genæ, and the lateral part of frons and the anteclypeus pale azure blue; rest of head velvety black, with a small oval spot, obscurely blue, on each side of vertex between the lateral ocelli and bases of antennæ, and an elongate, pyriform, pale yellow postocular spot on each side of occiput; beneath head black, except for an irregular pale blue border against the eyes; eyes olivaceous beneath and at sides, black above. *Prothorax* velvety black, with a broad border on each side confluent with an elongate spot on the side of the middle lobe pale azure-blue; posterior lobe simple, rounded. *Thorax* velvety black on dorsum, where are seen two large pale azure-blue spots on each side, a smaller upper spot near the antealar sinus and upper part of humeral suture extending slightly less than one-third the length of dorsum, and a larger, broader, lower spot extending for the lower two-thirds of dorsum, bordering the mid-dorsal carina inwardly, and with a tongue-like projection on the outer side near the humeral suture, giving a bifid appearance to the spot; laterally pale blue, with the postero-lateral suture narrowly outlined in black; beneath palest blue. *Legs* white, extensor surfaces of all femora black, as also the distal end of the flexor surface of hinder pair; tibiæ black on the flexor surface; tarsi entirely black. *Wings* hyaline; pterostigma blackish-brown, paler at margins, covering 1 to 1½ cells, strongly braced; 19 to 20 postnodal nervures to fore-wings, 19 in the hind; *Riv+v* arising together with *IRiii* from the subnode, or *IRiii* slightly distal to the level of that nervure; rarely *Riv+v* a shade proximal to the subnode. *Abdomen* black, marked with azure blue as follows:—Segment 1 broadly blue at the sides and more narrowly so at the apex, the black dorsal spot just reaching the apical border at a point; segment 2 broadly blue at the sides, and with a narrow mid-dorsal stripe extending for three-fourths the length of the mid-dorsal carina; segment 3 with a narrow basal annule narrowly interrupted on the mid-dorsal carina, and a very broad subapical annule, also narrowly interrupted on the mid-dorsum, this latter annule confluent with the blue of the sides and under surface; segments 4 to 6 very similar, but the basal annules reduced to two small basal spots which gradually diminish in size towards segment 6; segment 7 similar, but the basal spots absent; segment 8 with a broad blue apical annule resting on the apical border and covering about the apical half of the segment more subdorsally; segment 9 entirely blue, whilst 10 is blue, with a large trilobed spot of black on the apical border, the middle lobe of the spot extending to base of the segment. *Anal appendages* pale creamy yellow, exactly similar to those of *C. didyma* except for the basal spine of superiors, which is absent. Penis similar in shape to that of *C. didyma* (fig. 74); apex very broad and a little sinuous, very shallowly excavate, curling up over the stem of organ and deeply cleft into two long slim branches which curl down and forwards, and at the extreme apices backwards again; a strong dorsal spine within the curling arch of apex.

Female.—Abdomen 41 mm. Hind-wing 27 mm.

Differs in several respects from the male; labrum bright ochre, its base narrowly black; postclypeus black, with a rounded yellow spot on each side; bases of mandibles and genæ bright yellow; the isolated spots on vertex replaced by a complete transverse stripe, or with but a slight interruption at middle of head, this stripe very sinuous and traversing vertex between the ocelli; elongate spots behind eyes bright yellow; eyes bordered with yellow beneath. *Prothorax* black, broadly yellow at the sides and beneath, posterior lobe reduced to a small median tongue-like projection directed backwards. *Thorax* black, marked with bright yellow as follows:—A very narrow antehumeral stripe on each side of the dorsum and a tiny upper humeral spot; the sides and beneath all yellow save for a narrow black oblique stripe mapping out the postero-lateral suture. *Legs* yellow, marked similarly to the male, but the tibiæ entirely yellow, as also the trochanters and coxæ; tarsi black. *Wings* similar to the male in most respects; pterostigma reddish-brown between black nervures; 19 to 20 postnodal nervures in the fore-wings, 17 in the hind; *Riv+v* arising slightly proximal to the subnode: *IRiii* more distal, but the two arising very close together, as in the male. *Abdomen* black, with a bright yellow stripe running the whole length of the sides; segment 3 with a pair of basal subdorsal spots confluent with this yellow stripe; segment 1 broadly yellow, its dorsum with a broad triangular black spot which reaches the apical border at a point only; segments 2 to 6 with the mid-dorsal carina finely yellow; segment 8 with nearly its apical half bright ochreous, but this marked by a small triangular apical dorsal black spot with its point bifid and directed basalwards; segments 9 and 10 entirely bright ochreous. *Anal appendages* short, conical, pointed; bright ochreous, as also a conical protuberance between them; vulvar scales ochreous, extending to end of abdomen.

Distribution.—Upper Burma only. The *type*, in the author's collection, is from Gokteik, Northern Shan States;

other specimens were taken at Maymyo during June and July.

The species can only be confounded with *C. didyma*, from which, however, it is easily distinguished by the shape of the lower dorsal thoracic spot, by the conspicuous subapical blue annules on the abdomen, and by the blue markings at the end of abdomen extending on to segment 8 as well as the whole of segment 9. The origins of *Riv+v* and *IRiii* are also different, and the nodal index is higher.

61. **Cœliccia loogali** Laidlaw. (Fig. 75.)

Cœliccia loogali Laid., Rec. Ind. Mus. vol. xxxiv, pp. 12, 26–28 (1932).

Male.—Abdomen 42–43 mm. Hind-wing 28 mm.

Head: labium pale yellow; labrum and postclypeus glossy black, the latter with a small spot at each side pale azure blue; bases of mandibles, genæ, and outer part of frons pale azure blue; rest of head black, the vertex with an obscure triangular spot lying between the outer ocelli and the bases of antennæ; the occiput with a transversely elongate spot on each side behind the eyes; eyes dark olivaceous green at the sides and beneath, black above. *Prothorax* black, with a broad pale azure-blue stripe on each side below; posterior lobe rounded, simple. *Thorax* velvety black on dorsum to well beyond level of humeral suture, marked with a pair of antehumeral pale azure blue narrow curved stripes with the convexity facing outwards and extending the whole length of dorsum; a small upper humeral spot of the same colour, occasionally quite large, but sometimes obsolete; laterally pale blue, with a small black spot at the upper end of the postero-lateral suture and a similar spot below; beneath pale azure blue. *Wings* hyaline; pterostigma blackish-grey in a paler frame and thick black nervures, braced, covering 1½ cells; *Riv+v* arising well proximal to the subnode, *IRiii* slightly distal to that structure or occasionally at the same level; 19 to 23 postnodal nervures to fore-wings, 18 to 19 in the hind. *Legs* pale yellow as well as the trochanters and coxæ, femora lined with black on extensor surfaces, tibiæ on the flexor, tarsi black. *Abdomen* black on dorsum, pale blue to yellow on the lower part of sides of segments 3 to 6, segment 1 broadly blue on the sides, segment 2 more extensively so than the following segments; remaining segments black. *Anal appendages* black; superiors tipped with bright yellow at apex, and the apex of a ventral spine also tipped with the same colour; seen from above conical, with an inner ventral subapical spine which gives a bifid appearance to the apex of appendage; inferiors one-fourth longer than superiors, which are about as long as segment 10, tapered, very sinuous, and with the apical end turned strongly inward and downward. Penis broadly dilated at apex and rather deeply notched at the centre of this border, the end curling strongly over the stem of the organ and ending in two broad lobes which embrace the stem closely on either side.

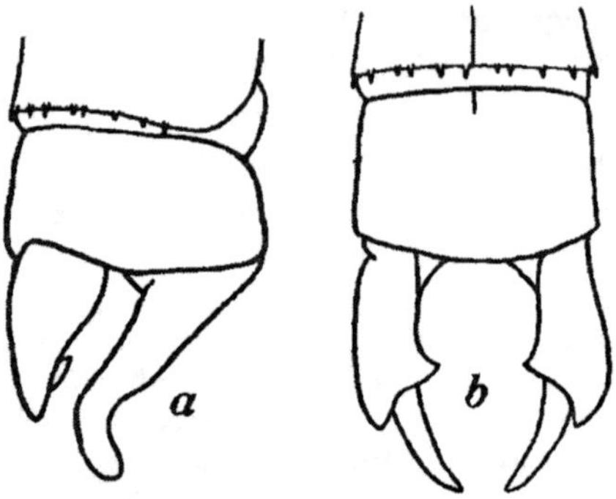

Fig. 75.—Anal appendages of *Cœliccia loogali* Laid., male. *a.* Right lateral view. *b.* Dorsal view.

Female.—Abdomen 40–41 mm. Hind-wing 28 mm.

Differs in but few respects from the male, much less than in other species of the genus. The labrum broadly bordered with yellow in all but very old specimens; the outer spots on the postclypeus yellow and very small or almost obsolete in old specimens; the blue of the genæ extending inwards across frons to a greater degree; a narrow zigzag transverse stripe across the vertex made up of six small blue spots behind the antennæ and linking up the ocelli in a chain; blue beneath head more extensive; *prothorax* and *thorax* similar to the male, but the posterior lobe of former reduced to a tiny median point and a small outer isolated lobe; *wings* entirely similar to the male, but palely enfumed in adults. *Abdomen* similar to the male except for the last three segments; segment 8 with a broad apical yellow annule covering slightly more than one-third the length of segment and confluent below, with a yellow border; segment 9 dark reddish-brown, with the apical border yellow or ochreous, gradually merging into the darker colour; segment 10 very short, black. *Anal appendages* black, conical, rather shorter than segment 10; vulvar scales robust, yellow tipped with black, extending a little beyond end of abdomen.

Distribution.—Maymyo, BURMA, during June and July.

A number of males and three females in the author's collection. *Type* and paratype female deposited in the British Museum collection. The species is easily distinguished from others by the curved antehumeral stripes on dorsum of thorax, by the lateral blue spots on the postclypeus, by the very black abdomen of the male, almost unrelieved by markings, especially the terminal segments, and by the black anal appendages tipped with bright yellow spots. The female is determined by the lateral spots on the postclypeus and the broad terminal rings of yellow on segments 8 and 9, followed by a black unmarked segment 10.

62. **Cœliccia poungyi** Fraser. (Fig. 76.)

Cœliccia poungyi Fras., J. Bombay Nat. Hist. Soc. vol. xxx, pp. 52, 53 (1926); Laid., Rec. Ind. Mus. vol. xxxiv, pp. 12, 24–25 (1932).

Male.—Abdomen 40–43 mm. Hind-wing 23–25 mm.

Head: labium black, white at base; labrum, bases of mandibles, and lower half of genæ, ante- and postclypeus glossy black; upper half of genæ palest blue, this colour encroaching slightly on frons, which is black; vertex and occiput mat black, the former with two oval spots, with their bases on the posterior ocelli and apex directed towards the bases of antennæ, the latter with an elongate pyriform creamy yellow spot on each side behind eyes; beneath head black, narrowly bluish against the eyes; eyes black above, olivaceous green at sides and below. *Prothorax* velvety black, the lower part of sides pale blue; posterior lobe rounded, simple. *Thorax* velvety black in the humeral area, pale azure blue on dorsum and laterally from about the middle of mesepimeron; dorsum almost entirely blue, the mid-dorsal carina finely black, as also the antealar sinus above and the small triangular area below, in which the carinal ridge ends; a small upper humeral spot sometimes present; laterally the postero-lateral suture heavily mapped out in black; beneath palest blue. *Legs* black; coxæ and trochanters posteriorly pale blue. *Wings* hyaline; pterostigma black or very dark brown, covering one cell, braced; 17 to 18 postnodal nervures in fore-wings, 16 in the hind; *Riv+v* arising close to *IRiii*, the former usually very slightly proximal to the subnode, but occasionally at its level, the latter generally slightly distal or rarely at the level of the subnode. *Abdomen* black; segment 1 with the apical half of its sides pale blue, extending up along the apical border and nearly confluent over dorsum; segment 2 with a narrow longitudinal stripe on each side at the ventral border, also a narrow linear stripe along the mid-dorsal carina, the latter pale creamy white; segment 3 with a pair of small subdorsal bluish spots at base and its mid-dorsal carina finely yellow, this latter continued on for a short distance on to segment 4; segments 9 and 10 bright chrome yellow. *Anal appendages* bright yellow; superiors as long as segment 10, narrow at base, broadening at middle and again slightly tapered to apex, which is bluntly conical, a small black-tipped ventral spine just beyond the middle and a smaller short spine near the base, the latter spine seen with difficulty; inferiors one-third longer than superiors, broad at base, then ungulate and very sinuous, blunt and rounded at apices, which are curved slightly inwards. Penis differing broadly from that of other species, very broad at apex, which is a little sinuous, the end curling strongly over and terminating in three processes, a median tongue-shaped lobe directed straight back over the stem of the organ, and two very thin, long, greatly curled branches, which curve down like a watch-spring on each side.

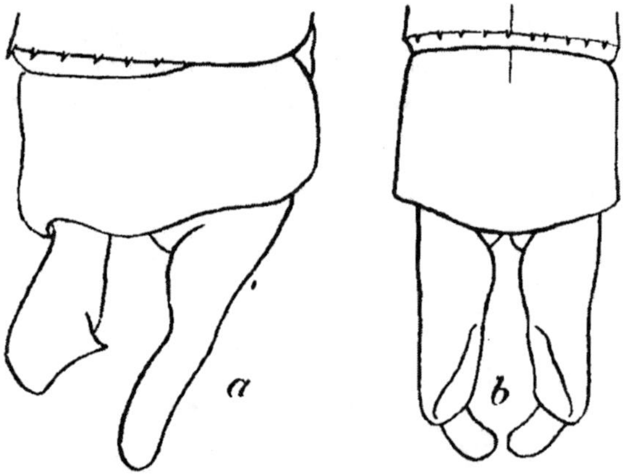

Fig. 76.—Anal appendages of *Cœliccia poungyi* Fras., male. *a*. Right lateral view. *b*. Dorsal view.

Female.—Abdomen 40 mm. Hind-wing 26 mm.

Differs rather broadly from the male, the blue markings replaced by bright ochre. Labium bright yellow; labrum yellow, with two large black basal spots narrowly joined across the middle; bases of mandibles, genæ, and a large spot on each side of postclypeus bright ochreous; the oval spots on vertex replaced by a continuous transverse and very sinuous bright yellow stripe which crosses the vertex between the ocelli; elongate spots behind eyes bright yellow, beneath eyes almost entirely bluish-green. *Prothorax* black, broadly yellow at the sides; posterior lobe almost obsolete, a mere median vestige left. *Thorax* black on dorsum to nearly as far as the antero-lateral suture, marked with a narrow bright yellow ante-humeral stripe on each side of dorsum; laterally and beneath yellow, with a short vestige of a narrow black stripe on the upper part of the postero-lateral suture. *Legs* yellow, posterior surfaces of femora and anterior of tibiæ black; tarsi black.

Wings hyaline; pterostigma black at centre, framed in pale yellow and thick black nervures; 19 postnodal nervures in fore-wings, 17 in the hind; *Riv+v* arising widely proximal to the subnode, *IRiii* in continuation of the subnode or very slightly distal to its level. *Abdomen* black, with a broad yellow stripe extending along its whole length, more extensively so on segment 1; segment 8 with rather more than its apical third bright ochreous, whilst segments 9 and 10 are entirely of this colour. *Anal appendages* short, conical, pointed, yellow, as also a conical protuberance beneath and between them.

Distribution.—UPPER BURMA.

The *type*, a male in the British Museum, is from Maymyo; paratype female and several males in the author's collection, taken during June and July. The species is determined from others by the whole of the dorsum of thorax being pale blue, whilst the female differs by having the whole of the under surface of the eyes pale greenish-blue and segments 8 to 10 yellow, with rather less than the basal two-thirds of the former black and the apical border unmarked.

63. **Cœliccia chromothorax** (Selys). (Fig. 77.)

Calicnemis chromothorax Selys, Ann. Mus. Civ. Genova, (2) vol. x (xxx), pp. 70–1 (1891); Laid., Rec. Ind. Mus. vol. xiii, p. 327 (1917); Fras., J. Bombay Nat. Hist. Soc. vol. xxix, p. 74 (1923).
Cœliccia chromothorax Laid., Rec. Ind. Mus. vol. xxxiv, pp. 12, 21–22 (1932).

Male.—Abdomen 42–43 mm. Hind-wing 26–29 mm.

Head: labium yellow, tipped with black; labrum, bases of mandibles, genæ, clypeus, vertex, and occiput black, the vertex with a small oval yellow spot lying between the posterior ocelli and antennæ on each side, the occiput with an elongate transverse pale yellow stripe on each side behind the eyes and a small point of yellow bordering the eyes beneath; eyes black on top, olivaceous laterally and beneath. *Prothorax* black, sides of middle and anterior lobes broadly yellow; posterior lobe rounded, simple. *Thorax* bright sulphur-yellow on dorsum, the mid-dorsal carina finely black, as also the antealar sinus and the small mesothoracic triangle below at end of carinal ridge; the humeral area broadly velvety black as far as the antero-lateral suture and with a small yellow point above just behind the upper end of humeral suture; laterally and beneath pale yellow, with a narrow black stripe mapping out the postero-lateral suture, but incomplete below. *Legs* yellow, with extensor surfaces of femora and flexor surfaces of tibiæ black as well as tarsi. *Wings* hyaline, palely enfumed in adults; pterostigma black, very finely framed in yellow, covering rather less than two cells, braced; *Riv+v* arising variably, often fused with *IRiii* or arising well proximal to it slightly before or at the level of subnode ; *IRiii* arising at or slightly distal to the level of subnode ; 19 to 22 postnodal nervures in fore-wings, 18 to 22 in the hind ; origin of anal bridge variable, arising at or a short distance proximal to *ac*. *Abdomen* black, the ventral border yellow as far as segment 6, broadly so on segment 1, where it passes up on dorsum and is narrowly confluent at the apical border ; segment 2 with its mid-dorsal carina finely yellow, as also segments 3 and 4 ; a pair of small bright yellow baso-dorsal spots on segments 3 to 5, that on 3 being confluent with the yellow on side ; the apical border of segment 9 and the whole of segment 10 bright chrome yellow. *Anal appendages* bright yellow ; superiors about as long as segment 10, broad at base, broad and shallowly notched at apex as seen in profile, obtuse at apex as seen from above, a small black-tipped spine

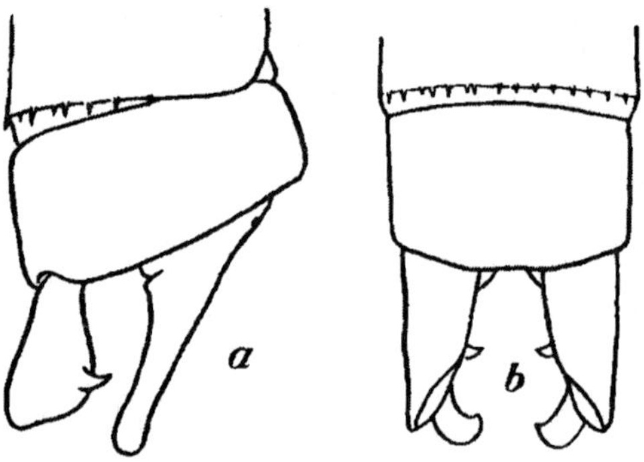

Fig. 77.—Anal appendages of *Cœliccia chromothorax* (Selys), male. *a*. Right lateral view. *b*. Dorsal view.

on the ventral inner surface about the middle of appendage ; inferiors broad at base, ungulate and very sinuous thereafter as far as apex, which is rounded as seen in profile, but curled strongly inwards and acute as seen from above. Penis with very blunt apex, its border concave, ending in two very narrow branches which curl strongly back to embrace the stem of the organ, and then turn abruptly back at ends, these branches springing from the centre of anterior border ; a robust spine on the dorsum of the stem of the organ lying within the arching apex.

Female.—Abdomen 40 mm. Hind-wing 28 mm.

Differs in several respects from the male, shorter and more robust ; bases of mandibles, genæ, anteclypeus except for a small spot on each side, and the anterior border of frons finely yellow ; 2nd joint of antennæ tipped with yellow ; oval spots on vertex replaced by a beaded transverse stripe traversing the vertex between the ocelli; elongate spots behind eyes larger; eyes bordered narrowly with yellow beneath. *Prothorax* very broadly yellow laterally; posterior lobe almost absent, merely a small black lobe on each side, the broad notch between bright yellow. *Thorax* black on dorsum, marked with a narrow bright yellow antehumeral stripe on each side which replaces the broad sulphur-yellow fascia seen in the male; laterally similar to the male. *Legs* and *wings* similar to the male; 20 to 21 postnodal nervures in fore-wings, 19 in the hind; *Riv+v* and *IRiii* arising a little proximal and a little distal to the subnode respectively; *ab* arising at level of *ac*. *Abdomen* black, with a yellow stripe running along the whole length of the ventral border; segment 1 entirely bright yellow save for a small dorsal black spot; segment 8 with a large apical dorsal spot covering nearly its apical half; segment 9 with a similar spot, but bifurcating into two lobes, which run slightly outwards and nearly as far as the base of the segment; segment 10 black, with the dorsum yellow, or more commonly a small round spot on each side of the mid-dorsal ridge at the middle of the segment. *Anal appendages* short, black, conical, pointed; vulvar scales robust, not extending quite to end of abdomen, yellow, blackish below.

Distribution.—UPPER BURMA. The *type* comes from Pwepoli and is in the Selysian collection. Occurs at Maymyo, Upper Burma, from June to August. Paratypes in the author's collection as well as the British Museum.

The male is easily distinguished from all others of the genus by having the whole of the dorsum of thorax bright sulphur-yellow; not infrequently it is found with the nervure *ab* arising proximal to the level of *ac*, thus differing from other species of the genus, and approximating to *Indocnemis* and *Calicnemis*; this character probably deceived Selys. The female is distinguished by the very distinctive yellow markings on the terminal abdominal segments.

64. **Cœliccia fraseri** Laidlaw. (Fig. 78.)

Cœliccia fraseri Laid., Rec. Ind. Mus. vol. xxxiv, pp. 11, 14–15 (1932).

Male.—Abdomen 37 mm. Hind-wing 23 mm.

Head: labium pale yellow; labrum dark brown, bordered with paler brown; bases of mandibles, genæ, anteclypeus, and the lateral part of frons pale blue; vertex and occiput and beneath eyes mat black, the former with a pair of round bluish spots lying in the triangle formed by the ocelli, and an oval spot on each side between the ocelli and base of antennæ; occiput with a vestigial pale yellow oblong spot on each side behind the eyes; the latter black above, olivaceous at the sides and below. *Prothorax* black, broadly blue or yellow laterally; posterior lobe rounded, simple. *Thorax* black on dorsum as far laterally as the anterior half of the mesepimeron, marked by very narrow dorsal blue or yellow stripes extending the whole length of dorsum, close to and parallel with the mid-dorsal carina; these stripes blue only in old adults and entirely absent in very old specimens; laterally and beneath pale azure blue, with a narrow black stripe mapping out the postero-lateral suture, incomplete below; a small spot, sometimes obsolete, on the upper part of humeral suture. *Legs* yellow, posterior surfaces of femora and the tarsi black. *Wings* hyaline: pterostigma reddish-brown framed in pale yellow and thick black nervures, covering $1\frac{1}{3}$ cells, braced; 16 to 17 postnodal nervures to fore-wings, 15 in the hind; *Riv+v* arising at a variable distance proximal

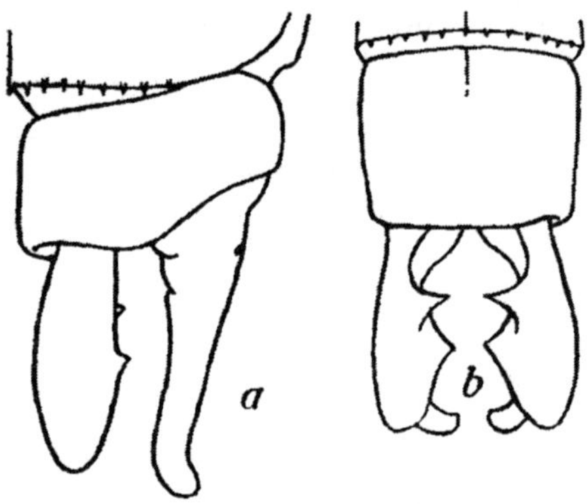

Fig. 78.—Anal appendages of *Cœliccia fraseri* Laid., male. *a.* Right lateral view. *b.* Dorsal view.

to subnode; *IRiii* arising at or a little distal to the level of subnode. *Abdomen* blackish-brown on dorsum, bluish near base and yellow elsewhere on the sides; segments 3 to 7 with small paired basal dorsal spots confluent with the yellow on the sides and beneath; segments 7 to 9 black on dorsum; segment 10 black, changing to reddish-brown and then broadly yellow at the sides (some specimens with the apical border of segment 9 broadly pale yellow, the base black, this extending as two dorsal points apically; segment 10 in the same specimens wholly pale yellow). *Anal appendages* pale yellow, inferiors black on dorsum and at apices; superiors narrow at base and hollowed out as seen from above, then broadening and club-shaped, very obtuse at apex, and with two inner points or broad spines at their middles; inferiors broad at base and then tapered, and ending in an obtuse or rounded apex which is slightly curled inwards as seen from above; penis closely similar to that of *C. chromothorax*, but the branches forming a watch-spring spiral and not directed back at the ends.

Female.—Abdomen 33 mm. Hind-wing 24 mm.

Closely similar to the male, far more so than is usual in the genus; differs as follows:—Labrum golden yellow, marked with a tiny black point at the middle line of base; blue of genæ almost meeting across the frons; spots on vertex confluent to form a complete slightly beaded stripe extending from eye to eye between the ocelli. *Prothorax* bright yellow, anterior and posterior lobes deep black, as also a fine mid-dorsal longitudinal line on the middle lobe; posterior lobe trilobate, projecting markedly over front of thorax. *Wings* hyaline; pterostigma ochreous, the central portion brownish; 17 to 18 postnodal nervures to fore-wings, 15 to 16 in the hind; *Riv+v* and *IRiii* arising variably as in the male. *Abdomen* similar, but segment 8 with a broad apical yellow annule occupying one-third of its length; segment 9 variable, dark brown or obscurely marked with yellow on dorsum and apical border; segment 10 reddish-brown, as also the *anal appendages*, which are short, conical, and pointed.

Distribution.—Khasia Hills, ASSAM. The *type*, in the British Museum, comes from Shillong, 5,000 ft. The species is stated by Mr. T. Bainbrigge Fletcher to be common in jungle near streams during the months of June and July, extending into August.

Distinguished from other species by the parallel blue or yellow lines on the dorsum of thorax lying in close apposition to the mid-dorsal carina, but, as mentioned above, these may be absent in very old specimens, when the determination must be made by the shape of the superior anal appendages, which are quite characteristic. The female invariably has the paired mid-dorsal thoracic stripes, even in the full adult, and these will serve to distinguish it from other females of the genus.

Genus **CALICNEMIS** Selys (1863). (Fig. 79.)

Calicnemis Selys, Bull. Acad. Belg. (2) vol. xvi, p. 159 (1863); id., Mém. Cour. vol. xxxviii, p. 130 (1886); Kirby, Cat. Odon. p. 131 (1890); Selys, Ann. Mus. Civ. Genova, (2) vol. x (xxx), p. 502 (1891); Laid., Rec. Ind. Mus. vol. xiii, pp. 322, 326 (1917); Munz, Amer. Ent. Soc. Mem. no. 3, p. 54 (1919); Kennedy, J. Sci. Ohio, vol. xxi, no. 1, p. 27 (1920); Fras., J. Bombay Nat. Hist. Soc. vol. xxix, p. 745 (1923); id., ibid. vol. xxxv, p. 646 (1932).

Zygopterous dragonflies of rather small size and of moderately slender build, with abdomen less than twice the length of hind-wings; coloured red or bright chrome yellow marked with black, or more rarely black partially obscured by pruinescence. Wings hyaline, rounded at apices, never falcate, petiole moderately long; discoidal cell elongate, costal side about three-fourths the length of posterior, distal side at least half as long again as the basal, and very oblique, so that the cell is acutely pointed at its lower distal angle; sectors of arc arising from lower end of arc and divergent from origin; no accessory basal postcostal nervure present; the nervure *ac* situate slightly nearer to the level of the distal antenodal nervure; *ab* always present and complete, continued on as *IA*, which latter extends for about three-fourths the length of wings; petiolation beginning well proximal to the level of *ac*; *Cuii* extending for a short distance beyond the end of *IA*, 20 cells or more in length; *Riv+v* arising well proximal to the level of subnode, *IRiii* a little distal; pterostigma slightly variable in the species, subquadrate, diamond-shaped, braced; cells of wings mainly quadrangular; 3 cells between the discoidal cell and nervure descending from subnode.

Head narrow; eyes relatively smaller than in species of *Platysticta*; occiput simple; posterior lobe of prothorax

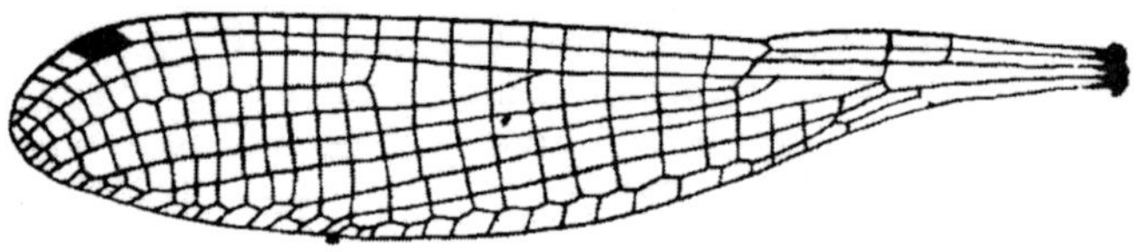

Fig. 79.—Wing of *Calicnemis miniata* Selys, male.

rounded and simple in both sexes; abdomen cylindrical, moderately robust, and short as compared with species of *Cœliccia*; anal appendages very homogeneous, superiors nearly as long as inferiors, ending in two long branches, one of which simulates a long robust sub-basal ventral spine, the other directed straight back to form the apex of appendage; inferiors long, tapered, sinuous, closely resembling those of *Cœliccia*. Genitalia: lamina narrowly and deeply cleft; hamules broad quadrate plates meeting broadly across the middle line; lobe flask-shaped, deeply grooved on its ventral surface; penis closely similar to that of *Cœliccia*, the apex dividing into two long, curling branches which curve down to embrace the stem of the organ; vulvar scales short and robust.

Genotype, *Calicnemis eximia* Selys.

Distribution.—The genus is almost entirely an Indian one, most species being confined to N.E. India and Burma, but one or more extending into Indo-China.

The genus is closely related to *Cœliccia* and *Indocnemis*, differing from the former by the shorter petiolation and broader and rounded apices of wings, and from the latter by its broader and shorter wings, by possessing only 3 cells between the discoidal cell and nervure descending from the subnode, and by the more open reticulation of the wings.

Key to the Species of Calicnemis.

Males.

1.	Beneath head entirely black	2.
	Head black, marked with yellow beneath.	3.
2.	Head and abdomen black, without markings	*pulverulans* Selys, [p. 175.
	Head black, with citron-yellow markings; abdomen with segment 1 yellow, segments 2 to 6 red, 7 to 10 black	*erythromelas* Selys, [p. 181.
3.	Abdomen bright vermilion-red, without markings; a small but conspicuous upper posthumeral spot on thorax; face and head as far back as posterior ocellus red, without markings.........	*eximia* Selys, p. 174.
	Abdomen black, more or less marked with red or ferruginous; posthumeral yellow spot absent or very tiny; front of head black or black traversed with red or citron-yellow bands	4.
4.	Abdomen black, with segments 1 and 2 and base of 3 brick-red or bright ochreous; front of head and face black.	*mortoni* Laid., p. 183.
	Abdomen black from segment 6 to the end, basal segments red; not more than 14 to 15 postnodal nervures in fore-wings.	*miles* Laid., p. 178.
	Abdomen red, with black markings on the three end segments only; 19 to 20 postnodal nervures to fore-wings..........	*miniata* Selys, p. 177.

Females.

1.	Beneath head entirely black	2.
	Beneath head black marked with yellow .	3.
2.	Abdomen robust, black, dorsum of segments 1 and 2 with yellow markings only; labrum black, broadly bordered with yellow	*pulverulans* Selys, [p. 176.
	Abdomen slim, yellow from segments 1 to 6, black from 7 to 10; labrum entirely black	*erythromelas* Selys, [p. 182.
3.	Not more than 11 to 13 postnodal nervures in fore-wings	*miles* Laid., p. 180.
	Not less than 17 to 19 postnodal nervures in fore-wings	4.
4.	Head broadly yellow beneath; a small but conspicuous upper posthumeral yellow spot present..................	*eximia* Selys, p. 174.
	Head black beneath, marked with a small wedge-shaped yellow spot; posthumeral yellow spot absent	5.
5.	Labrum yellow, with a large medio-basal black spot	*mortoni* Laid., p. 183.
	Labrum yellow, unmarked	*miniata* Selys, p. 177.

65. **Calicnemis eximia** Selys.

Calicnemis eximia Selys, Bull. Acad. Belg. (2) vol. xvi, p. 160 (1863); id., Mém. Cour. vol. xxxviii, p. 131 (1886); Kirby, Cat. Odon. p. 131 (1890); Selys, Ann. Mus. Civ. Genova, (2) vol. x (xxx), pp. 502–7 (1891); Mart., Mission Pavie (Neurop.), p. 18 (1904); Laid., Rec. Ind. Mus. vol. xiii, pp. 327, 330, 331, 338; (1917); Fras., J. Bombay Nat. Hist. Soc. vol, xxix, p. 745 (1923); Fras., J. As. Soc. Bengal, n. s. vol. xix, p. 462 (1923).

Calicnemis atkinsoni Selys (*eximia* ♀, nec *atkinsoni*), Mém. Cour. vol. xxxviii, p. 131 (1886); id., Ann. Mus. Civ. Genova, (2) vol. x (xxx), pp. 503–5 (1891); Laid., l. c. p. 327 (1917).

Male.—Abdomen 30–34 mm. Hind-wing 21–23 mm.

Head: labium pale ochreous; labrum, bases of mandibles, genæ, and upper surface of head as far backwards as the level of posterior ocelli bright vermilion-red, including antennæ; occiput velvety black, with a pale yellow streak behind each eye; beneath occiput black, with a broad quadrate patch pale ochre against the eye; the latter red above, paling to yellow beneath. *Prothorax* velvety black on dorsum, bright yellow laterally; posterior lobe rounded, simple, black. *Thorax* black on dorsum as far as the middle of mesepimeron laterally, marked with a pair of bright citron-yellow or bright vermilion antehumeral stripes with straight inner and curved outer border, extending into antealar sinus; a small yellow spot just behind the upper end of humeral suture; laterally bright yellow, with a narrow black oblique stripe mapping out the postero-lateral suture, tapering to a point below; beneath yellow. *Wings* hyaline; pterostigma reddish-brown, framed finely in yellow, braced, slightly longer than broad; 14 to 17 postnodal nervures to fore-wings, 13 in the hind; 3 cells between the nervure descending from subnode and outer end of discoidal cell in fore-wing, 2 to 3 in the hind. *Legs* vermilion, unmarked. *Abdomen* bright vermilion-red throughout, but becoming a little darker or brownish in old specimens, entirely without markings. *Anal appendages* bright yellow, slightly longer than segment 10, and of equal length; superiors of the generic shape, acute at apex, and with a long ventral sub-basal spine; inferiors a little sinuous, apex turned abruptly downwards.

Female.—Abdomen 28–30 mm. Hind-wing 23–25 mm.

Differs in several respects from the male; a more robust insect, the wings extending almost to end of abdomen; ground-colour bright citron-yellow instead of vermilion. *Head* differing from the male by a black band traversing the vertex and frons; *wings* rather broad, 17 to 18 postnodal nervures to fore-wings, 16 in the hind; pterostigma of the same colour as the male, usually broader, about one-fourth as long as broad, but variable in this respect, even in the wings of individual specimens. *Legs* yellow or ochreous, femora black on the extensor surface, especially at distal ends. *Abdomen* dark enfumed ochreous, changing to black on the terminal segments, segment 1 yellow, with a small wedge of dark colour at base; segment 2 with the mid-dorsal carina narrowly or broadly black, according to age, or blackish-brown in very old specimens; some obscure apical paired spots on segments 7 to 9 in some specimens, but these obscured in the fully adult. *Anal appendages* yellow or dark brown, conical, short, pointed; vulvar scales robust, extending to end of abdomen, yellow.

Distribution.—N.E. INDIA, BENGAL, SIKKIM, and ASSAM. I found this species very common at Mangpu, Darjeeling District, during May and June. It breeds in small brooks flowing through marshes, has a weak and short flight, and is usually gregarious.

Type probably in the Selysian collection; paratypes in most museums, including the Indian, Pusa and British. The generalized and vivid red colour of this insect serves to determine it at once from all other species of the genus. The female bears a close resemblance to those of *C. miniata* and *C. pulverulans*, and is distinguished from the former by being broadly yellow beneath the head instead of almost wholly black, whilst from *C. pulverulans* the same character, as well as the presence of the upper humeral yellow spot, will serve to separate it.

66. **Callcnemis pulverulans** Selys.

Calicnemis pulverulans Selys, Mém. Cour. vol. xxxviii, p. 133 (1886); Kirby, Cat. Odon. p. 131 (1890); Laid., Rec. Ind. Mus. vol. xiii, pp. 322, 329, 331 (1917); Fras., J. Bombay Nat. Hist. Soc. vol. xxix, p. 745 (1923).

Male.—Abdomen 31 mm. Hind-wing 25 mm.

Head: labium white, lateral palps black; labrum blackish-brown, broadly bordered with yellow; bases of mandibles, genæ, and clypeus glossy black; rest of head mat black, pruinosed in part as follows:—A broad fascia including the basal joints of antennæ extending from eye to eye from front of frons to a line drawn through the interocellar space, after which is a broad black belt extending from eye to eye, followed by a second and narrower belt of pruinescence across occiput dilated at both ends, and through which pale yellow stripes may occasionally be seen; beneath eyes black, unmarked. *Prothorax* black, the middle lobe pruinosed purplish, the posterior lobe black, elongate, rounded, projecting back over front of thorax. *Thorax* black, but appearing blue from pruinescence except the dorsum, which is black as far back as the middle of mesepimeron, and encloses narrow ante-humeral pruinosed blue stripes; laterally the whole of the side bluish except the postero-lateral suture, which does not share in the pruinescence; rounded yellow spots on the anterior parts of mes- and metepimeron; beneath black.

Legs and *abdomen* black, thinly pruinosed, and thus appearing bluish, unmarked. *Wings* hyaline; pterostigma dark reddish-brown framed in thick black nervures, covering a little more than one cell, braced; 18 postnodal nervures in fore-wings, 16 in the hind; 3 cells between the discoidal cell and nervure descending from subnode. *Anal appendages* black, inferiors pale yellow at base and apex; closely similar to those of *C. eximia*, superiors slightly longer than segment 10, inferiors slightly longer than superiors, the apex turned down and a little inwards, approaching a right angle.

Teneral males approximate in colour and markings to the female, and it is to be noted that, in the fully adult stage, all pale markings become obscured by dense pruinescence, thus perpetuating the markings in a different colour.

Female.—Abdomen 30 mm. Hind-wing 25–26 mm.

Resembles the young male and differs markedly from the adult stage of that sex. Labium and labrum similar to the male; bases of mandibles, genæ, and a narrow wavy stripe on vertex pale yellow, the latter marking shaped like a cupid's bow and extending from the base of one antenna to the other only; a pair of pale yellow postocular stripes on back of occiput similar to those occasionally seen in the male; *beneath head entirely black*; eyes black above, dark olivaceous brown below. *Prothorax* black, with a large hook-shaped pale yellow spot on each side of middle lobe; posterior lobe very small, rounded, black. *Thorax* black on dorsum as far back as the middle of mesepimeron, with a narrow antehumeral pale yellow stripe on each side of dorsum; laterally pale bluish-green or pale greenish-yellow with a moderately broad black stripe on the postero-lateral suture; beneath pale yellow. *Legs* yellow, the extensor surfaces of all femora and the flexor surfaces of anterior pair of tibiæ black, as also the tarsi. *Wings* hyaline; pterostigma blackish-brown, nearly half as long again as broad, covering 1½ cells; 16 postnodal nervures in fore-wings, 14 only in the hind. *Abdomen* deep black throughout except for segment 1, which is pale yellow with a small basal dorsal black spot, and segment 2, which is mottled or striped with yellow laterally; segments 3 and 4 have also small vestigial baso-dorsal paired yellow spots. *Anal appendages* black, conical, as long as segment 10; vulvar scales robust, extending to end of abdomen, dark brown; the abdomen very heavily built in this species.

Distribution.—N. BENGAL and SIKKIM only, at about 4,000 to 6,000 ft. Moderately common in the hills below Darjeeling from May to September.

The male is easily distinguished from all others by its extreme melanism, relieved only by extensive pruinescence. The female, more difficult in this respect, is determined by having the head entirely black beneath, the abdomen black beneath as well as on the dorsum, and by the absence of the small humeral yellow spot present in most species.

Type, an incomplete male in the Selysian collection ; allotype female in the Indian Museum collection ; paratypes in the British Museum and many private collections.

67. **Calicnemis miniata** Selys. (Fig. 79.)

Calicnemis miniata Selys, Mém. Cour. vol. xxxviii, p. 132 (1886) ; Kirby, Cat. Odon. p. 131 (1890) ; Selys, Ann. Mus. Civ. Genova, (2) vol. x (xxx), p. 504 (1891); Laid., Rec. Ind. Mus. vol. xiii, pp. 322, 328, 329, 331 (1917) ; Fras., J. Bombay Nat. Hist. Soc. vol. xxix, p. 745 (1923).

Calicnemis eximia Selys (*C. miniata* ♀, nec *C. eximia*), Bull. Acad. Belg. (2) vol. xvi, p. 160 (1886).

Male.—Abdomen 30–32 mm. Hind-wing 25–26 mm.

Head : labium pale brown ; labrum, bases of mandibles, genæ, frons, clypeus, and vertex as far back as the level of posterior ocelli bright brick-red, the latter traversed from eye to eye at the level of the antennæ by a narrow black stripe ; occiput black, marked by a narrow pale yellow postocular stripe on each side ; beneath head black, with a very small wedge of pale yellow against eyes ; the latter dark brown above, reddish below. *Prothorax* black, with a large reddish spot on each side of middle lobe ; posterior lobe rounded, simple, black. *Thorax* black, the dorsum bearing narrow bright red antehumeral stripes, broadening somewhat below, the sides with a narrow yellow or ferruginous stripe in the region of the spiracle and a yellow stripe on the lower half of border of metepimeron, these stripes in adults appearing purplish from overlying thin pruinescence ; beneath pale yellow clouded with olivaceous anteriorly. *Legs* dark reddish-brown, flexor surfaces of tibiæ ferruginous, trochanters and coxæ yellow, as also extreme proximal ends of femora. *Wings* hyaline ; 19 to 20 postnodal nervures in fore-wings, 17 to 18 in the hind ; pterostigma blackish-brown, finely framed in yellow and thick black nervures, markedly diamond-shaped and very acute at both ends, braced, covering one cell ; 3 or 4 cells between the discoidal cell and nervure descending from subnode in fore-wings, 3 in the hind. *Abdomen* blood-red as far as the basal third of segment 7, remaining segments black ; segments 5 and 6 with fine apical black rings. *Anal appendages* black, shaped very similarly to those of *C. eximia*, but the inferiors considerably longer than superiors, very sinuous, and with apices sloping inwards and meeting in the middle line.

Female.—Abdomen 27–29 mm. Hind-wing 23–24 mm.

Differs in a few respects from the male, the ground-colour or markings bright citron-yellow on head and thorax, ferruginous on abdomen. *Head* with the transverse black stripe on vertex much broader, extending forwards on to frons, involving the base of postclypeus more or less, rarely the whole of clypeus black; the stripe enclosing a pale yellow, broadly zigzagged one, often broken into two spots and incomplete on either side, situated at the level of the anterior ocellus; markings on occiput and beneath head similar to those of the male. *Prothorax* and *thorax* with similar markings, but the latter with sides bright citron-yellow, marked with a narrow oblique black stripe on the postero-lateral suture. *Legs* yellow, hinder surface of all femora black and distal ends of anterior pair brownish; tibiæ yellow on flexor surface, tarsi black. *Wings* with 17 to 19 postnodal nervures in the fore, 15 to 16 in the hind; pterostigma longer and less diamond-shaped than in the male, covering rather more than one cell, coloured similarly to the male; 2 to 3 cells between discoidal cell and level of subnode in hind-wings, 3 to 3½ in the fore. *Abdomen* much stouter than in the male and dilated at end, ferruginous as far as the middle third of segment 6, black thereafter as far as end, except laterally; segments 3 to 6 with fine apical black rings; vulvar scales robust, extending to slightly beyond end of abdomen, ferruginous. *Anal appendages* black, very minute, pointed at apices.

Distribution.—BENGAL and SIKKIM and probably NEPAL and the borders of Tibet. I found it moderately common during May and June at Mangpu in the Teesta Valley, Darjeeling District, 4,000 ft., and have specimens from Gangtok and Pashoke.

The male is distinguished by having the whole of the face and upper surface of the head bright red, except for a narrow black stripe traversing vertex, and the abdomen bright blood-red except for the last four segments, which are black. The female is determined with greater difficulty, but may be known by the following complex:—Labrum wholly yellow, superior posthumeral yellow spot absent, or very tiny and inconspicuous, and by the head being black beneath with but a small wedge-shaped yellow spot against the eyes, which is almost obsolete in adults.

68. **Calicnemis miles** Laidlaw. (Fig. 80.)

Calicnemis eximia race *atkinsoni* Selys (*C. eximia* ♀, nec *atkinsoni*), Mém. Cour. vol. xxxviii, pp. 131, 132 (1886).

Calicnemis atkinsoni Selys, Ann. Mus. Civ. Genova, (2) vol. x (xxx), pp. 503–5 (1891).

Calicnemis miles Laid., nom. nov., Rec. Ind. Mus. vol. xiii, p. 330 (1917); Fras., J. Bombay Nat. Hist. Soc. vol. xxix, p. 745 (1923).

Male.—Abdomen 33 mm. Hind-wing 25 mm.

Head: labium pale brown; labrum, bases of mandibles, genæ, clypeus, anterior border of frons narrowly, and the vertex bright brick-red, the latter traversed by a broad black stripe extending from eye to eye at the level of a line drawn through the ocellar space, and forwards at its middle two-fourths nearly to anterior border of frons; antennæ black; occiput and under surface of head black, the former with a short postocular pale yellow stripe and the latter with a diffuse pale greenish-yellow stripe extending in from border of eyes; eyes reddish-brown above, paler below. *Prothorax* black, with a small yellow spot on each side of middle lobe; posterior lobe rounded, simple, black. *Thorax* black on dorsum as far back as the middle of mesepimeron, marked with narrow bright brick-red antehumeral stripes of almost even thickness throughout; no upper humeral spot; laterally pale yellow, with a broad black stripe overlying the postero-lateral suture; beneath pale yellow. *Legs* ferruginous, posterior surfaces of femora darker, flexor surfaces of tibiæ

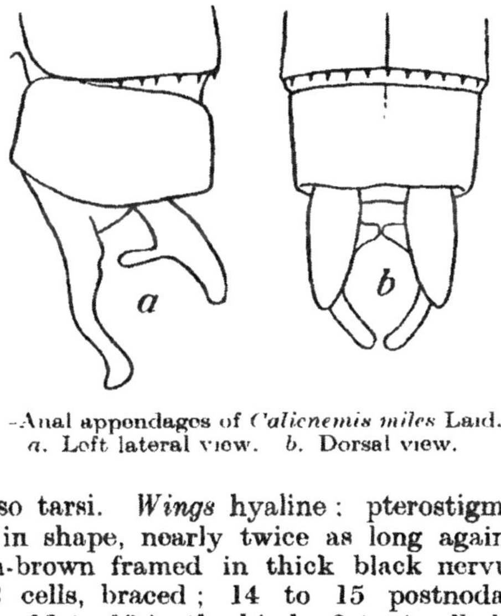

Fig. 80.—Anal appendages of *Calicnemis miles* Laid., male.
a. Left lateral view. *b*. Dorsal view.

black, as also tarsi. *Wings* hyaline: pterostigma elongate, rhomboidal in shape, nearly twice as long again as broad, pale reddish-brown framed in thick black nervures, covering $1\frac{1}{2}$ to 2 cells, braced; 14 to 15 postnodal nervures to fore-wings, 12 to 13 in the hind; 3 to 4 cells between the discoidal cell and nervure descending from subnode in fore-wings, 3 in the hind. *Abdomen* blood-red except for segment 1, which is pale greenish-yellow with a large brown spot on dorsum, and segments 8 to 10; segment 8 with a narrow lateral black stripe; 9 with a large subdorsal black spot at its base extending nearly to apical border, and 10 with a similar, but much shorter, black spot bordering its base; all segments with narrow black apical annules, broadening on segment 7, narrow on 8 to 10. *Anal appendages* dark reddish-brown, the inferiors paler at base; superiors about as long as segment

10 and markedly shorter than inferiors, shaped as for the genotype, the inferiors with the apices turned down abruptly and rather sinuous.

Female.—Abdomen 31 mm. Hind-wing 25 mm.

Differs markedly from the male as follows :—Labium bright yellow ; labrum with a tiny black point at the centre of its base, the *head* otherwise similar to the male, save that the markings are bright citron-yellow instead of brick-red, and the yellow spot beneath head is quadrate and larger. *Prothorax* and *thorax* with the ground-colour on sides and the antehumeral stripes bright citron-yellow, the lateral spot on prothorax larger, and the black stripe on sides of thorax narrower and incomplete below ; posterior lobe of prothorax similar in shape to that of the male. *Legs* bright yellow, only the femora marked with black posteriorly. *Wings* very similar to the male, the pterostigma paler, and with a fine frame of yellow lining the surrounding black nervures within ; 13 postnodal nervures to fore-wings, 11 in the hind ; 3 cells between discoidal cell and the nervure descending from subnode. *Abdomen* bright yellow, clouded with brown or reddish-brown on dorsum, becoming black on segments 8 to 10, more broadly so at base of 8 and 9 ; segment 1 with a small black dorsal spot at extreme base ; segments 2 to 7 with narrow bright yellow basal annules narrowly interrupted on mid-dorsum, and with the mid-dorsal carina blackish-brown and confluent on each segment, with a narrow black apical ring. *Anal appendages* very small, barely as long as segment 10, dark reddish-brown, conical, surmounting a yellow conical protuberance ; vulvar scale yellow, extending for a short distance beyond end of abdomen.

Distribution.—UPPER BURMA. The *type* comes from SIKKIM, and is presumably in the Selysian collection. It is a male named by Selys as *C. atkinsoni*, and thought by him to be the co-type of a female which he considered as belonging to a race of *C. eximia*. The female in question, however, turned out to be the female of *eximia* itself, hence a new name became necessary for the male specimen, which was called *C. miles* by Laidlaw.

The above descriptions have been made from a pair taken *in cop.* at Gokteik, Upper Burma, at the end of May, and now in my collection. The male closely resembles that of *C. miniata*, but is easily distinguished by the larger yellow spot beneath the head and by the red colour of the abdomen extending its whole length, with some black spots on the last three segments only. A ready means of distinguishing the two species, however, is the much lower postnodal index, 14 to 15 only in *C. miles*, as contrasted with 18 to 21 in *C. miniata*. Females are distinguished by the dorsal black markings being much reduced and restricted to the last three segments, the lower nodal index, and the very large quadrate pale yellow spot beneath the eyes. The elongate pterostigma in both sexes is also very characteristic when compared with the lozenge-shaped organ of *C. miniata*.

69. **Calicnemis erythromelas** Selys. (Fig. 81.)

Calicnemis erythromelas Selys, Ann. Mus. Civ. Genova, (2) vol. x (xxx), pp. 505, 506 (1891); Laid., Rec. Ind. Mus. vol. xiii, p. 331 (1917); Fras., J. Bombay Nat. Hist. Soc. vol. xxix, p. 745 (1923).

Male.—Abdomen 28 mm. Hind-wing 22·5 mm.

Head: labium with the middle lobe bright yellow, lateral lobes jet-black; labrum glossy black: bases of mandibles bright citron-yellow narrowly bordered with black; genæ bright citron-yellow, with a large central black spot confluent on its inner side with the clypeus which is also black; frons and vertex bright citron-yellow, with a very irregular crenated

Fig. 81.—Head of *Calicnemis erythromelas* Selys, male, showing markings.

bar of velvety black traversing the latter from eye to eye at level of antennæ, the yellow bar behind it very narrowly interrupted in the middle line; occiput and beneath head mat black, the occiput with narrow postocular citron-yellow stripes on each side; no markings beneath; eyes black above, olivaceous below. *Prothorax* velvety black, with a large citron-yellow spot on the subdorsum of middle lobe and another on each side; posterior lobe rounded, simple. *Thorax* velvety black on dorsum as far lateral as the middle of mesepimeron, and marked with narrow, bright, citron-yellow antehumeral stripes on each side; laterally yellow to greenish-yellow, with a moderately broad, complete, black, oblique stripe overlying the postero-lateral suture; tergum spotted with yellow, beneath palest yellow. *Legs* black, the inner surfaces of femora and outer surfaces of tibiæ yellowish. *Wings* hyaline; pterostigma dark reddish-brown finely framed in yellow and thick black nervures, inner border straight, outer oblique, nearly twice as long as broad, poorly braced, covering 1½ cells; 18 to 20 postnodal nervures in fore-wings, 17 in the hind; 3 cells between the discoidal cell and nervure descending from subnode; petiolation beginning much nearer the anal crossing than in other species, especially in the hind-wings. *Abdomen* bright blood-red from segments 2 to 6; segment 1 citron-yellow, with a large black dorsal spot extending to apical border; segments 7 to 10 entirely black, the others very finely ringed with black at apical end. *Anal appendages* black; superiors acutely pointed at apex, which is bevelled strongly within, and with a robust ventral sub-basal spine directed straight down, equal in length to segment 10; inferiors of the same length as superiors, curving outwards and then slightly inwards, apex with a minute tooth within.

Female.—Abdomen 30 mm. Hind-wing 23 mm.

Very similar to the male, of very slender build compared with other females of the genus, the abdomen, however, markedly dilated at segments 8 to 10. The stripe on vertex more extensive, and partially obscuring the yellow band behind it, so that this latter tends to become broken up into spots. *Prothorax*, *thorax*, and *legs* not differing in any respect from the male. *Wings* with 18 to 19 postnodal nervures in the fore, 16 to 17 in the hind; 2 or 3 cells between the discoidal cell and nervure descending from node in all wings; pterostigma similar to that of the male. *Abdomen* with segments 2 to 6 bright ochreous, tinted with red on dorsum; segment 1 and segments 7 to 10 yellow and black respectively, as in the male; the mid-dorsal carina of all segments from 2 to 6 narrowly black, but broader on the latter, that of the former with a short cross-bar near the apical border, and all with narrow apical black annules; vulvar scales robust, jet-black, extending to slightly beyond end of abdomen. *Anal appendages* black, shorter than segment 10, conical, pointed.

Distribution.—BURMA. *Type* in the Selysian collection, a male from Leito (? Leiktho, Toungoo Dist.); others from Copabo and Mt. Karen, May and June. The above descriptions are made from a pair in the author's collection from the Chin Hills, taken in November.

This species differs from others by its extreme slimness and narrower wings; it is quite the most beautiful of the genus. It is distinguished by the alternating bars of black and citron-yellow on the head, by the head unmarked beneath, by the isolated yellow spot on prothorax, by segment 7 being entirely black, by the shape of the pterostigma, and lastly, by the anal appendages of the same length. The female is at once distinguished by its graceful slimness, which contrasts so strongly with the robust build of other females in the genus. Other features are the head entirely black beneath, the alternating black and yellow bars on head, isolated spots on prothorax, and the striking coloration of the abdomen. The shape of the pterostigma appears to ally it to *C. miles*, also from Burma.

70. **Calicnemis mortoni** Laidlaw.

Calicnemis mortoni Laid., Rec. Ind. Mus. vol. xiii, pp. 526, 527 (1917); Fras., J. Bombay Nat. Hist. Soc. vol. xxix, p. 745 (1923).

Male.—Abdomen 33–36 mm. Hind-wing 26–29 mm.

Head: labium black, bases of palps bright yellow; labrum, bases of mandibles, genæ, and clypeus glossy black, rest of head mat black, the area between the ocelli and antennæ behind and the front border of frons in front purplish from a thin overlying pruinescence, a small oval ferruginous spot on each side lying between the posterior ocelli and roots of antennæ; a short, narrow, pale yellow postocular stripe on each side of occiput and a rather broad flame-shaped yellow fascia running inwards from the border of the eye on under side of head. *Prothorax* black, with a broad, thinly pruinosed purplish fascia on each side of middle lobe and two small ferruginous spots above this. *Thorax* black as far back as the antero-lateral suture, beyond which the sides are purplish black from thin pruinescence; a narrow slaty blue antehumeral stripe on each side of dorsum confluent with the lateral purplish area on prothorax; the anterior ends of mes- and metepimeron olivaceous yellow; beneath black, with a small geminate spot at the middle and a continuation of the yellow of anterior end of metepimeron crossing the anterior end. *Legs* black; *wings* hyaline; pterostigma blackish-brown, finely bordered with yellow and thick black nervures, covering 1½ to 2 cells, braced; 20 to 22 postnodal nervures to fore-wings, 16 to 20 in the hind. *Abdomen* black, segments 1 and 2 and the base of 3 with dorsum bright ochreous or brick-red, this colour extending partially round base of segment 2 as a narrow ring; segments 3 and 4 with obscure paired apical ferruginous spots. *Anal appendages* black, the inferiors tipped with yellow internally; superiors digitate, slightly longer than segment 10, directed straight back and furnished with an elongate tooth or spine near the base; inferiors much longer, broad at base, then cylindrical, tapering, slender and sinuous, the apices turned abruptly down at a right angle.

Female.—Abdomen 30 mm. Hind-wing 25 mm.

Differs considerably from the male as follows:—Labium wholly yellow: labrum yellow, with a large medio-basal black spot; anteclypeus yellow, with the outer ends and the postclypeus black; genæ, bases of mandibles, and frons bright yellow; a broad black stripe traversing the vertex at level of antennæ, narrowing to outer side of these organs and broadening forwards between them so as partially to overlap frons; this stripe followed posteriorly by a narrow yellow stripe at level of ocellar space and narrowly interrupted at its middle; occiput and under side of head black, with the usual postocular stripes and a broad quadrate yellow spot against the eyes beneath head; eyes dark brown above, paler and olivaceous below, the two areas sharply limited at the level of the black stripe crossing vertex. *Prothorax* black, middle lobe broadly yellow at sides; posterior lobe broadly rounded. *Thorax* black on dorsum as far lateral as the antero-lateral suture, marked with narrow citron-yellow antehumeral stripes; the sides yellow, with a broad black stripe lying along the anterior border of the postero-lateral suture incomplete below; beneath yellow. *Legs* black, coxæ and trochanters and the proximal two-thirds or three-fourths of flexor surfaces of femora bright yellow. *Wings* similar to male; pterostigma covering 2 cells, usually 2 rows of cells between the costa and radius after the end of pterostigma; 19 to 21 postnodal nervures in fore-wings, 18 to 19 in the hind. *Abdomen* black, segments 1 to 5 bright ochreous on dorsum, deepening to ferruginous and finally black at apical end of segment 5, the mid-dorsal carina finely black on all these segments, and narrow black apical annules on segments 2 to 5; segments 3 to 6 with small paired sub-dorsal pale yellow basal annules. *Anal appendages* short, conical, pointed, black; vulvar scales robust, extending to beyond end of abdomen, black, tipped with yellow.

Distribution.—Sikkim: Pashok, 5,500 ft., and Gangtok, 5,000 to 6,000 ft., during May and June.

The *type*, a male in the Indian Museum, comes from the former place, and has lost most of its markings and the whole of the pruinosed ones from greasing, which accounts for the differences between the original and the present descriptions, the latter having been made from a pair from Gangtok which have retained the original colours and markings in great perfection. Distinguished from other species by having the abdomen wholly black save for the dorsum of the basal three segments, which are brick-red. The female resembles that of *C. pulverulans* closely, but the higher nodal index and the yellow marking beneath head will serve to distinguish them. Related to *C. erythromelas* and *C. miniata* by the high nodal index, close venation, and dark colouring. Allotype female in the author's collection, as well as a paratype male.

Genus INDOCNEMIS Laidlaw (1917). (Fig. 82.)

Indocnemis Laid., Rec. Ind. Mus. vol. xiii, p. 325 (1917); Fras., J. Bombay Nat. Hist. Soc. vol. xxix, p. 745 (1923); id., ibid. vol. xxxv, p. 646 (1932).

Zygopterous dragonflies of rather large size and slender build, with abdomen less than twice the length of hind-wings; coloured black, marked with pale azure blue. Wings hyaline, subfalcate at apices, petiolation moderately long; discoidal cell elongate, costal side about three-quarters the length of posterior in fore-wing, four-fifths in the hind, distal side at least half as long again as the basal, so that the cell is acutely pointed at its lower distal angle and the distal side very oblique; sectors of arc arising from lower end of arc, the sectors divergent from origin; no accessory basal postcostal nervure present; the nervure *ac* situate midway between the two antenodal nervures; *ab* complete, always present, continued on as *IA*, which latter extends distally for more than half the length of wings; petiolation beginning slightly proximal to *ac*;

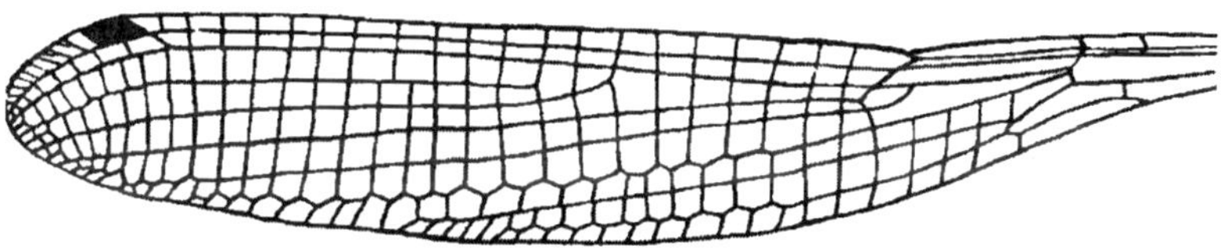

Fig. 82.—Wing of *Indocnemis kempi* Laid., male.

Cuii extending for a short distance beyond the end of *IA*, about 25 cells in length; *Riv+v* arising well proximal to the level of the nervure descending from subnode; *IRiii* arising in continuation of that nervure; pterostigma slightly longer than broad, proximal side very oblique, distal side much less so, strongly braced; 4 *cells between the discoidal cell and nervure descending from subnode.*

Head narrow; eyes relatively smaller than in species of *Platysticta*; occiput simple; posterior lobe of prothorax rounded and simple in the male (unknown in the female); abdomen cylindrical, long and slender, broadening somewhat towards the anal end; anal appendages closely similar to many species of *Cœliccia*, superiors blunt at apex, hollowed out within and with two inner spines, one subapical, the other sub-basal; inferiors considerably longer than superiors, cylindrical, broad at base, forcipate, the apices curling gently inwards.

Genotype, *Indocnemis kempi* Laidlaw.

Distribution.—Assam only; only one species is known to occur within Indian limits, from Cherrapunji.

The genus differs from *Cœliccia* by its shorter petiolation and by the larger number of cells between the discoidal cell and the nervure descending from the subnode, and from *Calicnemis* by the latter character, by the longer, narrower wings, and by the denser venation. Although it possesses intermediate characters, the falcate wings suggest a more archaic origin. In colouring and general facies it shows a remarkably close resemblance to species of the group *Indoneura*, but its appendages and venation are very different.

71. **Indocnemis kempi** Laidlaw. (Figs. 82 & 83.)

Indocnemis kempi Laid., Rec. Ind. Mus. vol. xiii, p. 325 (1917); Fras., J. Bombay Nat. Hist. Soc. vol. xxix, p. 745 (1923).

Male.—Abdomen 51 mm. Hind-wing 38 mm.

Head: labium whitish; labrum, bases of mandibles, and clypeus glossy black, the latter with a steely blue reflex, rest of head, including under surface, mat black: an obscure

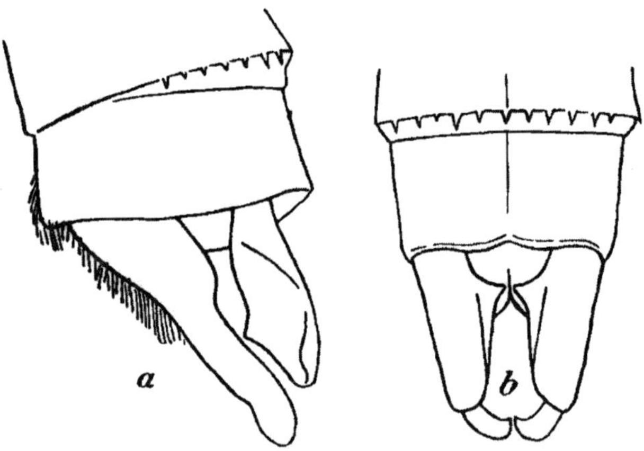

Fig. 83.—Anal appendages of *Indocnemis kempi* Laid., male. *a*. Left lateral view. *b*. Dorsal view.

bluish spot on each side between the posterior ocelli and bases of antennæ, and a narrow pale blue postocular stripe on each side of occiput. *Prothorax* and *thorax* black, the latter with narrow azure blue antehumeral stripes not extending as far up as the antealar sinus; the posterior halves of the mes- and metepimeron azure blue. *Legs* entirely black. *Wings* hyaline, the whole of the venation clouded with blackish-brown, the cell-middles clear; pterostigma black, nearly half as long again as broad, covering 1½ cells; 22 postnodal nervures in fore-wings, 18 to 20 in the hind. *Abdomen* black, except segment 10 and the dorsum of the apical half of 9, which are azure blue. *Anal appendages*: superiors slightly longer than segment 10, digitate as seen from above, and with the inner border near base prolonged inwards as a short robust spine, apex very obtuse; seen in profile deeply excavate below and within, the inner side prolonged into two robust spines, a sub-basal (also visible from above) and a subapical; inferiors bright ochreous tipped with black, about one-fourth longer than the superiors, broad at base, irregularly cylindrical, the apex curling inwards and pointed.

Distribution.—Known only from Cherrapunji, ASSAM, and seemingly a very rare and local insect.

Type in the Indian Museum, taken in December. The insect bears a very close resemblance to the larger species of the group *Indoneura*, e.g., *Phylloneura westermanni* and *Esme mudiensis*, both, however, from the Western Ghats, and to be distinguished by the shape of the anal appendages and venation.

Genus **PLATYCNEMIS** Charpentier (1840). (Fig. 84.)

Platycnemis Charp., Lib. Europ. p. 21 (1840); Ramb., Ins. Névrop. p. 239 (1842); Selys, Rev. Odon. p. 164 (1854); id., Bull. Acad. Belg. (2) vol. xvi, p. 162 (1863); id., Mém. Cour. vol. xxxviii, p. 127 (1886); Munz, Mem. Amer. Ent. Soc. no. 3, p. 54 (1919); Fras., J. Bombay Nat. Hist. Soc. vol. xxix, p. 744 (1923); id., ibid. vol. xxxv, p. 646 (1932).

Dragonflies of small size and rather slender build, with abdomen less than twice the length of wings: coloured creamy white or pale blue, marked with black or brown.

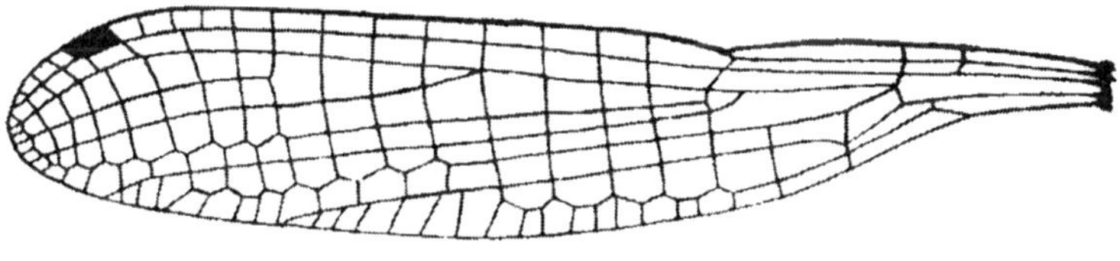

Fig. 84.—Wing of *Platycnemis latipes dealbata* Selys, male.

Wings hyaline, moderately rounded at apices, never falcate, petiolated to level of the proximal antenodal nervure; discoidal cell markedly elongate, with costal and posterior sides nearly equal in length, distal side about half as long again as basal, so that the outer end of cell is dilated and acutely angulated at its lower end; sectors of arc arising from lower end of arc and markedly divergent from origin: no accessory basal postcostal nervures present; the nervure *ac* situated about midway between the two antenodal nervures;

ab always present and complete, continued on as the nervure *IA*, which latter extends for about half the length of wings: petiolation beginning well proximal to *ac*; *Cuii* extending 5 or 6 cells beyond the end of *IA*, about 12 cells in length; *Riv+v* arising well proximal to the level of subnode, *IRiii* at the level of the nervure descending from it; pterostigma small, very oblique, diamond-shaped, braced, covering about one cell; cells of wings mainly quadrangular; 2 cells between the discoidal cell and nervure descending from subnode.

Head narrow; eyes smaller than in species of *Platysticta*: 3rd segment of antennæ as long as the 1st and 2nd together; occiput simple, deeply hollowed out; prothorax with posterior lobe simple, rounded, and without appendages in both sexes: thorax and abdomen moderately robust, the latter cylindrical and of about even thickness throughout; anal appendages rather homogeneous, superiors shorter than inferiors and with a small ventral spine; inferiors slightly sinuous, tapered gradually, and curved inwards nearly to meet at apices. Genitalia: lamina deeply and narrowly cleft; hamules as broad quadrate plates meeting across the middle line; penis with apical end curved up but not bifurcated, furnished at each side with a long, fine, curling branch directed somewhat backwards and outwards; lobe flask-shaped. Female with rather short vulvar scales, not extending as far as end of abdomen.

Legs with tibiæ broadly dilated in the two hinder pairs in both sexes; short stiff black hairs springing from the sides of the dilatations.

Genotype, *Platycnemis pennipes* Pall.

Distribution.—Throughout Europe and the British Isles, Asia Minor, West and Central Asia, Persia, N.W. INDIA, KASHMIR, and North Africa. Only one species found within Indian limits.

72. **Platycnemis latipes dealbata** Selys. (Figs. 84 & 85.)

Agrion dealbata Klug, MS. (1849).
Platycnemis syriaca Hagen, MS. (1849).
Platycnemis acutipennis var. *dealbata* Selys, Rev. Odon. p. 165 (1850).
Platycnemis latipes race *dealbata* Selys, Bull. Acad. Belg. (2) vol. xvi, p. 167 (1863); id., Mém. Cour. vol. xxxviii, p. 128 (1886); Morton, Trans. Ent. Soc. Lond. pt. ii, p. 306 (1907).
Platycnemis dealbata Selys, Rev. Odon., Add. et Corrig. p. 388 (1850); Kirby, Cat. Odon. p. 129 (1890).
Platycnemis latipes dealbata Morton, Ent. Month. Mag. (3) vol. vi, p. 84 (1920).

Male.—Abdomen 26 mm. Hind-wing 18 mm.

Head: labium pale bluish-white; labrum, bases of mandibles, genæ, and frons pale blue, limited posteriorly by a broad black stripe which traverses the vertex from eye to eye, broadening forwards on to frons at the middle, but narrowing to a point against the eyes; this stripe followed by a narrow pale blue one at level of middle ocellus, and again by a very broad black fascia limited behind by the posterior occipital line; beneath head pale bluish; eyes blue, capped with black; the blue ground-colour sometimes replaced by pale rusty red. *Prothorax* broadly black on dorsum, broadly blue on the sides, with a very irregular large black spot. *Thorax* pale blue on dorsum and sides, becoming yellowish below and beneath, or the whole suffused with pale ferruginous, marked with black as follows:—The dorsum broadly bronzed black, marked with narrow antehumeral pale yellow or bluish stripes, followed

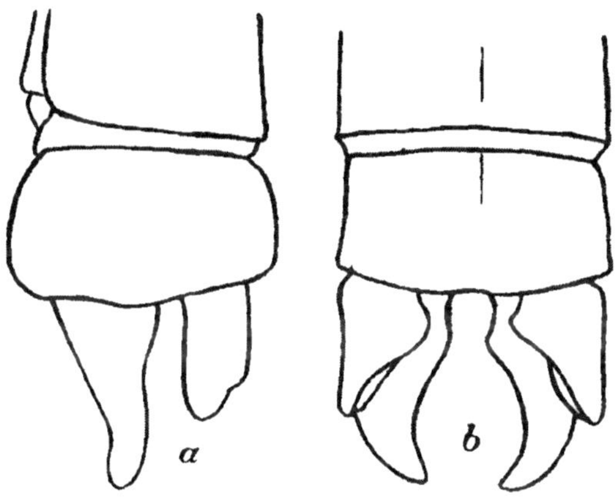

Fig. 85.—Anal appendages of *Platycnemis latipes dealbata* Selys, male.
a. Left lateral view. *b*. Dorsal view.

outwardly by two very narrow parallel black lines lying close together; the mid-dorsal carina and margins of antealar sinus very finely yellow; laterally a narrow, sometimes incomplete or interrupted black stripe on the postero-lateral suture. *Legs* bluish, the extensor surface of femora marked with a narrow black stripe limited on the hind pair to the distal half of limb; tibiæ unmarked, very broadly dilated. *Wings* hyaline; pterostigma small, subreniform, ochreous, the borders paler and framed in thick black nervures; 10 to 11 postnodal nervures to fore-wings, 9 in the hind. *Abdomen* bluish, or white in subadults and tenerals, marked with black and yellow on segments 6 to 10 only as follows:—Segment 6 with only the apical vestigial end of a stripe present as two large bronzed black spots; segments 7 to 10 broadly bronzed black on dorsum, the base of 7 and its mid-dorsal carina finely yellow, 8 with the carina more broadly yellow, and 9 with this same marking expanded fusiformly at the apical half of segment, whilst on 10 the spot is scutellate; all these segments yellow laterally, with a small rounded black spot on the centre of each of segments 8 and 9. *Anal appendages* yellow or bluish; superiors roughly triangular as viewed from above, with straight outer border and bevelled inner, hollowed out within, and with a small ventral spine situated near the base; inferiors broad at base, tapering to apex, which is a little rounded, turned down and curved inwards; seen from the side very sinuous.

Female.—Abdomen 27 mm. Hind-wing 18 to 19 mm.

Closely similar to the male, differing in the following particulars:—The transverse black stripe on vertex much narrower; dorsal black on *prothorax* narrower, the posterior lobe notched in the middle, the lateral portions rounded and depressed, the black spot on the sides reduced to a short longitudinal stripe or elongate spot. *Thorax* with the dorsal bronzed black more restricted and with a narrow rusty line bordering the mid-dorsal carina on each side; the posterior of the two posthumeral stripes is often vestigial or incomplete above and below, whilst the black stripe on the postero-lateral suture is very fine. *Wings* with 11 postnodal nervures to the fore, 10 in the hind, otherwise similar to the male. *Legs* with a vestige of a brown or blackish stripe on the basal portions of the tibiæ as well as on all the femora. *Abdomen* pale greenish-blue or tinted with ferruginous on dorsum, and with the mid-dorsal carina narrowly brownish-black from segment 1 to the middle of 9, this line often vestigial or split into fine parallel lines by the finely yellow mid-dorsal carina. *Anal appendages* pale, short, pointed; vulvar scales pale, rather short.

Distribution.—Asia Minor, Egypt, Persia, KASHMIR, and N.W. INDIA.

It is distinguished from other species by the absence of a black line behind the occiput, by the reduced dark markings of the femora and tibiæ, and, in the female, by the parallel reddish lines bordering the mid-dorsal carina on thorax.

Type in the Selysian collection.

The synonymy of this species is very confusing; named by Klug as a separate species, it was thought by Selys to be a variety of *P. acutipennis*, and to be identical with Hagen's species *P. syriaca*. Later, however, he modified this opinion on finding that the tibiæ of both sexes were dilated, and placed it as a race of *P. latipes*, after previously citing it as a separate species belonging to the group *P. pennipes* (*vide* Rev. Odon. p. 388). Morton has cited it both as a race and a subspecies of *P. latipes*; the latter is probably correct.

Genus **COPERA** Kirby (1890). (Fig. 86.)

Platycnemis, pars, Ramb. Ins. Névrop. p. 240 (1842).
Psilocnemis Selys, Bull. Acad. Belg. (2) vol. xvi, p. 168 (1863); id., Mém. Cour. vol. xxxviii, p. 121 (1886); Laid., Fasc. Malay. (Odon.) pt. ii, p 7 (1907).
Copera Kirby, Cat. Odon. p. 129 (1890); Laid., Rec. Ind. Mus. vol. xiii, p. 336 (1917); Munz, Mem. Amer. Ent. Soc. no. 3, p. 54 (1919); Fras., J. Bombay Nat. Hist. Soc. vol. xxix, p. 744 (1923); id., ibid. vol. xxxv, p. 646 (1932).
Pseudocopera Fras., Mem. Dept. Agric. India (Ent.), vol. vii, no. 7, p. 56 (1922).

Small or medium-sized dragonflies, with characters similar to those of the genus *Platycnemis*, differing by the 2nd segment of the antennæ being as long as or even longer than the 3rd; sexual differences more marked, the tibiæ in the females never being dilated, and the colour and markings differing,

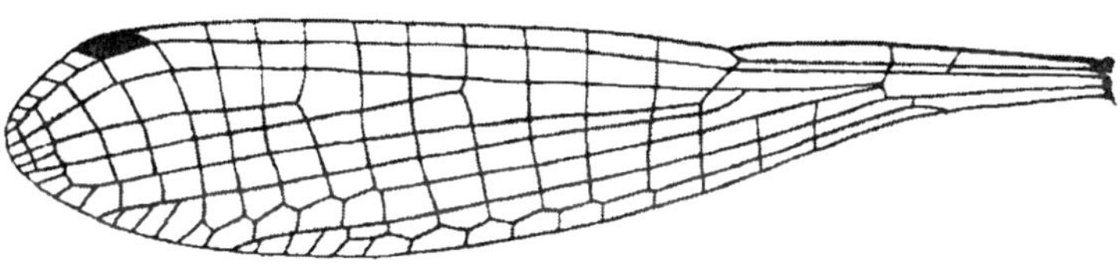

Fig. 86.—Wing of *Copera marginipes* (Ramb.), male.

sometimes very markedly, from those of the male; anal appendages of the male less homogeneous; venation entirely similar.

Genotype, *Copera marginipes* (Ramb.).

Distribution.—Throughout INDIA, CEYLON, and BURMA, save in the desert areas; Malaysia, Java, Sumatra, Borneo, Indo-China, China, and Formosa. One species is found in Madagascar and another in continental Africa.

The genus may be divided up into two groups—the *marginipes* group, with the legs coloured and tibiæ moderately or but little dilated, which breed in streams; and the *annulata* group, which have the legs white, tibiæ greatly dilated, and which breed in ponds and lakes. They are found from sea-level to some 4,000 ft. altitude, usually lying up in dark jungly shady spots. In the early stages all species are pure white, the dark markings developing comparatively slowly; thus, even in the darkest recesses, teneral forms may be seen threading their way stealthily through the undergrowth, often in great numbers, in the neighbourhood of streams.

Key to the Species of Copera.

1. The two hind pairs of tibiæ white and widely dilated .... 2.
   The two hind pairs of tibiæ yellow, reddish or brownish, only moderately dilated or not at all .... 3.

2. Legs very long, hind femora extending nearly to end of segment 2; 2nd abdominal segment wholly black on dorsum; inferior anal appendages strongly arched, black at apices ...... *annulata* Selys, p. 203
   Legs very short, hind femora extending only to end of thorax; 2nd abdominal segment with two dorsal blue spots; inferior anal appendages quite straight, white at apices .... *superplatypes* Fras., [p. 206.

3. Superior anal appendages only one-fourth the length of inferiors; female with posterior lobe of prothorax without spines .... *marginipes* Ramb., [p. 192.
   Superior anal appendages at least half the length of inferiors .... 4.

4. Legs very long, tibiæ not dilated, brownish; posterior lobe of prothorax of female without spines .... *assamensis* Laid., [p. 201.
   Legs shorter, tibiæ slightly dilated, citron-yellow or brick-red; posterior lobe of prothorax of female with a pair of divergent, slender, forwardly directed spines .... *vittata* Selys, p. 198.

73. **Copera marginipes** (Ramb.). (Fig. 86.)

Platycnemis marginipes Ramb. Ins. Névrop. p. 240 (1842).
Psilocnemis marginipes Selys, Bull. Acad. Belg. (2) vol. xvi, p. 168 (1863); id., Mém. Cour. vol. xxxviii, p. 123 (1886); Krug., Stett. Ent. Zeit. p. 102 (1898); Selys, Ann. Mus. Civ. Genova, (2) vol. x (xxx), p. 501 (1891).
Platycnemis lacteola Selys, Bull. Acad. Belg. (2) vol xvi, p. 167 (1863).
Psilocnemis striatipes Selys, Bull. Acad. Belg. (2) vol. xvi, p. 169 (1863).
Copera marginipes Kirby, Cat. Odon. p. 129 (1890); id., J. Linn. Soc. Lond., Zool. vol. xxiv, p. 560 (1894); Mart., Mission Pavie, sep. p. 18, (1904); Laid., Fasc. Malay. (Odon.), pt. ii, p. 8 (1907); Ris, Suppl. Ent. no. v, p. 18 (1916); Laid., Rec. Ind. Mus. vol. xiii, pp. 322, 337, pl. xiv, fig. 2 (1917); Fras., J. Bombay Nat. Hist. Soc. vol. xxvii, p. 543 (1921); id., ibid. vol. xxix, p. 744 (1923); id., Rec. Ind. Mus. vol. xxvi, pp. 428, 498 (1924); Laid., Spolia Zeylan. vol. xii, pp. 364, 365 (1924); id., J. Malay. Br. Roy. As. Soc. vol. iv, pp. 218, 230 (1926); Fras., Rec. Ind. Mus. vol. xxxiii, p. 448 (1931).
Copera acutimargo Krug., Stett. Ent. Zeit. p. 117 (1898).

Male.—Abdomen 28–31 mm. Hind-wing 16–18 mm.

Head: labium pale brownish-white; labrum, bases of mandibles, genæ, ante- and postclypeus pale greenish-yellow, the latter with a small black spot on each side the middle line; frons and vertex with a broad bronzed black fascia extending from eye to eye, followed by a moderately broad greenish-white stripe, equally wide, which wraps round the sides of the posterior ocelli and runs back for some distance as a narrow border to the eyes to become nearly confluent with a narrow, similarly coloured stripe which traverses the occiput behind the eyes, these two pale stripes enclosing a broadly elliptical stripe on the occiput; eyes black above, greenish at sides and beneath, and with a narrow equatorial black band encircling them; beneath the occiput broadly greenish-white from eyes inwards. *Prothorax* bronzed black on dorsum and lower part of sides, which are traversed by a broad pale greenish-yellow stripe invaded at two points by the dorsal black; lower border narrowly pale yellow; posterior border very finely yellow; posterior lobe shallow and very broadly rounded. *Thorax* bronzed black on dorsum for rather more than half way to the humeral suture, the mid-dorsal carina and borders of antealar sinus very finely lined with yellow; a narrow pale greenish-yellow humeral stripe in continuation of the prothoracic one, split in two above by a small black upper humeral spot; this stripe followed by a broad black fascia, narrowing above, on the middle of mesepimeron, peppered with small pale yellow spots; the sides beyond this pale greenish-yellow as well as below thorax, marked with an irregular black stripe on the anterior border of postero-lateral suture, and a second curved and narrower black stripe on the middle portion of metepimeron. *Legs* variably bright orange to dull reddish, the posterior two pairs of tibiæ moderately broadly dilated and about equal in length to the same femora. *Wings* hyaline; pterostigma brown, with a fine frame of yellow and thick black nervures, covering one cell, poorly braced; 12 postnodal nervures to fore-wings, 9 in the hind; 5 cells after the pterostigma between costa and radius. *Abdomen* bronzed black as far as the middle third of segment 8, from which to the end it is pale bluish-white or pale creamy white, according to age; segment 1 with the apical border narrowly and the sides broadly pale greenish-white, but the latter with a short dark stripe; segment 2 with a narrow pale greenish-white mid-dorsal stripe not extending quite to apical border, the sides also pale, clouded with brownish; segments 3 to 6 with a pale stripe along the sides confluent with a narrow pale greenish-white annule at the base of each segment; the lower border of segment 9 and the basal third of 8 narrowly black. *Anal appendages* pale yellow to white, the inferiors tipped with black beneath; superiors half the length of segment 10, vestigial, tiny rounded tubercles, with the apex pointed and hooked strongly ventralwards; inferiors at least four times the length of superiors, broad at base, tapering but slightly to a rounded obtuse apex, slightly separated at base, apices sloping in and finally meeting, inner surfaces hollowed out.

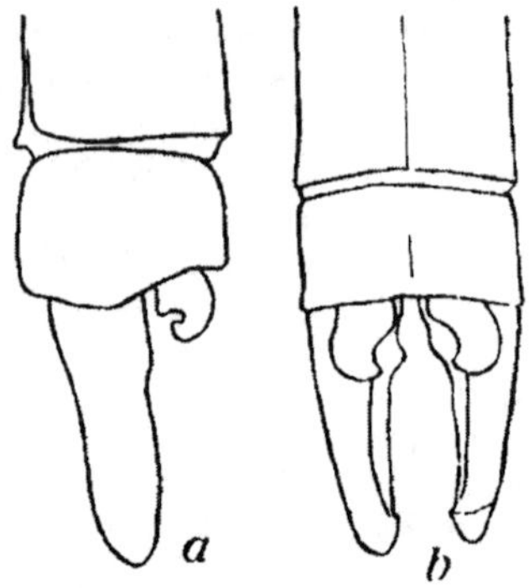

Fig. 87.—Anal appendages of *Copera marginipes* (Ramb.), male. *a*. Left lateral view. *b*. Dorsal view.

Female.—Abdomen 29–30 mm. Hind-wing 20 mm.

Differs from the male in several respects: the abdomen (very slim in the male) is comparatively stout and cylindrical, the ground-colour is drab and of varying shades of brown, whilst the black and darker markings are more restricted.

Head: labrum with a tiny median basal black point; genæ and bases of mandibles and clypeus palest brown, the latter with two black points as in the male; frons warm brown; the broad black fascia on vertex replaced by a broken black line, which broadens into a large bronzed black spot against the eyes; 2nd segment of antennæ and the distal end of 3rd creamy white; rest of upper surface of head and occiput pale brown or fawn, with a narrow interrupted black stripe just behind the ocelli, not extending out as far as the eyes, and a second longer, equally narrow black stripe on occiput behind the eyes; beneath head dirty white; eyes similar to male, but the polar cap dark olivaceous brown. *Prothorax* and *thorax* violaceous brown on dorsum, with a bronzed black fascia on the mid-dorsum like that in the male, pale brown laterally, with the black markings largely obsolete, the upper humeral spot present, but the posthumeral black stripe reduced to a lower zigzagged marking, whilst the stripes on the postero-lateral suture and metepimeron are mere black lines; posterior lobe of prothorax deeply notched, a small triangular lobe arising from centre of notch, and projecting forwards slightly over dorsum of middle lobe. *Legs* brownish.

white or carneous, femora with a speckled band or striated black stripe running their whole length, most conspicuous on the hinder pair, tibiæ not dilated. *Wings* similar to the male; 12 to 13 postnodal nervures in fore-wings, 11 in the hind; pterostigma a little paler and rather longer. *Abdomen* warm brown on dorsum, deepening to broad apical annules on segments 3 to 6 and the greater part of 7; markings similar to the male on segments 1 to 7; the apical half of segment 8 and the whole of 9 and 10 pale brownish-white or creamy white in subadults. *Anal appendages* shorter than segment 10, conical, pale brownish-white; vulvar scales brown, barely reaching end of abdomen.

Distribution.—This, the commonest species of the genus, is widely distributed throughout southern Asia and the Sondaic Islands. The *type*, which I have been unable to examine, came from Java, and formed part of the old Serville collection, now, I believe, in the Selysian collection. Java specimens resemble those from Siam and LOWER BURMA; the above description has been made from specimens from the latter region.

C. marginipes shows an infinite number of varieties, not only according to the age of specimens, tenerals being white with but few black markings, but also according to their locality. Unfortunately even in each locality varieties occur, so that it is difficult to say dogmatically that a number of races exist, and it seems safer, with our present knowledge and material, to speak of them as varieties only. Several species have been described which appear to be of not more than racial value or, possibly, only age varieties; among these is *C. acutimargo* Kruger, which Laidlaw records from Mergui. I possess specimens of *Copera* taken by the collector of Laidlaw's specimens, but after careful examination am unable to appreciate sufficient differences to separate them from others from Western India which I estimate as mere races or varieties of *C. marginipes*. All differ from the Burmese form by having segment 8 entirely black.

CEYLON forms:—Abdomen 34 mm. Hind-wing 18 mm.

Unusually long, and with very slender abdomen almost double the length of hind-wings; postclypeus and frons black, forming a continuous black area with the vertex as far back as the level of antennæ; 2nd and 3rd segments of latter black, tipped with white; the broad black fascia on occiput divided transversely into a broad anterior stripe, broadened at its outer ends and shaped like the propeller of an aeroplane, and a narrow posterior stripe, the postoccipital stripe lying well behind this. Lateral markings of *thorax* reduced, the posthumeral stripe present below only, and broken up into spots in its upper part, the stripes on the postero-lateral suture and metepimeron absent. *Legs* bright brick-red. *Wings* with 13 postnodal nervures to the fore, 11 in the hind; pterostigma black, framed finely in yellow. *Abdomen* black above on segments 3 to 8, with sharply defined pale blue basal annules on segments 3 to 7, very narrow on the latter; segment 2 has the mid-dorsal pale stripe heavily bordered with black on each side, and there is an apical line, the ventral ends of which tend to wrap round the ends of the stripes; segment 8 has the apical border and an apical small diamond-shaped spot pale blue, whilst segments 9 and 10 and the superior anal appendages are palest blue, the latter tipped with black at the apices. *Anal appendages* exactly the same shape as in the type.

Female.—Abdomen 30 mm. Hind-wing 18·5 mm.

Labrum and bases of mandibles bright yellow; postclypeus with rather larger black spots, but the black stripe on front of vertex even more reduced; only 2nd segment of antennæ pale; occipital black markings absent except for a vestige of the postocular stripe. Posterior lobe of *prothorax* with the middle portion longer and with the lateral portions forming small rounded lobes, recurved forwards. *Legs* similar, the dark stripes on femora like a chain of beads. *Abdomen* very similar to the type, the black apical marking on segment 2 recurved forwards as far as base, forming a letter U, which embraces the medial white and bordering black stripes; segments 9 and 10 and *anal appendages* creamy white.

Collected in Kandy, 2,000 ft., during August and September. Described from adults taken *in cop.*

Coorg forms:—Male: Abdomen 32 mm. Hind-wing 18 mm.

Black markings on *head* more extensive, the postclypeus, frons, and vertex forming one confluent area, the pale stripe separating this area from the occiput very narrow and almost obsolete, the dark areas of occiput extending out as far as eyes, markedly bronzed, almost dark metallic green, the post-orbital black stripe very broad. *Prothorax* and *thorax* with equally extensive melanism, the former with a narrow lateral pale stripe only, the mid-dorsal bronzed area on thorax extending to the humeral suture above, and nearly so elsewhere; the posthumeral black stripe extending broadly upwards as far as base of wings, and with the dorsal black enclosing a narrow pale humeral stripe which crosses the humeral suture obliquely from below upwards; the black stripes on the sides more extensive and made up of stippled spots irregularly confluent. *Legs* bright brick-red, the tibiæ moderately dilated. *Wings* with 12 to 14 postnodal nervures in the fore, 11 to 12 in the hind; pterostigma very oblique, very narrow, almost twice as long as broad, dark reddish-brown framed finely in pale brown. *Abdomen* similar to Ceylon forms. *Anal appendages* as for the type, inferiors palely ochreous.

WEST COAST forms (Poona and Mahableshwar, Bombay Presidency) :—These are intermediate between the type-form described above and Coorg forms. The patterns of the black head markings show much variation, but the pale stripe crossing vertex is moderately broad in all, and the occipital fascia just falls short of the eyes ; postclypeal markings vary from large spots to complete blackness ; pterostigma narrow and long, as in the Coorg forms ; segment 9 with its sides, especially at basal end, broadly black ; inferior anal appendages black as far as base on the outer sides.

DECCAN forms (Poona and Mhow) :—Abdomen 26 mm. Hind-wing 14 mm.

These show a reduction of melanism, especially in the females, the *heads* of which are often entirely without black markings and the whole a pale khaki-brown. Males from the Mhow area have the postclypeus unmarked and the frons narrowly bordered with pale brown ; the occipital markings are restricted to two large pyriform spots finely united by a median line, the postorbital stripe being nearly obsolete. The lateral markings of the *thorax* are obsolete, save for the lower part of the post-humeral marking. *Legs* carneous ; pterostigma remarkably narrow. They represent the smallest forms of the species.

Females of the same form have no head markings ; the mid-dorsal bronzed area of *thorax* is very narrow and the lateral markings obsolete, whilst the abdominal are reduced to a mere mid-dorsal dark brown stripe extending as far as segment 8 ; 9 has two subdorsal basal streaks and 10 is unmarked. The posterior lobe of *prothorax* is as in the type. The *wings* in adults are often deeply enfumed.

ASSAM forms :—Males closely resemble those from the West Coast, whilst the females have the dark stripes on *head* broken up into isolated bronzed green stripes, recalling strongly those of *Sympycna*. The lateral thoracic markings are much reduced ; the posterior lobe of *prothorax* is similar to that described for the Ceylon forms.

Margherita, Assam, during May.

BENGAL forms :—A pair from Jalpaiguri show the black markings much reduced, the occipital markings in the male tending to become broken up into isolated lateral oval spots, and in the female represented only by tiny black points lying within a pale area. Here again no modification of the posterior lobe of prothorax is found.

Jalpaiguri, Bengal, during March.

74. Copera vittata (Selys). (Figs. 88 & 89.)

Psilocnemis vittata Selys, Bull. Acad. Belg. (2) vol. xvi, p. 170 (1863); id., Mém. Cour. vol. xxxviii, p. 121 (1886).
Psilocnemis serapica Selys, Bull. Acad. Belg. (2) vol. xvi, p. 171, ♀ (1863).
Copera vittata Kirby, Cat. Odon. p. 129 (1890); Krug., Stett. Ent. Zeit. p. 117 (1898); Laid., Rec. Ind. Mus. vol. viii, pp. 342, 343 (1914); Fras., J. Bombay Nat. Hist. Soc. vol. xxix, p. 744 (1923); id., Rec. Ind. Mus. vol. xxvi, pp. 428, 498 (1924).
Copera vittata deccanensis Laid., Rec. Ind. Mus. vol. xiii, pp. 322, 337–9 (1923); Fras., J. Bombay Nat. Hist. Soc. vol. xxix, p. 744 (1923); id., Rec. Ind. Mus. vol. xxxiii, p. 448 (1931).

Male.—Abdomen 28–34 mm. Hind-wing 16–18 mm.

Head: labium pale reddish-brown; labrum bright greenish-yellow; genæ and bases of mandibles pale bluish-green;

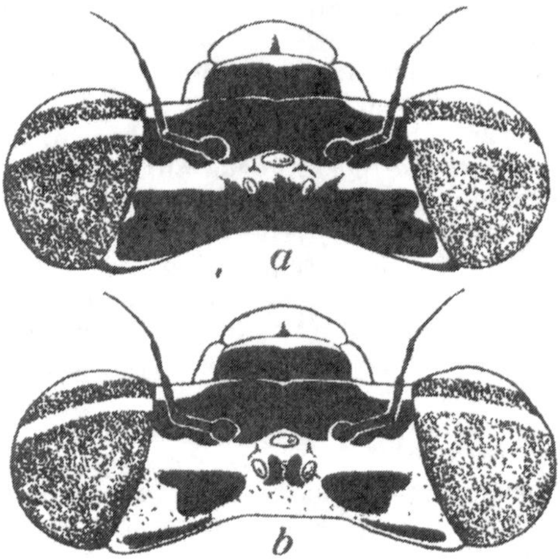

Fig. 88.—Heads of *Copera vittata* (Selys), male. *a*. From Bengal and Burma. *b*. From the Western Ghats and Nilgiris.

anteclypeus pale; postclypeus, frons, and vertex jet-black, the latter traversed from eye to eye by a moderately broad pale bluish-green stripe which passes between the anterior and posterior ocelli; behind the occiput a postocular narrow creamy-white stripe; eyes capped with black above, below this olivaceous green, marked with a narrow black equatorial belt. *Prothorax* black, with a narrow ochreous stripe on each side; posterior lobe broadly rounded. *Thorax* bronzed black on dorsum to nearly as far as humeral suture, the space between the humeral suture and first lateral being dark chocolate-brown or blackish, stippled with a few coarse yellow spots, the two dark areas separated by a narrow humeral ochreous stripe, the sides, posterior to the antero-lateral suture, bluish or yellowish; an obscure vestigial brownish stripe anterior to the suture, and a second, more pronounced, on the middle of metepimeron; yellow beneath thorax. *Legs* reddish, the two posterior pairs of tibiæ very slightly dilated. *Wings* hyaline; pterostigma much less elongated than in *C. marginipes*, squared outwardly, dark reddish-brown, paler at circumference, framed in black nervures, covering one cell, very poorly braced; 10 to 13 postnodal nervures to fore-wings, 10 in the hind. *Abdomen* black on dorsum and sides as far as segment 10, which latter is pale blue; segment 9 has also usually a small apical blue spot, sometimes extending along dorsum half way to base; segments 3 to 7 with very narrow pale blue basal annules, under surface white, expanding subapically on to the sides; segments 1 and 2 warm reddish-brown, the former with the dorsum largely black,

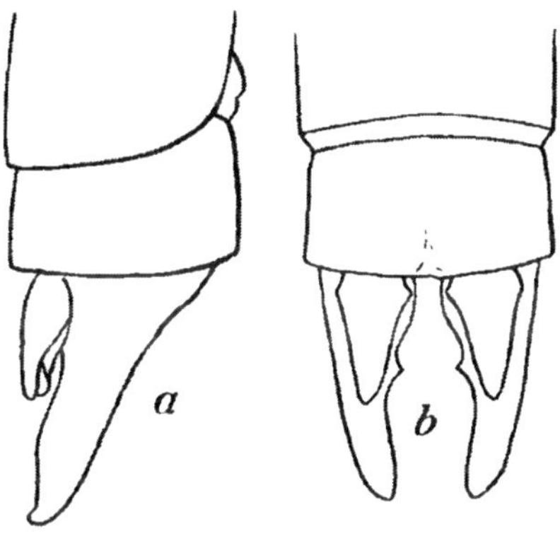

Fig. 89.—Anal appendages of *Copera vittata* (Selys), male.
a. Right lateral view. *b*. Dorsal view.

the latter with the mid-dorsal carina finely blue bordered with black. *Anal appendages*: superiors pale blue or creamy white, as long as segment 10, conical, depressed, acute and tapered at apex; inferiors nearly twice the length of superiors, broad at base, tapering to apex and directed obliquely medialwards so as to meet at apices, pale within, blackish-brown externally, a small obtuse spine at the middle of inner border. Penis broadly blunt at apex, the apical end curling over to embrace the stem of the organ, broadening at end, where it is shallowly cupped or grooved to embrace the stem, the outer angles pointed but not branched.

Female.—Abdomen 28–30 mm. Hind-wing 18 mm.

Differs from the male in several respects, the abdomen much more robust and cylindrical, the black markings more restricted, etc. *Head*: labrum bright yellow, with a small median basal black spot and its base narrowly black; anteclypeus brown;

genæ and bases of mandibles greenish-yellow; frons and vertex purplish-brown, with a moderately broad transverse black band extending from eye to eye at level of antennæ; the transverse black fascia on occiput absent, but the post-ocular stripe present; ocellar space paler, forming an ill-defined pale stripe; 2nd segment of antennæ white, the 3rd pale brown. *Prothorax* blackish-brown, with a pale stripe on each side; posterior lobe deeply notched, with a small median lobe lying within it, angles of lobe prolonged as fine spines, strongly divergent forwards. *Thorax* pale brown, marked very similarly to the male: the mid-dorsal bronzed area narrower, mid-dorsal carina and antealar sinus very finely yellow; a large black spot on the upper part of humeral suture, the suture below it finely black; posthumeral black band extending for only two-thirds up thorax, very ragged and broken up by large pale spots: the band bordering the postero-lateral suture very broad; two narrower stripes on metepimeron, one just posterior to the suture interrupted above, and another, shorter and narrower, on the lower border. Beneath pale brown *marked with three large black spots. Legs* yellow, all femora with the speckled beaded band on extensor surface. *Wings* hyaline, more or less palely enfumed; pterostigma subquadrate, shaped very much as in the male, but rather, shorter blackish-brown framed in pale yellow and black nervures, covering one cell, poorly braced; 14 postnodal nervures in fore-wings, 12 in the hind. *Abdomen* dark purplish-brown, changing to black on segments 7 to 10; segments 1 and 2 similar to the male; segments 3 to 7 with narrow basal pale yellow or whitish annules; segment 9 with a broad brownish-white T-shaped marking on dorsum, whilst segment 10 is entirely of that colour, except laterally. *Anal appendages* brownish-white, conical, pointed, surmounting a large conical protuberance; vulvar scales dark brown, robust, extending to the end of abdomen.

Distribution.—Almost identical with that of *C. marginipes*, if we consider as mere races or colour varieties all the forms which have been described as separate species. The colouring seems to have been the only factor considered in forming new species or subspecies; but varying degrees of melanism are found even in the same localities, and depend largely on the age of the specimens or the time of the year; wet- and dry-season forms undoubtedly occur. *C. atomaria*, from Borneo, represents the maximum melanism in the species, whilst dry-season forms from the West Coast of India show remarkably little of this. I have examined specimens from S. India (Malabar, Coorg, Nilgiris, Anaimalais), Bengal, Assam, Lower Burma, and Siam, and find that they all show an infinite variety of markings, often confluent, as often discrete. On the other hand, none differs so far as the shape of the anal appendages, the legs of the male, and the posterior lobe of prothorax of the female are concerned. The West Coast forms are described as a subspecies by Laidlaw under the name of *C. vittata deccanensis.* Although during life, and in spirit, specimens appear to be more brightly coloured than those described from elsewhere by other authors, after death, and when dried, much of this bright colouring is replaced by more drab browns and reds. Thus *C. vittata deccanensis* has the legs bright citron-yellow and the sides of the thorax bright greenish-yellow, whilst the humeral stripes and pale abdominal markings are pale blue, as also the anal appendages; these colours give place to drab shades after drying. *C. serapica*, from the Nicobars, is most certainly the female of *C. vittata*; the shape of the posterior lobe of prothorax at once determines it.

Males of *C. vittata* are readily distinguished from other species by the straight anal appendages, the superiors being just half the length of the inferiors, and by the very slight dilatation of the tibiæ. Females are most easily determined by the divergent spines on the posterior lobe of the prothorax, absent in other species.

The *type* of the species is in the Selysian collection, and has been erroneously described as having the labrum black; paratypes are to be found in most national collections and many private ones. The next species to be described is certainly distinct, though it was classed as a subspecies of *C. vittata* by its author.

75. **Copera assamensis** Laidlaw. (Fig. 90.)

Copera vittata assamensis Laid., Rec. Ind. Mus. vol. viii, pp. 342, 343 (1914); id., ibid. vol. xiii, p. 338 (1917).

Male.—Abdomen 32–34 mm. Hind-wing 17–18 mm.

Head: labium palest brown; labrum yellow; bases of mandibles and genæ greenish-yellow; anteclypeus pale brown, postclypeus, frons, and vertex black, the latter traversed from eye to eye by a moderately broad black stripe at the level of the ocellar space; distal end of 2nd segment of antennæ yellow; occiput reddish-brown, marked in full adults anteriorly and posteriorly by a diffuse black border; a narrow postorbital black stripe lying behind an equally narrow pale stripe on each side of the back of occiput; eyes dark brown above, olivaceous below, with a dark brownish-black equatorial belt. *Prothorax* brown, with the posterior lobe, three short parallel stripes on middle lobe, and a broad lateral stripe black. *Thorax* brown or carneous, marked irregularly with black as follows:—A broad bronzed-black band on mid-dorsum extending for rather more than half way to humeral suture, which latter is finely black, but expands into a small black spot above;

mid-dorsal carina and borders of antealar sinus finely yellowish or ferruginous; on the sides a continuation of the lateral black belt seen on the prothorax, which extends for but a short distance on to the mesepimeron, and breaks up into two obscure mottled fasciæ; postero-lateral suture finely black; lastly, a short black stripe on metepimeron; beneath yellowish. *Legs* reddish or reddish-brown, the tibiæ not dilated (Laidlaw states that the hind pair only show a trace of dilatation, but in an example before me I am unable to detect the slightest sign of this). *Wings* hyaline; 13 postnodal nervures to the fore-wings, 11 in the hind; pterostigma dark reddish-brown, framed in pale yellow and thick black nervures, the distal side nearly straight, the proximal very oblique, subquadrate, barely longer than broad, poorly braced, covering 1 cell. *Abdomen* black, ringed with pale blue basal annules as follows:—Segment 1 ochreous, with a curved black stripe on each side

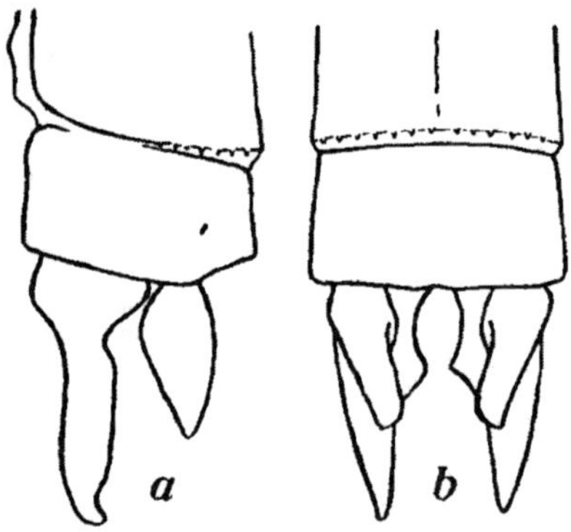

Fig. 90.—Anal appendages of *Copera assamensis* Laid., male. *a*. Left lateral view. *b*. Dorsal view.

and a short apical subdorsal black streak; segment 2 with a pale creamy-white stripe on mid-dorsum bordered heavily with black and not extending to the apical border of segment, and a narrow crenate black subapical line embracing the ends of the mid-dorsal stripes; segments 3 to 8 black, 3 to 7 with narrow blue basal annules and the black broadening as apical annules, the sides of 3 to 6 pale; segment 9 with its basal half black or entirely pale creamy white, as is also segment 10, except the sides. *Anal appendages* very similar to those of *C. vittata*; superiors creamy white, triangular as seen from above, conical and subcylindrical seen in profile, hollowed out within; inferiors nearly twice the length of superiors, dark reddish-brown, outer surface black, especially at apex, extending straight back, obtuse at apex, broad at base, subcylindrical thereafter. Penis not differing from that of *C. marginipes*.

Female.—Abdomen 30 mm. Hind-wing 17 mm.

Differs from the male in the black markings being rather more extensive and the more robust build of the abdomen. *Head* marked as in the male, but the occiput is entirely black, save for the postocular pale stripe, and the black fascia on vertex is rather broader. *Prothorax* coloured similarly; posterior lobe shaped as in *C. vittata*, the same two divergent spines being present. *Thorax* with the humeral and lateral black markings more extensive and mottled with coarse yellow spots. *Wings* not differing from the male. *Legs* yellowish, femora with the characteristic beaded black band on extensor surfaces. *Abdomen* blackish-brown to black on terminal segments, the 9th and 10th bearing palest brown dorsal spots. *Anal appendages* pale brown, conical, pointed; vulvar scales robust, dark brown.

Distribution.—From ASSAM to Indo-China.

The *type*, in the Indian Museum, has the upper surface of head mainly brown, with no black markings, save for the transverse stripe on vertex. A specimen from Siam in the author's collection has the markings more advanced, as described above. The male is easily distinguished from *C. vittata* by its longer legs, without dilatation of the tibiæ, the latter character separating it from all other species of the genus. The female described by Laidlaw may not be that of *C. assamensis*, and does not differ materially from the female of *C. vittata*. Förster has stated that the females of the genus are generally dimorphic, but after examining many scores I am unable to corroborate this; it would be more true to say that they are polychroic, according to age.

76. **Copera annulata** (Selys). (Figs. 91 & 92.)

Psilocnemis annulata Selys, Bull. Acad. Belg. (2) vol. xvi, p. 172 (1863); id., Mém. Cour. vol. xxxviii, p. 124 (1886).

Psilocnemis subannulata Selys (♀), Mém. Cour. vol. xxxviii, p. 125 (1886).

Copera subannulata Kirby, Cat. Odon. p. 129 (1890); Laid., Rec. Ind. Mus. vol. xiii, p. 337 (1917).

Copera annulata Kirby, Cat. Odon. p. 129 (1890); Laid., Rec. Ind. Mus. vol. viii, pp. 341, 342 (1914); Ris, Suppl. Ent. no. v, pp. 17, 18 (1916); Laid., Rec. Ind. Mus. vol. xiii, pp. 322, 337 (1917); Fras., J. Bombay Nat. Hist. Soc. vol. xxix, p. 744 (1923).

Copera annulata stevensi Laid., Rec. Ind. Mus. vol. viii, pp. 341, 342 (1914); id., ibid. vol. xiii, p. 337 (1917).

Copera arachnoides Fras., Mem. Dept. Agric. India (Ent.), vol. vii, no. 7, pp. 56, 57, pl. vii, fig. 4 (1922).

Male.—Abdomen 35–37 mm. Hind-wing 22–23 mm.

Head: labium white; labrum, genæ, bases of mandibles, and clypeus pale blue, the former with a tiny median black point at base, the latter with a variable transverse black mark confined to the middle of postclypeus, or in others extending by a short median stalk to its base ; frons and vertex in front and laterally pale blue, with a broad bronzed black triangular area which begins near front margin of frons and extends to back of occiput, touching the eyes only at a point in line with the posterior ocelli, from which point it recedes again to leave a clear pale blue area against the eyes ; this dark area marked with a small oval or triangular white spot between the posterior ocelli and bases of antennæ, and on back of occiput by two large triangular white spots near the middle line (in some specimens there is a second smaller triangular spot lying to the outer side of these) ; occiput beneath eyes pale bluish ; eyes black above, olivaceous at sides and below. *Prothorax* bronzed black, with a broad lateral pale blue stripe extending its whole length. *Thorax* dark bronzed blackish-green on dorsum, marked on each side with narrow, complete,

Fig. 91.—Head of *Copera annulata* (Selys), male.

pale blue antehumeral stripes lying in juxtaposition to the humeral suture ; a small upper posthumeral white spot (Burmese specimens) or a white sickle-shaped mark (Bengal or Assam forms) ; laterally and beneath palest blue, the postero-lateral suture rather broadly mapped out in black. *Legs* (fig. 94, *a*) pure white, the distal ends of all femora and the extreme proximal ends of tibiæ black ; hind femora extending to the end of second segment of abdomen ; tarsi black ; tibiæ widely dilated. *Wings* hyaline ; pterostigma reddish-brown, framed in white and thick black nervures, poorly braced, covering one cell, more or less diamond-shaped, less so in forms from Assam ; 13 to 16 postnodal nervures in fore-wings, 12 to 14 in the hind, the greater number in Assam forms ; petiolation beginning just distal to the level of proximal antenodal nervure. *Abdomen* bronzed black on dorsum, except segments 9 and 10, which are pale blue (Assam forms have the basal third or more of segment 9 black, whilst some Burmese forms have the whole of segment 9 black, and also a small diamond-shaped apical dorsal spot on segment 8); segments 1 and 2 with the sides broadly blue, but narrowing apically on the latter, whilst all segments from 3 to 6 have the lower part of sides bluish; segment 2 with its base narrowly blue, segments 3 to 6 with narrow pale blue basal annules confluent laterally with the blue on sides. *Anal appendages* pale blue or white, the apical half of inferiors black; superiors conical, triangular, acute at apex, as long as segment 10; inferiors broad at base, tapering to an acute apex, curved strongly downwards and very slightly inwards.

Female.—Abdomen 33–37 mm. Hind-wing 22–24 mm.

Differing in some respects from the male, as follows:—*Head* with an extension of the pale blue into the apex of the

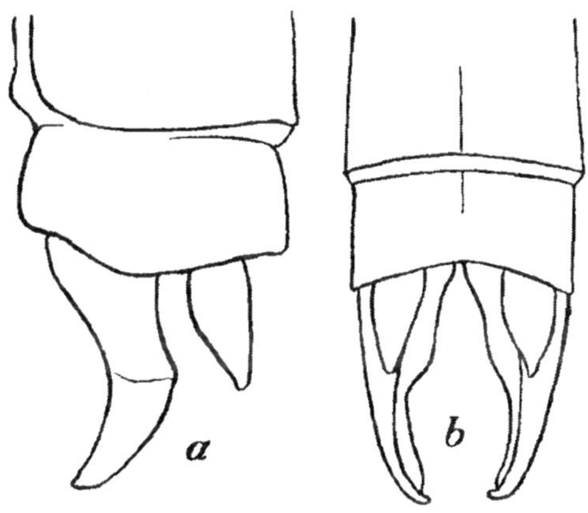

Fig. 92.—Anal appendages of *Copera annulata* (Selys), male.
a. Left lateral view. *b.* Dorsal view.

black triangle, this very variable (Assam forms are exactly similar to the males, save for a pair of obscure spots, one on each side of the anterior ocellus); the frons broadly blue, and more or less confluent with a series of spots on the ocellar space,—a small triangular spot on each side of anterior ocellus, another between the posterior ocellus and base of antennæ as in the male, a median small triangular spot between the posterior ocelli, and a spot on the outer side of each postocular spot often confluent with the latter. In some specimens the whole of the ocellar space blue, except for a narrow black bordering to the inner sides of posterior ocelli. *Legs* of great length, reddish or of the same colour as the males, but with none of the tibiæ dilated. *Prothorax* and *thorax* exactly similar to the male; posterior lobe of former simple, projecting back at middle, and with a slightly crenate border. *Abdomen* similar to the male except for segments 9 and 10, the former with two large subdorsal, bronzed, greenish-black spots broadly confluent at base and along basal half of mid-dorsum, its apical border and the whole of segment 10 pale blue. *Anal appendages* very short, pale blue, conical; vulvar scales pale, not extending to end of abdomen.

Distribution.—ASSAM, BENGAL, Malaysia, Indo-China and China. Selys also gives Japan, on the authority of a specimen, labelled from Yokohama, in the Zurich Museum. The species named *Copera ciliata* from a female by Selys, from Malacca, is probably not more than a variety of this species, and Selys expresses such an opinion in the 'Revision des Syn. Agrionines.'

This species is quite easily distinguished from all others except *C. superplatypes* by its pronounced black colour with pale blue markings and its broadly dilated white tibiæ, and from all other species, including the one cited, by the shape of its inferior anal appendages. The shape of the posterior lobe of the prothorax and the long hind-legs will determine the female from others.

Type in the Selysian collection; paratypes in most national collections.

Contrary to *C. marginipes* and *C. vittata*, both of which breed in running streams, this species breeds in weedy ponds and small lakes. It is common in the Company Gardens, Calcutta, breeding in most tanks there.

77. **Copera superplatypes** Fraser. (Figs. 93 & 94.)

Copera superplatypes Fras., Rec. Ind. Mus. vol. xxix, pp. 88, 89, figs. 6 *b* & 7 *b* (1927).

Male.—Abdomen 26 mm. Hind-wing 15 mm.

Head: labium white; labrum, anteclypeus, and genæ palest blue, rest of head black, save for a moderately broad pale blue stripe traversing the vertex from eye to eye at level of ocellar space. *Prothorax* black, with a pale bluish-white stripe and a spot of the same colour below it on each side. *Thorax* bronzed black as far lateral as the anterior suture, marked with pale blue, antehumeral stripes bordering the humeral suture anteriorly; laterally pale blue, the postero-lateral suture mapped out in black, a black spot on the mesepimeron confluent with the black in the humeral region, and another large black spot on the centre of metepimeron; beneath blue bordered with black. *Wings* hyaline; pterostigma black, covering one cell, diamond-shaped, slightly longer than broad, poorly or not braced; 12 to 13 postnodal nervures to fore-wings, 11 in the hind; petiolation beginning well proximal of *ac*, which lies about mid-way between the level of the antenodal nervures. *Legs* very short, hind femora extending only to end of thorax, pure creamy white with black spines, but the distal halves of all femora black ; the two hind

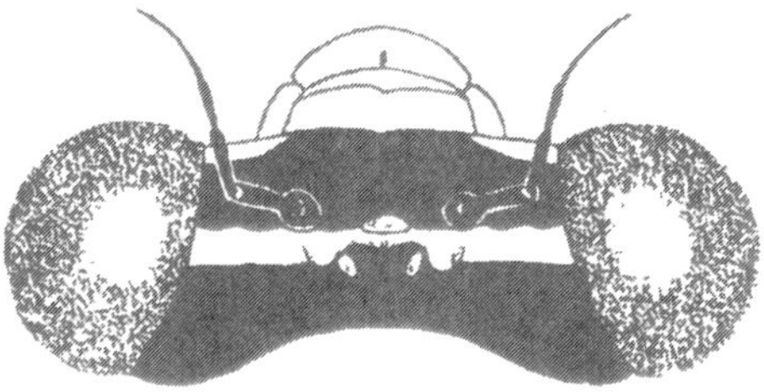

Fig. 93.—Head of *Copera superplatypes* Fras., male.

pairs of tibiæ enormously dilated, about two and a half times the width of hind femora. *Abdomen* black, the sides of segment 1 bluish-white ; segment 2 with a pair of oval blue dorsal sub-basal spots (similar to those found in a similar situation in several species of the genus *Agriocnemis*), the sides of the

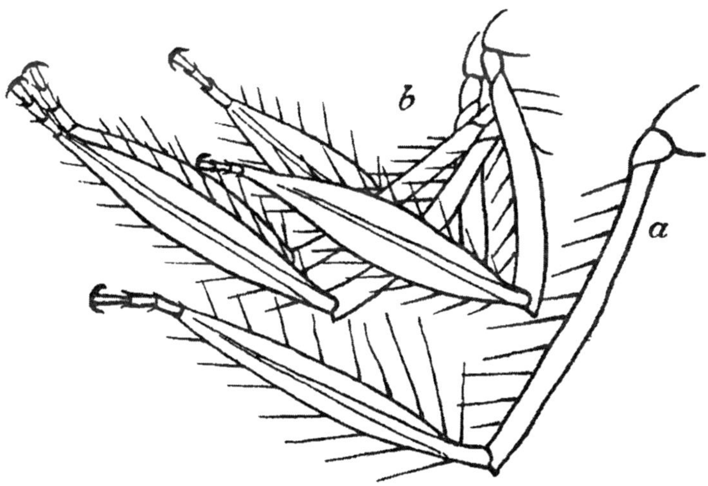

Fig. 94.—Legs of :—*a. Copera annulata* (Selys), male ; *b. Copera superplatypes* Fras., male, showing the dilated tibiæ of the two hind pairs of legs.

same segment bluish, broadly so at the apical end ; segments 3 to 5 with the lower parts of sides bluish and confluent with narrow pale blue basal annules narrowly interrupted on the mid-dorsum ; segments 6 to 8 similar, but the basal annules more widely interrupted, and the lateral blue extending subapically for a short distance up the sides; segment 9 unmarked; segment 10 with the dorsum palest blue. *Anal appendages* creamy white, the inferiors blackish at base; superiors slightly shorter than segment 10, conical, depressed, acute at apices; inferiors one-third longer than superiors, very broad at base, cylindrical, abruptly tapered and ending in a sharp point.

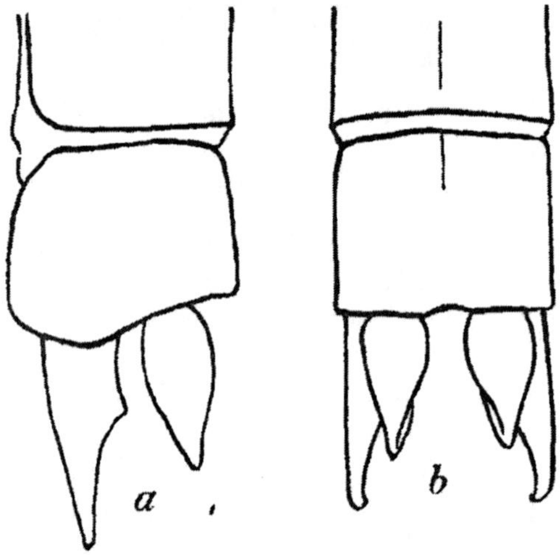

Fig. 95.—Anal appendages of *Copera superplatypes* Fras., male. *a.* Left lateral view. *b.* Dorsal view.

Female unknown.

Type in the Fraser collection.

Distribution.—A single male from Hasimara, Duars, BENGAL.

This species, which belongs to the *annulata* group, is easily distinguished by the following characters:—The black vertex traversed by a blue stripe instead of the broad black triangle spotted with white seen in *C. annulata*; the anal appendages, of which both pairs are directed straight back and are very similar in shape: the enormous dilatation of the tibiæ, which are ridiculously out of proportion to the small size of the insect; and lastly, by the very characteristic marking on the dorsum of segment 2.

Subfamily PROTONEURINÆ.

Genus *Alloneura* Solys, Bull. Acad. Belg. (2) vol. x, p. 441 (1860); id., Rev. Syn. Agr. p. 159 (1886).

Legion *Protoneura* Laid., Rec. Ind. Mus. vol. xiii, pp. 323, 343, 344 (1917).

Protoneurinæ Fras., J. Bombay Nat. Hist. Soc. vol. xxix, p. 742 (1923); id., ibid. vol. xxxiii, p. 834 (1929).

Dragonflies of small size and slender build, coloured usually black marked with blue, less commonly black marked with red or yellow, rarely metallic; resting with wings closed over dorsum. Wings narrow, long, and pointed at apex, or shorter, broader, and rounded at apex, hyaline, less commonly tipped or barred with black; the nervure *ac* situated midway between the two antenodal nervures or slightly nearer the distal antenodal; no accessory basal postcostal nervures ever present; petiolation beginning slightly proximal to the level of *ac*; the nervure *ab* variable in the genera, either present and vestigial or complete or entirely absent; *IA* absent or rudimentary; *Cuii* variable, often reduced; discoidal cell elongate, about four times longer than broad, the ends more or less squared, and costal and posterior sides approximately equal: sectors of arc arising from the lower end of arc and separate from their origins; individual cells mostly four-sided; *Riii* arising nearer the node than pterostigma; origin of *Riv+v* well proximal to the level of the oblique nervure descending from the subnode; *IRiii* in continuation of that nervure; pterostigma diamond-shaped, costal and posterior sides equal, distal and proximal sides oblique; no intercalated sectors save *IRii*.

Abdomen of moderate length, slender, never equal to twice the length of the hind-wing, usually considerably shorter than this; segments 8, 9, and 10 becoming progressively shorter from 8 to 10. Anal appendages very homogeneous; superiors shaped like the hammer of a revolver, acute at apex, and with a robust ventral spine; inferiors broad at base, tapering rapidly to an acute point sloped downwards and inwards. Vulvar scales robust, not extending beyond end of abdomen: no ventral spine at apex of segment 8 in the female, and no dorsal spine to the vulvar scales as in the PLATYSTICTINÆ. Penile organ variable, but the apical portion always curled strongly over and embracing the stem of the organ, branched or unbranched at the end.

Distribution.—Cosmopolitan. Within Indian limits only seven genera are known, of which two are represented in CEYLON and BURMA, and the whole in Continental India. Of these, four genera form a natural group confined to the Western Ghats of India, with the exception of one species found in the Eastern Ghats.

All species breed in running water, and are mostly confined to montane and submontane tracts at altitudes varying from 1,500 to over 7,000 ft., in heavily wooded country; no species are known from dry zones or desert areas. Very little is known about their life-histories, and the larvæ of the majority still remain unknown.

Key to the Genera of the Protoneurinæ.

1.	Wings tipped or barred with black in the male; the anal bridge (*ab*) present and complete	CHLORONEURA Laid., [p. 244.
	Wings hyaline in both sexes; *ab* variable.	2.
2.	*ab* present and complete	3.
	ab present, but incomplete and vestigial. .	5.
	ab entirely absent	MELANONEURA Fras., [p. 259.
3.	*Cuii* extending distally to beyond the middle of wing; not more than 10 to 12 postnodal nervures in hind-wings	DISPARONEURA Selys, [p. 228.
	Cuii not extending as far as middle of wing; usually about 20 postnodal nervures or more in the hind-wings	4.
4.	Large species, with abdomen about 50 mm. in length; fore-wings with about 30 postnodal nervures; segments 7 to 10 marked with azure blue	PHYLLONEURA Fras., [p. 268.
	Smaller species, with abdomen about 40 mm. in length; fore-wings with about 20 postnodal nervures; only segments 8 to 10 marked with azure blue.	ESME Fras., p. 262.
5.	Small species, with *Cuii* extending to middle of wing, and not more than 12 postnodal nervures	CACONEURA Kirby, [p. 210.
	Larger species, with *Cuii* not extending as far as middle of wing; fore-wings with 20 or more postnodal nervures	INDONEURA Laid., [p. 250.

Genus **CACONEURA** Kirby (1890). (Fig. 96.)

Caconeura Kirby, Cat. Odon. p. 132 (1890); Kruger, Stett. Ent. Zeit. vol. lix, p. 25 (1898); Ris, Nova Guinea, vol. ix (Zool.), 3, p. 478 (1913); id., Abh. senckenb. naturf. Ges. vol. xxxiv, pp. 508–11 (1913); id., Nova Guinea, vol. xiii (Zool.), pp. 90–1 (1914); Laid., Rec. Ind. Mus. vol. xiii, p. 344 (1917); id., Proc. Zool. Soc. Lond. pp. 338, 339 (1920); Tillyard, Ins. Aust. & New Zeal. p. 76 (1926).

Alloneura Selys (*nom. præoc.*), Bull. Acad. Belg. vol. x, p. 444 (1860); id., Mém. Cour. vol. xxxviii, pp. 176, 273 (1884).

Dragonflies of small size and slender build, with abdomen less than twice the length of wings; coloured black, marked with bright red, orange, blue, or yellow, or rarely without markings. Wings hyaline, narrow, moderately rounded at apices, never falcate, with a long petiole; discoidal cell elongate, costal and posterior sides equal, and four times as long as base or distal end ; sectors of arc arising from lower end of arc, divergent from origins ; no accessory basal post-costal nervure present ; the nervure *ac* situated about midway between the two antenodal nervures ; *ab* absent, or present and then always incomplete, extending in a strong curve from *ac* near the point where it meets the posterior margin of wing, and ending on the same margin a short distance from *ac* ; *IA* absent ; *Cuii* much reduced, only 1 to 4 cells in length ; *Riv+v* arising well proximal to the oblique nervure descending from the subnode ; *IRiii* in continuation of that nervure, or very slightly distal to it ; pterostigma small, about 1½ times as long as broad, diamond-shaped, distal and proximal ends very oblique, the latter strongly braced ; cells of wings mainly quadrangular.

Head narrow ; eyes smaller than in species of *Platysticta* ; occiput simple, not ridged posteriorly ; prothorax with the posterior lobe simple and rounded in the male, but armed with one or two pairs of strongly curved hooks in the female , abdomen cylindrical, slim, broadened at base and again

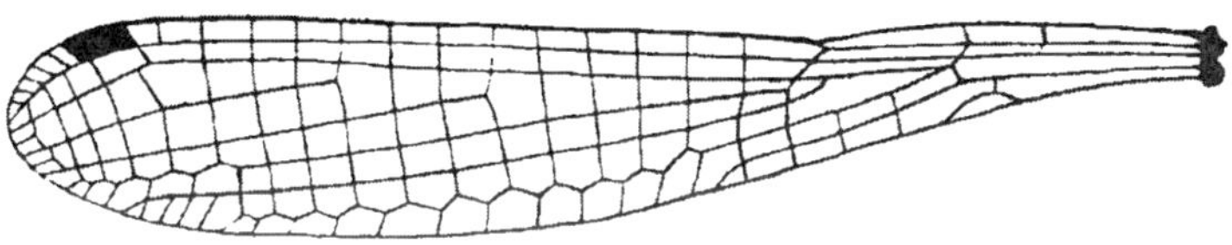

Fig. 96.—Wing of *Caconeura verticalis annandalei* Fras., male.

towards the terminal segments, about half as long again as the wings ; anal appendages very homogeneous, superiors shaped like the hammer of a revolver, tumid at base, tapering and pointed at apex and with a broad ventral spine : inferiors simple, strongly bevelled from above downwards, very broad at base, tapering to apex, which is bent strongly inwards and ends in an obtuse spine. Genitalia : lamina narrowly and deeply cleft, hamules broad foliate plates with an acute terminal spine, lobe very long, narrow, depressed, slipper-shaped ; penis with apex ending in two strongly curled branches which embrace the stem of the organ.

Female with robust vulvar scale extending to end of abdomen.

Distribution.—Peninsular India, except in desert tracts ; Ceylon, Burma, Malaysia, the Sondaic Archipelago, and Borneo.

No definition of the genus was given by its author, but the name *Caconeura* merely replaced the Selysian *Alloneura*, which was found by Kirby to be preoccupied. Kirby gives *C. dorsalis* Selys as the genotype, and includes under it all those species which Selys placed in his subgenus *Alloneura*.

We must, therefore, accept the definition of this subgenus as that of the genus *Caconcura*. Unfortunately, subsequent writers have included under the latter name species which clearly fall into the Selysian subgenus *Disparoneura*, more especially the 3rd group of the 2nd division of that subgenus, on the score that they possess a rudimentary anal bridge (*ab*). Although these were originally included by Selys under the subgenus *Alloneura*, he subsequently (Rev. Agrion.) removed them to the subgenus *Disparoneura*. The name *Caconeura* is employed here in its wider sense to include not only all those species originally included by Kirby under *Caconeura*, but also those included by Selys under *Disparoneura* in which the anal bridge is rudimentary; it would seem best to give generic rank to this last group, for which I would suggest *Cycloneura*. I would also suggest the inclusion in this new genus of all those species which fall into the 1st group of the old subgenus *Alloneura*, as these are clearly not congeneric with those of the 2nd. On the contrary, they bear the closest relationship to those in which the anal bridge is rudimentary, this characteristic being a very variable one, and quite occasionally breaking down even in the wings of an individual specimen. Thus there can be no doubt of the cousinship of such forms as *C. dorsalis auricolor*, *C. delicatula*, and *C. dorsalis dorsalis* (=*Alloneura dorsalis* Selys), although the anal bridge is present, but incomplete, in the first two and quite absent in the latter.

Key to the Species of Caconeura.

1.	Anal bridge (*ab*) absent in all wings	2.
	Anal bridge (*ab*) present in all wings	3.
2.	Thorax without any markings	*o'doneli* Fras., p. 225.
	Thorax with azure blue markings	*theebawi* Fras., p. 220.
3.	Thorax without any markings	4.
	Thorax black, marked with brick-red dorsal stripes	5.
	Thorax black, marked with pale yellow dorsal stripes	*sita* Kirby, p. 221.
	Thorax black, marked with broad pyriform blue dorsal stripes	*botti* Fras., p. 218.
	Thorax broadly golden yellow on dorsum.	*dorsalis auricolor* Fras., [p. 226.
4.	15 postnodals in fore-wings; anal appendages entirely black	*nigra* Fras., p. 225.
	13 postnodals in fore-wings; inferior anal appendages tipped with white	*autumnalis* Fras., [p. 223.
5.	A red transverse stripe traversing head from eye to eye	*verticalis verticalis* [Selys, p. 213.
	Head entirely black	6.
6.	13 to 14 postnodals in fore-wings, 12 in the hind	*verticalis annandalei* [Fras., p. 214.
	16 postnodals in fore-wings, 14 in the hind........................	*verticalis andamanensis* [Fras., p. 218.

78. Caconeura verticalis verticalis Selys.

Alloneura verticalis Selys, Bull. Acad. Belg. (2) vol. x, p. 453 (1860).

Disparoneura verticalis Selys, Mém. Cour. vol. xxxviii, p. 176 (1886); id., Ann. Mus. Civ. Genova, (2) vol. vii (xxvii), p. 484 (1889); Kirby, Cat. Odon. p. 134 (1890); Kruger, Stett. Ent. Zeit. vol. lix, p. 293 (1898); Laid., Fasc. Malay. (Odon.) pt. ii, p. 14 (1907); id., Rec. Ind. Mus. vol. xi, p. 388 (1915).

Male.—Abdomen 29 mm. Hind-wing 18·5 mm.

Head: labium pale brown; labrum, bases of mandibles, and genæ reddish-brown; rest of head velvety black, traversed from eye to eye at level of anterior ocellus by a narrow blood-red stripe: eyes black above, olivaceous brown beneath. *Prothorax* black, with a red stripe on each side, broadest on the middle lobe, but only just tipping the outer ends of the posterior. *Thorax* velvety black marked with blood-red and yellow, as follows:—Narrow blood-red antehumeral stripes broadening at the middle, tapering to a fine point which extends as high as the antealar sinus; an oblique blood-red stripe traversing the whole length of the middle of mesepimeron; an oblique irregular pale yellow stripe bordering the metepimeron ventrally. Beneath pale yellow, broadly bordered with black and with two small central black points. *Legs* black, bases of femora reddish: extensor surface of tibiæ bright ochreous. *Wings* hyaline; pterostigma dark brown, covering one cell, pointed at both ends, braced, about one-fourth as long as broad: 15 to 17 postnodal nervures to fore-wings, 13 to 14 in the hind; *Cuii* only 3 cells long in fore-wings, 4 cells in the hind; the nervure *ab* of moderate length, strongly curved, shorter in the hind-wings. *Abdomen* black: segment 1 with a cordate ferruginous spot on each side; segment 2 with a broad stripe along the ventral border and a linear spot on the mid-dorsal carina extending the whole length of segment, broadened in its middle two-fourths; segments 3 to 6 or 7 with small paired baso-dorsal yellow spots, remaining segments unmarked. *Anal appendages* of the generic shape; superiors reddish on dorsum, the ventral spine and apex black; inferiors black, paler within, curled strongly and abruptly inwards at the apices.

Female.—Abdomen 30 mm. Hind-wing 20 mm.

Head: labium palest brown or yellow; labrum, bases of mandibles, and genæ pale ochreous, the former with a median point of black: anteclypeus and postclypeus palest brown, the latter with a median and two lateral brown spots along its anterior border; rest of head mat black, with a narrow pale ochreous band extending from eye to eye in the same situation as a similar band in the male. *Prothorax* black, marked similarly to the male, but pale ochre instead of blood-red; posterior lobe narrowly and deeply cleft, and with a short, robust, spine-like process on each side of the fissure strongly projected forward. *Thorax* black, with similar markings to those of the male, but ochreous in colour; the lateral black stripes beneath thorax narrower, and the two medial points of black absent. *Abdomen* with similar markings to those of the male, but broader and pale ochreous; segment 1 with a V-shaped marking on each side, the upper arm of the V corresponding to the cordate marking of the male; segment 2 with two parallel stripes on each side confluent at the base and hook-shaped at the apical end of the segment; segments 3 to 7, and usually 8 to 10, with the sides broadly pale ochreous, except at the apical ends of 3 to 6, this colour confluent with the baso-dorsal paired spots, which are creamy white: segments 8 to 10 with the mid-dorsal carina finely pale yellow, expanding on the first two segments at the apical end into small oval spots. *Anal appendages* short, conical, pointed at apex, pale yellow; vulvar scales robust, extending to the end of abdomen, black, yellow above.

Distribution.—Borneo, whence the *type*. in the Selysian collection, comes, and Lower Burma. I have seen specimens from Mergui, Lower Burma, which do not differ greatly from others from Lio Natu, Borneo. This species seems to be dominant within the genus, if we include with it the closely related subspecies *annandalei*, *burmanensis*, and *andamanensis*, which extend its area to Western India.

79. **Caconeura verticalis annandalei** Fraser. (Figs. 96 & 97.)

Caconeura annandalei Fras., J. Bombay Nat. Hist. Soc. vol. xxvii, pp. 543, 544 (1921); id., ibid. vol. xxix, p. 743 (1923); id., Rec. Ind. Mus. vol. xxvi, pp. 503, 504 (1924); id., ibid. vol. xxxiii, p. 449 (1931).

Male.—Abdomen 29 mm. Hind-wing 19 mm.

Head: labium pale brown: labrum, genæ, bases of mandibles, and anteclypeus ochreous or reddish-brown; rest of head velvety black, including upper half of eyes. (In teneral specimens the head is traversed by a pale yellowish band at the level of the anterior ocellus representing a similar band seen in *C. verticalis*, but it becomes obsolete well before the adult stage.) *Prothorax* velvety black, with a small comma-like ochreous spot on each side of the middle lobe. *Thorax* velvety black, marked with very narrow antehumeral brick-red stripes not extending as far as the antealar sinus. Laterally two broad oblique stripes, one over the mesepimeron ochreous tinged with red along its anterior border, the second, lying along the ventral border of metepimeron, pale yellow; *beneath yellow*, bordered with black, and pulverulent throughout in the adult stage. *Legs* black, bases of all femora ringed narrowly with yellow; extensor surfaces of tibiæ greenish-yellow. *Wings* hyaline, enfumed palely at apices, especially in the hind; pterostigma dark reddish-brown, acutely pointed at each end, braced; *Cuii* 3 or rarely 4 cells long in the fore-wing, 4 or rarely 5 in the hind; *ab* vestigial, but always present; 13 to 14 postnodal nervures in fore-wings, 12 in the hind. *Abdomen* black, marked with pale yellow as follows:—Segment 1 with a small spot on each side; segment 2 with its ventral border narrowly yellow, and with a linear spot on the mid-dorsal carina not extending to either end of segment; segments 3 to 7 with small paired baso-dorsal spots; the mid-dorsal carina of segments 3 and 4, and sometimes 5, finely yellow. *Anal appendages* black, inner surfaces of inferiors yellowish; superiors and inferiors of the usual

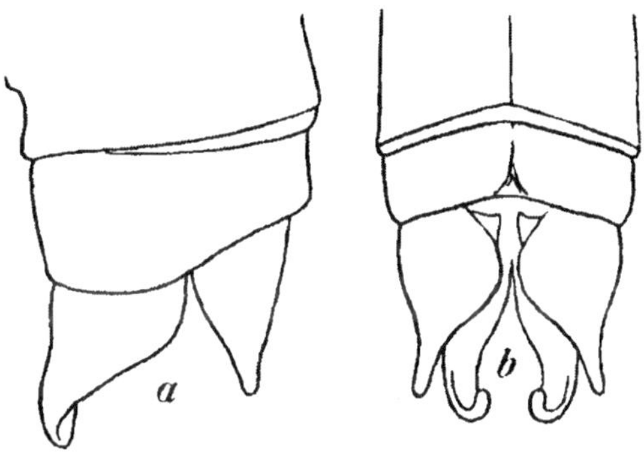

Fig. 97.—Anal appendages of *Caconeura verticalis annandalei* Fras. *a*. Left lateral view. *b*. Dorsal view.

Caconeura shape, and not differing from those of *C. verticalis verticalis*.

Female.—Abdomen 30 mm. Hind-wing 20 mm.

Of more robust build than the male, and the markings pale creamy white instead of bright red or ochreous. *Head*: labium pale yellow; labrum, genæ, bases of mandibles, clypeus, and the anterior margin of frons narrowly pale yellow, postclypeus with a median raphe and a small spot on each side dark brown; rest of head black, traversed by a yellow stripe at the level of anterior ocellus extending from eye to eye. Eyes dark olivaceous above, pale grey below. *Prothorax* black, traversed on each side by a broad yellow stripe, and marked on the mid-dorsum of middle lobe with a small geminate spot; posterior lobe deeply and rather narrowly cleft, a short, forwardly directed spine springing from each side of the latter. *Thorax* black on dorsum, with moderately narrow yellow antehumeral stripes extending nearly or quite up to the antealar sinus, which latter bears two small yellow spots; a tiny upper humeral spot on each side; laterally broadly yellow, with an oblique black stripe of variable width mapping out the postero-lateral suture; beneath similar to the male. *Legs* yellow, extensor surface of femora and flexor surface of tibiæ, as well as tarsi and spines, black. *Wings* hyaline; pterostigma yellow, dark olivaceous at the centre and framed in black nervures; 12 to 13 postnodal nervures to fore-wings, 11 in the hind; *Cuii* 3 cells long in fore-wings, 4 in the hind. *Abdomen* similar to the male, but the sides of all segments rather broadly yellow, clouded with dark brown nearly as far as apex, and segment 2 with two parallel stripes on each side. Segments 2 and 8 with the mid-dorsal carina finely yellow, extending to the base on the former segment; segment 9 with a spear-shaped stripe on the mid-dorsal carina, whilst the dorsum of 10 bears a saddle-shaped marking; the ventral borders of segments 8 to 10 narrowly bright yellow. *Anal appendages* shortly conical, black beneath, bright yellow above; vulvar scales robust, but not extending beyond end of abdomen.

Distribution.—WESTERN GHATS up to 4,000 ft. The *type*, in the British Museum, comes from Mahableshwar, where this form is common; it occurs in Coorg and throughout the Malabar and Nilgiri Wynaad, and also sparingly in the low hills south of Madras.

It differs from *C. verticalis verticalis* by the total absence of the stripe on the vertex, by the restricted markings of the prothorax, and by having the thorax yellow beneath. The female differs by the much shorter spines on the posterior lobe of prothorax, etc.

80. **Caconeura verticalis burmanensis,** subsp. nov.

Male.—Abdomen 31 mm. Hind-wing 19 mm.

Head: labium brownish-black; labrum, bases of mandibles, genæ, and anteclypeus reddish-ochreous; frons, postclypeus, vertex, and occiput velvety black, the vertex traversed by a narrow bright red band extending from eye to eye of anterior ocellus; eyes black above, olivaceous brown below. *Prothorax* black, with a narrow bright red stripe on each side, broadest on middle lobe, but only just tipping the outer ends of posterior lobe; the latter simple, rounded. *Thorax* velvety black, marked as follows:—A pair of bright red antehumeral stripes extending the whole length of dorsum, broadened at the middle, tapering gradually above, but rather abruptly below; two oblique stripes on each side, one on the mesepimeron yellow, tinted with red anteriorly, the second narrower, bordering the lower part of mesepimeron, yellow; small red points at the upper parts of humeral and postero-lateral sutures; *beneath wholly black*, pulverulent white in adults. *Legs* black, thinly white with pulverulescence; coxæ, tibiæ on the extensor surface, and trochanters greenish-yellow. *Wings* hyaline; pterostigma dark brown, finely framed in yellow and thick black nervures, outer end a little dilated, acutely pointed at both ends, strongly braced; *Cuii* 5 cells long in fore-wing, 6 in the hind; 14 to 15 postnodal nervures in fore-wings, 12 to 13 in the hind; *ab* vestigial but always present, shorter in the hind than in the fore-wings. *Abdomen* black, marked with pale yellow as follows:—Segment 1 with a lateral triangular spot and the ventral border narrowly, segment 2 with a fine mid-dorsal stripe extending nearly to end of segment and its ventral borders rather broadly yellow; segments 3 to 7 with small paired baso-dorsal and large subapical lateral spots. *Anal appendages* not longer than segment 10, of the usual *Caconeura* shape, superiors red on dorsum, the ventral spine black; inferiors black, apex very short and curled inward.

Female.—Abdomen 30–31 mm. Hind-wing 19–20 mm.

Differs in several respects from the male, of more robust build, the markings pale ochreous instead of red; labium yellow; labrum, bases of mandibles, genæ, and clypeus pale ochreous, the former with a small median point of black, the latter with three small black points on the anterior border of the postclypeus; the anterior border of the frons narrowly yellow, whilst the stripe on the vertex is broader and pale ochreous; eyes olivaceous green above, yellow below, a black equatorial belt separating the two areas. *Prothorax* with the posterior lobe deeply and narrowly cleft, and with a rather broad robust spine projecting on each side of the latter. *Thorax* and *abdomen* very similar to the male, but the markings pale ochreous; segments 2 to 6 have broad lateral yellow stripes confluent with the subapical spots, the first segment bearing two parallel stripes on each side; segment 9 with a mid-dorsal pale yellow spot shaped like the head of a spear, the point directed analwards; segment 10 pale yellow on dorsum. *Anal appendages* very short, conical, yellow; vulvar scales robust, but not extending beyond end of abdomen, black.

Distribution.—Maymyo, UPPER BURMA, during May, June, and July.

Differs from *C. verticalis verticalis* only by the greater length of *Cuii*. The female of the latter has not been described; a specimen before me from Ratuh and another from Natu, Borneo, differ only from the present form by the shorter *Cuii* and the higher nodal index—16 to 17 postnodals. From *C. verticalis annandalei* this form differs by the presence of the red stripe on vertex and by having the under surface of thorax entirely black (yellow bordered with black in *C. verticalis annandalei*). From *C. verticalis andamanensis* the presence of the red stripe on vertex will readily separate it, as also the greater length of *Cuii*.

81. **Caconeura verticalis andamanensis** Fraser.

Caconeura verticalis andamanensis Fras., Rec. Ind. Mus. vol. xxvi, pp. 412, 503, 504 (1924).

Differs from *C. verticalis verticalis* in the following characters :—Transverse stripe on vertex absent, the upper surface of head being entirely mat black ; the nervure *Cuii* considerably longer in both fore- and hind-wings ; thoracic markings more restricted.

The subspecies differs from *C. verticalis annandalei* by its higher nodal index—16 postnodal nervures in fore-wings, 14 in the hind ; the nervure *Cuii* is also one or two cells longer ; otherwise the two are very closely related.

Female unknown.

Distribution.—Andaman Is. only ; taken by the late Dr. N. Annandale in the neighbourhood of Mt. Harriet.

Type in the Indian Museum.

82. **Caconeura botti** Fraser.

Caconeura botti Fras., Mem. Dept. Agric. India (Ent.), vol. vii, pp. 41, 42 (1922) ; id., J. Bombay Nat. Hist. Soc. vol. xxix, p. 743 (1923).

Male.—Abdomen 27–31 mm. Hind-wing 17–19 mm.

Head : labium palest brown ; labrum pale blue, narrowly bordered with black, and with a small medio-basal point of the same colour ; genæ and bases of mandibles pale blue ; rest of head velvety black, but the vertex traversed by a narrow irregular blue stripe extending from eye to eye at the level of anterior ocellus ; eyes pale bluish-grey, broadly capped with black above. *Prothorax* black, the whole of posterior lobe and a large rounded spot on each side of middle lobe azure blue : posterior lobe rounded, simple. *Thorax* velvety black, marked with azure blue as follows :—A broad pyriform spot on each side of mid-dorsal carina, very broad below, where it extends from the mid-dorsal carina to the humeral suture, tapering above to a fine point at the antealar sinus near the upper end of humeral suture ; laterally two rather narrow stripes, one extending obliquely along the whole length of mesepimeron, the other bordering the metepimeron ventrally ; beneath palest blue, bordered on the outer side narrowly with black. *Legs* black, coxæ and trochanters spotted with yellow, bases of femora narrowly ringed with yellow ; extensor surface of tibiæ pale greenish-blue. *Wings* hyaline, palely enfumed in adults, especially towards the apices ; pterostigma blackish-brown, finely framed in pale brown, and beyond this by thick black nervures, short, but slightly longer than broad, strongly braced, covering 1 to 2 cells ; 15 to 17 postnodal nervures in fore-wings, 14 to 15 in the hind ; the nervure *ab* entirely absent ; *Cuii* only 1 to 2 cells long in fore-wings, 3 cells in the hind. (Abnormally, and very rarely, a vestige of *ab* present.) *Abdomen* black ; segment 1 with a small triangular blue spot on each side near the apical border ; segment 2 with its ventral border narrowly blue, as also the basal joint ; segment 3, and sometimes 4, with a pair of minute basal blue points on dorsum ; remaining segments unmarked. *Anal appendages* of the usual *Caconeura* shape ; superiors pale blue on dorsum, the ventral spine black, the apex acuminate ; inferiors creamy white tipped with black, the apices angulated inward and slightly upward.

Female.—Abdomen 29 mm. Hind-wing 19 mm.

More robust than the male, and differing markedly by having the antehumeral stripe very narrow instead of dilated below, and also by its falling well short of the antealar sinus. *Prothorax* with posterior lobe black, its outer edge only edged with blue ; deeply and narrowly cleft at its middle and bearing on each side of the fissure a long needle-like spine directed obliquely forward. *Wings* similar to those of the male ; 16 postnodal nervures in fore-wings, 14 in the hind ; *ab* invariably absent. *Abdomen* with the following additional distribution of blue :—An angulate spot on each side of segment 1, the sides of semgents 3 and 4, and, more narrowly, the ventral borders of segments 8 and 9. *Anal appendages* small, conical, black ; vulvar scales robust, extending well beyond end of abdomen, black, bordered with white above.

Distribution.—King Island, Mergui, Lower Burma, during June.

The description of the male has been amplified and that of the female given for the first time since more material has come to hand. This insect, along with the following, belongs to a group of very closely related species scattered throughout Malaysia, Borneo, and the Philippines. It is distinguished from *C. notostigma* Selys, from Singapore, by having the whole of the posterior lobe blue, a blue band on vertex and the dorsum of the superior anal appendages blue. From *C. interrupta*, from the same locality, it may be distinguished by the blue posterior lobe of prothorax and by the terminal abdominal segments being without markings. From *C. collaris* Selys, from Malacca and Borneo, it differs by the complete absence of the anal bridge, by the antehumeral bands extending right up to the antealar sinus, by the posterior lobe of prothorax being wholly blue, and by *Cuiii* being of greater length. Lastly, from *C. integra* Selys, from the Philippines, which resembles it very closely, it is to be distinguished by the complete absence of the anal bridge, by having the posterior lobe of prothorax entirely blue and the terminal abdominal segments without markings.

83. **Caconeura theebawi** Fraser.

Caconeura theebawi **Fras.**, Mem. Dept. Agric. India (Ent.), vol. vii, pp. 40, 41 (1922); id., J. Bombay Nat. Hist. Soc. vol. xxix, p. 743 (1923).

Male.—Abdomen 28–29 mm. Hind-wing 17·5–19 mm.

Head: labium yellow; labrum blue, broadly bordered with black, and with a medio-basal spot of the same colour: genæ, bases of mandibles, and a narrow transverse stripe extending from eye to eye on the vertex at the level of the anterior ocellus bluish-purple; anteclypeus dark brown, rest of head velvety black; eyes pale grey, broadly capped with black. *Prothorax* black, with a bluish purple stripe on each side, broadest on the middle lobe; posterior lobe rounded, and with a narrow moderately deep notch at the middle. *Thorax* black, with bluish-purple markings as follows:—A narrow antehumeral stripe not extending up as far as the antealar sinus, an oblique stripe on each side traversing the middle portion of mesepimeron, and a similar stripe, but broader and paler, on the ventral border of metepimeron; beneath yellow, bordered outwardly by a narrow well-defined black stripe. *Legs* black, bases of all femora narrowly yellow; coxæ and trochanters mottled with yellow; tibiæ pale bluish-green on the extensor surface. *Wings* hyaline; pterostigma pointed at both ends, nearly half as long again as wide, braced, blackish-brown, finely framed in yellow and thick black nervures; 14 to 15 postnodal nervures in fore-wings, 12 to 13 in the hind; the nervure *ab* very short, in some specimens almost obsolete, or more rarely entirely absent in all wings; *Cuii* 3 to 4 cells long in fore-wings, 4 to 5 in the hind. *Abdomen* black, marked with blue as follows:—A large cordate spot on each side of segment 1, and the apical half of the same segment narrowly; segment 2 with its ventral border pale yellow, a bluish spot near the apical end of each side and a fine pale yellow line on the mid-dorsal carina; segments 3 to 6 with small paired, blue, baso-dorsal spots; segment 7 unmarked; segment 8 with an azure blue trefoil-shaped spot at the apical end of the mid-dorsum; segment 9 with a broad triangular azure blue spot extending the whole length of dorsum, with its base resting on the apical border of the segment; segment 10 with a broad saddle-shaped, dorsal, azure blue spot. *Anal appendages* of the usual *Caconeura* shape, not differing markedly from those of *C. verticalis verticalis*, bright chrome-yellow, the ventral spine of the superiors black; inferiors curved abruptly inward at apices, which are acute and tipped with black.

Female.—Abdomen 26–30 mm. Hind-wing 17–18·5 mm.

Markings almost exactly similar to those of the male; labrum more narrowly bordered with black; the transverse stripe on vertex broader and crenate behind. Posterior lobe of *prothorax* deeply cleft at its middle into two minor lobes, on the outside of which springs a long thin spine directed strongly forwards; the lateral stripes on thorax broader. *Wings* with 13 to 16 postnodal nervures to fore-wings, 12 to 13 in the hind. *Abdomen* similar to the male as far as segment 7, but the sides of segments 2 to 6 with broad longitudinal yellow stripes clouded with brown and, on segment 2, split into two parallel stripes; segment 8 with a minute apico-dorsal spot of blue; 9 with its mid-dorsal carina narrowly blue, broadening out as a triangle at the apex of segment; segment 10 similar to the male. *Anal appendages* short, conical, pointed, black; vulvar scales black, very robust, extending beyond the end of abdomen.

Distribution.—King Island, Mergui, LOWER BURMA; found along the banks of small streams in rubber plantations.

The above description has been corrected and amplified with the aid of more material which has come to hand since the original description was compiled. The species is to be distinguished from *C. botti* Fras., *C. interrupta* Selys, *C. notostigma* Selys, *C. integra* Selys, and *C. collaris* Selys, all of which have the antehumeral stripes broadened below, especially so in *C. botti* and *C. collaris*. From *C. botti* it is also distinguished by the presence of an anal bridge. It bears a close resemblance to *D. campioni* Fras., but the complete anal bridge in the latter will prevent any confusion.

Type in the British Museum; paratypes in the Pusa and author's collections.

84. **Caconeura sita** Kirby.

Disparoneura sita Kirby, J. Linn. Soc. Lond., Zool. vol. xxiv, p. 563 (1894).
Caconeura sita Laid., Rec. Ind. Mus. vol. xiii, pp. 323, 348 (1917); Fras., J. Bombay Nat. Hist. Soc. vol. xxix, p. 743 (1923); Laid., Spolia Zeylan. vol. xii, p. 367, figs. 14 A, 14 B (1924).
Caconeura canningi Fras., J. Bombay Nat. Hist. Soc. vol. xxvi, p. 877 (1919); id., Rec. Ind. Mus. vol. xxvi, p. 504 (1924).

Male.—Abdomen 29 mm. Hind-wing 19 mm.

Head: labium white tipped with black; labrum and anteclypeus dark brown to ochreous; bases of mandibles black; genæ pale bluish; postclypeus, frons, vertex, and occiput mat black, the vertex traversed from eye to eye at the level of the anterior ocellus by a narrow partially interrupted stripe, bluish in colour, consisting of a narrow medial portion, followed by a small triangular spot lying midway between the posterior ocelli and the bases of antennæ, and finally, a narrow outer portion running to the eye; eyes dark olivaceous brown, capped with black above, paler below, and with a dark brown equatorial narrow belt. *Prothorax* black, with a narrow lateral subdorsal pale blue stripe on each side; posterior lobe rounded, simple. *Thorax* bronzed black on dorsum, pale bluish laterally, marked by a broad oblique black stripe on the postero-lateral suture; dorsally a pair of pale bluish antehumeral stripes tapering to a fine point near the antealar sinus; beneath whitish, bordered with black outwardly. *Legs* pale ochreous, marked broadly with black on the hinder surface of femora and flexor surface of tibiæ, the latter bright ochreous on extensor surface; tarsi black. *Wings* hyaline; pterostigma dark brown, framed finely in pale brown or yellow and by thick black nervures, braced, covering 1 to 2 cells; *Cuii* 4 cells long in fore-wings, 5 in the hind; 13 to 14 post-nodal nervures in fore-wings, 12 in the hind; *ab* vestigial, very short, a tiny arc from the top of which the nervure *ac* springs. *Abdomen* black, marked with yellow laterally; segment 1 with a large triangular spot on each side; segment 2 with its mid-dorsal carina finely and its sides broadly yellow; segments 3 to 6 with small paired baso-dorsal bluish-white spots and large diffuse pale brown subapical spots on each side; remaining segments unmarked. *Anal appendages* of the usual generic shape; superiors longer than inferiors, creamy white on dorsum, the ventral spine robust, bent strongly analwards; inferiors pale white or yellow, short, obtuse at apex, which is curled slightly upwards.

Female.—Abdomen 30–32 mm. Hind-wing 19 mm.

Very similar to the male, but more robust and with broader markings, as follows:—Segment 1 with a broad cordate spot on each side; segment 2 with the mid-dorsal carina well defined in ochreous, this line continued on to the following two or three segments; laterally segment 2 with a broad lateral stripe angulated upwards apically, and confluent here over the dorsum in tenerals; segments 3 to 6 with small paired baso-dorsal bluish-white spots as in the male, but broader and confluent laterally, with a broad yellow stripe running the length of each segment; segments 8 to 9 narrowly yellow ventrally, and with a fine yellow line on the mid-dorsal carina, broadening on 9 and ending as a large triangular spot on 10. The transverse stripe on vertex complete, and the lateral one on the prothorax broader than in the male. *Prothorax* with two robust black spines on the posterior lobe, strongly angulated and sloping forwards. *Wings* similar to the male, but the pterostigma a warmer brown, and covering a little more than one cell ; *Cuii* only 3 cells long in fore-wings, 4 in the hind. *Legs* similar to the male. *Anal appendages* pale yellow, very short, conical, pointed ; vulvar scales robust, extending a little beyond the end of abdomen, black laterally, yellow above.

Distribution.—Found widely distributed throughout the low country of CEYLON. Senior-White records it from the banks of the Suduganga River on scrub during August. Kirby records it from Trincomalee during October and September. It is the only species of the genus so far recorded from Ceylon.

The adult type-specimen is in the British Museum, where also a co-type teneral male is lodged. The female co-type, in the author's collection, is from the Suduganga River; this sex has not been described before. As in other species of the genus, teneral specimens of the male approach the female in colouring and markings, the present description thus differing from the original, which was made from a teneral male. Closely allied to *C. autumnalis* Fras., from which it differs by its more extensive markings, shorter and paler anal appendages, etc.

85. **Caconeura autumnalis** Fraser.

Caconeura autumnalis Fras., Mem. Dept. Agric. India (Ent.), vol. vii, p. 43 (1922); id., J. Bombay Nat. Hist. Soc. vol. xxix, p. 744 (1923).

Male.—Abdomen 31 mm. Hind-wing 19 mm.

Head : labium brown ; labrum glossy black ; anteclypeus dark brown, rest of head, including eyes, black. *Prothorax* and *thorax* black, unmarked in adults, but pruinosed on the sides and beneath ; rudiments of a pale blue lateral stripe in subadults on the anterior half of mesepimeron, and a creamy white stripe bordering the lower part of metepimeron. *Legs* black, extreme base of femora narrowly ringed with creamy white. Posterior lobe of prothorax rounded, simple. *Wings* hyaline, palely enfumed in adults ; pterostigma dark reddish-brown framed in black nervures, narrow, pointed at both ends, the distal and hinder borders forming a continuous arc, strongly braced, covering one cell ; 13 to 15 postnodal nervures to fore-wings, 12 to 13 in the hind ; *Cuii* 2 to 3 cells long in fore-wing, 4 to 6 in the hind ; the nervure *ab* vestigial, of variable length, almost obsolete in some, but extending to the base of discoidal cell in others. *Abdomen* velvety black ; segments 3 to 6 or 7 with small paired creamy white baso-dorsal spots; segments 3 to 6 with an apico-lateral diffuse brownish-white spot on each side; segment 1 in adults often pruinosed, and with a pale latero-apical bordering of white. *Anal appendages* black, of the usual *Caconeura* shape, superiors as long as segment 10, broad at base, conical and tapered to the apex, which is acute, the ventral spine very robust; inferiors pale brown internally, curled in rather abruptly at apex.

Female.—Abdomen 31–34 mm. Hind-wing 20–23 mm.

A more robust insect than the male. *Head* boldly marked with creamy white as follows:—Labrum, bases of mandibles, genæ, anteclypeus, penultimate segment of antennæ, and a narrow stripe traversing the vertex from eye to eye at the level of the anterior ocellus. Postclypeus variable, dark brown in some, with two broad basal yellow spots, or pale ochreous with three small anterior brown spots or a single medial linear spot. Eyes black above, olivaceous beneath. *Prothorax* black, with a broad lateral creamy white stripe on each side, broadest on the middle lobe; posterior lobe very deeply and narrowly cleft at its middle almost to its base, thus forming two small almost separate lobes, on the outer side of which is a short recurved spine. *Thorax* bronzed black, marked with a narrow creamy white antehumeral stripe in continuation of that on the prothorax and extending almost up to the antealar sinus, which latter bears a small creamy spot on each side. Laterally two broad oblique stripes traversing the whole length of the middle of mesepimeron and the lower border of metepimeron; a small white spot situated between the upper ends of these two stripes. Beneath creamy white, bordered outwardly with black. *Legs* yellow, with black spines; distal ends of all femora on the extensor surface, and flexor surface of the anterior pair of tibiæ brown. *Wings* with 14 to 15 postnodal nervures to fore-wings, 12 to 14 in the hind. *Abdomen* black, marked with pale yellow as follows:—The lower border and a cordate spot on each side of segment 1; two parallel stripes on each side of segment 2 confluent at the base; paired baso-dorsal spots on segments 3 to 6, as in the male, but larger; a broad lateral fascia on the sides of segments 3 to 6, becoming rather obscure on 4 to 6; segment 8 with the mid-dorsal carina finely yellow, its ventral border broadly; 9 with a yellow spot on the mid-dorsum shaped like a spear-head, the point directed apically; segment 10 with its mid-dorsum and apical border narrowly yellow, its ventral border, as also that of 9, broadly so. *Anal appendages* short, conical, pointed at apex, yellow; vulvar scales robust, extending slightly beyond the end of abdomen, dark brown, paler above.

Distribution.—Assam, Burma, Tonkin, and Java during June to August. Common at Maymyo, Upper Burma. I have seen a single male from Hoa Minh, Tonkin, which, except for its smaller size (abdomen 29 mm., hind-wing 17 mm.) and rather more reduced markings, does not differ from the type. Java specimens differ even less from those of Assam and Burma.

Type in the British Museum; paratypes in the Pusa and author's collections.

86. **Caconeura nigra** Fraser.

Caconeura nigra Fras., Mem. Dept. Agric. India (Ent.), vol. VII, pp. 42–3 (1922).

Male.—Abdomen 28·5 mm. Hind-wing 18 mm.

Head, prothorax, and *thorax* velvety black, without markings; eyes pale grey beneath, black above. *Legs* black. *Wings* hyaline; pterostigma black, about half as long again as broad; 15 postnodal nervures in fore-wings, 13 in the hind; the nervure *ab* greatly reduced, but a vestige always present; *Cuii* 5 cells long in fore-wings, 5 to 6 in the hind. *Abdomen* black; segment 1 with a small whitish spot on each side; segment 2 narrowly bordered with white ventrally; segments 3 to 6 with tiny paired, white, baso-dorsal spots, remaining segments unmarked. *Anal appendages* black, of the usual *Caconeura* shape, but rather broader at the base and shorter; ventral spine of superiors with very broad base: inferiors with apices strongly angulated inwards.

Female unknown.

Distribution.—King Island, Mergui, LOWER BURMA, during June.

This species is very closely related to *C. autumnalis* and *C. o'doneli*; indeed, it is possible that the three may be but local races of a single species, but more material is needed before we can pronounce on this. From the first of the above species *C. nigra* differs by its extreme melanism, there being no markings, save on the basal abdominal segments; from the second the presence of the nervure *ab* is sufficient to distinguish it.

87. **Caconeura o'doneli** Fraser.

Caconeura o'doneli Fras., J. Bombay Nat. Hist. Soc. vol. XXIX, p. 743 (1923).

Male.—Abdomen 29 mm. Hind-wing 17·5 mm.

Very closely allied to the two former species, from which it differs in the following respects:—*Thorax* with a more or less complete ochreous oblique stripe traversing the middle of mesepimeron, but often obscured above; pterostigma black, finely framed in yellow and thick black nervures; 12 to 13 postnodal nervures only in fore-wings, 11 to 12 in the hind; ***the nervure ab always entirely absent in all wings.***

Female.—Abdomen 29 mm. Hind-wing 19 mm.

Differs from the female of *C. autumnalis* (that of *C. nigra* being unknown) in the following characters :—Labrum with a medio-basal spot of blackish-brown ; postclypeus black ; the transverse creamy white stripe on vertex much narrower. *Thorax* with the antehumeral stripes absent or represented by a mere chain of minute dots ; the dorsal marking on segments 8 and 9 much reduced, that on the latter expanded ever so slightly apically ; segment 10 with only a minute dorsal spot. *Wings* with 12 to 13 postnodal nervures in the fore, 11 only in the hind ; pterostigma pale brown, finely framed in yellow and thick black nervures ; *the nervure ab entirely absent in all wings*, as in the male. *Anal appendages* black, not differing in shape from those of *C. autumnalis*.

Distribution.—Duars, BENGAL. Found along the banks of the Silli Torsa River by Mr. H. V. O'Donel.

Type in the Fraser collection. The total absence of markings in the males, save on the thorax, and the entire absence of the anal bridge (*ab*) in all wings, which seems to be a very constant character, will serve to distinguish it from all other species of the genus.

88. **Caconeura dorsalis auricolor** Fraser.

Caconeura dorsalis auricolor Fras., Rec. Ind. Mus. vol. xxix, p. 90 (1927).

Male.—Abdomen 29–30 mm. Hind-wing 18 mm.

Head : labium black ; labrum ochreous, marked with a medio-basal point of black ; bases of mandibles and genæ blue ; anteclypeus olivaceous brown ; postclypeus black, or in tenerals orange, marked with three small black spots along its anterior border, and with its base finely black ; rest of head velvety black, with a broad transverse bright orange stripe extending from eye to eye on the vertex at the level of the anterior ocellus ; eyes olive-green, darker above, the two zones being separated by a narrow equatorial line of black. *Prothorax* velvety black, marked with bright orange as follows :—A stripe running the whole length of each side and broadest at the middle lobe, a pair of juxtaposed triangular spots on the dorsum of the middle lobe and the whole of the posterior lobe. *Thorax* velvety black, with the whole of the dorsum between the humeral sutures bright golden yellow, the mid-dorsal carina being very finely black ; laterally two pale blue or pale greenish-blue oblique stripes, one extending along the centre of the mesepimeron, the other along the ventral border of the metepimeron ; beneath black, the central part white and often pruinosed. *Legs* black, femora white on the flexor surface, tibiæ yellow on the extensor. *Wings* hyaline ;

pterostigma blackish-brown, finely framed in yellow and thick black nervures, covering 1 to 1½ cells, lozenge-shaped; the nervure *ab* always present, of variable length, but always incomplete; *Cuii* 4 cells long in fore-wings, 5 in the hind; 14 to 16 postnodal nervures in fore-wings, 13 to 14 in the hind. *Abdomen* black, marked with palest blue as follows:—Segment 1 with a large spot on each side; 2 with a narrow ventro-lateral stripe and a narrow fusiform streak on the mid-dorsal carina; segments 3 to 7 with small paired baso-dorsal and subapical ventro-lateral spots, segments 8 and 9 unmarked, 10 broadly blue on dorsum. *Anal appendages* black, superiors blue on dorsum, inferiors carneous within; not differing in shape from those of *C. verticalis*.

Female.—Abdomen 30 mm. Hind-wing 20 mm.

Resembling the male closely in regard to markings, which are, however, palest ochre instead of orange or golden yellow. *Head*: labrum, genæ, and bases of mandibles palest ochre, anteclypeus ochreous enfumed with brown; postclypeus black. *Prothorax* without the mid-dorsal marking on middle lobe, and the posterior lobe marked with yellow at the tips of the sides only; this lobe deeply cleft at its middle, and with a broad robust flattened spine directed rather sharply forwards. *Thorax* with narrow, complete, ochreous antehumeral stripes, which replace the broad golden fascia seen in the male; laterally and beneath similar to the male. *Legs* black, coxæ, trochanters, and femora creamy yellow, the latter with a broad beaded black stripe on the extensor surface; tibiæ bright yellow on extensor surface. *Wings* similar to the male; 13 to 15 postnodal nervures to fore-wings, 12 to 13 in the hind; *Cuii* 4 cells long in fore-wings, 5 in the hind, the nervure *ab* always present. *Abdomen* black, marked with pale ochreous as follows:—Segment 1 with its apical border finely and a large spot on each side; segment 2 with two parallel lines on each side, the upper angulated apically; segments 2 to 6 with the mid-dorsal carina finely yellow and with paired linear baso-dorsal spots; segment 7 unmarked, whilst 8 has its mid-dorsal carina finely, and its ventral border, as well as that of 9, broadly yellow; segment 9 with a spear shaped marking on the dorsum, the head of the spear directed apically; segment 10 broadly yellow on dorsum. *Anal appendages* and a protuberance between and below them pale yellow; vulvar scales robust, short, dark brown.

Distribution.—Maymyo, UPPER BURMA, during August. Appears to be common locally, and is probably found in other localities throughout Burma.

Closely related to *C. delicatula* Lieft., from which it is only distinguished by having the epistome glossy black instead of bright orange, and segment 10 marked with blue. Both forms are related to *C. dorsalis* Selys from Borneo, in which the transverse orange stripe on the vertex is represented by two triangular spots, the labrum black instead of orange, and the nervure *ab* entirely absent.

Genus **DISPARONEURA** Selys (1860). (Fig. 98.)

Disparoneura Selys, Bull. Acad. Belg. (2) vol. x, p. 443 (1860); id., Mém. Cour. vol. xxxviii, p. 160 (1886); Kirby, Cat. Odon. p. 133 (1890); Laid., Rec. Ind. Mus. vol. xiii, pp. 344, 345 (1917); id., Proc. Zool. Soc. Lond. pp. 338, 339 (1920); Ris, Odon. S. Afr. p. 293 (1920).

Dragonflies of small size and slender build, with abdomen less than twice the length of wings: coloured black, marked with bright red, orange, yellow or blue, less rarely without markings or pruinosed. Wings hyaline, narrow, moderately rounded at apices, and with a long petiole; discoidal cell elongate, costal and posterior sides equal or slightly subequal, and four times as long as the base and distal end; sectors

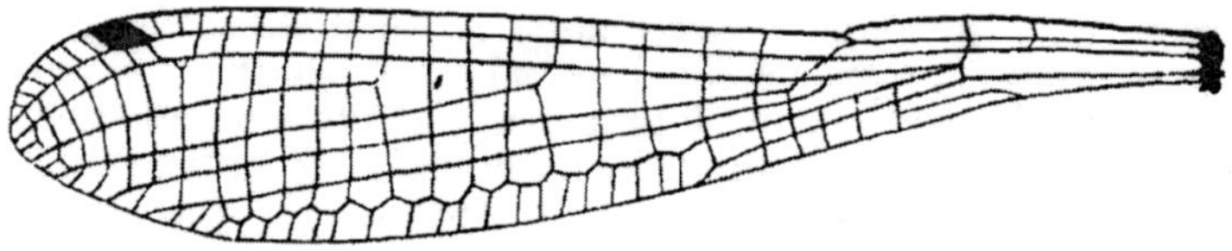

Fig. 98.—Wing of *Disparoneura atkinsoni* Selys, male.

of arc arising from the lower end of arc, divergent from origin; no accessory basal postcostal nervure present; the nervure *ac* situated about midway between the two antenodal nervures; *ab always present and always complete*; *IA* absent; *Cuii* much reduced, rarely more than 4 cells in length; *Riv+v* arising well proximal to the nervure descending from the subnode; *IRiii* in continuation of that nervure or arising very slightly distal to it; pterostigma small, about $1\frac{1}{2}$ times as long as broad, diamond-shaped, distal and proximal ends oblique, the latter strongly braced; areolets of wings mainly quadrangular.

Head narrow; eyes smaller than in *Platysticta* or *Protosticta*; occiput simple, not ridged posteriorly; prothorax with the posterior lobe simple and rounded in the male, armed with one or two pairs of hooks in the female; abdomen cylindrical, slim, broadened at base and again towards the terminal segments, about half as long again as the wings; anal appendages resembling those of *Caconeura* and very homogeneous, superiors shaped like the hammer of a revolver, tumid at base, tapering and pointed at apex, and with a broad ventral spine near the base; inferiors simple, strongly bevelled from above down, very broad at base, tapering to apex, which is bent strongly inwards and finally a little upwards, obtuse or acute at extreme apex. Genitalia: lamina narrowly and deeply cleft; hamules broad foliate plates with an acute terminal spine; lobe elongate, narrow, depressed, slipper-shaped; penis with apex ending in two long strongly curled branches which embrace the stem of the organ. Female with robust vulvar scales extending to end or slightly beyond end of abdomen; anal appendages very short, conical, pointed at apex.

Distribution.—Similar to that of *Caconeura*, but wider in that it has succeeded in colonizing the continent of Africa. Species are found throughout INDIA and CEYLON save desert tracts, BURMA, the Malay Peninsula, Sumatra, and Borneo.

So far as I know, no single specimen of any species of *Disparoneura* has been taken in which the anal bridge (*ab*) has been found incomplete as in *Caconeura*, so that the separation of the two genera, although resting on this single character, is sound. The genotype is in doubt, as the one given by Selys—*Agrion glaucum* Burm.—has been shown by Calvert to be an *Enallagma*—*E. glaucum* (Burm.). It seems convenient, therefore, to take the next Selysian species—*D. mutata*—for the generic type, especially as its author remarks that it is, perhaps, only a race of "*glauca*" (*E. glaucum*). The genus as it originally stood has been somewhat cut up by Laidlaw and myself, and the following genera have been derived from it: *Chloroneura* Laid., *Indoneura* Laid., *Phylloneura* Fras., *Esme* Fras., and *Melanoneura* Fras.

Key to the Species of Disparoneura.

1.	Black species, without markings	2.
	Black species, marked with bright red	3.
	Black species, marked with azure blue	*campioni* Fras., p. 242.
2.	Thorax thinly pruinosed; superior anal appendages bifid at apex as seen in profile	*tetrica* Laid., p. 235.
	Thorax not pruinosed; superior anal appendages truncate at apex seen in profile	*nigerrima* Laid., p. 236.
	Thorax densely pruinosed white on dorsum.	*cæsia* (Selys), p. 240.
	Thorax steely metallic purple	*centralis* Hagen, p. 238.
3.	Very large species, with abdomen nearly 40 mm. long; Ceylon only	*tenax* (Hagen), p. 241.
	Smaller species, with abdomen about 30 mm. long; Continental India only	4.
4.	Red markings confined to head and thorax	*atkinsoni* Selys, p. 230.
	Red markings extending on to abdomen as far as the base of segment 3	*souteri* Fras., p. 232

NOTE.—*D. cæsia* and *D. centralis* are from Ceylon only, *D. campioni*, *D. tetrica*, and *D. nigerrima* from Continental India and Burma.

89. **Disparoneura atkinsoni** Selys. (Figs. 98 & 99.)

Disparoneura atkinsoni Selys, Mém. Cour. vol. xxxviii, p. 168 (1886); Kirby, Cat. Odon. p. 133 (1890); Laid., Rec. Ind. Mus. vol. xi, p. 388 (1915); id., ibid. vol. xiii, pp. 323, 347 (1917); Fras., J. Bombay Nat. Hist. Soc. vol. xxix, p. 743 (1923).
Disparoneura fletcheri Fras., J. Bombay Nat. Hist. Soc. vol. xxvi, p. 876 (1919); id., ibid. vol. xxix, p. 743 (1923).

Male.—Abdomen 31 mm. Hind-wing 22 mm.

Head: labium pale brown; labrum, bases of mandibles, genæ, clypeus, frons, and vertex as far back as a line drawn posterior to the anterior ocellus bright blood-red, traversed by a narrow black stripe at the level of antennæ extending from eye to eye and encroaching somewhat on the frons; labrum with a tiny median black point at base; anteclypeus suffused with brown, and postclypeus with three small black spots; a small black oval spot also on each cheek confluent with the brown of anteclypeus; base of antennæ reddish; occiput mat black, with a minute point of red on each side at back of head; eyes dark olivaceous above, paler beneath, capped with black, and with an equatorial black belt confluent with the black stripe on vertex. *Prothorax* black, the lower border of middle lobe narrowly ochreous, the anterior border of anterior lobe narrowly red, a narrow blood-red stripe traversing the middle lobe on each side, a small geminate spot on the mid-dorsum of middle lobe, a tiny point on each side just in front of the latter, and a fine mid-dorsal streak on the posterior lobe all of the same colour. *Thorax* velvety black on dorsum as far back as the middle of the mesepimeron, marked with broad blood-red antehumeral stripes, which are continuous in front with the lateral stripe on prothorax and narrowly separated behind from a spot of the same colour on the antealar sinus; on the upper part of the mesepimeron a bright ochreous or reddish elongate spot shaped like the human leg, and above this two small spots or red points just posterior to the upper end of humeral suture. Laterally bright ochreous, suffused partly with red and traversed by an irregular stripe of black on the postero-lateral suture, which sends off two prolongations on the metepimeron posteriorly, an upper angulated spot and a lower elongate one, which traverses the metepimeron from before back in its anterior half; finally, a black stripe bordering the posterior half of the hinder lower border of the metepimeron; beneath pale olivaceous, marked with two black spots which taper to a point posteriorly. *Legs* black, coxæ and trochanters irregularly marked with red, as also the flexor surfaces of femora distally; extensor surfaces of tibiæ ochreous (some specimens have the femora ochreous spotted and striped with black). *Wings* hyaline, palely enfumed in adults; pterostigma diamond-shaped, dark ochreous framed in black nervures, braced, covering one cell; 17 postnodal nervures to fore-wings, 14 in the hind; *ac* situated nearer the distal antenodal, especially in hind-wing; *Cuii* 4 cells long in fore-wings, 5 in the hind. *Abdomen* black, marked with blood-red as follows:—A triangular spot on each side and the lower border of segment 1 narrowly (these markings sometimes yellow); a narrow mid-dorsal stripe not quite extending to ends of segment and broadened medially, the ventral border also narrowly red; segments 3 to 7 with small paired baso-dorsal spots and larger lateral subapical obscure reddish spots; segments 8 and 9 with narrow apical rings, whilst 10 has small basal and apical red spots. *Anal appendages* black, rather longer than segment 10, of the usual generic shape; superiors ending in an acute apex, and with a broad acutely pointed ventral spine;

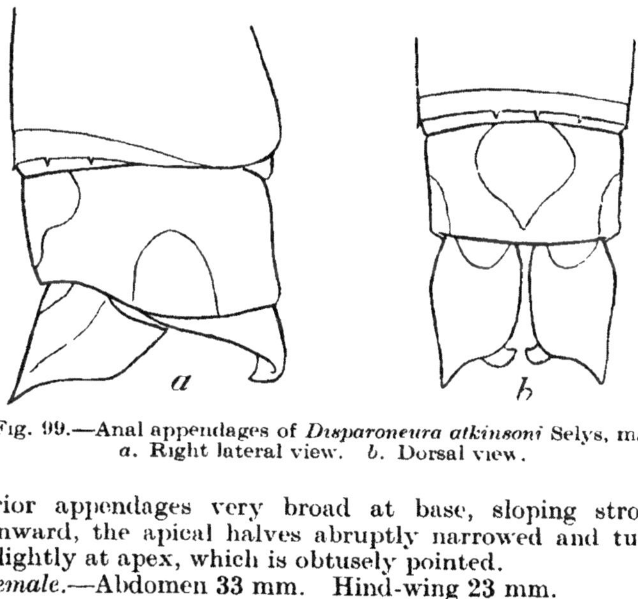

Fig. 99.—Anal appendages of *Disparoneura atkinsoni* Selys, male. *a*. Right lateral view. *b*. Dorsal view.

inferior appendages very broad at base, sloping strongly downward, the apical halves abruptly narrowed and turned up slightly at apex, which is obtusely pointed.

Female.—Abdomen 33 mm. Hind-wing 23 mm.

Somewhat similar to the male in markings, but the ground-colour pale ochreous or yellow, and the general colouring one of marblings of black and brown on a yellow background. Differs from the male as follows:—The transverse black stripe on vertex broader and ill-defined behind, where it fades into the yellow stripe; the small spots on back of occiput absent. *Prothorax* marked similarly, the posterior lobe deeply bifid at its middle, and furnished on either side of the cleft with a robust spine, which slopes strongly forward over the dorsum of middle lobe and has a small ventral process below it; seen laterally, these spines appear as if springing from a common base, and strongly angulated away from one another.

Thoracic markings black and dark brown on a yellow ground, the dark markings shaped as in the male. *Legs* pale yellow, mottled and speckled with blackish-brown, especially on the femora. *Abdomen* with more conspicuous markings, the subapical lateral spots of greater extent and confluent with broad ventral yellow stripes on segments 3 to 6 or 7, thus appearing more as a constriction of the dorsal black; the baso-dorsal paired spots creamy white and more conspicuous; segments 8 to 10 with a mid-dorsal ochreous or yellowish white stripe, very narrow, and tapering away towards the base on the first of these segments, much broader, and broadening towards the apex of 9 and of even breadth on segment 10; segments 8 and 9 with the ventral border narrowly palest yellow. *Wings* with 16 to 17 postnodal nervures in fore-wings, 14 to 15 in the hind; pterostigma blackish-brown at centre, framed in pale yellow and thick black nervures; *Cuii* 3 cells long in fore-wings, 4 in the hind. *Anal appendages* black, as long as segment 10, which is very short, conical, and pointed, surmounting a conical brown protuberance; vulvar scale robust, extending to the end of abdomen, brown.

Distribution.—Confined to the Khasia Hills, ASSAM, from where the type was originally taken.

Type in the Selysian collection; paratypes in the Indian Museum, Pusa Museum, and the author's collection. Mr. T. B. Fletcher has taken it along the borders of small montane brooks, and remarks that the red markings are very conspicuous as the insect hovers over the water. On the wing it bears a resemblance to *C. verticalis*, and is quite indistinguishable from *D. souteri*, a closely related species which, however, is only found in S. India, and has more extensive red markings, these extending broadly on to the basal abdominal segments. The markings show some slight variations, and the ground-colour varies widely according to the age of specimens, being in tenerals a wedgewood blue, then ochreous, and finally red. It was this difference in the ground-colour which led me to describe the same species as new under the name of *D. fletcheri*. The type of the latter is in the British Museum.

90. **Disparoneura souteri** Fraser. (Fig. 100.)

Disparoneura souteri Fras., Rec. Ind. Mus. vol. xxvi, pp. 521, 523 (1924); id., ibid. vol. xxxiii, pp. 449, 468 (1931).

Male.—Abdomen 30 mm. Hind-wing 18 mm.

Head black, marked with bright cherry-red; labium brown; labrum and anteclypeus dark brown; postclypeus and cheeks pale red, this colour forming a broad band across the face from eye to eye and most intensely red on the clypeus; rest of head velvety black, traversed by a broad band of bright cherry-red, with its hinder border at level of posterior ocelli. Eyes reddish-brown above, changing gradually to greenish-yellow beneath. *Prothorax* velvety black, with a fine point of cherry-red at middle of posterior lobe, and a subdorsal stripe of the same colour in continuation of the humeral stripes. *Thorax* velvety black, marked with a broad cherry-red humeral stripe on each side of dorsum, these stripes being of the same breadth as the intervening black; laterally a stripe of citron-yellow, limited in front by the first lateral suture, and with its anterior border bright cherry-red; finally, the hinder half of the metepimeron primrose-yellow. *Legs* black, femora pale on inner side, tibiæ yellow on extensor surface. *Abdomen* black, segment 1 marked with bright cherry-red at its base, and with an angulated yellow marking

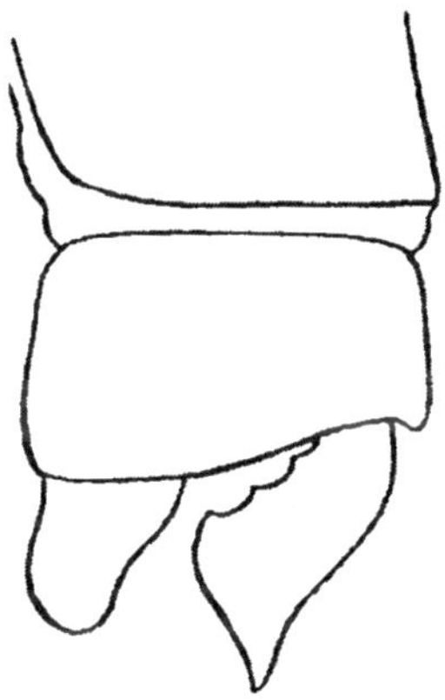

Fig. 100.—Anal appendages of *Disparoneura souteri* Fras., male, viewed from the left side.

on the sides; segment 2 broadly cherry-red on the dorsum, this marked subapically with two moderately large spots of black and a narrow black apical ring, laterally and apically the red changing to yellow, the sides entirely black below; segment 3 with its basal sixth cherry-red, changing to yellow on the sides, the red part marked with a narrow sub-basal black annule broadly incomplete on the dorsum; also a diffuse subapical yellowish spot on the sides which is repeated on segments 4 and 5, but more obscurely so; rest of abdomen black, unmarked. *Wings* hyaline, palely enfumed towards the apices in adults; pterostigma black, with a fine frame of yellow lining the inner side of the enclosing black nervures, covering one cell; 17 postnodal nervures in fore-wings, 15 in the hind; *Cuii* 3 cells long in the fore-wings, 5 in the hind. *Anal appendages*: the superiors black, marked above with a large dash of cherry-red, directed straight back, constricted at base, ending in a point; seen in profile they dilate rapidly towards the apex, and are furnished below with two robust spines, one medial, one sub-basal; inferiors black, slightly longer than superiors, directed straight back, tapering to an obtuse point, apices curled inward rather abruptly and nearly meeting at the middle line.

Female.—Abdomen 28 mm. Hind-wing 18 mm.

Head: eyes dark brown above, greenish-yellow in the lower half, the two zones of colour sharply limited from one another; labium, labrum, and anteclypeus pale brown, cheeks greenish-yellow and connected across the postclypeus by a golden yellow stripe; the red band on vertex of head replaced by a golden yellow one. *Prothorax* and thoracic markings similar to the male, but bright greenish-yellow instead of red, these stripes margined finely with golden yellow. *Abdomen* black, marked with golden-brown and olivaceous; segment 1 with an olivaceous angulated mark on the sides; 2 with a fine golden yellow line on the mid-dorsal carina, which is continued on to segment 3 nearly as far as its apical border; laterally segment 2 has a hasp-shaped mark of yellowish-brown, and beneath it a streak of yellow on the ventral border; 3 to 5 have the sides broadly golden brown, the basal portion nearly cut off by an invasion of the black, while apically the golden-brown expands into a diffuse olivaceous spot; 6 to 7 have this marking very obscure; 8 to 10 have a lateral greenish-yellow stripe continuous from segment to segment; the dorsal carina on these segments is golden yellow, this colour expanding broadly on segment 9 and less so on 10. *Anal appendages* brownish-yellow, shortly conical. *Wings* hyaline, pterostigma dark brown; nodal index: 15 postnodals in fore-wing, 13 in the hind. *Legs* black, the yellow markings brighter and more extensive than in the male.

Distribution.—Confined to the WESTERN GHATS of India, and so far reported from only two localities, Sampaji, Coorg, and near Tamaracherri, South Malabar.

This beautiful and conspicuously coloured insect (contrary to the rule in heliochromes) is very retiring, hiding in shaded spots beneath overhanging bamboo, cane or bushes on the banks of submontane streams. When on the wing the whole of the head, thorax, and base of abdomen appear uniformly red, especially when the insect emerges into the sunlight. The only species with which it may be confused is *D. atkinsoni* from Assam, which has less extensive red markings and none on the basal abdominal segments; the lateral thoracic markings of this species are also very complex, but simple in *D. souteri*.

Type in the British Museum; paratypes in the author's collection.

91. **Disparoneura tetrica** Laidlaw.

Disparoneura tetrica Laid., Rec. Ind. Mus. vol. xiii, pp. 323, 345, 346 (1917); Fras., J. Bombay Nat. Hist. Soc. vol. xxix, p. 743 (1923); id., Rec. Ind. Mus. vol. xxvi, pp. 429, 503 (1924); id., ibid. vol. xxxiii, p. 449 (1931).

Male.—Abdomen 27–28 mm. Hind-wing 18–19 mm.

Head: labium brown; labrum dark brown; genæ and anteclypeus palest blue; rest of head velvety black; eyes capped with black, pale blue beneath, marked with a black equatorial belt. *Prothorax* black, slightly pruinosed at the sides and more so beneath. *Thorax* velvety black, the anterior border of mesepimeron for the front half, the upper half of the hinder border of the postero-lateral suture, and the lower border of metepimeron creamy white; beneath pale yellow, heavily bordered with black, this part and the lower part of sides thinly pruinosed white. *Legs* black, the flexor surfaces of all tibiæ bright yellow, the femora all pruinosed white. *Wings* hyaline, the outer fifth more or less palely enfumed; pterostigma dark blackish-brown, covering from 1 to 1½ cells, braced; 12 to 13 postnodal nervures in fore-wings, 10 to 11 in the hind; *Cuii* 3 cells long in fore-wings, 4 in the hind; *ab* complete, but not uncommonly ending on the posterior border of the wing, thus approaching the condition seen in the genus *Caconeura*. *Abdomen* black, thinly pruinosed on the basal segments in adults; segment 1 with an obscure cordate lateral marking heavily pruinosed; segment 2 with its ventral border narrowly pale yellow; segments 3 to 6 or 7 with minute baso-dorsal paired pale bluish spots, the sides and beneath yellowish, this colour encroaching on the black subapically. *Anal appendages* black, the base and tips of superiors paler; superiors of the usual Disparoneurine shape, about as long as segment 10, the robust ventral tooth bifid at its apex as in *D. souteri*; inferiors very broad at base, apices obtuse and slightly inturned.

Female.—Abdomen 26–28 mm. Hind-wing 18 mm.

Head: labium pale brown; labrum and anteclypeus pale ochreous, the former with a tiny median black point; genæ, anterior border of frons, and postclypeus pale greenish-blue, the latter broadly bordered with blackish-brown; rest of the head blackish-brown, traversed at level of anterior ocellus by a narrow pale yellow stripe extending from eye to eye and involving bases of antennæ; eyes similar to those of the male, but the ground-colour pale olive green. *Prothorax* dark brown, middle lobe with a large pale greenish-white spot on each side, below and behind which is a smaller yellowish spot; posterior lobe deeply and narrowly cleft, and with a robust bright yellow horn on each side of the cleft strongly sloped forwards. *Thorax* dark blackish-brown as far as the antero-lateral suture, beyond which the sides are bright yellow with a greenish tinge, marked with a narrow black stripe on the postero-lateral suture and a chain of three spots on the centre of metepimeron, the central one elongate; dorsally a rather broad ochreous antehumeral stripe on the lower two-thirds of each side, whilst the underneath of thorax is similar to the male. *Wings* similar to the male, but usually less enfumed, and usually an extra postnodal nervure in each wing and quite often an extra cell-length to *Cuii* in hind-wings; pterostigma rather paler, and with a fine yellow frame inside the bordering nervures. *Abdomen* black on dorsum, pale ochreous on the sides, this latter paling to whitish near the apex of segments 3 to 6; baso-dorsal spots much larger than in the male and yellowish in colour; on segments 3 to 6 the dorsal black makes complete broad apical black rings round each segment, very conspicuous as viewed from the side; segment 9 with a small apico-dorsal yellow diamond-shaped spot, whilst 10 has a short linear baso-dorsal yellow stripe. *Legs* yellow, hinder surfaces of femora, flexor surfaces of tibiæ, and the tarsi black. *Anal appendages* very short, conical, black: vulvar scales black, yellowish above, very robust, extending well beyond the end of abdomen.

Distribution.—WESTERN GHATS of India; the type, from Talewadi, North Kanara, was taken in October. I found this insect quite common on most streams in Coorg during May and June, and less commonly after the monsoon was over.

Teneral males approach the female in colour and markings, whilst very adult males lose almost all their markings beneath pruinescence. A shy retiring insect, frequenting dark shady spots on submontane streams, usually in heavy jungle. Related closely to the next species, which it much resembles, and distinguished from it by the larger size and higher nodal index (only 11 to 12 in the fore-wings and 9 in the hind of *D. nigerrima*).

Type in the Indian Museum; paratypes in the British Museum and author's collections.

92. **Disparoneura nigerrima** Laidlaw.

Disparoneura nigerrima Laid., Rec. Ind. Mus. vol. xiii, pp. 346, 347 (1917); Fras., J. Bombay Nat. Hist. Soc. vol. xxix, p. 743 (1923); id., Rec. Ind. Mus. vol. xxvi, pp. 429, 503 (1924); id., ibid. vol. xxxiii, p. 449 (1931).

Male.—Abdomen 23–25 mm. Hind-wing 15–17 mm.

Head: labium palest brown; labrum pale blue, paler at border; genæ pale blue; rest of head blackish-brown or black; occiput behind the head paler brown and often pruinosed; eyes capped with black, olivaceous to bluish below, and with a narrow equatorial belt of black. *Prothorax* and *thorax* black, the latter paling to yellow at the sides and beneath ; a narrow oblique stripe of black bordering the postero-lateral suture anteriorly. *Legs* black, coxæ and trochanters shaded with yellow ; tibiæ yellow on the distal half of flexor surfaces ; femora pruinosed in adults. *Wings* hyaline ; pterostigma black, framed finely in yellow and thick black nervures, very small, covering but little more than half a cell, braced, usually imperfectly so ; 10 to 12 postnodal nervures to fore-wings, 9 in the hind ; *Cuii* 3 to 4 cells long in fore-wings, 5 in the hind ; *ac* meeting *ab* well distal to the origin of latter. *Abdomen* black, segments 1 and 2 with the lower parts of sides pale yellow ; segments 3 to 7 with small paired white basal spots. *Anal appendages* blackish-brown, rather longer than segment 10, of the generic shape, the superiors having the ventral spine bifid as in *D. tetrica* ; inferiors longer than superiors, apices obtuse and curled slightly upward and inward.

Female.—Abdomen 27 mm. Hind-wing 15–16 mm.

Differs considerably from the male in colour and markings. *Head* : labrum darker blue, clouded with brownish ; genæ and bases of mandibles pale greenish-blue ; vertex black, traversed from eye to eye by a narrow pale azure blue band at the level of anterior ocellus ; occiput pale yellow behind and below ; eyes dark olivaceous brown above, pale blue beneath this area, and with two narrow black equatorial bands, the upper of which is continued across the genæ and confluent with the black of the epistome, whilst the narrow belt of blue on the eyes is continuous with the narrow blue band on vertex. *Prothorax* black, with a tiny geminate mid-dorsal spot on the anterior lobe and an irregular stripe on each side pale blue ; posterior lobe narrowly and deeply cleft at middle, and with two stout short spines directed rather obliquely forwards. *Thorax* black on dorsum, palest yellow laterally, marked with very narrow pale blue antehumeral stripes, followed by an irregular pale brown posthumeral stripe, incomplete above ; laterally the black extends as far as the middle of mesepimeron, and there is a narrow black stripe on the postero-lateral suture ; beneath whitish. *Legs* yellow, the femora blackish on extensor surface, spotted at proximal ends, and the tibiæ blackish on flexor surface. *Wings* similar to the male, but the pterostigma palest brown, and an extra postnodal nervure in all wings ; *Cuii* is also usually 1 or 2 cells longer. *Abdomen* black on dorsum, pale brown to yellow laterally, the dorsal black expanding subapically on segments 2 to 7, and only reaching the base of segments at a point, so that narrow pale yellow annules slightly interrupted on the mid-dorsum are left here ; segments 7 to 9 with a mid-dorsal stripe of pale yellow gradually broadening backwards, forming a sagittate marking on 9; segment 10 pale yellow, with two small subdorsal basal brown triangles. *Anal appendages* pale yellow, very short, conical; vulvar scales robust, pale yellow, extending to the end of abdomen.

Distribution.—Confined to CENTRAL and WESTERN INDIA on the eastern side of the watershed, and not further south than Satara.

Frequents the banks of streams, hovering over runlets or settling on grass-stems at the river-side. On the wing bears a close resemblance to *Pseudagrion hypermelas*, with which it often keeps company, but is easily distinguished by an examination of the anal bridge and discoidal cells. From *D. nigerrima* it is distinguished by its pale blue labrum, and by the double equatorial belt on the eyes. The distribution of the former insect is also to the west of the watershed and more southerly.

93. **Disparoneura centralis** Hagen.

Disparoneura centralis Hagen, Verh. zool.-bot. Ges. Wien, vol. ix, p. 207 (1859); Selys, Bull. Acad. Belg. (2) vol. x, p. 449 (1860); id., Mém. Cour. vol. xxxviii, p. 167 (1886); Kirby, Cat. Odon. p. 133 (1890); id., J. Linn. Soc. Lond., Zool. vol. xxiv, p. 562 (1894); Laid., Rec. Ind. Mus. vol. xi, p. 388 (1915); id. (*pruinosa*), ibid. vol. xiii, p. 323 (1917); Fras., J. Bombay Nat. Hist. Soc. vol. xxix, p. 743 (1923); Laid., Spolia Zeylan. vol. xii, p. 366 (1924).

Alloneura centralis Selys, Bull. Acad. Belg. (2) vol. x, p. 449 (1860).

Disparoneura oculata Kirby (♀), J. Linn. Soc. Lond., Zool. vol. xxix, pp. 562, 563 (1894).

Caconeura mackwoodi Fras., J. Bombay Nat. Hist. Soc. vol. xxvi, pp. 876, 877 (1919); id., ibid. vol. xxix, p. 743 (1923).

Disparoneura cæsia Laid. (*centralis*, nec *cæsia*) Spolia Zeylan. vol. xii, pp. 365, 366 (1924).

Male.—Abdomen 30–33 mm. Hind-wing 20–24 mm.

Head: labium blackish-brown; labrum, bases of mandibles, and genæ glossy black, but paler in tenerals; rest of the head velvety black, with a dark purplish sheen; eyes black, with purple reflex. *Prothorax* and *thorax* velvety black, with a beautiful dark purple reflex, unmarked in adults, but with vestigial pale lateral bands in subadults and tenerals; beneath and the legs black. *Wings* hyaline; pterostigma black, with a very fine yellow line bordering the inner side of the framing nervures, strongly braced, covers one cell; 17 to 19 postnodal nervures to fore-wings, 14 to 16 in the hind; *Cuii* 4 to 5 cells long in fore-wings, 5 to 6, or more rarely 7, in the hind; *ac* meets *ab* well distal to the origin of the former, which ends at the hinder end of the nervure descending from the outer end of discoidal cell, or more rarely on the posterior border of the wing near this nervure. *Abdomen* velvety black, with a purplish reflex, unmarked. *Anal appendages* black, rather longer than segment 10, of the generic shape, the ventral spine of superiors not bifid, as in the two preceding species; inferiors longer than superiors, sloping downwards, tapered rather abruptly at the apex, which is turned slightly inwards.

Female.—Abdomen 31–32 mm. Hind-wing 21–22 mm.

Differs very considerably from the male, which accounts for Kirby describing it as a separate species under the name of *D. oculata*. *Head*: labium brown, black at apices; labrum black; anteclypeus, genæ, extreme bases of mandibles, and the base of postclypeus broadly creamy white or pale yellow; postclypeus bordered with black anteriorly, the hinder border of this festooned; rest of head steely black, with a small rounded dark ochreous spot between the root of antennæ and anterior ocellus on each side; eyes dark brown, capped with black. *Prothorax* black, with a pale yellow stripe on each side, broken at the spot where it passes from the middle to the posterior lobe; the latter rounded, cleft at its middle, and with two lappet-like hooks which are obtusely pointed, nearly parallel and, contrary to what is usually found in the genus, directed straight back towards the thorax. *Thorax* black, with a very narrow antehumeral stripe on each side of the dorsum, a broad oblique stripe bordering the postero-lateral suture posteriorly, and a third stripe coursing along the lower border of the metepimeron all pale yellow; beneath yellow, broadly bordered with black outwardly. *Legs* black, extreme bases of all femora and trochanters and coxæ yellow. *Wings* as for the male; pterostigma pale brown, framed narrowly in yellow and thick black nervures; 15 to 16 postnodal nervures to fore-wings, 14 in the hind; *Cuii* 4 cells long in fore-wings, 5 to 6 in the hind. *Abdomen* black, marked with creamy white as follows:—Segment 1 with a cordate spot on each side; segment 2 with a pair of parallel stripes on each side, often reduced in adults to a small spot at the base, and another at the apical end of segment; segments 3 to 6 with small paired baso-dorsal spots; segments 8 and 9 with the ventral border narrowly, whilst 9 and 10 have each a small diamond-shaped dorsal spot. *Anal appendages* very short, conical, pointed, black.

Distribution.—The hilly tracts of CEYLON; common round about Kandy during September and October and around Banderawela during May and June.

Kirby has identified the British Museum specimens as *D. cæsia* in error, and Laidlaw, misled by this, has described the female as for the same species. Kirby described the female as a separate species under the name of *D. oculata*. The beautiful purple reflex of the head and thorax of the male will serve to distinguish this insect from all others of the same genus, and the total absence of pulverulence, even in the most adult specimens, will distinguish it from *D. cæsia*. Females are distinguished with more difficulty, especially if teneral; adults have the pale markings very restricted, and all specimens show only small dorsal spots on segments 9 and 10 instead of the usual dorsal stripes on segments 8 to 10 seen in the females of other species.

Type in the Hagen collection (Harvard Mus. Comp. Zoöl.); paratype female (*D. oculata*) in the British Museum collection.

94. **Disparoneura cæsia** (Selys).

Alloneura cæsia Selys, Bull. Acad. Belg. (2) vol. x, p. 450 (1860).
Disparoneura cæsia Selys, Mém. Cour. vol. xxxviii, p. 167 (1886); Kirby, Cat. Odon. p. 133 (1890); id., J. Linn. Soc. Lond., Zool. vol. xxiv, p. 562 (1894); Laid., Rec. Ind. Mus. vol. xi, p. 388 (1915); id., ibid. vol. xiii, p. 323 (1917); Fras., J. Bombay Nat. Hist. Soc. vol. xxix, p. 743 (1923); Laid., Spolia Zeylan. vol. xii, p. 366 (1924).

Male.—Abdomen 35–36 mm. Hind-wing 22–23 mm.

Head: labium black; labrum, ante- and postclypeus steely blue-black; genæ and bases of mandibles glossy black or dark brown or paler; rest of head velvety black, without purple reflex; eyes black above, olivaceous brown below. *Prothorax* and *thorax* black, the whole of dorsum and greater part of sides densely pruinescent; laterally an anterior vestige of an oblique pale yellow stripe on the mesepimeron, absent in very adult specimens, and a second similar stripe bordering the lower part of metepimeron and nearly always present; beneath yellow, broadly bordered with black, the yellow portion often obscured by pruinescence. *Legs* black, coxæ and trochanters pruinescent. *Wings* hyaline, palely enfumed in adults; pterostigma black, very finely framed in yellow and thick black nervures, covering nearly one cell, braced; 17 to 20 postnodal nervures to fore-wings, 17 to 18 in the hind; *Cuii* 3 to 4 cells long in fore-wings, 6 to 7 cells in the hind; *ac* meeting *ab* well distal to origin of the latter, which ends on the hinder part of the nervure descending from the distal end of discoidal cell ("medio-anal link") by inclining sharply towards the hinder border of the wing (occasionally connected to the hinder border of wing by one or more accessory short nervures). *Abdomen* black, with a steely blue reflex, unmarked. *Anal appendages* black, about as long as segment 10; superiors with an unusually long, thin, and slightly sinuous ventral spine; inferiors with the apices curled strongly inward and acutely pointed.

Female unknown.

Distribution.—Hilly tracts of CEYLON. I have taken it at Dyatalawa after the S.W. monsoon was over, and possess some males taken by Col. F. Wall at Kandy, 2,000 ft., during September. Kirby has reported it from Gamboda and Pundu-loya in May. Distinguished from other species by the dense white pruinescence of the dorsum of thorax, which is somewhat similar to that seen in the male of *Mnais earnshawi*.

Type in the Hagen collection.

95. **Disparoneura tenax** (Hagen).

Agrion tenax Hagen, Verh. zool.-bot. Ges. Wien, vol. viii, p. 478 (1858).
Alloneura tenax Selys, Bull. Acad. Belg. (2) vol. x, p. 449 (1860).
Disparoneura tenax Selys, Mém. Cour. vol. xxxviii, p. 168 (1886); Kirby, Cat. Odon. p. 133 (1890); Laid., Rec. Ind. Mus. vol. xi, p. 388 (1915); id., ibid. vol. xiii, p. 323 (1917); Fras., J. Bombay Nat. Hist. Soc. vol. xxix, p. 743 (1923); Laid., Spolia Zeylan. vol. xiii, p. 365 (1924).

Male.—Abdomen 35–39 mm. Hind-wing 25–26 mm.

Head: labium light brown, with apices black; labrum bright orange, bordered narrowly with dark brown, and with a small median black point; genæ and bases of mandibles pale ochreous; anteclypeus brown; postclypeus bright reddish-orange, bordered anteriorly with black, very narrowly so at the middle, but broadly so at either side; vertex and occiput velvety black, the former traversed from eye to eye at the level of anterior ocellus and antennæ by an irregular moderately broad bright reddish-orange stripe, slightly interrupted at its middle; eyes reddish, paler on summit. *Prothorax* velvety black, traversed on each side by an irregular orange stripe interrupted narrowly at the junction of middle and posterior lobes. *Thorax* velvety black, marked with narrow brick-red antehumeral stripes, and yellow stripes on the mesepimeron at level of spiracle and on lower border of metepimeron, the former stripe on mesepimeron tinted with reddish; beneath yellow at centre, broadly black at each side. *Legs* black, some reddish marks on coxæ and trochanters, and the flexor surfaces of tibiæ ferruginous. *Wings* hyaline, palely enfumed in adults; pterostigma blackish-brown, narrowly framed in reddish and thick black nervures, covering rather more than one cell, braced; 19 to 21 postnodal nervures in fore-wings, 18 to 19 in the hind; *Cuii* 7 cells long in fore-wings, 9 in the hind; *ac* arising from *ab* well distal to the origin of the latter, which ends on the nervure descending from the discoidal cell rather near the hinder border of the wing. *Abdomen* black, marked with bright chrome-yellow as follows:—Segment 1 with a small triangular spot on each side; segment 2 with a linear spot on the mid-dorsum, expanded subapically, a small spot on each side near apical border and a narrow stripe along the ventral border; segments 3 to 6 with small paired baso-dorsal spots. *Anal appendages* black, as long as segment 10; superiors of the usual generic shape, but the apices curling out and divaricate as seen from above, the ventral spine robust and directed straight down; in some specimens the base of these appendages reddish; inferiors with apices curled strongly inwards.

Female.—Abdomen 35 mm. Hind-wing 26–27 mm.

Rather similar to the male, but the markings bright chrome-yellow instead of bright brick-red or orange. Differs as follows:—Postclypeus entirely black. The lateral spot on segment 1 larger; mid-dorsal marking on segment 2 absent, but an additional spot at the base on each side and the ventral border narrowly yellow; segment 3 with an ochreous stripe running along the whole length of each side, and a vestige of a similar stripe on segment 4; segment 9 with a prominent diamond-shaped spot of yellow on the mid-dorsum, whilst 8 and 9 have a bright yellow stripe along the ventral border not quite extending to the base of segments. *Prothorax* with the posterior lobe furnished with a pair of very broad, robust, horn-like processes directed obliquely forwards, and with a smaller pair between them directed straight back. *Anal appendages* very short, conical, black; vulvar scales robust, black, tipped with yellow.

Distribution.—CEYLON only, in montane areas. I found it not uncommon on all streams on the grassy uplands around Diyatalawa, and as far up as Hakgalla. It is quite the largest of all the Disparoneuras, equalling in size some of the smaller *Indoneura* group, so that there is no difficulty in distinguishing it from all others of the genus.

96. **Disparoneura campioni** Fraser. (Fig. 101.)

Disparoneura campioni Fras., Mem. Dept. Agric. India (Ent.), vol. vii, no. 7, pp. 43, 44 (1922); id., J. Bombay Nat. Hist. Soc. vol. xxix, p. 743 (1923).

Male.—Abdomen 31 mm. Hind-wing 29 mm.

Head: labium brownish-white; labrum azure blue, with a tiny median black point; anteclypeus, genæ, and bases of mandibles azure blue; postclypeus black, with two elongate azure blue spots at its base; vertex and occiput mat black, the former traversed by a narrow azure blue band, extending from eye to eye at level of anterior ocellus; eyes black above, azure blue below, traversed by a narrow equatorial belt of black. *Prothorax* black, with a lateral stripe of blue extending from margins of posterior lobe to anterior lobe. *Thorax* velvety black on dorsum as far back as level of antero-lateral suture; laterally pale sky blue, traversed by a broad oblique black stripe on the postero-lateral suture; beneath palest blue, bordered broadly with black. *Legs* black, coxæ and trochanters spotted with blue ; femora carneous on the extensor surface, tibiæ bluish on flexor surface. *Wings* hyaline ; pterostigma black, finely framed in ochreous and thick black nervures, braced, covering one cell ; 15 to 16 postnodal nervures to fore-wings, 13 to 14 in the hind ; *Cuii* 3 cells long in fore-wings, 4 to 5 in the hind ; *ab* invariably complete and arising well distal to the level of *ac*. *Abdomen* black, marked with azure blue as follows :—Segment 1 with a small cordate spot on each side ; segment 2 with a medio-dorsal stripe shaped like the head of a javelin, and not quite reaching the extremities of the segment, a small triangular latero-apical spot, and the ventral border narrowly ; segments 3 to 7 with small paired

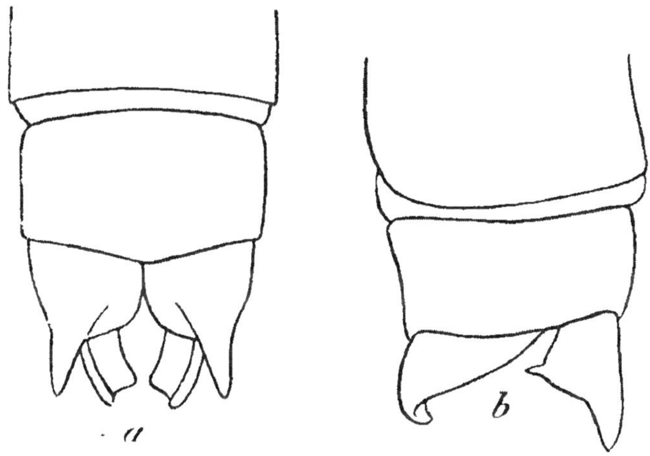

Fig. 101.—Anal appendages of *Disparoneura campioni* Fras., male. *a*. Dorsal view. *b*. Left lateral view.

baso-dorsal spots, and 3 to 5 with additional lateral larger pale brown or dirty bluish spots on each side subapically, the segments being dark reddish-brown basal to this spot and black apically, giving the appearance of broad apical black annules : segments 8 to 10 with broad triangular azure blue spots, the one on 8 extending half-way to base of the segment, that on 9 just tipping the base, and that on 10 extending broadly to the base. *Anal appendages* about as long as segment 10, of the usual generic shape ; superiors broadly azure blue on dorsum, the apex prolonged, the ventral spine black at apex and with a tiny imbricated hook at this point ; inferiors blue or carneous at base, black at apex, which is acute and curled upwards at the extremity.

Female unknown.

Distribution.—ASSAM and UPPER BURMA. Col. F. Wall found it not uncommon at Maymyo, Upper Burma. The type comes from Margherita, Assam.

The only dragonflies likely to be confused with this insect are species of the genus *Caconeura* with blue markings, and from these *D. campioni* is easily distinguished by the complete anal bridge.

Type in the British Museum; paratypes in the author's collection.

97. **Disparoneura campioni cacharensis**, subsp. nov.

Differs from *D. campioni* by the much broader transverse stripe on vertex: by the character of the antehumeral stripes, which are moderately broad below and taper quite abruptly in the upper third, a character paralleled in *Caconeura interrupta* Selys, from Singapore: the lateral spot on segment 1 larger and confluent with a narrow apical annule; the dorsal spot on segment 2 broader, and the ventral border of blue more extensive: segment 3 with the basal spots confluent to form a complete basal annule: segment 8 with a pair of blue semilunar spots at apical border, whilst 9 and 10 are azure blue on the whole of the dorsum and greater part of the sides: superior anal appendages nearly entirely blue, whilst the inferiors are carneous.

Distribution.—Cachar, ASSAM.

Type in the author's collection. Female unknown.

Genus **CHLORONEURA** Laidlaw (1917). (Fig. 102.)

Disparoneura, 2me Groupe, Selys, Bull. Acad. Belg. (2) vol. x, p. 445 (1860); id., Mém. Cour. vol. xxxviii, p. 161 (1886).

Chloroneura Laid., Rec. Ind. Mus. vol. xiii, pp. 323, 344 (1917); Fras., J. Bombay Nat. Hist. Soc. vol. xxix, p. 742 (1923).

Dragonflies of small size and slender build, with abdomen less than twice the length of wings: coloured black or red with blackish markings. Wings hyaline, barred with black or dark brown, moderately rounded at apices, and with a long petiole; discoidal cell elongate, costal and posterior sides equal, and about four times as long as the distal and proximal ends; sectors of arc arising from its lower half, divergent from origin; no accessory basal postcostal nervure present; the nervure *ac* situated about midway between the two antenodal nervures; *ab* always present and complete; *IA* absent; *Cuii* extending for more than half the length of wings; *Riv+v* arising well proximal to the level of subnode; *IRiii* in continuation of the subnode; pterostigma acutely diamond-shaped, sides very oblique, braced; areolets of wings mainly quadrangular.

Head narrow; eyes smaller than in *Platysticta* or *Protosticta*; occiput simple; prothorax with the posterior lobe rounded in the male, armed with robust hooks in the female; abdomen cylindrical, broadening towards the hinder segments, about half as long again as the wings; anal appendages resembling those of the genus *Caconeura* and very homogeneous, superiors shaped like the hammer of a revolver, tumid at base, tapering towards the apex, and with a very robust ventral spine shaped like a briar-thorn; inferiors simple, short, and broad, bevelled from above downwards, apices curled strongly in and obtuse. Genitalia very similar to those of the genus *Caconeura*, but with the penis throwing off two terminal branches on each side which embrace the body of the organ, and with a dorsal spine.

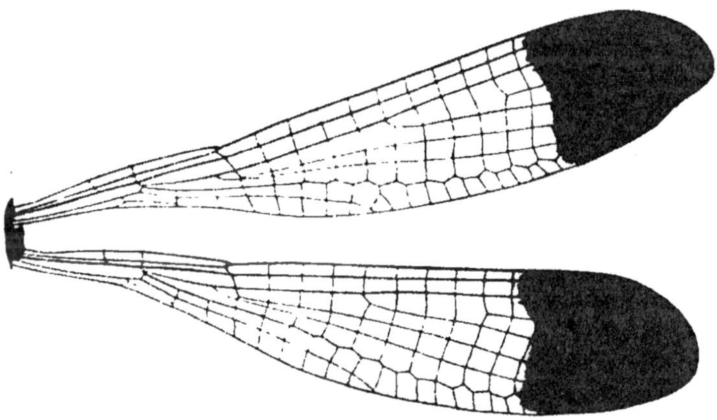

Fig. 102.—Wings of *Chloroneura apicalis* Fras., male.

Distribution.—CENTRAL INDIA, the DECCAN and COORG only. Only two species are known, the two being found in company on the lower reaches of the Cauvery River, Coorg, one being confined exclusively to this locality, the other, *C. quadrimaculata*, with a more extensive range from the Western Ghats and Deccan as far as the Central Provinces.

The genus differs from the two preceding by the species being larger and more robust, by the coloured wings, and by the greater length of *Cuii*.

Genotype, *C. quadrimaculata* (Ramb.).

Key to the Species of Chloroneura.

Wings of male black at apices	*apicalis* Fras., p. 248.
Wings of male broadly barred with blackish-brown across the middle	*quadrimaculata* [(Ramb.), p. 246.

98. **Chloroneura quadrimaculata** (Ramb.). (Fig. 103.)

Argia quadrimaculata Ramb., Ins. Névrop. p. 255 (1842).
Disparoneura quadrimaculata Selys, Bull. Acad. Belg. (2) vol. x, p. 446 (1860); id., Mém. Cour. vol. xxxviii, p. 163 (1886); Laid., Rec. Ind. Mus. vol. xi, pp. 388, 391 (1915).
Chloroneura quadrimaculata Laid., Rec. Ind. Mus. vol. xiii, pp. 323, 344 (1917); Fras., ibid. vol. xvi, p. 466, fig. 3, pl. xxxvi, and fig. 4, pl. xxxvii (1919); id., J. Bombay Nat. Hist. Soc. vol. xxvii, p. 44 (1921); id., ibid. vol. xxix, p. 742 (1923); id., Rec. Ind. Mus. vol. xxvi, pp. 429, 501 (1924); id., ibid. vol. xxxiii, p. 449 (1931).

Male.—Abdomen 32 mm. Hind-wing 22 mm.

Head: labium pale whitish-brown; labrum, clypeus, vertex, and occiput dull brick-red, genæ and under surface of head pale ochreous; labrum with a small median black spot; eyes bright brick-red, with two horizontal black equatorial lines coursing round them, and continued on to the vertex as a black stripe running inwards as far as the base of antennæ, and more indefinitely so across the frons; another obscure stripe running in from the eyes behind the level of antennæ, and a small black oval spot behind and to outer side of posterior ocelli; occiput traversed behind by two transverse black stripes, one of which separates the red and yellow areas, the other lying more posteriorly in the yellow area; antennæ reddish. *Prothorax* dull brick-red, the sutures and sulci between the lobes mapped out in black; posterior lobe broadly rounded. *Thorax* bright brick-red on dorsum, paler at the sides, marked with black as follows:—The mid-dorsal carina narrowly, a narrow antehumeral stripe incomplete above and below, a more or less broken humeral stripe, sometimes made up of three elongate spots, sometimes broader and complete; a broader irregular stripe on the mesepimeron, often broken up into two or more spots, or broad, complete, and, rarely, partially confluent with the humeral stripe; a short stripe descending from the antero-lateral suture, a complete narrow stripe mapping out the postero-lateral suture; and finally an incomplete stripe, sometimes absent, on the lower border of metepimeron; beneath pale ochreous, bordered outwardly with black and often thinly pruinosed. *Legs* pale brown, femora speckled outwardly with black. *Wings* hyaline, the fore pair traversed by a broad blackish-brown fascia, which extends from near the node to half-way from that structure to the pterostigma, the hind pair traversed by a similar fascia which begins about 5 cells distal to the node and extends to within 1 or 2 cells of the pterostigma; the latter organ bright ochreous framed in thick black nervures, covering one cell, braced; 15 postnodal nervures to fore-wings, 12 to 13 in the hind; *Cuii* 9 to 12 cells long in fore-wings, 10 to 12 in the hind. *Abdomen* brick-red, marked with white, brown, and black as follows:—A cordate spot and the ventral border of segment 1 red, dorsum broadly black; segment 2 with a large ring of black on dorsum, extending at one point to the base, but falling just short of the apical border, a narrow longitudinal stripe of black on each side; segments 3 to 6 each with a pair of small dorsal whitish spots at base, narrowly bordered with black posteriorly, a narrow lateral stripe on each side, not extending to apex or base of segment, and finally broad black apical annules; segments 7 to 9 black for the greater part, the mid-dorsal carina palely red; segment 10 largely reddish, the base and sides changing to blackish-brown. *Anal appendages* of the *Caconeura* shape; superiors carneous, the broad ventral spine tipped with black, the apex long and tapering to a point; inferiors very short and obtuse

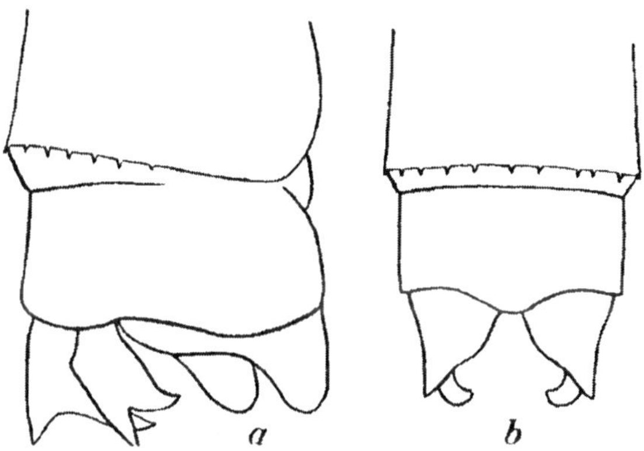

Fig. 103.—Anal appendages of *Chloroneura quadrimaculata* (Ramb.), male. *a*. Right lateral view. *b*. Dorsal view.

as seen from the side, apices black and curled strongly inwards, extending slightly beyond the apices of superiors as seen from the dorsum.

Female.—Abdomen 29–30 mm. Hind-wing 22 mm.

Differs considerably from the male, and more robust in build; black markings better defined and rather more extensive. *Head*: the black stripes on occiput much broader, and the oval spot on vertex very elongate and sometimes joined to its fellow by a narrow black line on the hinder border of occiput. *Prothorax* with the sides broadly black, marked with a small rounded ochreous spot; dorsum pale ochreous, with fine median and broader submedian black lines mapping out the contours of lobes; posterior lobe broadly and shallowly notched behind, on either side of which springs a very long, slightly sinuous, spine-like process shaped like a cow's horn, directed very obliquely forwards, yellow in colour; at the root of each spine a second shorter and thicker black spine projects horizontally back. *Thorax* broadly black on dorsum, the sides olivaceous green shaded with ochreous and marbled with black as in the male, but the markings more confluent and more sharply defined; beneath and the lower parts of sides more thickly pruinosed than in the male. *Legs* pale reddish-brown, femora speckled with black outwardly. *Wings* hyaline, without any black markings, although very adult specimens may have the costal and medial portions markedly enfumed with brownish; 16 postnodal nervures to fore-wings, 13 to 14 in the hind; *Cuii* as in the male, as also the pterostigma. *Abdomen* pale olivaceous, tinted with ochreous dorsally, and marked and shaded with blacks and browns; segments 3 to 8 with small paired white spots at the base, each spot framed heavily in black; segments 3 to 7 with broad black apical annules, in front of which the olivaceous ground-colour forms a broad pale ring; the mid-dorsal carina finely yellow from segment 2 to the middle of 6 and again on segments 8 and 9, where it expands to form marks shaped respectively like an arrow-head and a spear-head; segment 10 pale olivaceous shaded with brown laterally. *Anal appendages* pale olivaceous, very short, barely half the length of segment 10, pointed, conical; vulvar scales thick and robust, not extending to the end of abdomen, black tipped with yellow.

Distribution.—CENTRAL INDIA in submontane areas, the DECCAN on its western side only, and the western side of the watershed of the WESTERN GHATS as far south as South Kanara and Coorg. I found it fairly common on the tributaries of the Cauvery River in Coorg and the upper reaches of the same river in Mysore during June, July, and September; at Mahableshwar and Poona during May to July. Dr. Gravely records it from Satara in June, and the insect has been taken in the Panch Mahals, Bombay Pres. It does not appear to breed below 1,500 ft., so that the locality Bombay, given for the type, is an obvious error.

Type in the Serville collection; paratypes in most museums and private collections.

99. **Chloroneura apicalis** Fraser. (Fig. 102.)

Chloroneura apicalis Fras., Rec. Ind. Mus. vol. xxvi, pp. 429, 501–3, fig. 4 (1924); id., ibid. vol. xxxiii, p. 449 (1931).

Male.—Abdomen 31 mm. Hind-wing 20 mm.

Head: eyes reddish-brown, changing to olivaceous beneath; labium whitish; labrum brownish-yellow, genæ ochreous; rest of head black, traversed from eye to eye at the level of middle ocellus by a brick-red stripe. *Prothorax* black, with a geminate spot of reddish-brown at the middle of dorsum of mid-lobe. *Thorax* black on dorsum, with a dull green metallic sheen, pruinosed on the sides and beneath in adults ; a narrow antehumeral brick-red stripe is present in moderately young specimens, but this is soon lost in the adult ; there is also a brownish-white stripe on the sides between the lateral sutures which is almost, or quite, obsolete in full adults. *Legs* black, femora pruinosed on the flexor surface, tibiæ bright yellow on extensor surface. *Wings* hyaline, the apices for rather more than the outer fourth of wings deep blackish-brown ; 15 postnodal nervures in fore-wing, 13 in the hind ; *Cuii* 7 to 8 cells long in fore-wings, 8 in the hind ; pterostigma dark reddish-brown, well braced, covering about $1\frac{1}{2}$ cells. *Abdomen* reddish-brown, black on dorsum, with small paired dorsal, basal, whitish spots on segments 3 to 7. *Anal appendages* black ; superiors tipped with white, trigger-shaped, triangular seen from above, pointed, and with a broad, short, robust spine beneath as seen from the side : inferiors, viewed laterally, broad at base, tapering and sloping downwards, apices curling strongly inward.

Female.—Abdomen 31 mm. Hind-wing 18 mm.

Head : eyes bluish above, whitish-green below, with a small polar cap above and an equatorial belt of warm brown ; labium white; labrum, cheeks, and clypeus pale greenish-white (or in some the genæ reddish-brown) ; a narrow belt of black connecting the equatorial belts of eyes across vertex, and a broader belt of the same colour running between the polar caps on occiput, between which lies a pale yellowish stripe corresponding to the red stripe seen in the male. *Prothorax* black, marked with a longitudinal subdorsal and a lower lateral stripe on each side ; below the latter two small spots of yellowish ; the processes on the posterior lobe long, yellow, sloping obliquely forward, and between them a pair of small black divaricate spines. *Wings* hyaline, slightly enfumed, apices unmarked ; pterostigma brown, 14 postnodal nervures in fore-wings, 12 to 13 in the hind. *Thorax* black on dorsum, pale greenish-white laterally, an antehumeral stripe of the same colour ; the black, posterior to the latter stripe, marked with a sinuous upper ochreous spot, and, in some specimens, two warm brown spots about its middle ; postero-lateral suture finely mapped out in black, its upper part more broadly so. *Abdomen* greenish-white, dorsum broadly blackish-brown ; on segment 1 the ground-colour present as an apical ring and two parallel yellowish-green stripes, on 2 the black on dorsum expanding broadly apicalward and joining a fine apical black ring, its dorsum finely yellow along the carina, segments 3 to 6 somewhat similar, but the black expanding basalward as well as apicalward, the ground-colour on these segments passing up at the base and nearly confluent over the dorsum (seen from above this appears as paired spots); segment 7 broadly black except the sides; 8 and 9 with the mid-dorsum and latero-ventral borders narrowly whitish-yellow; 10 whitish on mid-dorsum, apical border, and ventrum. *Anal appendages* very short, whitish yellow, conical; vulvar scales robust, extending to the end of abdomen, black at base.

Distribution.—COORG only, in the restricted area of Fraserpet, on the upper reaches of the Cauvery River, before and after the S.W. monsoon, especially in May and September.

Shy insects keeping to the shade of overhanging trees, settling on grasses lining the banks. Females keep solely to scrub adjoining the river, except when actually ovipositing; males when hovering, as they constantly do in flight, exhibit the thin black body lying in the centre of a thick black circle produced by the black bars on the rapidly whirring wings (a similar effect, but with a broader wall to the circle, is produced by *C. quadrimaculata* when in flight). The broad black apices to the wings of this insect will distinguish it from all others of the same group.

Genus **INDONEURA** Laidlaw. (Fig. 104.)

Indoneura Laid., Rec. Ind. Mus. vol. XIII, p. 347 (1917); Fras., J. Bombay Nat. Hist. Soc. vol. XXIX, p. 743 (1923); id., Rec. Ind. Mus. vol. XXVI, pp. 506–9, fig. 5 (1924).

Zygopterous dragonflies of moderate size and slender build, species of the genus being considerably larger than those of the two preceding genera; abdomen less than twice the length of wings; colouring invariably black, marked with blue in both sexes. Wings hyaline, narrow, very long and subfalcate at apices, with a moderately long petiole; discoidal cell elongate, costal and posterior sides equal and four to five times longer than the base or distal end; sectors of arc arising from the lower end of arc, divergent from origin; no accessory basal postcostal nervure present; the nervure *ac* situated about midway between the two antenodal nervures or slightly nearer the distal one; *ab* always present but incomplete, curving down to meet the posterior margin of wing as in the genus *Caconeura*; *IA* absent; *Cuii* reduced, not extending to half the length of wings, usually 8 to 9 cells in length; *Riv+v* arising well proximal to the subnode; *IRiii* arising at the level of the subnode; pterostigma comparatively small, diamond-shaped, proximal and distal sides very oblique, well braced; areolets of wings nearly entirely quadrangular.

Head narrow; eyes smaller than in species of *Platysticta*; occiput simple, not ridged posteriorly; prothorax with *posterior lobe simple in both sexes,* but hook-like processes on the anterior lobe of females; abdomen cylindrical, long, slim, broadened at base and again towards the terminal segments, about half as long again as the wings; anal appendages very homogeneous, superiors shaped like the butt end of a revolver, obtuse at apex, and with a long ungulate ventral process; inferiors simple, strongly bevelled from above down, very broad at base, tapering rapidly to apex, which is obtuse or pointed, directed straight back or curled a little upwards. Genitalia: lamina narrowly but deeply cleft: hamules broad quadrate plates, pointed within; penis with the apical end curling strongly over dorsum, its end expanding

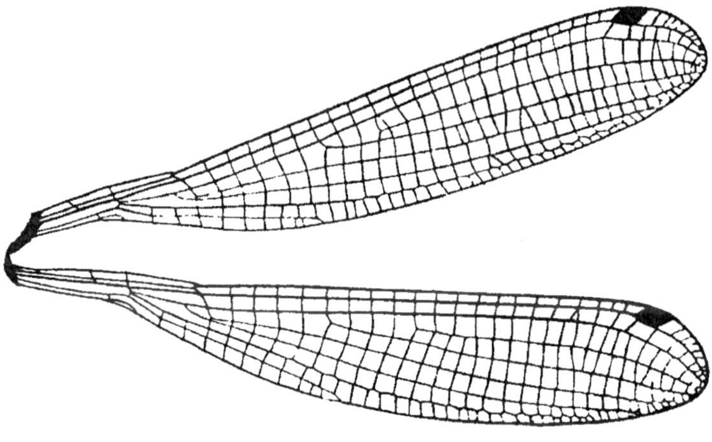

Fig. 104.—Wings of *Indoneura ramburi* Fras., male.

into two long branches on each side, spreading widely: vulvar scales robust, but not extending beyond the abdomen.

Distribution.—Confined to PENINSULAR INDIA and, with the exception of one species, to the WESTERN GHATS of India, including the Nilgiri, Palni, and Shevaroy Hills. Most species found at an altitude of about 3,500 to 5,500 ft., but one at least at an altitude of over 7,000 ft., and one not far above sea-level.

The genus differs markedly from *Caconeura*, which has a similar formation of the anal bridge, by the sexes being almost similar in colour and markings, by the female being without appendages to the posterior lobe of prothorax, and by the greater length of *Cuii*, etc. Four or five species are at present known, and this number is not likely to be exceeded.

Genotype, *Indoneura gomphoides* (Ramb.).

Key to the Species of Indoneura.

1.	Pterostigma bright ochreous in both sexes; dorsum of 2nd abdominal segment marked with blue	*gomphoides* (Ramb.), [p. 252.
	Pterostigma black; 2nd abdominal segment unmarked on dorsum	2.
2.	Under surface of head bright blue; pterostigma subquadrate; superior appendages with a tiny sub-basal tooth	*risi* Fras., p. 258.
	Under surface of head black; pterostigma longer than broad; superior appendages without a basal tooth................	3.
3.	Head and body without blue markings ..	*obscura*, sp. n., p. 257.
	Head and body with azure blue markings.	4.
4.	Segments 8 and 9 uniformly blue on dorsum	*ramburi* Fras., p. 254.
	Segments 8 and 9 with blue T-shaped markings	*T-cœrulea*, sp. n., p. 256.

100. **Indoneura gomphoides** (Ramb.). (Figs. 104 & 105.)

Argia gomphoides Ramb., Ins. Névrop. p. 256 (1842).
Alloneura gomphoides Selys, Bull. Acad. Belg. (2) vol. x, p. 448 (1860).
Disparoneura gomphoides Selys, Mém. Cour. vol. xxxviii, p. 171 (1886); Kirby, Cat. Odon. p. 134 (1890).
Indoneura gomphoides (*ramburi*, nec *gomphoides*) Laid., Rec. Ind. Mus. vol. xiii, pp. 347, 348, text-fig. 7, pl. xv, fig. 7 (1917); Fras., ibid. vol. xxiv, pp. 1, 2 (1922); id., J. Bombay Nat. Hist. Soc. vol. xxix, p. 743 (1923); id., Rec. Ind. Mus. vol. xxvi, pp. 429, 505 (1924); id., ibid. vol. xxxiii, p. 449, text-fig. 6, iii (1931).

Male.—Abdomen 34–35 mm. Hind-wing 26–27 mm.

Head: labium dirty white or yellow at the base, blackish-brown at the apex; labrum and base of mandibles azure blue, the former narrowly bordered with black at the base, the latter along the upper border; anteclypeus dark brown, post-clypeus steely metallic blue-black; genæ and frons traversed by a broad azure blue band, divided on the former by a moderately broad black band, which is confluent with the black of clypeus; rest of head, including the under surface, mat black; eyes deep marine blue, capped with black. *Prothorax* black, anterior lobe blue for its middle two-fourths only, middle lobe with a large lunar-shaped spot on each side, and a pair of short parallel stripes on the posterior half of the mid-dorsum. *Legs* black, coxæ and trochanters marked irregularly with blue and pale brown. *Thorax* velvety black on dorsum, azure blue on the sides, which latter are traversed obliquely by a moderately broad complete black stripe situated over the postero-lateral suture; dorsum marked by a very narrow antehumeral azure blue stripe, on each side, extending nearly up to the antealar sinus; beneath white or palest blue. *Wings* hyaline, evenly enfumed in the adult stage, petiolated as far as well proximal to *ac*, which lies much nearer the distal antenodal nervure ; *ab* short, strongly arched ; 23 postnodal nervures to fore-wings, 21 in the hind ; pterostigma light reddish-brown, framed in thick black nervures, costal and outer sides longer than posterior and inner, markedly diamond-shaped, strongly braced ; commonly two rows of cells between the costa and radius distal to the pterostigma. *Abdomen* black, marked with azure blue as follows :—Segment 1 broadly so at the sides and narrowly along the apical border ; segment 2 with a broad stripe on each side, a narrow complete annule at the base, and a fine linear mid-dorsal stripe extending along the basal three-fifths of segment ; segments 3 to 6 with small paired basal spots : segments 8 to 10 almost entirely azure blue, the ventral borders of 8 and 9 broadly, and the apical border of 10 narrowly black. *Anal*

Fig. 105.—Anal appendages of *Indoneura gomphoides* (Ramb.), male, viewed from the left.

appendages rather longer than segment 10 and about equal to one another ; superiors of the generic shape, apex moderately long, ventral process at least four times as long as apex, sinuous, turned down rather abruptly and inclined a little inward, its apex somewhat tumid ; inferiors tapering rapidly from base, curling gently up and slightly inwards, acute at apex as seen in profile.

Female.—Abdomen 36–37 mm. Hind-wing 28 mm.

Differing but slightly from the male ; *wings* with 20 to 22 postnodal nervures to fore-wings, 17 to 18 in the hind ; pterostigma bright ochreous framed in black nervures ; nearly always two rows of cells distal to the pterostigma. *Abdominal* markings differing only in the hind segments, segment 8 with the ventral border narrowly blue, the dorsum and subdorsum broadly so ; segment 9 with a broad T-shaped mark.

the cross-bar of the T bordering the apical half of segment, a squarish spot on each side of this same segment; segment 10 with the apical border broadly black. *Anal appendages* short, conical, surmounting a conical protuberance of the same length as themselves; vulvar scales robust, extending to the end of abdomen, broadly blue, tipped and bordered with blackish-brown.

Distribution.—Confined to the grassy uplands of the NILGIRI HILLS at an altitude of 7,000 to 7,500 ft., from May to July.

Gregarious in habits, large colonies being found clinging to ferns on the banks of the small moor-streams in which the insect breeds. Flight short and weak. Distinguished from all other species by the pale colour of the pterostigma and by the linear mid-dorsal marking on segment 2. The insect also differs markedly from other species by its comparatively short and thick abdomen.

Type in the Selysian collection; co-type ♀ in the British Museum; paratypes of both sexes in the author's, Morton, and Laidlaw collections.

101. **Indoneura ramburi** Fraser. (Fig. 106.)

Indoneura gomphoides Laid. (*ramburi*, nec *gomphoides*) Rec. Ind. Mus. vol. xiii, pp. 347, 348, text-fig. 7, pl. xv, fig. 7 (1917).
Indoneura ramburi Fras., Rec. Ind. Mus. vol. xxiv, pp. 2, 3 (1922); id., J. Bombay Nat. Hist. Soc. vol. xxix, p. 743 (1923); id., Rec. Ind. Mus. vol. xxvi, pp. 429, 505 (1924); id., ibid. vol. xxxiii, p. 449, text-fig. 6, a (1931).

Male.—Abdomen 43–44 mm. Hind-wing 28–29 mm.

Head: labium pale yellow; labrum, anteclypeus, postclypeus and bases of mandibles steely bluish-black; genæ and a broad transverse stripe traversing frons azure blue, this blue area divided transversely by a black stripe which crosses the genæ and becomes confluent with the black of clypeus; eyes deep marine blue capped with black. *Prothorax* black, the whole of the anterior lobe blue, as well as a small comma-shaped spot on each side of the middle lobe. *Thorax* velvety black on dorsum, azure blue on the sides, the former marked with very narrow, straight, azure blue antehumeral stripes, a little dilated below and extending nearly up to the antealar sinus; laterally a rather narrow, oblique, complete black stripe mapping out the postero-lateral suture, occasionally partially obscured in adults by pruinescence; beneath palest blue, unmarked. *Legs* black, coxæ and trochanters rather broadly marked with blue; tibiæ bright yellow on extensor surfaces. *Wings* hyaline, palely enfumed in adults; pterostigma black or dark reddish-brown, diamond-shaped, longer than broad; 21 to 23 postnodal nervures in fore-wings, 19 to 20 in the hind; only a single row of cells distal to the pterostigma between the costa and radius. *Abdomen* black, marked with blue as follows :—Segment 1 broadly so at the sides, apical border unmarked with blue, as in *I. gomphoides* ; segment 2 with a broad stripe on each side, the dorsum unmarked ; segment 3 with a pair of small baso-dorsal spots ; segments 4 to 7 with larger complete baso-dorsal annules ; segments 8 to 10 blue, the latter with its apical border narrowly and its sides broadly black. *Anal appendages* black, rather longer than segment 10, and of about equal length ; superiors of the usual generic shape, apex rather short, tumid and obtuse, ventral process about three times as long as apical, more angulated to the base than in *I. gomphoides*, and angulated on themselves about the middle, apex slightly tumid and rounded ; inferiors shaped very similarly to those of *I. gomphoides*, terminating in an acute point at the apex as seen in profile.

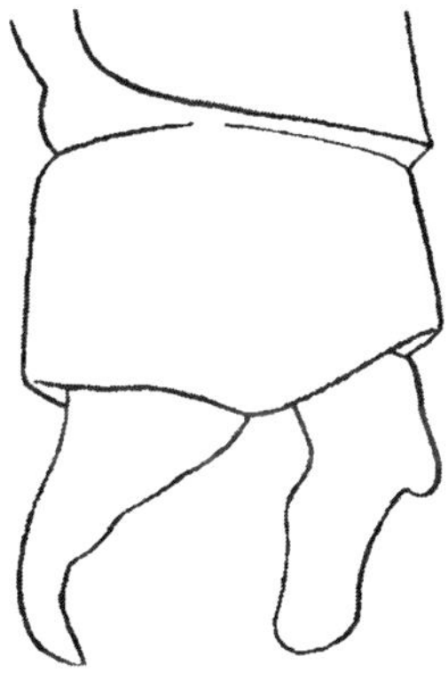

Fig. 106.—Anal appendages of *Indoneura ramburi* Fras., male, viewed from the left.

Female.—Abdomen 40–41 mm. Hind-wing 28 mm.

Closely similar to the male, but the markings more greenish-blue or even yellow, except in very old specimens, where the colour approximates to that of the male ; markings of *head* and *thorax* exactly similar to the male, those of the *abdomen* differing as follows :—The baso-dorsal annules interrupted on all segments so as to form paired spots as on segment 3 of the male ; segment 8 with a broad blue triangular spot, with its base on the apex of the segment and its apex extending for two-thirds the length of the segment mid-dorsally ; segment 9 with a T-shaped marking similar to that of *I. T-cœrulea* ; segments 8 and 9 with the ventral border blue ; segment 10 with a small saddle-shaped marking on the dorsum. *Anal appendages* black, short, conical, surmounting a conical black protuberance of their own length ; vulvar scales robust, not quite extending to the end of abdomen, bluish or pale brown throughout. *Wings* with 22 to 23 postnodal nervures to fore-wings, 20 in the hind; pterostigma ochreous, framed in black nervures; anal bridge more flatly arched than in the male.

Distribution.—Widely distributed throughout the WESTERN GHATS from Kanara to Malabar, including the hilly tracts of Mysore, Coorg, and the Nilgiris up to 6,000 ft. A comparatively rare insect south of the Palghat Gap, where it is replaced largely by *Esme*. I have specimens from the Shevaroy Hills, which seem to be the limits of the eastern distribution.

A larger and more slender insect than *I. gomphoides*, and breeding at a lower altitude; the conspicuous blue basal annules are sufficient to determine it, even when on the wing. The larger size, differently shaped pterostigma, and the absence of a ventral spine on the base of the superior appendages, as well as the head entirely black beneath, will distinguish it from *I. risi*.

Type in the British Museum; paratypes in many private collections. The two males indicated by Selys (Mém. Cour. vol. xxxviii, p. 171 (1886)), now in the MacLachlan collection, and determined by Selys as *I. gomphoides*, belong to this species, as also those indicated by Laidlaw from South Kanara

102. **Indoneura T-cœrulea,** sp. nov.

Indoneura ramburi Fras., Rec. Ind. Mus. vol. xxiv, p. 3 (1922) (var. ?).

Male.—Abdomen 42–44 mm. Hind-wing 30 mm.

Closely resembles the preceding species but differs constantly in the markings of labrum and terminal abdominal segments; labrum bordered with azure blue; segments 8 and 9 with very broad T-shaped azure blue markings on dorsum, the crossbar of the T lying along the apical border of the segments, the tail extending as a broad mid-dorsal stripe on segment 8 as far as its base, broader and shorter on segment 9, and with the ends of the crossbar curling round and extending nearly to the base of the segment like the flukes of an anchor, which the marking here, perhaps, more closely resembles than a T; segment 10 entirely blue on dorsum. *Anal appendages* similar to the last species.

Female unknown.

Type in the author's collection.

Distribution.—Appears to be confined to the south-eastern aspects of the NILGIRIS at rather low altitudes, the type being taken at Mattupaliyam Ghat, 2,500 ft., towards the end of July. It is, perhaps, but a local race of *I. ramburi*, but has been taken in company with it. On the wing it is impossible to distinguish the two species.

103. Indoneura obscura, sp. nov. (Fig. 107.)

Male.—Abdomen 44 mm. Hind-wing 28 mm.

Head: labium cinereous, tipped with black; labrum glossy black, unmarked with blue; clypeus steely blue-black; genæ and base of mandibles glossy black; rest of head, including under surface, mat black; eyes black above, dark bluish beneath. *Prothorax* and *thorax* velvety black, the former unmarked, the latter with the blue dorsal and lateral markings largely obscured, the antehumeral stripes present as a mere upper point of blue, those of the sides limited to the postero-inferior two-thirds of the metepimeron; beneath black posteriorly, pale bluish in front, obscured by dense white pruinescence. *Legs* black, extensor surface of tibiæ yellow.

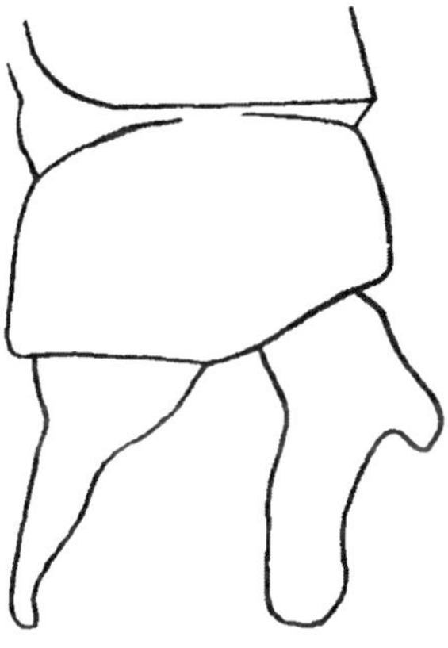

Fig. 107.—Anal appendages of *Indoneura obscura* Fras., male, viewed from the left.

Wings hyaline, palely enfumed; pterostigma markedly elongate, lozenge-shaped, proximal side shorter than the others, braced, dark blackish-brown finely framed in pale brown and thick black nervures; 20 to 21 postnodal nervures in fore-wings, 18 to 19 in the hind. *Abdomen* black, marked sparingly with blue as follows:—Segment 1 with a small spot on each side; segment 2 with a narrow stripe on each side; segments 3 to 6 with narrow obsolete blue annules at the base; remaining segments unmarked. *Anal appendages* almost exactly similar to those of *I. ramburi*.

Female unknown.

Distribution.—Eastern Ghats of India. The *type*, in the author's collection, is from Koraput, Vizagapatam, and was taken at an altitude of about 2,000 ft. during October. This species, remarkable for the almost complete absence of markings, is the only representative of its genus found outside the limits of its zoo-centre in the Western Ghats. Its isolation in the Eastern Ghats is a difficult problem to explain, as no species have so far been found in Central India.

104. **Indoneura risi** Fraser. (Fig. 108.)

Indoneura risi Fras., Rec. Ind. Mus. vol. xxxiii, pp. 449, 469–71, text-fig. 6, I (1931).

Male.—Abdomen 37–38 mm. Hind-wing 23–24 mm.

Head: labium white; labrum, ante- and postclypeus steely metallic blue-black; bases of mandibles brownish; genæ azure blue, marked with a large glossy black median spot; rest of head black, but the frons traversed by a broad azure blue band confluent with the blue of genæ; *head*

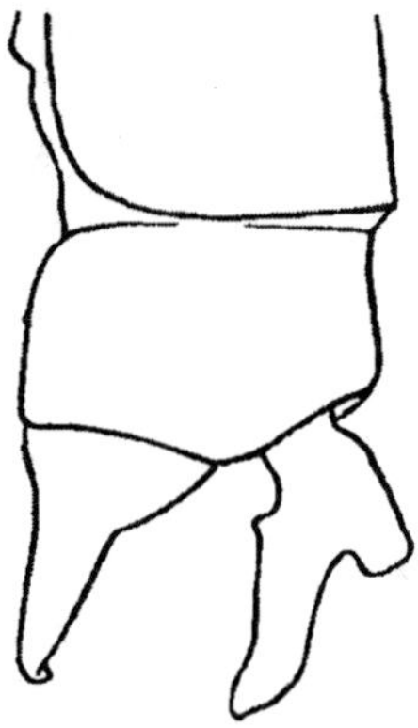

Fig. 108.—Anal appendages of *Indoneura risi* Fras., male, viewed from the left.

beneath occiput pale azure blue; eyes deep marine blue, capped with black. *Prothorax* black, the anterior lobe entirely azure blue, the middle lobe broadly blue laterally, this colour extending as a broad band for the whole length of lobe; dorsum and posterior lobe unmarked. *Thorax* velvety black on dorsum as far lateral as the middle of mesepimeron, and marked with a pair of azure blue antehumeral, narrow, slightly curved stripes, the inner border of which is slightly concave, the outer convex; laterally pale blue, marked with a fusiform black stripe which occupies the middle two-fourths of the posterior border of mesepimeron; beneath thorax pale greenish-blue, unmarked. *Legs* pale brown or dirty white, the femora mottled with black on the extensor surface, more broadly so at the distal ends; tibiæ with the extensor surface pale bluish; tarsi black; coxæ and trochanters pale blue. *Wings* hyaline, palely enfumed in adults, petiolated as far as a little proximal to *ac*; pterostigma black, *almost quadrate*, not diamond-shaped, braced; the nervure *ac* lying nearest the distal antenodal nervure or about half-way between the two antenodals in the hind-wing; *ab* feebly arched and short; 19 to 21 postnodal nervures in fore-wings, 19 in the hind. *Abdomen* black, marked with blue as follows:—Segment 1 almost entirely blue, marked on the dorsum with a moderately broad black band, extending the whole length of the segment; segment 2 broadly blue at the sides, especially apically; segments 3 to 7 with narrow blue basal annules, narrowest on the first and last of these segments; segments 8 to 10 entirely blue, except for a very narrow apical black border and the lower part of sides of segment 10. (The abdomen in this species is very narrow, and only about two-fifths longer than the wings.) *Anal appendages* rather longer than segment 10, of equal length; superiors black, of the usual generic shape, broad at base, lower border markedly angulated and bayonet-shaped, superior process about one-third the length of the inferior, which is directed straight back and of only moderate length, its extreme apex directed slightly downwards; inferior appendages brownish-black or paler, directed almost horizontally straight back or but slightly downward, broad at base, then tapered and tongue-like as seen in profile, the apex not visible from this direction, but turned horizontally and abruptly inward to end in an acute point.

Female unknown.

Distribution.—WESTERN GHATS; foothills of the Malabar Wynaad, near Tamaracherri, at about 500 ft., during the months of May to August.

Found amidst dense jungle in densely shaded streams, and, although local, is not a common insect. Distinguished from other species by its small size, by having the under surface of the head blue, and by the tiny spine near the base of the superior anal appendages beneath.

Type in the author's collection.

Genus **MELANONEURA** Fraser (1922). (Fig. 109.)

Melanoneura Fras., Mem. Dept. Agric. India (Ent.), vol. vii, p. 55 (1922); id., J. Bombay Nat. Hist. Soc. vol. xxix, p. 743 (1923); id., Rec. Ind. Mus. vol. xxvi, pp. 506, 508, fig. 5 (1924); id., ibid. vol. xxxiii, p. 469 (1931).

Zygopterous dragonflies of moderate size and slender build, with the characteristics of the genus *Indoneura*, differing only in the neuration of the wings, in which the anal bridge (*ab*) is entirely absent (as in some species of *Caconeura*), and by an abbreviated nervure *Cuii* which is never more than 5 to 6 cells in length.

Distribution.—WESTERN GHATS; the genus is monotypic, the only species being found solely in Coorg and the Malabar Wynaad. It is probably the most primitive of the *Indoneura* group, and is distinguished from the genus *Caconeura* mainly by its larger size and by the female being without processes

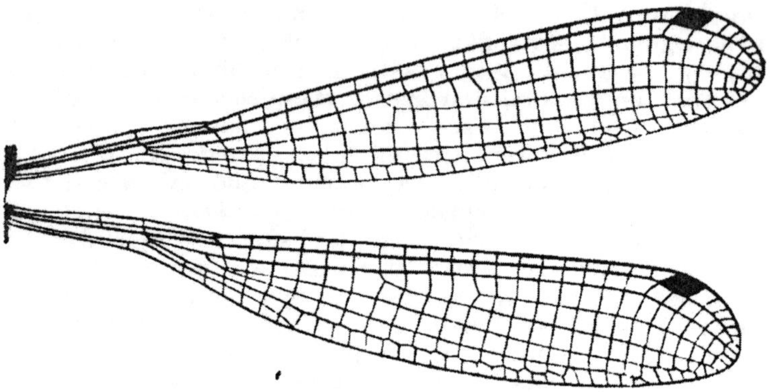

Fig. 109.—Wings of *Melanoneura bilineata* Fras., male.

to the posterior lobe of prothorax; the anal appendages also differ strongly from those of *Caconeura*, agreeing closely with those of *Indoneura*.

Genotype, *M. bilineata* Fras.

105. **Melanoneura bilineata** Fraser. (Figs. 109 & 110.)

Melanoneura bilineata Fras., Mem. Dept. Agric. India (Ent.), vol. vii, p. 55 (1922); id., J. Bombay Nat. Hist. Soc. vol. xxix, p. 743 (1923); id., Rec. Ind. Mus. vol. xxvi, pp. 429, 504, 505, 506, 508, text-fig. 5 (1924); id., ibid. vol. xxxiii, p. 449 (1931).

Male.—Abdomen 41–45 mm. Hind-wing 27–30 mm.

Head: eyes azure blue for the lower two-thirds, black for the upper third; labium ashy white; labrum and anteclypeus metallic blue-black; rest of head mat black, traversed by a moderately broad band across the postclypeus and cheeks. *Prothorax* black, the sides broadly azure blue, posterior lobe rounded, simple. *Thorax* mat black, marked with a narrow antehumeral azure blue stripe and two broad blue stripes on the sides, the posterior one covering the metepimeron. *Legs* black, coxæ and trochanters blue, tibiæ yellow on extensor surfaces. *Wings* hyaline, pterostigma black, over $1\frac{1}{2}$ cells; 19 to 22 postnodal nervures in fore-wing, 17 to 20 in the hind; anal bridge absent; position of *Riv+v* variable, usually arising opposite or distal to the line of node ; *Cuii* only 5 to 6 cells long in all wings. *Abdomen* black, marked with azure blue, segments 1 and 2 broadly on the sides, 3 to 6 with tiny basal lunules confluent over dorsum, 7 *unmarked*, 8 to 10 blue, with very narrow basal rings of black. *Anal appendages* black, equal, similar to those of *Indoneura* ; superiors directed at first straight back and then down and slightly back at a rather acute angle, a robust spine on dorsum springing from the point of angulation, the whole appendage shaped like a wrist and hand held in the attitude of clasping a ball, the spine representing the thumb, the distal two-thirds the fingers ; inferiors broad at base, conical, tapering to a point which is slightly recurved at the apex.

Fig. 110.—Anal appendages of *Melanoneura bilineata* Fras., male, viewed from the left.

Female.—Abdomen 37–43 mm. Hind-wing 27–31 mm.

Very similar to the male, the markings blue only in very adult specimens, usually citron yellow or pale bluish-green, especially the facial and antehumeral stripes, the sides of the thorax nearly always yellow ; segments 8 and 9 with apical fleur-de-lis-shaped markings, blue. Pterostigma pale brown ; 19 to 21 postnodal nervures in fore-wings, 17 to 18 in the hind.

Distribution.—COORG and MALABAR only, at altitudes of 3,000 to 4,000 ft. A fairly large colony has established itself near the head of the Sampaji River, where the insect shelters among cane-brakes, appearing only after 4 p.m. ; a smaller colony has made its home on a small stream at Santikupa in a coffee estate, and a few occasional specimens are found on the stream at Hallery, near Mercara. A third colony is established at the head of the Tamaracherri Ghat, Malabar. I failed to rediscover it at Sidapur, Coorg, where the type was taken.

M. bilineata is occasionally taken in company with *Indoneura ramburi*, but the two can always be distinguished by the absence of the conspicuous basal blue annules, so marked a feature in the latter. The total absence of the anal bridge will distinguish it from all other species of the group.

Genus **ESME** Fraser (1922). (Fig. 111.)

Esme Fras., Mem. Dept. Agric. India (Ent.), vol. vii, no. 7 (June 1922); id., Rec. Ind. Mus. vol. xxvi, pp. 506–9 (1924).

Zygopterous dragonflies of moderate size and slender build, with the same characters as the genus *Indoneura*, differing only in the neuration of the wings, which have a complete anal bridge (*ab*) instead of the vestigial curtailed structure found in *Indoneura*.

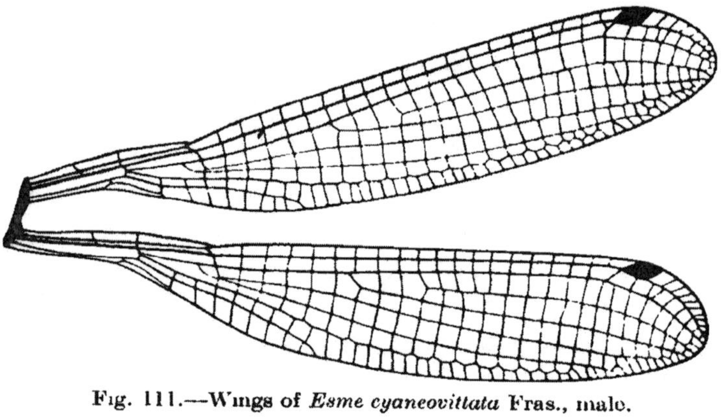

Fig. 111.—Wings of *Esme cyaneovittata* Fras., male.

Distribution.—Confined to the Western Ghats and, with the exception of one species that is found in South Kanara, restricted to the area south of the Palghat Gap and Palni Plateau.

Genotype, *E. cyaneovittata* Fraser.

Key to the Species of Esme.

1.	Labrum metallic blue-black, without blue markings	*mudiensis* Fras., p. 264.
	Labrum azure blue, bordered with black..	2.
2.	Legs black, unmarked; inferior anal appendage stout; blue on sides of prothorax interrupted anteriorly	*cyaneovittata* Fras., [p. 263.
	Legs black, femora and tibiæ marked with blue; inferior anal appendage slender; prothorax with the lateral blue stripe complete	*longistyla* Fras., p. 266

106. Esme cyaneovittata Fraser. (Figs. 111 & 112.)

Esme cyaneovittata Fras., Mem. Dept. Agric. India (Ent.), vol. vii, pp. 45, 46 (1922); id., Rec. Ind. Mus. vol. xxvi, pp. 429, 506, fig. 5 (1924); id., ibid. vol. xxxiii, p. 449, text-fig. 6, v (1931).

Male.—Abdomen 41 mm. Hind-wing 30–31 mm.

Head: labium with lateral lobes black, mid-lobe bluish-white; labrum azure blue, anterior border and a small median point black; anteclypeus dull black; postclypeus glossy steely blue-black; genæ and a broad fascia traversing the frons and confluent with the former azure blue, this blue area traversed by a black band which is confluent medially with the black of clypeus; rest of the head, including under surface, black; eyes deep sea-blue, capped above with black. *Prothorax* black, the greater part of anterior lobe azure blue,

Fig. 112.—Anal appendages of *Esme cyaneovittata* Fras., male, viewed from the left.

as well as the lower part of sides of middle lobe, the two areas not confluent, but only narrowly interrupted. *Thorax* velvety black on dorsum, azure blue laterally, the dorsum marked with narrow antehumeral blue stripes extending almost up to the antealar sinus, tapering above, squared below; laterally traversed by a moderately broad oblique black stripe situated over the postero-lateral suture; beneath palest blue. *Legs* wholly black, coxæ and trochanters with small spots of blue. *Wings* hyaline, more or less enfumed in adults; *ab* complete and often exhibiting aberrations in one or more wings of individual specimens, extra nervures connecting the anal bridge with the posterior border of wing, or the bridge continued distalwards as a vestigial nervure *IA*; pterostigma black, finely framed in pale brown and thick black nervures;

20 to 21 postnodal nervures in fore-wings, 18 in the hind; rarely a few double cells in the space between the costa and radius distal to the pterostigma. *Abdomen* black, marked with azure blue as follows :—Segment 1 entirely blue save for a small basal dorsal black spot; segment 2 with a broad stripe on each side, broadest apically; segments 3 to 7 with very narrow complete baso-dorsal annules; segments 8 to 10 blue, the apical border of 10 narrowly and the ventral borders of all segments broadly black. *Anal appendages* black, rather longer than segment 10, inferiors slightly longer than superiors, which are of the usual *Indoneura* type, broad at the base; apex a mere tiny spine, ventral process very long, narrow, and of even thickness throughout, sloping down and inward; inferiors very broad, conical as seen from the side, tapered, apices curled strongly in and upward.

Female.—Abdomen 40 mm. Hind-wing 30–32 mm.

Closely similar to the male, differing in the following points :—*Abdomen* of more robust build: labrum more broadly bordered with black, and its base also narrowly black: anterior lobe of *prothorax* with the blue area less extensive; segment 8 with the blue limited to the apical third of dorsum; segment 9 black, with a broad inverted T-shaped blue mark on dorsum; segment 10 blue, with its sides broadly black; *anal appendages* short, conical, black, surmounting a conical protuberance equal in length to the appendages. *Wings* with 20 postnodal nervures to fore-wings, 18 in the hind: vulvar scale robust, but not extending quite to the end of abdomen, black, marked with blue at apex.

Distribution.—Confined to the WESTERN GHATS south of the Palghat Gap; very common in the Palni Hills, where it appears to form colonies like those of *I. gomphoides*; less common in the Anaimalai and Mudis Hills, where it is more or less replaced by the next species, *E. mudiensis*, from which it is determined by its labrum marked with blue. From *E. longistyla* it may be known by its black legs unmarked with blue, by its stout inferior anal appendages, and by the broken blue lateral stripe on the prothorax.

Type in the British Museum; paratypes in the Indian and Pusa collections and many private collections.

107. **Esme mudiensis** Fraser. (Fig. 113.)

Esme mudiensis Fras., Rec. Ind. Mus. vol. xxxiii, pp. 449, 472, 473, text-fig. 6, iv (1931).

Male.—Abdomen 46 mm. Hind-wing 32 mm.

Head: labium white, palely tinted with blue; labrum glossy metallic blue-black, quite unmarked with blue; anteclypeus glossy black, postclypeus metallic blue-black; genæ pale azure blue, traversed by a broad black stripe which is continuous with the black of clypeus ; rest of the head velvety black, but the blue of genæ continued on to the frons. being only narrowly interrupted at the middle line ; antennæ black ; eyes dark blue during life ; the adjacent part of occiput, beneath head, black. *Prothorax* velvety black, the anterior lobe with a transversely oval blue spot at its middle, the middle lobe azure blue at its lower lateral border, more broadly behind than in front. *Thorax* velvety black on dorsum as far lateral as the middle of mesepimeron, marked with a narrow azure blue antehumeral stripe on each side, the outer border of the stripe convex, the inner border straight, its upper end tapering to an obtuse point which falls well short of the antealar sinus ; laterally azure blue, marked with an oblique complete black stripe on the postero-lateral suture and adjacent part

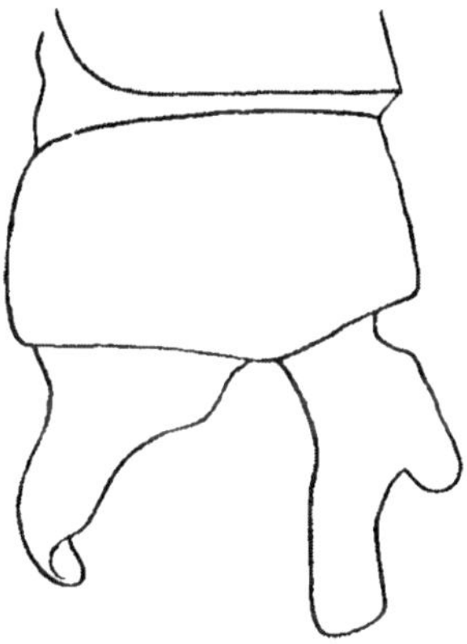

Fig. 113.—Anal appendages of *Esme mudiensis* Fras., male, viewed from the left.

of mesepimeron ; beneath palest blue, unmarked. *Legs* black, coxæ and trochanters azure blue. *Wings* hyaline, palely enfumed in adults, petiolated to well proximal to the level of *ac* ; pterostigma blackish-brown, framed finely in paler brown and thick black nervures, diamond-shaped, its sides equal in length, strongly braced ; 23 postnodal nervures in fore-wings, 21 in the hind ; *ac* situated slightly nearer the distal antenodal nervure and meeting *ab*, which is complete, well distal to its origin. *Abdomen* black, marked with azure blue as follows :—Segment 1 broadly blue at the sides, this colour extending dorsalwards along the apical border of the segment, and almost confluent with the blue on the opposite side ; segment 2 with a broad stripe on each side at its lower part, extending the whole length of the segment ; segments 3 to 6 with very small dorsal paired spots at base, obsolete on segments 4 to 6 in some specimens ; segment 7 entirely black; segments 8 to 10 azure blue, but the lower part of sides black, the black border broadening towards the base of each segment and forming a narrow basal annule on segment 8. *Anal appendages* black, slightly longer than segment 10, and of the usual *Indoneura* shape; superiors with a short upper process and a very long lower one, which latter is strongly curved inwards as seen from above, and at first down and then analwards as seen from the side, the apex slightly tumid and turned a little down; inferiors of the same length as superiors, broad at base, compressed laterally, tapering gradually to the apex, which is turned abruptly upwards and inwards.

Female.—Abdomen 44 mm. Hind-wing 29 mm.

Closely similar to the male, differing only in sexual characters and the shape of the blue markings on the terminal abdominal segments. Bases of mandibles blue, as in the male; the blue band traversing face broadly interrupted across the frons. *Wings* with 20 postnodal nervures to fore-wings, 18 in the hind; *ac* situated much nearer the distal antenodal nervure; sides of segments 1 and 2 with the blue marking more restricted; only segment 3 with basal paired spots; segment 8 with three small apical spots, a median pyriform spot, and an oval one situated subdorsally on each side of it; segment 9 with a broad triangular spot, its base resting on the apical border of the segment, its apex just reaching the base; segment 10 with a very small mid-dorsal basal spot. *Anal appendages* black, small, conical, surmounting a short conical protuberance; vulvar scales robust, black, tipped with yellow.

Distribution.—Mudis Hills, Travancore, and Anaimalais, WESTERN GHATS of South India south of the Palghat Gap.

A somewhat scarce and very local species, which seems to be confined to the Anaimalai group at an elevation of about 3,000 to 4,000 ft., emerging during May. Its habits do not differ from others of the *Indoneura* group. From other species of *Esme* it is easily distinguished by the labrum being entirely unmarked with azure blue.

Type in the British Museum; paratypes in the author's collection

108. **Esme longistyla** Fraser. (Fig. 114.)

Esme longistyla Fras., Rec. Ind. Mus. vol. xxxiii, pp. 449, 471, 472, text-fig. 6, vi (1931).

Male.—Abdomen 44 mm. Hind-wing 28–29 mm.

Head: labium pale brownish-white or white; labrum azure blue, its base moderately broadly bordered with black; ante- and post-clypeus black; bases of mandibles, genæ, and a broad fascia traversing frons azure blue, a large black spot on each cheek narrowly confluent with the black of clypeus; rest of head above and beneath mat black: eyes deep marine blue, capped with black. *Prothorax* black, with the whole of the anterior lobe and a broad stripe traversing the lower part of sides of middle lobe blue, the two blue areas being confluent. *Thorax* velvety black on dorsum as far back as the middle of mesepimeron, marked with very narrow blue antehumeral stripes, broadening below and falling well short of the antealar sinus above, laterally azure blue, traversed by a moderately broad black oblique stripe which overlaps the postero-lateral suture on both sides: beneath pale greenish-yellow or blue, pulverulent in adults. *Legs* black, with a bluish stripe on the outer and inner sides of hind pair of femora: tibiæ with extensor surfaces pale bluish; tarsi black: coxæ and trochanters pale blue. *Wings* hyaline.

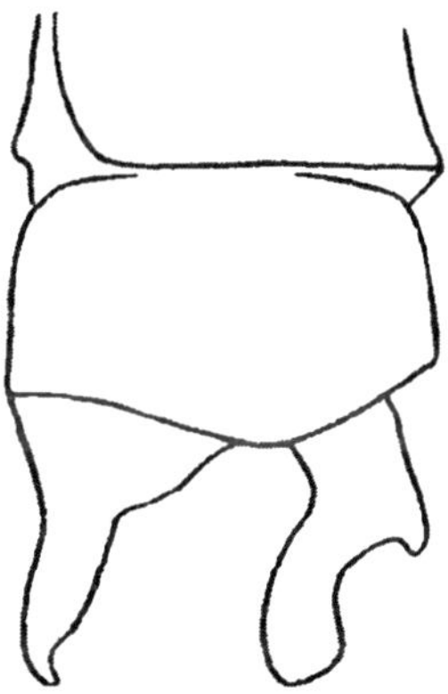

Fig. 114.—Anal appendages of *Esme longistyla* Fras., male, viewed from the left.

palely enfumed in adults, petiolated to a point well proximal to *ac*, which nervure lies nearest the distal antenodal nervure. *ab* complete; pterostigma with sides subequal, subquadrate in shape, black: 22 postnodal nervures in fore-wings, 20 in the hind. *Abdomen* long and slender, black, marked with blue as follows:—Segment 1 entirely blue, save for a large triangular black dorsal spot, the apex of which just reaches the apical border of the segment; segment 2 with a broad irregular blue stripe traversing each side, the upper border of the stripe with two broad indentations; segments 3 to 7 with narrow basal blue annules, narrowly bisected by the finely black mid-dorsal carina on segment 3, and partially so on the other segments: segments 8 to 10 wholly azure blue, but the latter segment with the sides and apical border narrowly black. *Anal appendages* black, superiors slightly longer than inferiors and segment 10; superiors of the usual *Indoneura* shape, strongly curled in towards each other, broad at base, inferior border shallowly concave, superior process obtuse, very short, a mere angular projection; inferior process very long, rather narrow, a little constricted at its middle, a little clubbed and obtuse at its apex; inferior appendages broad at base, after which there is a short constriction, then slightly tapered as far as apex, which is curled upward and inward and ends in a very acute point.

Female.—Abdomen 42–43 mm. Hind-wing 28 mm.

Remarkably similar in markings to the male, differing only by the anterior lobe of *prothorax*, which is blue at its middle only. *Wings* with 19 to 20 postnodal nervures in fore-wings, 18 in the hind; pterostigma warm brown, more diamond-shaped than in the male, especially in the hind-wings, where the costal and posterior borders are longer than the lateral. *Anal appendages* very short, conical, black, projecting over a conical blackish-brown protuberance of the same length as appendages; vulvar scales blue at the sides, long, robust.

Distribution.—South Kanara, S. India. A few pairs only of this rare insect were taken by me at Perage, Coorg, during September and October, settled on ferns and scrub beside a torrential stream. Its habits are similar to the rest of the group and, like most species, it appears to be very local. The species is considerably smaller than either *E. cyaneovittata* or *E. mudiensis*, the only two other known species of the genus; it differs from the former by the aborted superior process of the superior appendages, the slender shape of the inferior appendages, by the anterior lobe of the prothorax being wholly blue and the hind pair of legs marked with blue, this latter character serving to separate them at a glance. From *E. mudiensis* it is distinguished by the labrum, entirely blue save for the base, which is in striking contrast to the entirely blue-black metallic structure of that species.

Type and co-type in the British Museum; paratypes in the author's collection.

Genus **PHYLLONEURA** Fraser (1922). (Fig. 115.)

Phylloneura Fras., Rec. Ind. Mus. vol. xxiv, p. 3 (1922); id., Mem. Dept. Agric. India (Ent.), vol. vii, p. 45 (1922); id., J. Bombay Nat. Hist. Soc. vol. xxix, p. 743 (1923); id., Rec. Ind. Mus. vol. xxvi, pp. 506, 508 (1924); id., ibid. vol. xxxiii, p. 469 (1931).

Zygopterous dragonflies of moderate size, considerably larger than species of *Indoneura*, almost double that of the smallest species of that genus; of slender build, and with the characteristics of the genus *Indoneura*, but differing in the neuration of the wings, in which the network is denser, in having the nodal index higher, and the anal bridge complete, as in the genus *Esme*; the wings are also longer and considerably narrower and rather more pointed and falcate at the apices, and the azure blue markings are more extensive on the terminal abdominal segments. In many specimens vestigial remnants of intercalated sectors appear in the form of double cells,

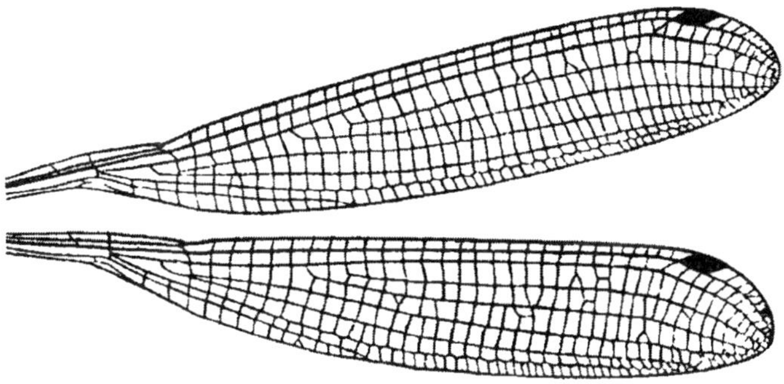

Fig. 115.—Wings of *Phylloneura westermanni* (Selys), male.

the sectors of which are joining up here and there to form broken longitudinal nervures.

Distribution.—WESTERN GHATS: this monotypic genus is restricted to the Nilgiri Hills, Coorg, and a few localities in the Malabar Wynaad, in similar situations to those in which species of *Indoneura*, *Esme*, and *Melanoneura* are found.

Genotype, *P. westermanni* (Selys).

109. **Phylloneura westermanni** (Selys). (Figs. 115 & 116.)

Alloneura westermanni Selys, Bull. Acad. Belg. (2) vol. x, p. 447 (1860).

Disparoneura westermanni Selys, Mém. Cour. vol. xxxviii, p. 171 (1886); Kirby, Cat. Odon. p. 134 (1890); Laid., Rec. Ind. Mus. vol. xi, p. 388 (1915); id., ibid. vol. xiii, pp. 323, 347 (1917).

Phylloneura westermanni Fras., Rec. Ind. Mus. vol. xxiv, pp. 3, 4 (1922); id., J. Bombay Nat. Hist. Soc. vol. xxix, p. 743 (1923); id., Rec. Ind. Mus. vol. xxvi, pp. 506–9 (1924); id., ibid. vol. xxxiii, p. 469 (1931).

Male.—Abdomen 41–51 mm. Hind-wing 28–37 mm.

Head: labium white, tipped with black; labrum, ante- and postclypeus steely blue-black; bases of mandibles black, with a central blue spot; genæ azure blue, traversed by a black band which is confluent with the black of clypeus; frons black, but the blue of the genæ spreading narrowly on to it at either side; rest of head velvety black, including under surface; eyes deep sea-blue, capped with black. *Prothorax* black, the greater part of the anterior lobe and a broad stripe on the lower part of sides of middle lobe azure blue. *Thorax* velvety black on dorsum, azure blue at the sides, the former marked with very narrow antehumeral blue stripes which do not extend to the alar sinus; laterally the blue traversed by a moderately broad oblique black complete stripe situated over the postero-lateral suture; a small pyriform posthumeral blue spot on each side; beneath bluish-white. *Legs* black, coxæ and trochanters blue; extensor surfaces of tibiæ yellow. *Wings* hyaline, more or less deeply enfumed in adults, but quite clear in younger specimens; anal bridge arched at origin,

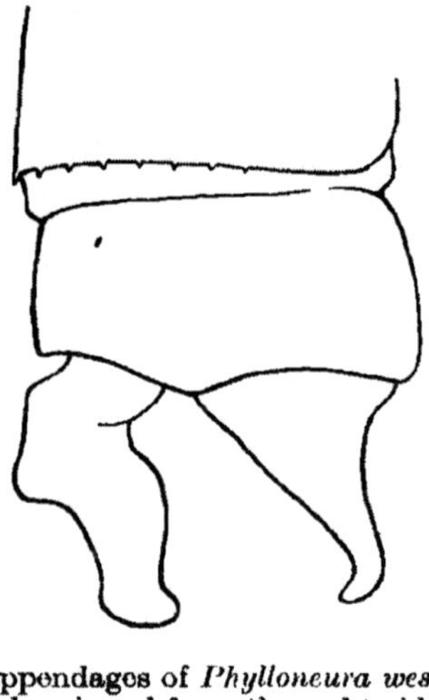

Fig. 116.—Anal appendages of *Phylloneura westermanni* (Selys), male, viewed from the right side.

but inclining towards the hinder border of wing distally; pterostigma black or dark reddish-brown, framed in black nervures, braced, narrower than in the rest of the *Indoneura* group, covering two cells; 28 to 31 postnodal nervures in fore-wings, 26 to 27 in the hind; *Cuii* 13 to 15 cells in length; many double cells between main nervures, especially, in some specimens, towards the apex of wings. *Abdomen* black, marked with blue as follows:—Segment 1 broadly blue at the sides, narrowly so along the apical border; segment 2 with the ventral border broadly blue; segments 3 to 5 with very narrow basal annules, almost obsolete on 5; segment 6 entirely black; segment 7 with its apical half azure blue; segments 8 and 9 entirely blue on dorsum, except for very narrow basal black annules; 10 blue on the dorsum, black laterally. *Anal appendages* black, half as long again as segment 10; superiors slightly longer than inferiors, of the usual *Indoneura* shape, broad at base, apex a rounded tubercle about one-fourth the length of the ventral process, which is a little sinuous, rounded, and tumid at apex; viewed in profile, this appendage has an upper outer obtuse tubercle at the base; inferiors broad at base, sloping downwards and tapering gradually to an acute point, which is directed inward and slightly upward.

Female.—Abdomen 45–46 mm. Hind-wing 33–35 mm.

Exactly similar to the male, except for sexual differences and the blue markings on terminal segments of abdomen; *abdomen* shorter and more robust; segments 8 and 9 with large triangular dorsal blue spots, the base of the triangles resting on the apical border of the segments, the apex tapering as far as the base of the segments, especially on segment 9; ventral border of segment 8 and dorsum of segment 10 blue. *Anal appendages* conical, short, black, surmounting a conical process of their own length and colour; vulvar scale robust, not extending quite to the end of abdomen, blue, tipped with black.

Distribution.—WESTERN GHATS: scattered colonies in the Nilgiri Hills, Coorg, and Malabar Wynaad; one large colony about half-way up the Gudalur Ghat, Nilgiris; another in the bed of the Burliyar River, in the same district; a third, smaller colony about half-way down the Sampaji Ghat, Coorg; found on the wing from May to September. The smallest specimens come from Coorg, the largest from the Nilgiris, whilst intermediate sizes are found in a colony on the Tamaracherri Ghat, Malabar. It is to be noted that the size of specimens in individual colonies is constant, so that it would appear that the colonies are of very ancient standing.

Type in the Hagen collection; paratypes in the Indian, Pusa and British Museum collections; paratype female in the British Museum, collected by me in the Nilgiris.

Subfamily CŒNAGRIINÆ.

Agrion, 5me Legion, Selys, Bull. Acad. Belg. (2) vol. xx, p. 377 (1865); id., ibid. vol. xliii, p. 97 (1877).

Cœnagrioninæ, pars, Kirby, Cat. Odon. p. 119 (1890).

Agrioninæ (minus genus *Lestes*) Lucas, British Dragonflies, pp. 30, 58 (1900).

Agrioninæ Tillyard, Biology of Dragonflies, pp. 279, 280 (1917).

Legions *Argia* and *Agrion* Laid., Rec. Ind. Mus. vol. xvi, pp. 169–72 (1919).

Cœnagrioninæ (minus group *Platycnemis*) Munz, Mem. Amer. Ent. Soc no. 3, pp. 21, 41 (1919).

Cœnagrionidæ Tillyard, Insects of Aust. & New Zeal. pp. 76, 77 (1926).

Cœnagrioninæ Fras., J. Bombay Nat. Hist. Soc. vol. xxxiii, p. 834 (1929); Needham, Zool. Sinica, ser. A, vol. xi, p. 238 (1930).

Dragonflies of small size and slender build, vari-coloured, non-metallic or occasionally partly metallic; resting with wings closed over dorsum; wings narrow, rounded at apex, hyaline, rarely coloured; situation of *ac* variable; no accessory basal postcostal nervures ever present; petiolation variable, but usually beginning at or a little proximal to the level of *ac*; *ab* always present and complete; *1A* present; *Cuii* variable in length; discoidal cell elongate, of variable length, di tal end very acute, costal side always much shorter than the posterior; sectors of arc arising from the lower end of arc and separated at origin; individual cells mostly five-sided; *Riii* usually arising about midway between the node and pterostigma; origin of *Riv+v* usually well proximal to the oblique nervure descending from the subnode; *IRiii* in continuation of that nervure; pterostigma generally diamond-shaped, but variable in the genera, sometimes variable in shape and colour in the fore- and hind-wings of the male, and often variable in the sexes; no intercalated sectors save *IRii*.

Abdomen of moderate length, or long and very slender, usually not greatly exceeding the length of hind-wing, and never equal to twice its length; segments 8 to 10 becoming progressively shorter, the latter segment sometimes bearing two dorsal apical tubercles and very often emarginate. Anal appendages very variable in the genera and species, superiors shorter or longer than inferiors, the latter often rudimentary; vulvar scales robust, but not generally extending beyond the end of abdomen; a ventral spine present or absent at the apical border of segment 8 in the female. Penile organ very variable, but the end portion always curled strongly over and embracing the stem of the organ, branched or unbranched at the extreme end, and sometimes furnished with a pair of dorsal spines.

Distribution.—Cosmopolitan. Within Indian limits a large number of genera known, mostly with a wide distribution, throughout CONTINENTAL INDIA, BURMA, and CEYLON. The majority of species breed in the still waters of lakes and tanks, but a number are only riverine in habit; they have a wide adaptability, being found in dry and wet zones, heavy jungles, and dry arid desert lands. The life-histories of many species have been worked out in Europe and America, but little is known of the Indian species in this regard.

Key to the Genera of the Cœnagriinæ.

1.	Are situated at the level of the distal antenodal nervure	2.
	Are situated distal to the level of the distal antenodal nervure	11.
2.	*ab* (anal bridge) arising from the hind border of wing at the point where *ac* meets it	3.
	ab arising from the hind border of wing more or less proximal to the point where *ac* meets it	6.
3.	A prominent ridge on the frons; no postocular coloured spots on head; head and thorax of a uniform colour, without any dark markings	CERIAGRION Selys, [p. 313.
	No ridge on frons; postocular coloured spots always present; head and thorax not of a uniform colour, and usually with black or dark markings	4.
4.	Pterostigma in fore-wing larger than that in the hind; abdomen usually very long and slender; female with an apical ventral spine on segment 8	ACIAGRION Selys, [p. 333.
	Pterostigma in fore- and hind-wings of the same size; abdomen not unduly long and slender; female without a ventral spine on apical border of segment 8	5.
5.	Pterostigma longer than broad, diamond-shaped, distal and proximal sides very oblique; 10 to 12 postnodal nervures in fore-wings	PSEUDAGRION Selys, [p. 274.
	Pterostigma almost square, distal side convex; 14 to 16 postnodal nervures in fore-wings	ARCHIBASIS Kirby, [p. 307.
6.	Pterostigma differing in shape and size in fore- and hind-wings of male; segment 10 of male with a pair of dorsal apical tubercles; postocular coloured spots present or absent in the adult stage	7.
	Pterostigma of the same colour and shape in fore- and hind-wings of male; segment 10 of male usually without dorsal apical tubercles; postocular coloured spots present or absent	8.

7.	Postocular coloured spots always present in adult stage; tubercles on apical border of segment 10 closely apposed.	ISCHNURA Charp., [p. 346.
	Postocular coloured spots absent in the adult stage; tubercles on apical border of segment 10 widely separated by a shallow notch	RHODISCHNURA Laid., [p. 368.
8.	Female with a robust ventral apical spine on segment 8; postocular coloured spots present	ENALLAGMA Charp., [p. 371.
	Female without a ventral apical spine on segment 8; postocular coloured spots present or absent	9.
9.	*ab* arising well proximal to the level of the proximal antenodal nervure; tarsal claw-hooks of equal length; postocular coloured spots absent	ONYCHARGIA Selys, [p. 416.
	ab arising well distal to the level of the proximal antenodal nervure; tarsal claw-hooks unequal in length; postocular coloured spots present (except in some very old pruinosed specimens).	10.
10.	Basal side of discoidal cell in fore-wing equal to the costal; 15 to 16 postnodal nervures in fore-wings; *ab* arising from hinder border of wing very slightly proximal to the level of *ac*	HIMALAGRION Fras., [p. 330.
	Basal side of discoidal cell in fore-wing shorter than the costal side; only 8 to 10 postnodal nervures in fore-wings; *ab* arising from hinder border of wing well proximal to the level of *ac*	CŒNAGRION Kirby, [p. 412.
11.	Moderately large species, with abdomen about 30 mm. in length; 10 to 12 postnodal nervures in fore-wings.....	ARGIOCNEMIS Selys, [p. 405.
	Smaller species, with abdomen not more than 20 mm. in length; only 6 to 8 postnodal nervures in fore-wings	12.
12.	The junction of *ab* and *IA* (medio-anal link) markedly angulated	AGRIOCNEMIS Selys, [p. 379.
	The junction of *ab* and *IA* not angulated, continued on in the same straight line	MORTONAGRION Fras., [p. 408.

Genus **PSEUDAGRION** Selys (1876). (Fig. 117.)

Pseudagrion Selys, Bull. Acad. Belg. (2) vol. xlii, p. 490 (1876); Kirby, Cat. Odon. p. 153 (1890); Selys, Ann. Mus. Civ. Genova, (2) vol. x (xxx), p. 80 (1891); Laid., Rec. Ind. Mus. vol. xii, pp. 21, 22 (1916); Ris, Odon. South Africa, pp. 297–9 (1918); Laid., Rec. Ind. Mus. vol. xvi, pp. 170, 191–2 (1919); Munz, Mem. Amer. Ent. Soc. no. 3, p. 62 (1919); Fras., J. Bombay Nat. Hist. Soc. vol. xxix, p. 748 (1923); id., Rec. Ind. Mus. vol. xxxiii, p. 465 (1931).

Dragonflies of small size and slender build, colours non-metallic and usually bright blue, marked with black or red, orange or green, etc., less commonly dull coloured, black, brown, or pruinosed; wings hyaline; pterostigma similar in the fore- and hind-wings, lozenge-shaped, narrow, oblique at both ends, but especially so at the distal extremity, braced, covering rather less than one cell; postnodals numbering 8 to 15 in the fore-wings, usually about 10: discoidal cell acutely pointed at the distal end, with the costal side less than half the length of posterior in fore-wing, and about half the length in the hind, basal side equal to the costal in the fore-wing, shorter than the costal in the hind, distal side very oblique; sectors of arc arising from the lower end of arc, divergent from origins; arc situated at the level of the distal antenodal nervure; *ab* (anal bridge) always present and complete, arising at the point where *ac* meets hinder border of wing, and continued on as the nervure *IA*, but an angulation at the junction of the two; *Riv+v* arising well before the nervure descending from the subnode; *IRiii* at that level; *ac* situated considerably nearer the level of the proximal antenodal nervure.

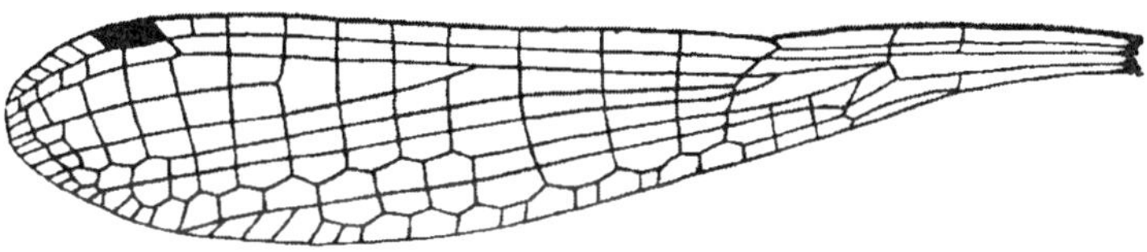

Fig. 117.—Hind-wing of *Pseudagrion spencei* Fras., male.

Head narrow, usually bearing postocular triangular coloured areas, but these absent in some species; eyes globate; thorax rather slender; posterior lobe of the prothorax of female with two small hooks or spines sloping strongly forwards on the dorsum of middle lobe; abdomen slender, cylindrical, not nearly twice the length of hind-wing; legs short, tibial spines of moderate length, 4 to 7 on the hind pair; anal appendages very variable, as long as or shorter than segment 10, the inferiors much shorter, superiors frequently forked or notched at apex and with or without spines at base; inferiors simple, conical. Genitalia: lamina narrowly and deeply cleft; hamules broad quadrate plates meeting across the middle line; lobe small, depressed; penis with broad apex, the end curled over the dorsum and bifid to embrace the stem of the organ; vulvar scales moderately robust, no basal spine present on segment 8 of the female.

Genotype, *Pseudagrion furcigerum* (Ramb.).

Distribution.—Restricted to the Old World, but very widely distributed therein. Continental Africa, Madagascar, INDIA; BURMA, and CEYLON, Malaysia, Java, Sumatra, New Guinea, Australia, South China, Japan, the Philippines, Fiji, and Samoa. Africa appears to be the headquarters of the genus, no less than 24 species having been listed by Dr. Ris from that continent, but 13 species are found within Indian limits, seven of which are restricted to the area. Some species breed both in still and running water but, for the greater part, the former is preferred. *P. indicum*, *P. williamsoni*, *P. hypermelas*, and *P. rubriceps* breed exclusively in streams. Some species are given to biennial migrations, especially *P. decorum* and *P. microcephalum*, which may be seen in myriads during September and October migrating up the coast of Western India.

It is impossible to give a hard and fast definition for the genus, and, if the original one is adhered to, several species, obviously true Pseudagrions, would be ruled out of it. Thus the postocular spots are not invariably present, the origin of *ab* is not always at the level of *ac*, and even the prothoracic spines of the female are subject to some varia ion. As, however, there are no well defined groups within the genus, it is convenient to adhere to the old definition until such time as the genus can be split up.

Key to the Indian Species of Pseudagrion.

1.	Face, frons, and vertex bright reddish-orange or dark ochreous	2.
	Face, frons, and vertex mainly black	*hypermelas* Selys, [p. 289.
	Face, frons, vertex, and occiput blue or green, marked with black	3.
2.	Thorax golden green on dorsum, azure blue on sides, sparingly marked with black; never pruinosed; a narrow black humeral stripe present	*rubriceps* Selys, p. 296.
	Similar to the last, but without the humeral stripe	*ceylanicum* (Kirby), [p. 299.
	Thorax black, densely pruinosed, especially on dorsum	*pruinosum* Burm., [p. 305.
3.	Thorax azure blue on dorsum and sides, marked with medial and humeral black stripes	4.
	Thorax green on dorsum, azure blue on sides, marked with medial and humeral black stripes	7.
	Thorax palest blue, with three fine black lines on mid-dorsal carinal ridge and a thicker black humeral stripe	*decorum* (Ramb.), [p. 286.
4.	Superior anal appendages as long as segment 10 and bifid at apex as seen in profile	5.
	Superior anal appendages shorter than segment 10 and bifid at apex as seen in profile; 11 postnodals in fore-wings	*bengalense* Laid., p. 282.
	Superior anal appendages shorter than segment 10, not bifid at apex as seen in profile	*malabaricum* Fras., [p. 284.

5.	Superior anal appendages expanded on the inner side and bearing 1 to 3 small spines near base on this border	6.
	Superior anal appendages not expanded on the inner side, but with a long robust spine near base within	*spencei* Fras., p. 292.
	Superior anal appendages without expansion on the inner side and without any spines on inner side or near base ..	*laidlawi* Fras., p. 294.
6.	Dorsal marking on segment 2 goblet-shaped; superior anal appendages with 2 or 3 small spines on inner border	*microcephalum*(Ramb.), [p. 278.
	Dorsal marking on segment 2 a narrow rectangle; superior anal appendages with only a single small basal spine....	*andamanicum* Fras., [p. 281.
7.	Superior anal appendages with a robust curled spine at base; segment 2 with a goblet-shaped marking on dorsum or one similar to that found in *Cœnagrion mercuriale*	*indicum* Fras., p. 303.
	Superior anal appendages without any spine at base; segment 2 with a broad dorsal stripe enclosing a small triangular or V-shaped blue marking near the base.	*williamsoni* Fras., [p. 300.

Key to the Females of the above.

1.	More than 12 postnodal nervures to fore-wings..............................	*pruinosum* Burm., [p. 306.
	Less than 12 postnodal nervures to fore-wings............................	2.
2.	Thorax with a single fine black line on mid-dorsum	3.
	Thorax with three fine black lines on mid-dorsum	5.
	Thorax with a single thick black stripe on mid-dorsum; segment 9 wholly black save for a fine blue apical ring...	*malabaricum* Fras., [p. 285.
3.	Head and upper surface of eyes with a large amount of bright orange; segment 9 with a bifid black marking	*microcephalum*(Ramb.), [p. 279.
	Head and upper surface of eyes without orange colouring	4.
4.	Segment 2 with a black arrow-head-shaped marking on dorsum; segment 9 with a quadrate black dorsal spot	*decorum* (Ramb.), [p. 288.
	Dorsal markings on segments 2 and 8 very broad, extending the whole length of segments, that on 2 broad at base	*bengalense* Laid., [p. 283.
	Segment 2 with a diamond- or cordate-shaped subapical spot on dorsum connected finely to base and apex	*andamanicum* Fras., [p. 281.
5.	Segments 2 to 7 with dorsal markings reduced to fine black lines............	*spencei* Fras., p. 293.
	Segments 2 to 7 with broad dorsal black markings	6.

6.	Segment 9 with the black dorsal marking bifid	*rubriceps* Selys, p. 298.
	Segment 9 with the black dorsal marking not bifid	7.
7.	Head and thorax uniform pale greenish-yellow	*hypermelas* Selys. [p. 291.
	Head and thorax uniform pale lilaceous..	*laidlawi* Fras., p. 295.
	Head and thorax dark green, with the black markings surrounded by an areola of golden yellow	8.
8.	Dorsal marking on segment 2 broad	*indicum* Fras., p. 304.
	Dorsal marking on segment 2 very broad	*williamsoni* Fras., [p. 302.

110. **Pseudagrion microcephalum** (Ramb.). (Fig. 118.)

Agrion microcephalum Ramb., Ins. Névrop. p. 259 (1842).
Pseudagrion microcephalum Selys, Bull. Acad. Belg. (2) vol. xlii, p. 504 (1876); Kirby, Cat. Odon. p. 153 (1890); Laid., Rec. Ind. Mus. vol. xii, pp. 21–4 (1916); Ris, Suppl. Ent. no. v, pp. 40–3 (1916); Laid., Rec. Ind. Mus. vol. xvi, pp. 172, 192, 193 (1919); Fras., ibid. vol. xvi, p. 467, pl. xxxvi, fig. 2 (1919); id., ibid. vol. xxiv, p. 311 (1922); id., J. Bombay Nat. Hist. Soc. vol. xxix, p. 749 (1923); Laid., Spolia Zeylan. vol. xii, p. 372 (1924); Fras., Rec. Ind. Mus. vol. xxvi, pp. 428, 497 (1924); id., ibid. vol. xxxiii, pp. 448, 465 (1931).

Male.—Abdomen 27 mm. Hind-wing 17 mm.

Head: labium white; labrum, bases of mandibles, genæ, frons, and fore part of vertex pale azure blue, the postclypeus with three small black spots at base; occiput blue behind, and marked with very large postocular azure blue spots, narrowly framed in black and bordered with dark brown or black anteriorly; basal joints of antennæ blue; eyes palest sky-blue beneath, deepening to dark azure blue above, and with a small cap of brown confluent with the brown on back of vertex; beneath black, bordered with blue against the eyes. *Prothorax* azure blue, the anterior lobe and five parallel narrow lines on middle lobe all confluent in front and behind, black. *Thorax* azure blue, paling to white below and beneath; dorsum with a moderately broad black stripe on the mid-dorsal carina, and rather narrower humeral black stripes; the lateral sutures finely mapped out in black, but broadening on the upper part of postero-lateral suture. *Wings* hyaline; pterostigma greyish, framed in black sutures, covering slightly less than one cell, braced; 10 postnodal nervures in fore-wings, 9 in the hind. *Abdomen* azure blue, marked with black as follows:—Segment 1 with a small quadrate black basal spot; segment 2 with a goblet-shaped dorsal black mark, the stem of the goblet very short and confluent with a narrow apical black ring, whilst from the centre of the cup part a fine black line runs along the mid-dorsal carina to connect with a narrow basal ring; segments

3 to 7 with broad dorsal markings not extending quite as far as the base of segments, but expanding subapically and then contracting to become confluent with apical narrow rings; on segment 7 the stripe expands gradually as far as the apex of the segment; segment 8 with a thick apical dorsal ring; segment 9 unmarked, save for a narrow black line and a row of black apical spines; segment 10 with a broad saddle-shaped dorsal mark extending from base to apex. *Anal appendages*: superiors black, blue within, as long as segment 10, digitate, slightly narrower at base, deeply cupped within, rounded externally and bifid at apex as seen in profile. A small black spine on one side of the apical notch and two small black spines crossing each other at the base on inner side; inferiors less than half the length of superiors, white, small cone-like

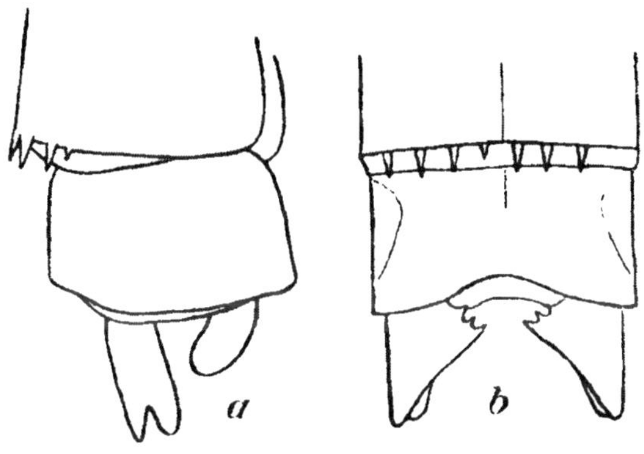

Fig. 118.—Anal appendages of *Pseudagrion microcephalum* (Ramb.), male.
a. Right lateral view. *b.* Dorsal view.

structures, pointed within. (These appendages show considerable variation, especially the superiors. Dr. Ris (*l. c.*) has illustrated this well, giving figures of several types ranging from Ceylon to Formosa.)

Female.—Abdomen 29 mm. Hind-wing 20 mm.

Differs in several respects from the male. Eyes palest blue beneath, olive-green in the upper half, deepening on the summit; postocular spots blue behind, graduating to olive-green in front, where they are tinted with orange; whole of face and vertex olivaceous suffused with orange; genæ paler and without the orange tinting; the fine transverse black line crossing vertex interrupted, finer, absent at the eyes; the transverse black stripe behind the postocular spots thicker and extending as far out as the eyes. *Prothorax* bluish-green, the black markings more complete and cutting up the ground-colour into a large mid-dorsal and large lateral spots; posterior lobe orange at its centre, which is curled strongly back and has a long spine-like process on either side projecting and slanting strongly forwards, shaped like the stalked eyes of a snail as seen from above and ochreous in colour, tipped with black. *Thorax* bluish-green, richly suffused with golden orange on dorsum, especially in the humeral area, azure blue laterally. Black markings similar to the male, but the parallel lines on mid-dorsum much interrupted and often a mere chain of fine dots; numerous blue spots on tergum: white and pruinosed beneath. *Legs* as for the male, but pruinosed. *Wings* with 10 postnodal nervures to fore-wings, 9 in the hind; pterostigma paler brown, otherwise similar; *ac* situated more proximally. *Abdomen* coloured similarly to the male, but the black markings differing as follows:—Segment 2 with a thick dumbbell-shaped dorsal mark extending from the base to apex, shortly stalked at the latter; segment 8 with a broad dorsal stripe extending the whole length of the segment; segment 9 with a bifid or two tongue-like spots at basal two-thirds, pointed apically: segment 10 unmarked, deeply and narrowly notched at apex; segment 9 has occasionally an additional apical transverse fine black line with a bordering of black spines. *Anal appendages* very small, conical, black: vulvar scales pale blue, rather short, not extending to the end of abdomen.

Distribution.—Throughout the plains of INDIA, BURMA, and CEYLON, and extending to Australia. It is entirely a creature of the plains, being replaced by *P. malabaricum* in the neighbouring hills of the West Coast and Ceylon, and by *P. bengalense* to the north-east and Burma. It migrates in vaste hordes on the West Coast, together with *P. decorum*, during October and September, and, like that species, breeds in lakes and marshes.

The male is distinguished from other species by the character of the dorsal marking on segment 2; the only species likely to be confused with it are *P. bengalense* and *P. malabaricum*, both marked very similarly on the thorax; it is distinguished from the former by the superior anal appendages being as long as segment 10, while they are only half that length in *P. bengalense*; this species has, moreover, no apical black marking on segment 8; from *P. malabaricum* also it is distinguished by this last feature, as well as by possessing no transverse black stripe on the vertex. Females are distinguished by the bifid black mark on segment 9 and by the orange suffusion on the vertex and upper part of the eyes.

Type in the Selysian collection, formerly in the Rambur collection. The male from Sylhet mentioned by Selys is probably *P. bengalense*; Singapore, Malacca, and Java are also given as localities by the same author.

111. Pseudagrion andamanicum Fraser.

Pseudagrion andamanicum Fras., Rec. Ind. Mus. vol. xxvi, pp. 411, 412, pl. xxiv, figs. 6, 7, 8, 10, 11 (1924).

Male.—Abdomen 25 mm. Hind-wing 17 mm.

Head: labium white; labrum, bases of mandibles, cheeks, and frons azure blue; vertex and postocular spots azure blue, marked with black as follows:—A small spot in front and one on each side of the middle ocellus; an irregular black stripe extending from eye to eye at the level of the posterior ocelli, and a narrow black border behind the postocular spots; eyes blue above, capped diffusely with brown, paling to greenish-yellow below. *Prothorax* azure blue, the middle lobe with a black crown-like spot bearing three small blue spots, one on each side and a bifid one at the centre. *Thorax* azure blue, paler at the sides; a narrow mid-dorsal black carinal stripe and somewhat narrower humeral ones which do not extend quite to the upper limit of the thorax in the case of the latter; laterally a small black spot at the middle of mesepimeron and another at the upper part of the postero-lateral suture; beneath pale blue. *Legs* pale blue, femora striped with black on extensor surfaces. *Wings* hyaline, faintly enfumed; pterostigma dark brown; 11 to 12 postnodal nervures in fore-wings, 8 to 9 in the hind. *Abdomen* azure blue, marked with black as follows:—A small quadrate spot on segment 1 not quite extending to apical border; segment 2 with a rectangular dorsal subapical spot which is connected to a narrow apical annule by a short linear neck, and to the base by a fine black line on the mid-dorsal carina; segments 3 to 6 with moderately broad dorsal stripes expanding subapically, and then contracting again to join narrow apical annules; segment 7 similar but expanding continuously as far as apex of segment; segments 8 and 9 blue, with moderately broad apical black annules, that on segment 8 at least double the width of that on 9; segment 10 broadly black on dorsum. *Anal appendages*: superiors as long as segment 10, black, blue within, bifid at apex as seen in profile, and with a small spine at base; inferiors shorter, broad at base, tapering to apex, sloping up towards superiors, pale bluish-white.

Female.—Abdomen 27 mm. Hind-wing 20 mm.

Head somewhat similar to that of the male, but the transverse band absent; three minute black points at base of labrum and two large spots on the occiput between the postocular spots; labrum, genæ, and frons bluish-green, vertex and postocular spots greenish. *Prothorax* bluish-green, with similar black markings to the male: posterior lobe with the usual forwardly sloping spines. *Thorax* azure blue to bluish-green, without the black dorsal bands of male, which are here replaced by golden brown areolæ bordering the sutures; small black points at the upper ends of humeral and postero-lateral sutures. *Wings* with 11 and 9 postnodal nervures to fore- and hind-wings respectively: pterostigma pale brown. *Abdomen* blue to bluish-green, marked with black somewhat as in the male: segment 2 with the subapical spot smaller, cordate or diamond-shaped, and surrounded by a golden halo; segment 8 broadly black and with a narrow apical blue ring: segment 9 similar or connected to an apical ring by a very narrow neck; segment 10 entirely blue. *Anal appendages* dirty blue, very small, conical; vulvar scales moderately robust, extending to end of abdomen.

Distribution.—Confined, so far as known, to the ANDAMAN IS. *Type* in the Indian Museum, one pair from Mt. Harriet, taken during November.

Closely related to *P. microcephalum*, from which it is distinguished by the characteristic marking on the dorsum of segment 2 and by the more restricted markings on the head. The dorsal marking on segment 2 will also serve to distinguish the female from others.

112. **Pseudagrion bengalénse** Laidlaw. (Fig. 119.)

Pseudagrion bengalense Laid., Rec. Ind. Mus. vol. xvi, pp. 192, 193 (1919); Fras., J. Bombay Nat. Hist. Soc. vol. xxix, p. 750 (1923).

Pseudagrion australasiæ Laid., Rec. Ind. Mus. vol. xii, pp. 21–3 (1916).

Male.—Abdomen 30–32·5 mm. Hind-wing 20–21 mm.

Head: labium palest yellow; labrum, bases of mandibles, clypeus, genæ, frons, and front portion of vertex pale greenish-blue, unmarked, vertex traversed by a broad black band from eye to eye at level of ocellar space; large postocular blue spots bordered behind with black: eyes pale blue below, deepening to azure blue above; beneath head pale blue, bordered inwardly with black. *Prothorax* pale blue, the middle lobe on dorsum with a black crown-shaped marking enclosing three blue spots, the middle one bifid. *Thorax* azure blue, the mid-dorsum with a broad black band and somewhat narrower black humeral stripes: laterally a small spot in the middle of mesepimeron and another at the upper end of postero-lateral suture; beneath palest blue. *Legs* pale blue, femora broadly black on extensor surfaces. *Wings* hyaline; pterostigma pale yellow to brown, very oblique outwardly, covering less than one cell; 11 postnodal nervures in fore-wings, 9 in the hind. *Abdomen* pale azure blue, marked with black as follows:—Segment 1 with a small dorsal black spot; segment 2 with an elongate vase-shaped spot extending from base to apex, broadly expanded at the middle two-fourths of segment, constricted markedly at both apex and base, and then expanding again to join narrow apical and basal annules; segments 3 to 6 with narrow mid-dorsal stripes pointed at base of segments, expanded subapically, and then contracted again to join narrow apical annules; segment 7 with a similar band expanding gradually as far as apex of segment; segments 8 and 9 entirely blue, save for an apical fringe of black spines; segment 10 with a coarse X-shaped dorsal marking covering most of the dorsum of segment. *Anal appendages*: superiors black, about half the length of segment 10, bifid at apex as seen in profile, with the apex hooked strongly inward and deeply excavate on inner side as seen from the dorsum, no basal spines; inferiors blue, very short conical objects, rounded at apex as seen from above.

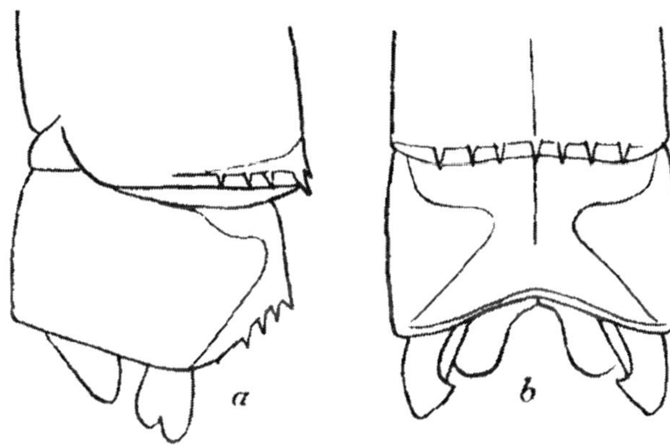

Fig. 119.—Anal appendages of *Pseudagrion bengalense* Laid., male. *a*. Left lateral view. *b*. Dorsal view.

Java specimens have the postclypeus black or with three small black basal spots, and there is usually a black crown-shaped spot on the middle of the vertex; the marking on segment 2 is quite different, being a short goblet-shaped marking not extending to the base of the segment, which, however, bears a narrow dorsal basal ring.

Female.—Abdomen 29 mm. Hind-wing 20 mm.

Closely similar to the male, differing as follows:—Labrum with a tiny medio-basal spot of black; postclypeus with three small basal black points; face and vertex more greenish-blue than blue; postocular spots blue, tinted with greenish. *Prothorax* with the blue spots smaller; posterior lobe and its spines yellowish. *Thorax* pale blue on dorsum, tinted with ferruginous, and with the same three black stripes, which are, however, much narrower in the humeral region; sides pale greenish-blue ; legs with black markings less in evidence, especially on the hind femora. *Wings* similar, but the pterostigma golden yellow, framed in heavy black nervures. *Abdomen* with the dorsal marking on segment 2 extending its whole length, broad at apex, where it is narrowly confluent with an apical annule, tapering but slightly as far as base ; dorsal markings on segments 3 to 7 broader, whilst 8 and 9 have similar broad markings extending the whole length of the segments ; segment 10 blue, unmarked. *Anal appendages* brownish, very minute, conical : vulvar scales palest blue, barely reaching the end of abdomen.

Distribution.—BENGAL, ASSAM, and BURMA within our limits, but probably extending into Malaysia. Mr. Lieftinck has sent me examples from Java which differ, so far as the males are concerned, from the type as mentioned above ; they may be a local race, and are certainly not *P. australasiæ*, nor a race of *P. microcephalum*. The female has not so far been taken within Indian limits, the description given above being made from Java specimens. *Type* in the Indian Museum ; paratype female (from Java) in the author's collection.

Distinguished from *P. microcephalum* by its short anal appendages and by the very characteristic marking on the dorsum of segment 2.

113. **Pseudagrion malabaricum** Fraser. (Fig. 120.)

Pseudagrion malabaricum Fras., Rec. Ind. Mus. vol. xxvi, pp. 428, 494–5 (1924) ; id., ibid. vol. xxxiii, pp. 448, 465 (1931).
Pseudagrion sp. Laid., Spolia Zeylan. vol. xii, pp. 370, 371 (1924).

Male.—Abdomen 33 mm. Hind-wing 20 mm.

Head : labium white ; labrum and epistome azure blue, with three minute points on the epistome and a small central point behind the frons black ; vertex blue, traversed by a broad black band from eye to eye ; occiput black, marked with two very large blue postocular spots ; eyes turquoise blue, with a black cap above. (In adult specimens often the whole of upper surface of head posterior to epistome is black, save for the postocular spots.) *Prothorax* black, with an anterior collar, the posterior lobe, two dorsal spots near hinder margin, and a spot on either side blue. *Thorax* azure blue, marked with three broad black bands on dorsum, a medial and humeral, and on the sides a small spot between the lateral sutures above and a second on the upper part of postero-lateral suture. *Legs* blue, femora with black stripe behind. *Wings* hyaline, pterostigma dark brown, postnodals in fore-wing 10 to 11, usually the latter number, 9 in the hind. *Abdomen* blue, marked with black as follows :—Segment 1 with a broad quadrate basal spot; 2 with an apical ring, a short goblet-shaped mark on the dorsum extending over three-fourths of the segment at the apical end, hollowed out basalwards, and with a very thin stem which joins the apical ring; lastly, a basal ring on dorsum and sides extending as a point along dorsal carina, which projects into mouth of goblet; segments 3 to 6 with broad dorsal stripes expanding apicalwards and connected with narrow apical rings; 7 with a similar stripe, which broadens steadily as far as the apical border; 8 and 9 with fine apical rings only; 10 broadly black on dorsum. *Anal appendages* black, superiors slightly shorter than segment 10, pale within, *not bifid*, diverging strongly, apices curling strongly inward as a robust tooth; seen from the side bluntly pointed and bevelled from apex below; inferiors very small, pointed obtusely.

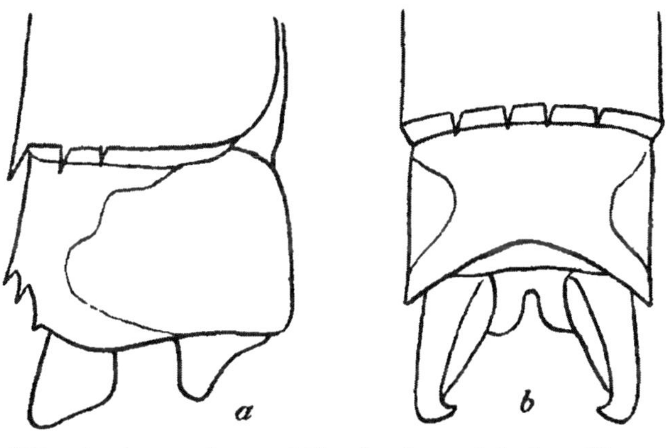

Fig. 120.—Anal appendages of *Pseudagrion malabaricum* Fras., male. *a.* Right lateral view. *b.* Dorsal view.

Female.—Abdomen 32 mm. Hind-wing 22 mm.

Labium white; labrum and epistome pale olive-green, deepening to black on vertex; postocular spots peacock-blue; eyes olivaceous, with dark brown cap. *Prothorax* as for the male, but anterior collar broader, and posterior lobe and processes yellow. *Thorax* peacock-blue, paling to pale azure blue on the lower part of sides and beneath, marked similarly to the male, but the humeral stripes narrower and surrounded with an areola of golden green. *Wings* similar to the male, but the pterostigma paler brown; 11 to 12 postnodal nervures to fore-wings, 10 in the hind. *Abdomen* dark azure blue, broadly marked with black as follows:—Segment 1 with a quadrate dorsal spot; segment 2 with a dorsal marking shaped like a chess-pawn, not quite extending to the base of segment; segments 3 to 7 similar to the male: segments 8 and 9 broadly black, with a narrow apical blue annule only; segment 10 entirely blue. *Anal appendages* bluish, very short, conical, pointed: vulvar scales robust, carneous or white, extending to the end of abdomen.

Distribution.—Throughout the WESTERN GHATS in submontane and montane areas from South Kanara to CEYLON. The type was taken at Kodaikanal, Palni Hills, in September, and I found it during the same month at Ootacamund, 7,500 ft., Nilgiris, where it was not uncommon on the Lovedale Lake. Coorg, however, is the locality in which it occurs in greatest numbers, some tanks there being literally blue with the vast numbers present.

This species is closely related to *P. microcephalum* and *P. bengalense*, and is distinguished from them by the superior anal appendages being non-bifid at apex. It was described by Dr. Laidlaw in the same month and year as my own description appeared, but without name, as he had only a single male to study, from Batticola, Ceylon. He notes its differences from *P. microcephalum*, and gives excellent figures of the anal appendages, which leave no doubt as to its being the same insect as *P. malabaricum*. Breeds in small lakes in the hills, and is never found over running water.

Type in the British Museum; paratypes in the Pusa and author's collections.

114. **Pseudagrion decorum** (Rambur). (Fig. 121.)

Agrion decorum Ramb., Ins. Névrop. p. 258 (1842).
Pseudagrion decorum Selys, Bull. Acad. Belg. (2) vol. xlii, p. 504 (1876); Kirby, Cat. Odon. p. 153 (1890); Laid., Rec. Ind. Mus. vol. xii, pp. 22, 24 (1916); id., ibid. vol. xvi, p. 172 (1919); Fras., J. Bombay Nat. Hist. Soc. vol. xxvii, p. 543 (1921); id., ibid. vol. xxix, pp. 750, 755 (1923); id., Rec. Ind. Mus. vol. xxvi, pp. 428, 497 (1924); id., ibid. vol. xxxiii, pp. 448, 465 (1931).

Male.—Abdomen 28–30 mm. Hind-wing 18–20 mm.

Head: labium white; labrum, genæ, bases of mandibles, clypeus, frons, and vertex as far as the level of posterior ocelli a beautiful pale bluish-green; occiput with a large triangular postocular spot on each side deep azure blue, paling somewhat behind occiput and confluent narrowly by a bridge of bluish-green; these large spots separated from the anterior bluish-green by a fine crenate black line which passes behind the posterior ocelli medially and borders the eyes outwardly; a short thick transverse stripe partially bordering these spots behind; eyes with a small semilunar patch of black at the upper back part changing to blue and then bluish-green beneath, and finally palest green at the lower and back part; a small black spot on each side of the anterior ocellus and three small black points on the postclypeus. *Prothorax* pale blue, posterior lobe bluish-green, middle lobe with base narrowly black, from which a short subdorsal black point passes forwards; anteriorly three small black points and a fine lateral black streak separating middle and anterior lobes; posterior lobe narrow, simple, rounded. *Thorax* a beautiful bluish-green on dorsum

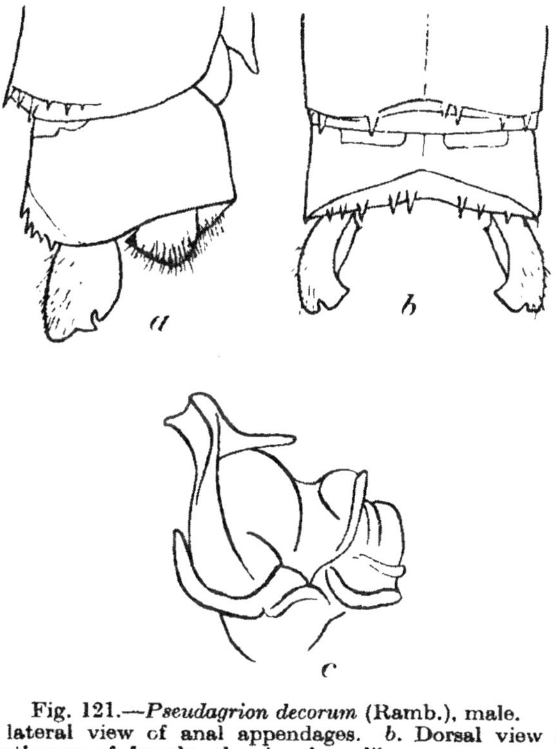

Fig. 121.—*Pseudagrion decorum* (Ramb.), male.
a. Right lateral view of anal appendages. *b.* Dorsal view of same. *c.* Prothorax of female, showing horn-like processes on posterior lobe.

to slightly beyond level of humeral suture; laterally azure blue; mid-dorsal carina very finely black, and with an equally narrow black line running close and parallel to it on each side; a narrow black humeral stripe and a short black point at upper part of postero-lateral suture; beneath white; tergum azure blue. *Wings* hyaline; pterostigma diamond-shaped, very narrow, braced, covering less than one cell; 10 postnodal nervures to fore-wings, 9 in the hind; *ac* nearly opposite level of basal antenodal nervure. *Abdomen* pale azure blue, except segments 8 to 10, which are deep azure blue, and segment 2, which is bluish-green on dorsum. Black markings as follows:—Segment 1 with a basal spot notched at its middle apically; segment 2 with a mid-dorsal stripe slightly broadened at base and again subapically, the intervening portions very narrow and apically confluent with a narrow apical annule; segments 3 to 7 with the dorsum steely or bronzed black nearly as far as base of segments, this band expanding subapically and then constricted again to join a narrow apical annule; segment 7 has this band broader than on the rest, the apical dilatation scarcely noticeable; segments 8 to 10 with only narrow apical black lines augmented by stout black spines. *Anal appendages* carneous, superiors changing to pale azure blue and tipped with black at apex, and with a small basal black spine, deeply cupped on inner surface, narrowly bifid at apex, the black portion forming a stout inwardly directed spine; inferiors short, barely half the length of superiors, with a minute black spine on inner side, rounded.

Female.—Abdomen 31 mm. Hind-wing 20 mm.

Differs rather markedly from the male; labrum, bases of mandibles, genæ, and clypeus palest green, changing on the frons to a golden green or shagreen on vertex and occiput; the transverse black belt on vertex absent; postocular spots very large, azure blue, finely framed in black; eyes pale green below, changing to olivaceous on the summit, where they are tinted more or less with ochreous. *Prothorax* palest green, with a diffuse golden yellow stripe on each side; black markings entirely similar to the male, a series of five parallel longitudinal black lines confluent anteriorly and posteriorly; posterior lobe with the usual pair of recurved spines. *Thorax* a beautiful pearly blue, with the mid-dorsal carina and humeral sutures bordered with golden yellow, this gradually vignetted into the neighbouring ground-colour. *Legs* palest green or blue, marked similarly to the male. *Wings* hyaline; pterostigma paler than in the male; 9 postnodal nervures to fore-wings, 8 in the hind. *Abdomen* pale blue, with a greenish tinge at base and on terminal segments, the dorsum of segment 2 bearing a black marking shaped like the head of an arrow, the shaft directed apically and confluent with a narrow apical black annule, the point not reaching as far as base, this marking with an areola of golden yellow similar to that bordering the sutures of thorax; segment 1 with a basal dorsal black spot; segments 3 to 10 with black dorsal stripes or spots, which fall just short of the base and expand subapically, and again contract to join narrow apical black rings on segments 3 to 6; on segment 7 the broad stripe broadening as far as apical border ; segment 8 with the dorsal marking extending broadly from the base, falling just short of the apical margin of segment: segments 9 and 10 similar to 8. *Anal appendages* nearly as long as segment 10, conical, pointed, pale bluish-green ; vulvar scales extending to end of abdomen, bluish-green.

Distribution.—Widely distributed throughout CONTINENTAL INDIA and BURMA, but I can find no records of it from Ceylon, although it is almost certain to occur throughout the plains there. It is essentially an insect of the plains, and the few records from montane areas are undoubtedly due to strays ; thus I have taken a single specimen at Ootacamund, Nilgiris, 7,250 ft., and a few specimens at 4,000 ft. in Coorg. During September this species takes to flight, the migration taking place in a north-westerly direction up the West Coast of India, when millions may be seen in company with *P. microcephalum*.

Distinguished easily from other species, even when on the wing, by the light blue ground-colour, with but few black markings on the thorax to relieve it, and the arrow-head marking on segment 2. Females distinguished with more difficulty, especially from those of *P. microcephalum*, *P. bengalense*, and *P. malabaricum* ; from the former it is determined by segment 9 being without the bifid marking, and segment 10 being not entirely unmarked, as in that species ; the dorsal marking on segment 2, shaped as in the male, will determine it from all other species.

Type in the Selysian collection ; paratypes in most national collections and the Indian Museum.

115. **Pseudagrion hypermelas** Selys. (Fig. 122.)

Pseudagrion hypermelas Selys, Bull. Acad. Belg. (2) vol. xlii, p. 519 (1876) ; Kirby, Cat. Odon. p. 153 (1890) ; Kruger, Stett. Ent. Zeit. vol. lix, p. 119 (1898) ; Morton, Trans. Ent. Soc. Lond. pp. 307, 308, pl. xxiv, figs. 11, 12, p. 306 (1907) ; Laid., Rec. Ind. Mus. vol. xii, pp. 21, 22, 25 (1916) ; id., ibid. vol. xvi, pp. 172, 184 (1919) ; Fras., J. Bombay Nat. Hist. Soc. vol. xxvii, p. 543 (1921) ; id., ibid. vol. xxix, pp. 749, 755 (1923) ; id., Rec. Ind. Mus. vol. xxvi, pp. 428, 497, 498 (1924) ; id., ibid. vol. xxxiii, p. 448 (1931).

Pseudagrion bidentatum Morton, Trans. Ent. Soc. Lond. p. 307, pl. xxiv, figs. 9, 10 (1907) ; Laid., Rec. Ind. Mus. vol. xii, pp. 21, 22, 25 (1916) ; Fras., J. Bombay Nat. Hist. Soc. vol. xxix, p. 749 (1923).

Male.—Abdomen 23–24 mm. Hind-wing 14–15 mm.

Head : labium white or palest brown ; labrum and anteclypeus blackish-brown ; genæ and bases of mandibles glossy brown ; rest of the head mat black, the frons and fore part of vertex thinly pruinosed in full adults ; eyes dark olivaceous brown above, paler below. *Prothorax* black, obscurely brown at sides, but this obscured by thin pruinescence. *Thorax* black on mid-dorsum and broadly so over the humeral sutures, enclosing a broad purplish-brown antehumeral stripe ; laterally purplish-brown, with obscure broad blackish fasciæ largely obscured by pruinescence ; beneath white with dense pruinescence. *Wings* hyaline ; 9 postnodal nervures in fore-wings, 8 in the hind ; pterostigma black, bordered finely with yellow and enclosed in thick black nervures, very narrow, twice as long as broad, with distal side very oblique, strongly braced, covering one cell. *Legs* black, overlaid with pruinescence ; tibiæ and tarsi ochreous. *Abdomen* violaceous brown, overlaid thinly with pruinescence and broadly marked with black on dorsum ; segment 2 with a lateral stripe confluent subapically with the dorsal black ;

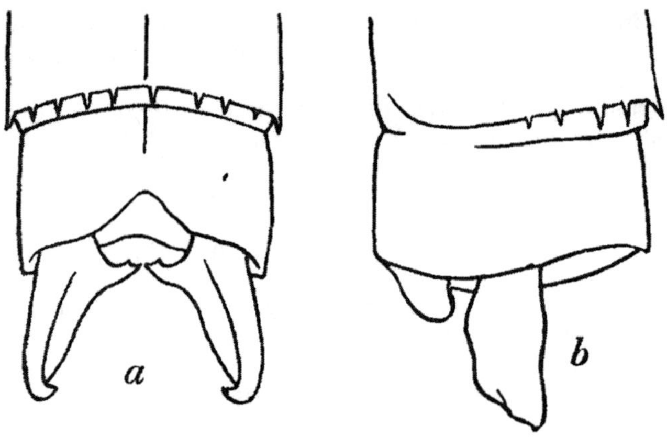

Fig. 122.—Anal appendages of *Pseudagrion hypermelas* Selys, male. *a*. Dorsal view. *b*. Left lateral view.

segments 3 to 7 with the black dorsal band contracted at base of segments, expanded subapically, then contracted again to become confluent with narrow black apical annules ; segments 8 and 9 with lateral black spots more or less confluent with the dorsal black. *Anal appendages* carneous, deepening to black at apices ; superiors bifid as seen in semi-profile, the upper branch shaped as a short, curved, crochet-like hook ; seen in full profile the apex curved like a parrot's beak, the fissure not visible, the lower border sinuous and bevelled ; seen from the dorsum the appendages end in a sharply recurved hook, and the inner border of the base is expanded into a very robust spine minutely bidentate at apex ; the basal spines of the two appendages overlapping and more or less hidden by the overhanging border of segment 10 ; inferior appendages short, conical, not half the length of superiors.

Female.—Abdomen 23 mm. Hind-wing 15–16 mm.

Differs strongly from the male, much paler in colour and the markings greatly restricted. *Head* : labium creamy white ; labrum and anteclypeus palest brown ; postclypeus, frons, vertex, and occiput pale olivaceous, the former with or without three small black basal points ; genæ and bases of mandibles yellow ; postocular spaces bordered in front and behind narrowly with black ; beneath the head palest ochre ; eyes warm brown above, pale green at sides and below, marked with a narrow blackish equatorial band. *Prothorax* palest ochreous, marked dorsally with three short parallel black stripes confluent in front and behind on middle lobe ; posterior lobe raised, broadly rounded, and with two short spines projecting forwards from its base. *Thorax* pale olivaceous, with the mid-dorsal carina and humeral sutures finely mapped out in black, the former also with a narrow black line running closely parallel to it on each side ; the mid-dorsal and humeral lines suffused laterally with bright ochreous. *Legs* carneous, with black spines. *Wings* similar to the male, but the pterostigma uniform pale yellow, framed in black nervures ; 9 to 10 postnodal nervures to fore-wings, 7 to 8 in the hind. *Abdomen* pale greenish-yellow on the basal half, palest blue on the distal segments marked dorsally with black as follows :—Segment 1 with a small geminate spot on mid-dorsum not extending to base or apex ; segment 2 with a narrow mid-dorsal stripe expanding abruptly subapically, then constricted again and joining a small apical spot ; segments 3 to 7 with similar mid-dorsal narrow stripes extending from base of segments, and markedly and abruptly expanded subapically on all segments, a narrow interval occurring between this and narrow apical annules ; segment 8 with a brown dorsal band slightly expanded at the apical end of segment ; segment 9 with a broad basal dorsal spot connected to a narrow mid-dorsal line which extends to the apical end of segment ; segment 10 unmarked. *Anal appendages* very short, conical, pale brown ; vulvar scales rather inconspicuous, pale brown or ochreous.

Distribution.—Central Provinces, Deccani areas of the Bombay Presidency, and S. Punjab.

Haunts the grassy banks of small streams and canals, often in company with *Disparoneura nigerrima*, which species it so closely resembles as to be indistinguishable on the wing. Some confusion appears to have arisen between this species and *P. bidentatum* Morton, the author of which latter has, it seems, mixed up the descriptions of two species. His figures of the anal appendages of *P. bidentatum* are certainly those of *P. hypermelas*, whilst the description of the colours and markings given for *P. bidentatum* are those of *P. spencei* mihi. The figures of the anal appendages given for *P. hypermelas* are, on the other hand, those of *P. spencei*.

Type a male in the Selysian collection; allotype female in the author's collection, and not hitherto described. Type of *P. bidentatum* in the Morton collection.

116. **Pseudagrion spencei** Fraser. (Figs. 117 & 123.)

Pseudagrion spencei Fras., Mem. Dept. Agric. India (Ent.), vol. vii, no. 7, pp. 47, 48 (1922); id., J. Bombay Nat. Hist. Soc. vol. xxix, pp. 755, 759 (1923).

Male.—Abdomen 22–24 mm. Hind-wing 15 mm.

Head: labium white; labrum, bases of mandibles, genæ, clypeus, frons, basal joints of antennæ, and the vertex as far

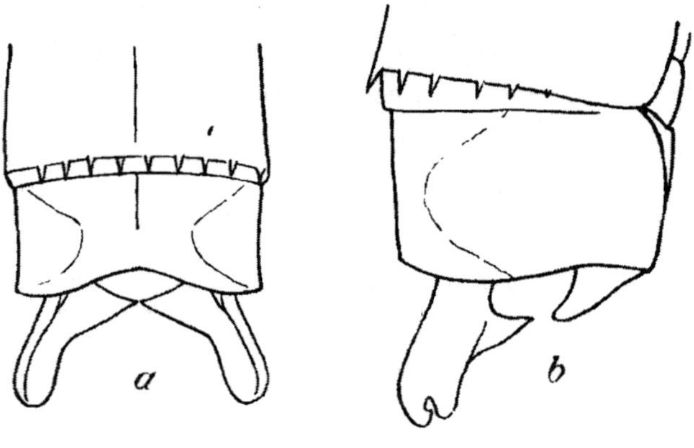

Fig. 123.—Anal appendages of *Pseudagrion spencei* Fras., male. *a.* Dorsal view. *b.* Right lateral view.

back as the anterior ocellus pale greenish-blue, rest of head black, save for very large postocular spots bluish in colour, and connected narrowly across the occiput: beneath head pale olivaceous; eyes dark brown above, bluish laterally and below. *Prothorax* black on dorsum, with a small geminate spot on mid-dorsum of middle lobe, as well as the lower part of sides, blue. *Thorax* pale azure blue, with the mid-dorsum broadly black, a narrow sinuous humeral stripe and the upper part of the postero-lateral suture black; beneath white. *Legs* pale carneous, femora rather broadly black on extensor surface. *Wings* hyaline; pterostigma very narrow, half as long again as broad, distal side very oblique, covering about half a cell, dark reddish-brown, framed in black nervures; 9 postnodal nervures to fore-wings, 8 in the hind. *Abdomen* azure blue, marked with black as follows:—Segment 1 with a quadrate dorsal spot extending to apical end of segment; segment 2 with a thistle-shaped spot on mid-dorsum connected by its stalk to a narrow black apical annule; segments 3 to 6 with broad dorsal bands, which expand subapically and then contract again to become confluent with narrow apical annules; segment 7 similar, but the band expanding steadily as far as apical border of segment; segments 8 and 9 wholly blue, save for a row of a few small black spines at apical border; segment 10 with a broad X-shaped dorsal marking. *Anal appendages*: superiors as long as segment 10, black, shaped very similarly to those of *P. hypermelas*, but bifid at apex as seen in profile, the upper arm of fissure bent towards the lower and shaped like a parrot's beak, the extreme base constricted, then broadly expanded within to form a long acutely pointed spine, *not bidentate* as in *P. hypermelas*; inferior appendages less than half the length of superiors, conical, sloping up to meet the basal spine of superiors, the latter partially hidden by the overhanging border of segment 10, as in *P. hypermelas*.

Female.—Abdomen 23–24 mm. Hind-wing 16 mm.

Differs in several respects from the male, as follows:—Labrum, bases of mandibles, and genæ pale ochreous, rest of head olivaceous, tinged with ochreous; the postocular areas finely bordered with black in front; eyes dark olivaceous above, paler below. *Prothorax* pale ochreous or greenish-blue, with black markings similar to those of the male, thinly pruinosed beneath; posterior lobe broadly arched, recurved forwards slightly, and with two small yellow spines projecting forwards from its base. *Thorax* pale bluish-green or pale ochreous, the mid-dorsal carina finely black and with a fine black line running close to and parallel with it on each side; the humeral suture finely mapped out in black and a small black spot on the upper part of the postero-lateral suture; beneath white with pruinescence. *Wings* and *legs* similar to the male, but the fore-wings with 10 postnodal nervures occasionally. *Abdomen* ochreous, the mid-dorsum in adults ferruginous, and marked sparingly with black as follows:—Segment 1 with a small quadrate spot touching the base; segment 2 with a fine mid-dorsal streak, which expands into a small triangle just short of apical end of segment; segments 3 to 6 with only small subapical black triangles; segment 7 with a dorsal band extending the whole length of the segment and slightly expanded at apical end; segment 9 with a broad belt on dorsum narrowing from base to apex; segment 10 unmarked. *Anal appendages* and vulvar scales small, the former carneous or tinged with ochreous dorsally, conical and obtusely pointed at apex.

Distribution.—CENTRAL PROVINCES, BIHAR, BENGAL, and ASSAM. *Type* a male, in the British Museum, from Shillong, Assam ; paratypes in the author's collection from Kajibthatkhana, Jalpaiguri, Bengal, and from Margherita and Gauhati, Assam, during March, May, and September.

As noted under the description of *P. hypermelas*, *P. bidentatum* Morton appears to be a partial description of this insect ; the type of the latter, a male, is described as "discoloured, colours very likely blue and bronzed black" ; allowing for this, the markings agree closely with those given above for the male of *P. spencei*, save for segments 8 and 9, which are said to be black, but were probably so from decomposition and post-mortem changes. The anal appendages described are, however, undoubtedly those of *P. hypermelas*, and it would appear that the descriptions of the two insects have been muddled together, as Mr. Morton was studying the two at the same time. In my original description of *P. spencei* I mentioned that there was no basal spine to the superior appendages, but I now find that this is incorrect, the spine being concealed beneath the overhanging border of segment 10, and thus seen with difficulty unless the appendage be removed for examination. Distinguished from *P. hypermelas* by the blue ground-colour of the male, and by the greatly restricted black markings of the abdomen in the case of the female.

117. **Pseudagrion laidlawi** Fraser. (Fig. 124.)

Pseudagrion laidlawi Fras., Mem. Dept. Agric. India (Ent.), vol. vii, no. 7, pp. 48–50 (1922) ; id., J. Bombay Nat. Hist. Soc. vol. xxix, pp. 750, 755 (1923).

Male.—Abdomen 25 mm. Hind-wing 16 mm.

Head : labium white ; labrum, bases of mandibles, genæ, clypeus, frons, and vertex as far back as anterior ocellus pale azure blue, with three small black spots finely confluent at base of postclypeus and a small median black point on frons ; a broad black band extending from eye to eye at the ocellar space ; occiput with large pale, lilaceous, postocular spots, narrowly bordered with black posteriorly and confluent across the middle line ; occiput behind and beneath pale blue ; eyes dark blue above, pale blue beneath. *Prothorax* lilaceous or bluish, black on mid-dorsum of middle lobe, the black enclosing a small bigeminate blue spot ; a narrow line of black on each side. *Thorax* lilaceous, paler at the sides and beneath ; the mid-dorsum broadly black, but often with a pale lilaceous line on each side of the mid-dorsal carina ; a rather sinuous narrow black humeral stripe and a small black spot on the upper end of the postero-lateral suture. *Legs* pale blue or creamy white; extensor surfaces of all femora broadly black. *Wings* hyaline; pterostigma diamond-shaped, covering less than one cell, braced, black, framed narrowly in yellow and thick black nervures; 8 postnodal nervures to fore-wings, 7 in the hind. *Abdomen* pale blue, with lilaceous tinge on mid-dorsum, especially on segment 10; segment 1 with a small apical and a large basal black spot on dorsum; segment 2 with a thistle-shaped black marking on dorsum, connected by its stalk to a narrow apical black annule; segments 3 to 6 with broad dorsal bands expanding subapically and then constricted again to become confluent with apical annules; segment 7 with a similar band, which, however, broadens gradually as far as the apical border of the segment; segments 8 and 9 deep azure blue, with a few black spines at the apical border; segment 10 with a narrow X-shaped dorsal marking.

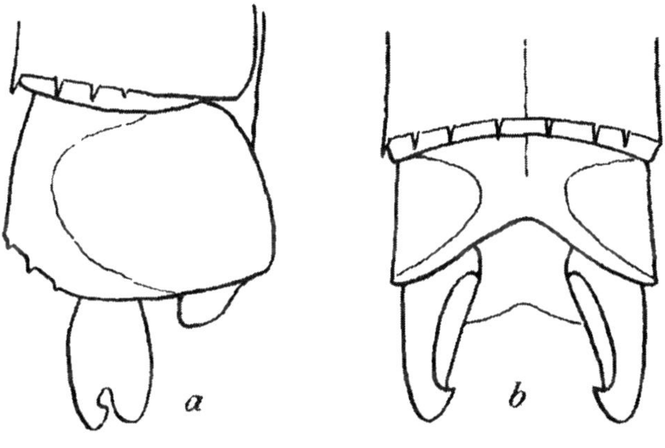

Fig. 124.—Anal appendages of *Pseudagrion laidlawi* Fras., male. *a*. Right lateral view. *b*. Dorsal view.

Anal appendages lilaceous, tipped with black externally and white within; superiors as long as segment 10, bifid as seen in profile, the outer arm of fissure ending in a short, sharply recurved spine, the base slightly constricted and without a spine, deeply hollowed out within; inferiors less than half the length of superiors, small conical processes.

Female.—Abdomen 25 mm. Hind-wing 16 mm.

Differs considerably from the male, both in body-colouring and markings; labrum, face, and upper surface of *head* pale violaceous or purplish-brown, without any markings: eyes bluish-green above, carneous below, and banded faintly with two lighter blue bands and a linear dark purplish equatorial band; beneath head carneous. *Prothorax* palely violaceous or pale lilaceous brown, the middle lobe with a linear black

V-shaped marking, the posterior lobe margined finely at the base with black; this lobe with the usual short spines at base, directed sharply forward. *Thorax* coloured similarly to the prothorax, but the sides palest blue; the mid-dorsal carina finely black, as also a narrow parallel stripe on each side of it; a very narrow humeral stripe usually present and a small spot on the upper part of postero-lateral suture; the antealar sinus finely mapped out in black, as also the anterior portion of thorax prior to the mid-dorsal carinal ridge; beneath pale blue to white. *Legs* pale, carneous, the anterior pair of femora with a black stripe on extensor surface. *Wings* as for the male, but the pterostigma uniformly pale yellow and an extra postnodal nervure usually present in fore-wings. *Abdomen* pale greenish-blue on dorsum and sides, tinted with lilaceous on segments 1, 2, and 10, marked dorsally with black except segment 10; segment 1 with a basal dorsal spot; segment 2 with a mid-dorsal stripe expanded at base and apex of segment, but narrowing again at the latter to join a narrow apical black annule; segments 3 to 7 similar to the male; segments 8 and 9 with broad dorsal bands broadening towards apex of segment on 8 and narrowing on 9. *Anal appendages* very short, conical, pale greenish-white; vulvar scales very short, coloured similarly.

Distribution.—Throughout LOWER SIND, and probably, also, the dry areas of neighbouring States; moderately common at Karachi, where I found it breeding in water-channels in the municipal sewage farm. *Type* in the British Museum; paratypes in the Pusa and author's collections.

The male is easily distinguished from other species by its pronounced lilaceous colouring, which is similar to that seen in some species of *Aciagrion*, and also by the shape of its superior anal appendages. The female is not unlike those of *P. hypermelas* and *P. spencei*, but the broad abdominal markings will distinguish it from the latter, and the absence of head-markings from the former.

118. **Pseudagrion rubriceps** Selys. (Fig. 125.)

Pseudagrion rubriceps Selys, Bull. Acad. Belg. (2) vol. xlii, p. 510 (1876); Kirby, Cat. Odon. p. 153 (1890); Selys, Ann. Mus. Civ. Genova, (2) vol. x (xxx), p. 83 (1891); Laid., Rec. Ind. Mus. vol. xii, pp. 22, 24, 25 (1916); id., ibid. vol. xvi, pp. 172, 193, 194 (1919); Fras., J. Bombay Nat. Hist. Soc. vol. xxix, pp. 750, 755 (1923); id., Rec. Ind. Mus. vol. xxvi, pp. 428, 497 (1924); Laid., Spolia Zeylan. vol. xii, p. 370 (1924); Fras., Rec. Ind. Mus. vol. xxxiii, p. 448 (1931).

Male.—Abdomen 29 mm. Hind-wing 18–20 mm.

Head: labium citron-yellow; labrum, bases of mandibles, genæ, clypeus, and frons bright orange; vertex and occiput dark olivaceous, but the posterior portion of the latter changing to pale blue, and marked above with a pair of large triangular dark blue postocular spots, which are finely bordered posteriorly with black; eyes olivaceous green above, changing rapidly to bright rich orange and then golden yellow, and faintly bluish beneath. *Prothorax* black, marked with pale blue as follows:—The sides broadly, the whole of anterior lobe, and three large spots on dorsum, the middle one of which contains a median geminate spot of paler blue. *Thorax* olivaceous green, with a golden tinge as far back as the first lateral suture, the sides azure blue; mid-dorsal carina finely black, and a somewhat thicker black line running close to and parallel with it on each side; a narrow black humeral stripe, somewhat sinuous and crossing the suture to its inner

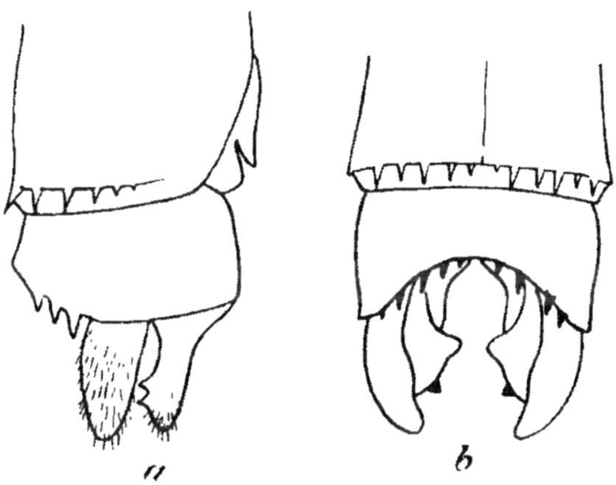

a b

Fig. 125.—Anal appendages of *Pseudagrion rubriceps* Selys, male. *a*. Right lateral view. *b*. Dorsal view.

side above; the upper half of the first lateral suture and the upper third of the second finely black; beneath white. (In Assam specimens the three median black lines may be confluent and the humeral very thick.) *Legs* yellow, femora black on the outer and posterior surface, tibiæ speckled with black on the flexor surface; spines 4 or 5 on femora, short, black (in the original description the author states that the claw-hooks are small or absent, but I have not found this to be the case in most specimens examined). *Wings* hyaline; pterostigma strongly braced, covering less than one cell, reddish-brown, diamond-shaped; 10 postnodal nervures in fore-wings, 9 in the hind; *ac* situated about midway between the two antenodal nervures. *Abdomen* marked broadly with black on dorsum; segments 1 and 2 olivaceous green above, azure blue laterally, the former with a quadrate dorsal spot not quite extending to apical border, which is ringed with blue; segment 2 with a goblet-shaped marking on dorsum, the cup with a shallow concavity and its stem very short and confluent with an apical black annule (in Coorg specimens the cup is deep and its sides connected by a cross-bar, thus enclosing a spot of the ground-colour, whilst in Assam specimens the black is even more extensive, completely blotting out the blue spot); segments 3 to 7 black, bronzed green on dorsum, pale greenish laterally, and with narrow blue basal annules: segment 8 with a broad black cone on dorsum, the apex of cone extending nearly to base of segment, the base resting on the apical border of segment; segments 9 and 10 azure blue, unmarked, or, in Assam specimens, with a broad black apical ring on 9 and a narrow one on 10, which latter is deeply excavate on the border, with spines along the concavity. *Anal appendages* black, the inferiors paler: superiors as long as segment 10, subconical, and with blunt apex as seen from the side, broad and deeply hollowed out on the inner side as seen from the dorsum, and with a thick, short, spine-like process on the inner side nearer base than apex; inferiors one-fourth shorter than superiors, ungulate as seen from the side, narrow at base and broadened at apex as seen from below, and with a small spine at the inner angle of apex, which is partly visible from the side. Genitalia as for the genus: penis with dorsal spine, branches long and twisted spiralwise near their centres, blunt at apices.

Female.—Abdomen 29 mm. Hind-wing 21 mm.

Differs in colour and markings from the male as follows:—Labrum, face, vertex, and the occiput a uniform dark olivaceous, with the postocular spaces finely margined in black; the antennæ and a small spot at base of postclypeus also black; eyes dark blue above, paling to azure blue below. *Prothorax* and *thorax* with markings similar to the male, but the ground-colour of dorsum a dull bluish-green; posterior lobe of prothorax shaped as in the preceding species, and with two short forwardly directed spines at base. *Wings* with an extra postnodal nervure and the pterostigma much paler in colour. *Abdomen* with dorsal markings broader, that on segment 2 of almost equal width from base to apex, but slightly constricted at the middle and slightly broadened subapically; segment 8 with a broad band extending the whole length of segment, and broadening towards the apical end; segment 9 with a basal bifid dorsal marking, the two points of the marking extending nearly to apical border of segment; segment 10 unmarked. *Anal appendages* shorter than segment 10, small, conical, white processes; vulvar scales bluish, robust, extending to end of abdomen.

Distribution.—Widely distributed throughout the plains and submontane areas of CONTINENTAL INDIA and BURMA except in desert tracts. Beyond our limits reported from Java, Indo-China, Malaysia and Formosa.

The male is easily distinguished from all others by the brilliant reddish-orange face, from which it derives its name, and which is very conspicuous, even when the insect is on the wing. The female presents more difficulty, as the bifid mark on segment 9 is shared by at least one other species; it resembles the female of *P. indicum* most closely, but this latter has the band on segment 9 complete from base to apex; the female of *P. williamsoni* is unknown, but probably closely resembles that of the present species; all have the dorsum of thorax dark olive-green and the sides azure blue. Breeds in and frequents the grassy banks of small streams: the male common, but the female is rarely seen except when *in cop.*

Type in the Selysian collection; paratypes in the Indian and British Museums and in most national collections

119. **Pseudagrion ceylanicum** (Kirby).

Archibasis ceylanica Kirby, Proc. Zool. Soc. Lond. p. 255, pl. xx, fig. 4 (1881); id., J. Linn. Soc. Lond., Zool. vol. xxiv, p. 564 (1894); Laid., Rec. Ind. Mus. vol. xvi, pp. 172, 194 (1919); Ris, Ark. Zool. Uppsala, Bd. 21 A, no. 31, pp. 15, 20–2 (1930).

Pseudagrion rubriceps Laid., Spolia Zeylan. vol. xii, p. 370 (1924); Ris, l. c. pp. 20–2 (1930).

Male.—Abdomen 32 mm. Hind-wing 20 mm.

Differs from *P. rubriceps* in the following characters:—The whole of the labrum, face, basal joints of antennæ, upper surface of eyes, and the vertex and occiput as far back as the hinder border of postocular areas rust-red, unmarked with black, the latter areas broadly bordered with black posteriorly: beneath head cinereous, bordered inwardly with black; beneath eyes olivaceous brown. *Prothorax* olivaceous green, with a small black dorsal spot on each side of middle lobe and a fine line of the same colour on each side. *Thorax* olivaceous, tinted with ferruginous on dorsum and changing to blue on the sides and beneath: the mid-dorsal carina finely black in its lower two-thirds or three-quarters, and without the bordering black lines seen in *P. rubriceps*: a small black spot on the upper ends of the humeral and postero-lateral sutures. *Wings* palely enfumed; 11 to 12 postnodal nervures in fore-wings, 9 to 10 in the hind: pterostigma dark reddish-brown, framed in black nervures, covering less than one cell. *Abdomen* with the dorsal marking on segment 2 very narrow and extending the whole length of the segment, or with a small bluish spot on its mid-dorsum near the base, or in others with an elongate goblet-shaped marking, into the cup of which dips the point of a basal black triangle, the two markings nearly confluent ; segment 8 broadly black on dorsum, the marking extending the whole length of the segment ; segments 9 and 10 with some black bordering spines and the apical borders finely black. *Anal appendages* similar to those of *P. rubriceps*.

Female.—Abdomen 29 mm. Hind-wing 21 mm.

The ground-colour almost exactly similar to the male, thus differing from the female of *P. rubriceps* ; the *head* and *thorax* entirely similar to the male. *Abdomen* differing by the markings on segment 2 being very narrow in the basal two-thirds, broadening markedly subapically, and then contracting again to join a narrow black apical annule ; segment 9 with the bifid dorsal black marking shorter and with a narrow black apical annule broadening abruptly at the mid-dorsum and nearly confluent with the points of the basal marking ; a row of very robust black spines situate on this apical ring. *Wings* with the pterostigma paler than in the male.

Distribution.—CEYLON only, throughout which it appears to be widely distributed ; Trincomalee during September, at sea-level, Kandy and Nalande during July and September at 2,000 ft.

Distinguished from *P. rubriceps* by the head and thorax being without black markings and by the much broader and more extensive marking on segment 8 ; the female differing by resembling the male in colouring and, for the greater part, in markings. *Type* in the British Museum ; paratypes in the author's collection. As a certain amount of variation occurs in the continental *P. rubriceps*, the present closely similar form is probably a race of *P. rubriceps* in process of becoming specialized.

120. **Pseudagrion williamsoni** Fraser. (Fig. 126.)

Pseudagrion williamsoni Fras., Mem. Dept. Agric. India (Ent.), vol. vii, no. 7, pp. 46, 47 (1922) ; id., J. Bombay Nat. Hist. Soc. vol. xxix, p. 750 (1923).

Male.—Abdomen 23 mm. Hind-wing 18 mm.

Head : labium carneous ; labrum yellow ; genæ glossy blackish-brown to black ; ante- and postclypeus, frons, front of vertex, and basal joints of antennæ pale golden olivaceous ; rest of the head posterior to this black, with very large postocular turquoise-blue spots connected by a short stripe of yellow ; eyes yellowish-green, capped with dark brown ; beneath head dull purplish-brown. *Prothorax* with the anterior and posterior lobes turquoise-blue, middle lobe black, with the lower part blue and enclosing a small quadrate spot on each side, and a geminate spot on the dorsum blue, the latter divergent anteriorly. *Thorax* greenish-yellow on dorsum, the mid-dorsal carina broadly black, and a black humeral stripe on each side almost equally broad; laterally azure blue, with a small round spot on the centre of the antero-lateral suture, and another point of black on the upper part of the postero-lateral suture; beneath white. *Legs* white or pale yellow, the femora black on extensor surface. *Wings* hyaline; pterostigma warm brown, finely framed in yellow and thick black nervures, covering less than one cell, very narrow, at least half as long again as broad; 11 postnodal nervures in fore-wings, 9 to 10 in the hind. *Abdomen* with the ground-colour

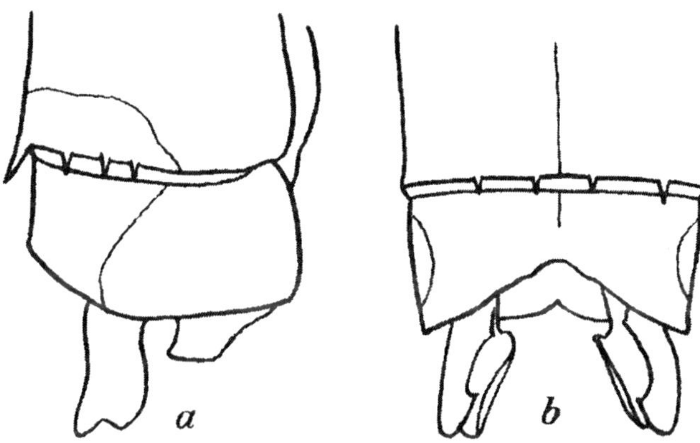

Fig. 126.—Anal appendages of *Pseudagrion williamsoni* Fras., male. *a*. Right lateral view. *b*. Dorsal view.

of basal segments azure blue, 3 to 7 greenish-blue, the remainder azure blue, marked with black as follows:—Segment 1 with a basal quadrate dorsal spot not extending to apex of segment, which has a fine black ring; segment 2 with a goblet-shaped marking on dorsum, not extending to base of segment, but connected by a short stem to a narrow apical ring; a rather broad basal ring not wider than the goblet, which may or may not be confluent with the edges of the goblet to enclose a small triangular or V-shaped blue spot on mid-dorsum; segments 3 to 7 with broad dorsal stripes expanding subapically, and then contracting again to join narrow apical black annules, except on segment 7, where the band expands gradually as far as the apical border of the segment; segments 8 and 9 with the basal third and fourth respectively black, the rest azure blue, the apical black rings subject to slight variations in depth; segment 10 with its dorsum broadly black, this narrowing towards apex, below which is a narrow blue shelf. *Anal appendages*: superiors black, as long as segment 10, slightly sinuous and truncate as seen in profile, and notched at the apex, deeply excavate on the inner surface as seen from the dorsum, the edges slightly curled and the apex bearing a robust hook to the outer side of notch, angulated but not spined near base; inferiors small, conical, white, bearing a minute black point at apex.

Female.—Abdomen 25 mm. Hind-wing 17 mm.

Differs somewhat from the male in colour and markings:—Labium cinereous: labrum and bases of mandibles yellow; genæ glossy black; rest of head olivaceous, with a golden sheen, but the postocular areas with blue centres and narrowly margined in front and broadly at the back with black: beneath head black; eyes olivaceous above, paling below. *Prothorax* olivaceous, with similar blue markings, framed in black, as seen in the male; posterior lobe with short yellow spines. *Thorax* ferruginous along the mid-dorsum and humeral areas, bluish-green between these and on the sides; mid-dorsal carina finely black, and small black points on the upper ends of the humeral and postero-lateral sutures, as well as a small round spot on the mesepimeron, the same as seen in the male; beneath white. *Legs* yellow, distal ends of all femora on extensor surface. *Wings* as for the male, but the pterostigma a paler brown. *Abdomen* greenish-blue at base, bluish at anal end, marked with black as follows:—Segment 1 with a small basal quadrate spot; segment 2 with a narrow mid-dorsal stripe extending the whole length of the segment, broadening abruptly subapically, then contracting to join a narrow apical annule: segments 3 to 7 similar to the male; segments 8 and 9 broadly black from base to apex, except on the sides, which are blue; segment 10 entirely pale blue. *Anal appendages* pale bluish, short, conical, pointed; vulvar scales carneous, extending to end of abdomen.

Distribution.—LOWER BURMA only; King Island, Mergui, in swamps and small streams in rubber plantations during June.

The broad golden olivaceous area of the front of the head, contrasting strongly with the black on the upper posterior part, as well as the marking on segment 2, will serve to distinguish this remarkably beautiful insect, perhaps the most beautiful of the genus. The colouring of the head and thorax, almost without black markings, is equally distinctive of the female.

Type in the British Museum; paratypes in the Pusa and author's collections.

121. **Pseudagrion indicum** Fraser. (Fig. 127.)

Pseudagrion indicum Fras., Rec. Ind. Mus. vol. xxvi, pp. 428, 495–7, fig. 3, i, ii, iv, v (1924); id., ibid. vol. xxxiii, pp. 448 465 (1931).

Male.—Abdomen 34 mm. Hind-wing 22 mm.

Head: labium white; labrum, face, and cheeks pale yellowish-green; eyes black above, greenish at sides and beneath; vertex and occiput black, marked with large azure blue postocular spots. *Prothorax* azure blue, marked with fine subdorsal and lateral longitudinal black lines confluent behind; posterior lobe blue. *Thorax* with mid-dorsum broadly black and somewhat narrower black humeral stripes, the area between grass-green, as also the posterior border of the humeral stripes; laterally azure blue, with the upper half of the antero-

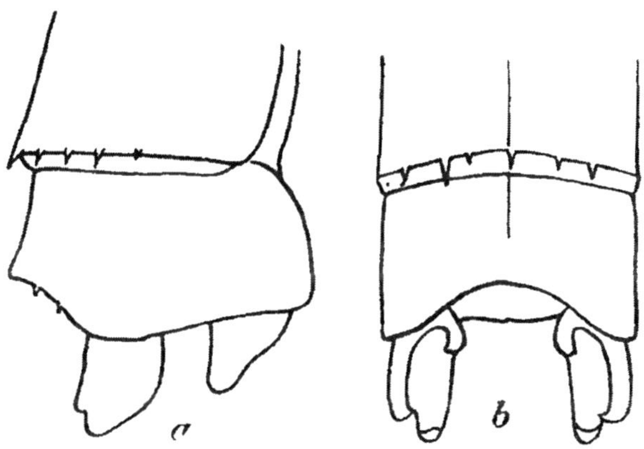

Fig. 127.—Anal appendages of *Pseudagrion indicum* Fras., male. *a*. Right lateral view. *b*. Dorsal view.

lateral suture and a small spot on the upper end of the postero-lateral suture black; beneath white. *Wings* hyaline; pterostigma very narrow and very acute at both ends, covering rather less than one cell, blackish-brown, framed in pale yellow and thick black nervures; 12 postnodal nervures in fore-wings, 10 in the hind. *Abdomen* azure blue on segments 1, 2, and basal half of 3; 3 to 7 greenish laterally; 8 to 10 azure blue; marked with black as follows:—Segment 1 with a quadrate basal dorsal marking, which may or may not reach apex of segment to become confluent with a narrow apical ring; segment 2 with a dorsal marking usually shaped similarly to that on the same segment of *Cœnagrion mercuriale*, but occasionally goblet-shaped, and with a small medio-basal spot; segments 3 to 6 with a broad dorsal stripe which expands subapically and again contracts to join a narrow apical black annule; segment 7 similar, but the dorsal stripe expanding continuously as far as apex of segment; segments 8 and 9 azure blue, with broad apical dorsal annules sometimes bifid at the mid-dorsal carina, or sometimes crenated; segment 10 broadly black on dorsum. *Anal appendages*: superiors black, shorter than segment 10, slightly and unequally bifid at apex as seen in profile, the inner lower branch of the bifurcation the longer as seen from above, and with its apex hooked over; a moderately robust, long, slightly curled spine near the base, the inner surface of appendage deeply excavate; inferiors paler, a little more than half the length of superiors, conical, bevelled from apex downward.

Female.—Abdomen 32 mm. Hind-wing 22 mm.

Differs widely from the male in colour and markings. *Head* pale green, the eyes dark emerald-green above, with or without a small black cap; the vertex and occiput are tinted with golden yellow, the black very restricted, and the postocular spots pale green. *Prothorax* pale green, finely marked with black. *Thorax* grass-green on dorsum, pale yellowish-green on sides: three fine parallel lines of black on mid-dorsal carina and a fine black humeral line (not amounting to a stripe); vestiges of a black stripe on the upper part of first lateral suture. *Wings* hyaline, pterostigma pale brown, 12 to 13 postnodal nervures in fore-wings, 10 to 11 in the hind. *Abdomen* with ground-colour similar to the male; markings on segments 1 to 7 similar, except that the goblet mark on 2 is filled in to form a thistle-head, which extends to base of segment; segments 8 and 9 are black, save for fine apical blue rings; 10 is all blue save for a very fine basal ring.

Distribution.—Known so far only from the NILGIRIS, COORG, and MALABAR, in submontane and montane tracts, from 2,000 to 4,000 ft., almost throughout the year.

Some variation is found in the markings, especially in those on segments 2, 8, and 9. The markings on the above-mentioned segments and the beautiful grass-green colour of the dorsum of thorax will determine this insect from all others of the genus. It appears to be related to *P. williamsoni* from Burma, and is distinguished by the colouring of the face and upper surface of the head and the marking on segment 2. Breeds in small streams.

Types from the Nilgiri Wynaad in the British Museum; paratypes in the author's collection, from the same locality, and Coorg, South Kanara, and Malabar.

122. **Pseudagrion pruinosum** Burmeister. (Fig. 128.)

Agrion pruinosum Burm., Handb. Ent. vol. ii, p. 821 (1839).
Pseudagrion pruinosum Selys, Bull. Acad. Belg. (2) vol. xli, p. 517 (1876); Kirby, Cat. Odon. p. 153 (1890).

Male.—Abdomen 37 mm. Hind-wing 25 mm.

Head: labium pale brown; labrum, bases of mandibles, genæ, and anteclypeus ochreous, the former with a minute medio-basal spot of black; postclypeus black, often with an ill-defined transverse spot of ferruginous at its middle; frons and front of vertex ferruginous as far back as the ocellar space, as also the 2nd joint of antennæ; rest of head blackish-brown to black, the postocular areas ferruginous at the centres, but obscured in very adult specimens (as well as in

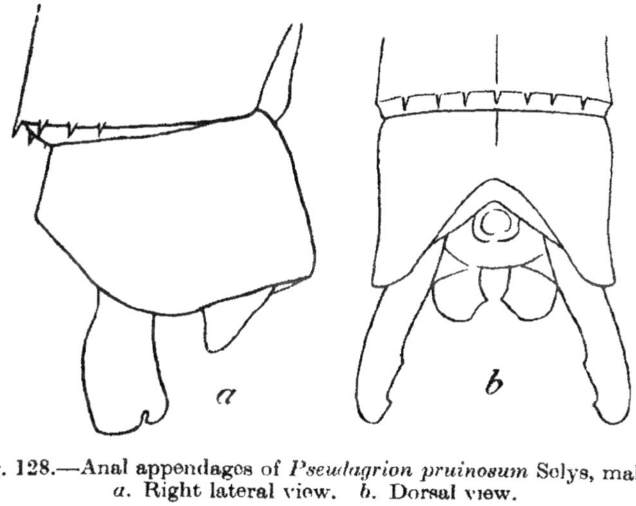

Fig. 128.—Anal appendages of *Pseudagrion pruinosum* Selys, male.
a. Right lateral view. *b*. Dorsal view.

Java specimens, which have the whole of vertex and frons also black); eyes dark brown, paler below; beneath head black, bordered narrowly and outwardly with ochreous (ochreous with an irregular line of black in Java specimens). *Prothorax* black, the middle lobe and sides densely pruinosed bluish-white. *Thorax* black, with some obscure brown and ochreous markings on the lower part of sides; dorsum thinly pruinosed, the sides densely so, and often chalky white, especially below. *Wings* hyaline; pterostigma blood-red to dark brown, usually the former in Burmese specimens, and framed in thick black nervures, covering rather less than one cell; 13 to 15 postnodal nervures in fore-wings, 11 to 12 in the hind. *Legs* black, pruinosed, coxæ and trochanters and proximal part of hind femora yellowish: tibiæ bright ochreous on extensor surfaces; tarsi black. *Abdomen* black, ochreous on lower part of sides; segment 1 and the dorsal three-fourths of 2 pruinosed white or bluish-white; segments 3 to 6 with obscure narrow, basal, ochreous annules, otherwise steely blue or greenish-black; segments 8 and 9 often pruinosed on the dorsum; segment 10 rather broadly ochreous laterally. *Anal appendages*: superiors black, as long as segment 10, minutely notched below as seen in profile, ungulate, rather squared at apex; long, narrow, and ungulate as seen from above, the inner side a little excavate, and the border with two short obtuse spines near apex, the distal of these formed by the hooking inwards of the apex; inferiors pale yellow, half the length of superiors, conical, sloping upwards.

Female.—Abdomen 32 mm. Hind-wing 24 mm.

Head: labrum, clypeus, frons, vertex as far back as posterior ocelli, and the antennæ dull olivaceous; bases of mandibles and genæ pale greenish-yellow; postocular areas ochreous, changing to bluish-green at the centres, and bordered all round narrowly with black; the posterior ocelli edged posteriorly with black; eyes a beautiful pea-green; rest of head as for the male. *Prothorax* dull olivaceous, with four longitudinal black lines on middle lobe; posterior lobe crenate, with the usual spines directed forwards from its base. *Thorax* dull olivaceous on dorsum, changing to pale bluish-green on the sides, marked with a small black spot on the upper ends of the humeral and postero-lateral sutures; beneath yellowish-green, thinly pruinosed. *Legs* yellow or ochreous, distal ends of femora on extensor surfaces and the flexor surfaces of tibiæ and whole of tarsi black. *Wings* hyaline: pterostigma palest yellow, framed in black nervures; 13 postnodal nervures to fore-wings, 11 in the hind. *Abdomen* similar to the male but without pruinescence, and with the yellow on sides of all segments much more extensive. *Anal appendages* black, not longer than segment 10, pointed, conical; vulvar scales yellow, extending to end of abdomen.

Distribution.—Extending from BURMA to Indo-China and Java. The *type* comes from Java, and is one of the Burmeister collection; paratypes in most national collections and the British Museum. I have specimens from Maymyo, Upper Burma, taken in August and June; from Java in September; from King Island, Mergui, Lower Burma, taken in September; and from Siam in October. It is the largest species of the genus taken within Indian limits, and is easily distinguished by the reddish face and pruinosed body, as well as by the blood-red pterostigma.

Genus ARCHIBASIS Kirby (1890). (Fig. 129.)

Stenobasis Selys, Bull. Acad. Belg. (2) vol. xliii, p. 107 (1877).
Archibasis Kirby, Cat. Odor. p. 156 (1890) ; Laid., Rec. Ind. Mus. vol. xvi, p. 194 (1919); Munz, Mem. Amer. Ent. Soc. no. 3, p. 63, tab. 17, fig. 124 (1919) ; Ris, Ark. Zool. Uppsala, Bd. 21 A, no. 31, pp. 15–16 (1930).

Dragonflies of small size and slender build ; colours non-metallic, blue marked with black ; wings narrow, subrotundate at apex, hyaline, but palely enfumed in adults ; pterostigma similar in the fore- and hind-wings, subquadrate, distal side slightly convex, proximal side oblique, braced, covering about one cell ; postnodal nervures numbering from 13 to 16 in the fore-wings ; discoidal cell acutely pointed at the distal end, its costal side about one-fourth the length of posterior in fore-wing, about one-half or less in the hind : arc in line with the distal antenodal nervure : *ab* arising from the posterior border of wing at the point where *ac* meets it, or fusing with the hinder border slightly distal to *ac* : *ab* continued

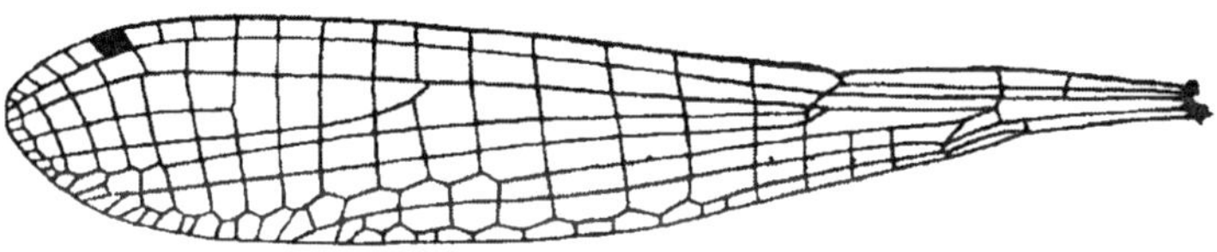

Fig. 129.—Wing of *Archibasis mimetes præclara* (Fras.), male.

on as *IA*, but the junction of the two nervures strongly angulated ; *Riv+v* arising slightly before the level of the nervure descending from subnode ; *IRiii* in continuation of that nervure ; *ac* situated much nearer the level of the distal antenodal nervure.

Head very small, with or without triangular postocular spots or areas ; eyes globate ; thorax moderately robust ; posterior lobe of prothorax crenate or arched and, in the female, furnished with two small spines sloping strongly forwards and lying almost prone on the middle lobe ; legs short, robust, tibial spines moderately long, about five in number on the hind pair of tibiæ ; claw-hooks without spines. Superior anal appendages about as long as segment 10, rounded or slightly notched at apex, hollowed out on the inner side ; inferior appendages half the length of superiors, small cone or club-shaped organs. Genitalia : lamina narrowly and deeply cleft ; hamules small quadrate plates meeting broadly across the middle line. penis simple, squared at apex, the end curled strongly over and hollowed out below to embrace the stem of the organ, unbranched ; lobe flask-shaped, depressed ;

vulvar scales robust, extending to end of abdomen ; no apical spine on the lower border of segment 8 in the female.

Genotype, *Archibasis melanocyana* (Selys).

Distribution.—WESTERN GHATS of India, ASSAM, BURMA, Indo-China, Sumatra, and Australia. One form found in southern India is not more than a race of an Australian species which so far has not been recorded from intervening territories.

Specimens of the genus are very uncommon, probably on account of the wild jungly nature of their habitats ; they breed in small, swiftly running brooks or streams near sea-level, and have habits somewhat similar to species of *Pseudagrion* found in similar haunts, such as *P. rubriceps* and *P. indicum*. The genus is readily separated from *Pseudagrion* by the square-shaped pterostigma, the higher nodal index, and by the total absence of teeth on the claw-hooks.

Key to the Species of Archibasis.

Labrum finely black at base ; segment 10 with a black dorsal spot ; inferior anal appendages slightly less than half the length of superiors and without a small spine on the inner side near apex	*melanocyana* [(Selys), p. 308.
Labrum entirely blue ; segment 10 entirely blue ; inferior anal appendages two-thirds the length of superiors, and armed with a small spine on the inner side near apex	*mimetes præclara* [(Fras.), p. 310.

123. **Archibasis melanocyana** (Selys). (Fig. 130.)

Stenobasis melanocyana Selys, Bull. Acad. Belg. (2) vol. xliii, p. 107 (1877).

Archibasis melanocyana Kirby, Cat. Odon. p. 156 (1890) ; Laid., Proc. Zool. Soc. Lond. p. 387, ex Selys (1902); Ris, Zool. Meded. Leiden, vol. x, p. 25 (1927) ; id., Ark. Zool. Uppsala, Bd. 21 A, no. 31 (1930).

Male.—Abdomen 30–32 mm. Hind-wing 21–23 mm.

Head : labium white ; labrum, bases of mandibles, genæ, anteclypeus, frons, and vertex as far back as the roots of antennæ azure blue, the labrum finely bordered with black at base ; postclypeus black, as well as vertex from the level of antennæ to back of occiput, the anterior border of this black area crenate, and the occipital area with large azure blue triangular postocular spots ; beneath head black, but with a broad pale yellow belt against the eyes ; eyes blue above, capped with black, olive-green below. *Prothorax* black, palest blue on the sides changing to white below, a mid-dorsal blue spot often obliterated in full adults ; posterior lobe rounded, simple. *Thorax* black on dorsum, with broad dark blue antehumeral stripes which are deeply indented by the humeral black near the top of thorax ; laterally pale azure blue, the upper part of postero-lateral suture black ; beneath thorax palest yellow, densely overlaid by pulverulent white. *Legs* pale yellow, femora broadly black on extensor surface ; tibiæ black on flexor surface. *Wings* hyaline ; pterostigma black, finely framed in pale brown and thick black nervures, covering rather less than one cell, poorly braced ; 13 to 15 postnodal nervures in fore-wings, 13 to 14 in the hind. *Abdomen* black, marked with blue as follows :—Segment 1 with a large black spot at base, the apical border pale blue ; segment 2 broadly black on dorsum, this slightly constricted at apical border, and bearing a large blue oval spot on the dorsal basal half, laterally pale blue ; segments 3 to 6 with narrow blue basal paired spots, the black dorsal band

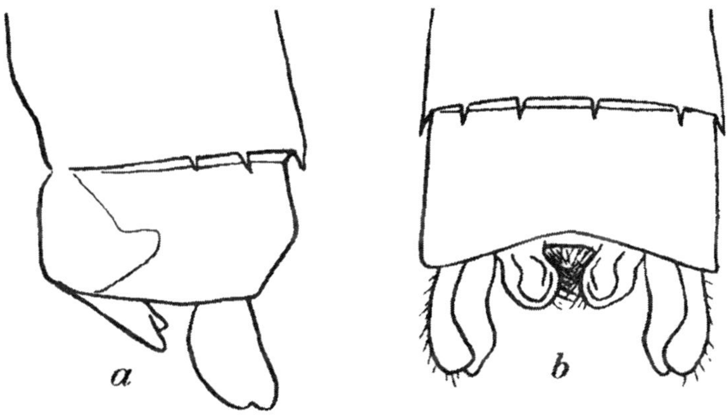

Fig. 130.—Anal appendages of *Archibasis melanocyana* (Selys), male. *a*. Left lateral view. *b*. Dorsal view.

slightly expanded subapically ; segment 7 with the black band broadening continuously to the apical border, its sides only pale greenish-blue ; segments 8 and 9 entirely blue, save for a row of black apical spines on 8 and a very narrow black apical border on 9 ; segment 10 with a black dorsal spot, the sides broadly blue. *Anal appendages* black ; superiors as long as segment 10, slightly notched at apex as seen in profile, hatchet-shaped as seen in a semi-dorsal view, a narrow notch separating the blade of the hatchet from the handle at the end ; seen from above sinuous, curled in slightly at apex and deeply hollowed out within ; inferiors nearly half the length of superiors, conical and sloping up towards superiors as seen from the side, roughly squared and angulated and hollowed out as seen from the dorsum.

Female.—Abdomen 35 mm. Hind-wing 28 mm.

Very similar to the male, but the blue markings rather more obscured, especially on the *head* and *abdomen*; posterior lobe of *prothorax* rounded, but slightly concave dorsally, and with two small low-lying spines directed strongly forward from its base and extending to the middle of middle lobe; *thorax* similar to the male, but the antehumeral stripe less indented and the postero-lateral suture black for its upper half. *Wings* deeply enfumed around all nervures, so that these appear thickened and blurred, the cell-middles being clear; pterostigma reddish-brown, framed in black; 16 postnodal nervures in fore-wing, 13 to 14 in the hind; *ab* (anal bridge) fusing with hinder border of wing slightly before the level of *ac*, so that the petiolation is rather longer than in the male. *Legs* as for the male. *Abdomen* without the blue oval spot on segment 2; segment 8 black, broadly blue on sides; segment 9 blue, broadly black at base; segment 10 blue. *Anal appendages* black, short, conical, pointed; vulvar scales bluish, robust, extending to end of abdomen.

Distribution.—From LOWER BURMA only within our limits; elsewhere from Malaysia, Sumatra, Banka, and Borneo.

The 10th segment of the type is given as blue, but in all specimens which I have examined I have been unable to detect this, although decomposition may have obscured it. The blue antehumeral stripes appear to be somewhat variable, being given as interrupted above in the type. In the teneral form the humeral black is made up of a lower and upper portion, so that variation in the stripe may be due to more or less incomplete fusion of these two parts as adult markings are developed. Distinguished from other species by the shape of the anal appendages and blue markings.

Type in the Selysian collection: paratypes in the Stockholm, Ris, and author's collections.

124. **Archibasis mimetes præclara** (Fraser). (Figs. 129 & 131.)

Pseudagrion præclarum Fras., Rec. Ind. Mus. vol. xxvi, pp. 428, 493, 494, fig. 3, vi (1924); id., ibid. vol. xxxiii, pp. 448, 465, 466 (1931).

Male.—Abdomen 36–39 mm. Hind-wing 24–27 mm.

Head: labium white; labrum, bases of mandibles, genæ, anteclypeus, frons, and vertex as far posterior as the middle ocellus, including basal joints of antennæ, turquoise-blue; postclypeus black, with two small basal blue spots; a broad black band crossing the head from eye to eye on top of vertex, behind which are large triangular dark blue postocular spots narrowly connected across the occiput by a blue stripe;

beneath head black, broadly white against the eyes; eyes blue above, pale greenish-blue to yellow below. *Prothorax* black, the anterior lobe, two large spots on the dorsum of middle lobe, and the lower part of the sides broadly blue. *Thorax* broadly black on dorsum to well beyond humeral suture, marked with moderately broad dark azure blue antehumeral stripes, strongly indented by the humeral black in the upper third; laterally blue, with a small round black spot in middle of mesepimeron, and another small spot on upper part of postero-lateral suture; beneath white or carneous, thinly pruinosed. *Legs* creamy white, with black spines; femora broadly black on extensor surface, tibiæ on the flexor surface.

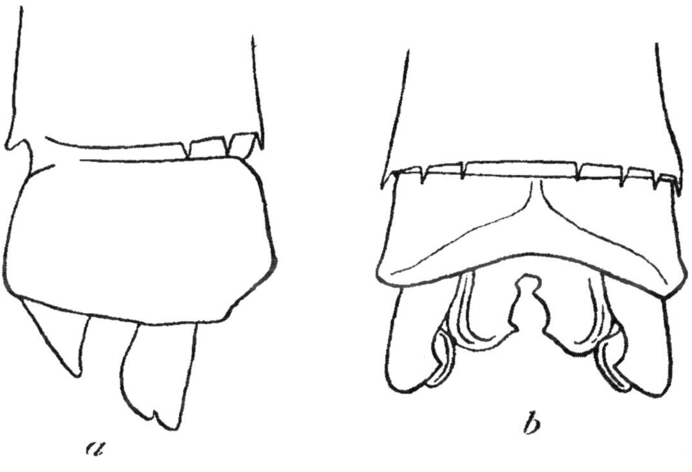

Fig. 131.—Anal appendages of *Archibasis mimetes præclara* (Fras.), male. *a.* Left lateral view. *b.* Dorsal view.

Wings hyaline; pterostigma subquadrate, blackish-brown, framed palely in yellow, covering less than one cell; 15 to 16 postnodal nervures in fore-wings, 14 in the hind; petiolation ceasing before level of *ac* in the fore-wings, at *ac* in the hind. *Abdomen* black on dorsum, blue to greenish-yellow on the sides; segment 1 with a broad black basal dorsal spot not extending to apical border of segment; segment 2 with a broad black goblet-shaped dorsal marking, with a very short apical stem connected to a narrow apical black ring, a large oval mid-dorsal blue spot extending along the basal half of segment, laterally broadly blue; segments 3 to 6 with broad black dorsal stripes expanding subapically and then contracting again to join narrow apical rings; segment 7 somewhat similar, but the apical dilatation continuing to the apical border; segments 8 and 9 with an apical ring of black spines only, whilst 10 is entirely blue. *Anal appendages* black; superiors as long as segment 10, with a tiny apical notch as seen in profile, narrow at base, expanded rather broadly on the inner side of the apical half, this expansion hollowed out; inferiors two-thirds the length of superiors, broad at base, then tapered rapidly to a fine point as seen in profile, rounded, and with a small tooth on the inner side as seen from above.

Female.—Abdomen 38 mm. Hind-wing 27 mm.

Differs in a few respects from the male, a more robust insect. *Head*: labium greenish-yellow, with a small medio-basal point of black; genæ, bases of mandibles, and anteclypeus bluish-green; postclypeus, frons, vertex, and occiput black, the latter with large greenish-blue postocular spots; eyes olivaceous green, paler below. *Prothorax* marked similarly to the male; posterior lobe with two short forwardly directed spines which lie closely apposed to the dorsum of middle lobe. *Thorax* with the humeral black irregular and often interrupted above and margined diffusely with golden ochreous; laterally and beneath as for the male. *Wings* evenly enfumed in adults; pterostigma violaceous brown; 16 postnodal nervures in fore-wings, 14 to 15 in the hind. *Abdomen* differing in the following respects:—Segment 2 with a rather narrow thistle-shaped mark on dorsum; segments 8 to 10 azure blue, 8 with a large cordate dorsal spot, its point prolonged narrowly along the mid-dorsal carina as far as base of segment, its base constricted and joining a narrow apical annule; segment 9 with two large triangular black spots, the base of the spots resting on the base of the segment; segment 10 wholly blue. *Anal appendages* small, conical, pointed, blue; vulvar scales robust, not extending as far as the end of abdomen, bluish.

Distribution.—Confined to the WESTERN GHATS of India; Coorg, South Kanara, South Malabar, and parts of the Wynaad. The type, described as *Pseudagrion præclarum*, is a slightly teneral male from Mukat, Coorg, and is the sole specimen known from that area. I found this species rather common on a small brook running through dense jungle at Tamaracherri, S. Malabar; it was present on this stream nearly the whole year round.

The anal appendages are identical with those of *A. mimetes mimetes* Till., an Australian form, from which it differs only by the higher nodal index and larger size, so that, in spite of the vast area separating the two, *A. mimetes præclara* cannot be considered of more than racial or subspecific value.

Type in the author's collection; paratypes in the British Museum.

Genus **CERIAGRION** Selys. (Fig. 132.)

Ceriagrion Selys, Bull. Acad. Belg. (2) vol. xlii, p. 525 (1876); Kirby, Cat. Odon. p. 154 (1890); Ris, Abh. Senckenb. Ges. vol. xxxiv, p. 519 (1913); Laid., Rec. Ind. Mus. vol. xii, p. 132 (1916); id., ibid. vol. xvi, pp. 187–9 (1919); Fras., J. Bombay Nat. Hist. Soc. vol. xxix, p. 748 (1923).

Dragonflies of small size and slender build; colours non-metallic, usually some shade of yellow, orange, or olivaceous, or, more rarely, blue, and rarely marked with black; wings hyaline; pterostigma lozenge-shaped, narrow, oblique at both ends, braced, covering about one cell; postnodal nervures numbering from 10 to 12 in fore-wings; discoidal cell acutely pointed at the distal end, with costal side about one-third the length of posterior side in fore-wing and one-half the length in the hind, basal side equal to costal in the fore-wing, only half that length in the hind, distal side very oblique; sectors

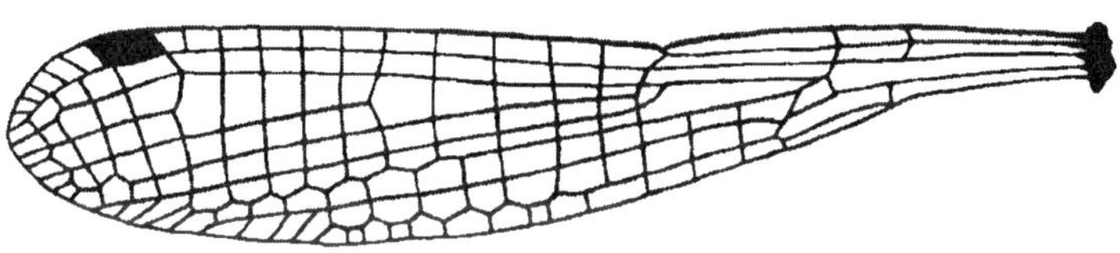

Fig. 132.—Wing of *Ceriagrion coromandelianum* (Fabr.), male.

of arc arising from lower end of arc, divergent from their origins; arc situated at the level of the distal antenodal or very slightly distal to that level; *ab* (anal bridge) always present and complete, arising at the point where *ac* meets hinder border of wing, *ac* being situated much nearer the proximal antenodal nervure; *ab* continued on as the nervure *IA*, but an angulation (medio-anal link) present at the junction of the two; *Riv*+*v* arising well before the oblique nervure descending from the node; *IRiii* in continuation of that nervure. (In some species *ab* arises slightly but distinctly proximal to the level of *ac*.)

Head narrow, *with a well defined frontal ridge* (resembling that found in most of the Anisoptera); postocular coloured spots always absent; eyes globate; thorax rather long and narrow; posterior lobe of prothorax simple, that of female entirely free of any armature; abdomen slender, cylindrical, not nearly twice the length of hind-wing; legs short, tibial spines short and thick, 4 to 6 or 7 in the hind and middle pairs; anal appendages variable in the species, but generally the superiors short and hooked, the inferiors longer, conical, and tapering to a point. No spine at base of segment 8 in the female.

Genitalia: lamina narrowly and deeply cleft; hamules quadrate plates slightly hollowed out or curled, meeting across middle line; lobe long, narrow, flask-shaped; penis blunt or emarginate at apex, its end curled over dorsum and bifid or hollowed out to embrace stem of penis.

Genotype, *Ceriagrion coromandelianum* (Fabr.)

Distribution.—Southern Europe, throughout the greater part of the African continent, Southern Asia, China, Japan, and the Sondaic Archipelago. Nearly all the species breed in still waters, but occasionally some take to streams through force of circumstances, e. g., *C. cerinorubellum*; weedy ponds and lakes are preferred by most species.

The genus is closely related to *Pseudagrion*, by the petiolation and structure of the wings, and to the Australian genus *Xiphiagrion*. About eight or, possibly, nine species are found within Indian limits, some widely scattered, others extremely local.

Key to the Species of Ceriagrion.

1.	Abdomen pale azure blue; segments 9 and 10 marked with black on dorsum..	*azureum* (Selys), [p. 328.
	Abdomen uniform dull olivaceous in both sexes	*olivaceum* Laid., p 324.
	Abdomen bright red at base and anal ends, black on dorsum in between	*cerinorubellum* (Br.), [p. 326.
	Abdomen bright citron-yellow, without markings	*coromandelianum* (Fabr.), [p. 315.
	Abdomen pale citron-yellow, with black markings on the end segments	2.
	Abdomen bright reddish-orange, without markings	3.
2.	Superior anal appendages more than half the length of segment 10	*fallax* Ris, p. 321.
	Superior anal appendages less than half the length of segment 10	*nigroflavum*, sp. n., [p. 323.
3.	Inferior anal appendages with a thick robust spine projecting back from their middle............................	*prætermissum* Lieft., [p. 319.
	Inferior anal appendages without such a spine	4.
4.	Superior anal appendages with no visible tooth as seen from above; inferior anal appendages shorter than superiors.....	*rubiæ* Laid., p. 318.
	Superior anal appendages with two minute teeth on inner border when viewed from above; inferior anal appendages longer than superiors	*erubescens* Selys, [p. 316.

125. **Ceriagrion coromandelianum** (Fabr.). (Figs. 132 & 133.)

Agrion coromandelianum Fabr., Ent. Syst. Suppl. p. 287 (1798).
Agrion cerinum Ramb., Ins. Névrop, p. 279 (1842).
Ceriagrion coromandelianum Selys, Bull. Acad. Belg. (2) vol. xlii, p. 528 (1876); Kirby, Cat. Odon. p. 154 (1890); id., J. Linn. Soc. Lond., Zool. vol. xxiv, p. 564 (1894); Martin, Mission Pavio (Névrop.), (sep.) p. 18 (1904); Morton, Trans. Ent. Soc. Lond. p. 358 (1907); Ris, Abh. Senckenb. Ges. vol. xxxiv, p. 519, pl. xxiii, figs. 13, 14 (1913); Laid., Rec. Ind. Mus. vol. viii, p. 345 (1914); id., ibid. vol. xii, p. 132 (1916); id., ibid. vol. xvi, pp. 188–90 (1919); Fras., J. Bombay Nat. Hist. Soc. vol. xxvii, p. 542 (1921); id., Rec. Ind. Mus. vol. xxiv, pp. 303, 309 (1922); id., J. Bombay Nat. Hist. Soc. vol. xxix, p. 748 (1923); id., Rec. Ind. Mus. vol. xxvi, pp. 428, 491 (1924); Laid., Spolia Zeylan. vol. xii, p. 369 (1924); Fras. J. Siam. Soc. Nat. Hist. vol. vii, p. 91 (1927); Needham, Man. Dragonflies of China, pp. 264–5 (1930); Fras., Rec. Ind. Mus. vol. xxxiii, p. 448 (1931).

Male.—Abdomen 28–30 mm. Hind-wing 18–20 mm.

Head: labium pale yellow: labrum, clypeus, genæ, bases of mandibles, and frons bright citron-yellow; vertex olivaceous: occiput and postocular area ochreous: basal three joints of

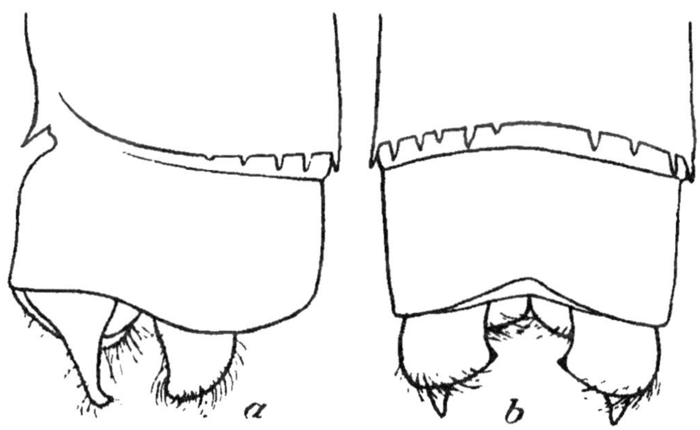

Fig. 133.—Anal appendages of *Ceriagrion coromandelianum* (Fabr.), male. *a*. Left lateral view. *b*. Dorsal view.

antennæ pale yellow; eyes olivaceous above, pale greenish-yellow below; beneath head pale yellow. *Prothorax* and *thorax* uniformly olive-green, the posterior lobe of the former and the dorsum of latter often tinted with ochreous; laterally citron-yellow; beneath more or less thinly pruinosed white. *Legs* citron-yellow, with short black spines. *Wings* hyaline, pterostigma golden yellow, framed in brown nervures covering one cell, braced; 11 to 12 postnodal nervures in fore-wings, 10 to 11 in the hind; *ab* arising at the level of *ac*. *Abdomen* uniformly citron-yellow; segment 10 deeply emarginate on dorsum. *Anal appendages* citron-yellow or ochreous, the inferiors tipped with black; superiors about half the length of segment 10, subquadrate as seen from above, but the corners gently rounded and the inner ending in a minute black point, the whole shaped like a pullet's head, the spine representing the small beak; seen in profile, subtriangular or briefly ungulate; inferiors about half as long again as superiors, sloped strongly upwards, broad at base, then abruptly narrowed and tapering rapidly to an acute point, which is slightly upturned: seen from below, the bases meeting but the apices widely separated and turned in slightly.

Female.—Abdomen 29–32 mm. Hind-wing 20 mm.

A larger, more robust and duller coloured insect. *Head*: face pale olivaceous, vertex olivaceous brown, rest of head as for the male. *Prothorax* and *thorax* golden olivaceous brown, paler at the sides and beneath, here usually thinly pruinosed. *Legs* as for the male. *Wings* similar to the male, but the pterostigma paler. *Abdomen* uniformly olivaceous, with an ochreous or golden brown tint on dorsum, which deepens somewhat on the hinder segments. *Anal appendages* and vulvar scales olivaceous or golden brown, the former conical, pointed, as long as segment 10, the latter robust, extending to end of abdomen.

Distribution.—Throughout INDIA, CEYLON, BURMA, Malaysia, Indo-China, and South China.

Easily distinguished from all others by its abdomen being entirely pale lemon-yellow, without markings. Breeds in weedy ponds and tanks, on the banks and borders of which it may be found for the greater part of the year, threading its way through the rank herbage or reeds. The female is commonly found in scrub-jungle some distance away from water. Whereabouts of type unknown; paratypes in most national and private collections.

126. **Ceriagrion erubescens** Selys. (Fig. 134.)

Ceriagrion coromandelianum race *erubescens* Selys, Ann. Mus. Civ. Genova, (2) vol. x (xxx), p. 516 (1891).

Ceriagrion erubescens Laid., Proc. Zool. Soc. Lond. p. 388 (1902); Ris, Abh. Senckenb. Nat. Ges. vol. xxxiv, pp. 519–20 (1913); id., Nova Guinea, vol. xiii, Odon. Fauna, p. 122 (1915); Laid., Sarawak Mus. Journ. no. 6, p. 274 (1915); id., Rec. Ind. Mus. vol. xii, p. 188 (1919); Fras., J. Siam. Soc. Nat. Hist. vol. iii, p. 461 (1919); id., J. Bombay Nat. Hist. Soc. vol. xxix, p. 748 (1923); Tillyard, Rec. Austral. Mus. vol. xv, no. 2, p. 158 (1926); Fras., J. Siam. Soc. Nat. Hist. vol. vii, p. 9 (1927); Lieftinck, Tijdschr. Ent. vol. lxxii, p. 117 (1929); Liu, Bull. Peking Soc. Nat. Hist. vol. iii, pt. iv, p. 16 (1929); Needham, Man. Dragonflies of China, pp. 262–3 (1930).

Male.—Abdomen 27–30 mm. Hind-wing 18–20 mm.

Head: labium pale yellow; labrum, face, and frons bright orange, genæ and base of mandibles citron-yellow; vertex and occiput dark ochreous, separated by a fine black line, and from the eyes by a similar black line; eyes olivaceous. *Prothorax* dark ochreous on dorsum, olivaceous laterally. *Thorax* olivaceous, paling low down on the sides and beneath, but with a golden tinge on the dorsum. *Wings* hyaline; pterostigma nearly twice as long as broad, narrow, golden olivaceous, framed in black nervures, covering one cell; *ab* arising at the level of *ac* or a mere fraction proximal to it; 12 postnodal nervures in fore-wings, 10 in the hind. *Legs* yellow, with black spines, 5 to 6 of these on hind pairs of tibiæ. *Abdomen* uniform rich vermilion or orange (Needham, *l. c.*, states that Chinese specimens differ in no way from *C. coromandelianum* in colouring. I think, therefore, that he has mixed up a new species with *erubescens*, as Selys says definitely, "the abdomen is clearly red or pale carmine"; the present

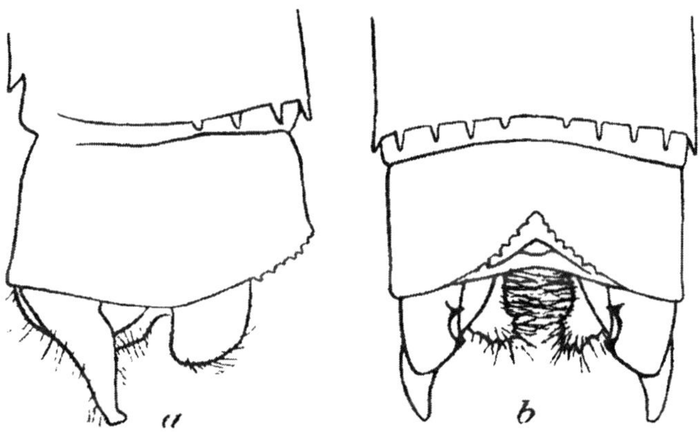

Fig. 134.—Anal appendages of *Ceriagrion erubescens* Selys, male. *a.* Left lateral view. *b.* Dorsal view.

description has been made from specimens from Burma, whence the type came, and is, therefore, more likely to be correct). *Anal appendages* dark olivaceous or orange tipped with black; superiors half the length of segment 10, short blunt processes with the end curled very slightly down and tufted with golden hairs; seen in profile they are hood-shaped, with two ventral short spines, one near apex, the other on a more superior plane near the middle of appendix; inferiors at least half as long again as superiors, broad at base, tapering to apex, which is minutely truncate, sloping up at about half a right angle to the long axis of abdomen.

Female.—Abdomen 28 mm. Hind-wing 20 mm.

A more robust and more dully coloured insect than the male. *Head*: labrum, clypeus, bases of mandibles, genæ, and frons citron-yellow, with a slight olivaceous tint; vertex as for the male, but a duller brown; occiput bright ochreous. *Prothorax* and *thorax* olivaceous yellow; abdomen reddish-brown on dorsum, deepening on the hinder segments and paling to a greenish-yellow on the sides of segments 1 and 2; vulvar scale very robust. *Anal appendages* conical, short, reddish-brown. *Wings* as for the male, but *ab* arising more distinctly proximal to *ac*; 13 postnodal nervures to fore-wings, 12 in the hind.

Distribution.—BURMA, Indo-China, and Malaysia. The forms described from China are certainly not identical with those from Burma or Siam, and are more likely to be a race of *C. coromandelianum*, as was conjectured by Selys.

Distinguished from all other Indian species, except *C. risi* and *C. rubiæ*, by the red colour of the abdomen, and from these two species by the superior anal appendages, which have two minute spines on the inner border as seen from above.

Type apparently in the Genoa Museum; paratypes in the author's collection, from Siam and Burma.

127. **Ceriagrion rubiæ** Laidlaw. (Fig. 135.)

Ceriagrion rubiæ Laid., Rec. Ind. Mus. vol. xii, pp. 132–3 (1916); id., ibid. vol. xvi, pp. 188, 190–1 (1919); Fras., J. Bombay Nat. Hist. Soc. vol. xxix, p. 748 (1923); id., Rec. Ind. Mus. vol. xxvi, pp. 428, 491–2 (1924); id., ibid. vol. xxxiii, p. 448 (1931).

Male.—Abdomen 26–29 mm. Hind-wing 17–18 mm.

Head: labium pale yellow; labrum, clypeus, and frons bright ochreous; bases of mandibles and genæ citron-yellow; vertex warm reddish-brown; occiput rich ochreous; eyes olivaceous, paler below; beneath head yellow. *Prothorax* and *thorax* bright orange, paling to yellow on the sides and beneath. *Legs* pale ochreous, with black spines. *Wings* hyaline; pterostigma amber-tinted, framed in paler yellow and reddish-brown nervures: 12 postnodal nervures in fore-wings, 10 to 11 in the hind; *ab* arising at the level of *ac*. *Abdomen* rich orange, or with a vermilion tint on dorsum paling to yellow beneath. *Anal appendages* reddish-brown, the inferiors paler and tipped minutely with black; superiors about half the length of segment 10, curled slightly downwards, blunt or squared at apex, on the lower angle of which is a minute black spine; seen from above broad at base, roughly triangular, apex rounded; inferiors of the same length, sloping very sharply up to meet the superiors, very broad at base, tapering to apex, which ends in a minute black spine.

Female.—Abdomen 30–31 mm. Hind-wing 20 mm.

More robust than the male, and the reddish or orange colour replaced by olivaceous. *Head*: labrum, anteclypeus and front of frons yellow ; postclypeus and upper surface of head olivaceous brown, but occiput similar to the male. *Prothorax* with the posterior lobe subtrilobate, broad and shallow. the anterior lobe with a small hook on each side, the nature of which is obscure. *Thorax* and *abdomen* olivaceous, the former pruinosed white beneath, as also are the under sides of the first two abdominal segments. *Legs* similar to the male. *Anal appendages* and end segments of abdomen reddish-brown ; vulvar scales moderately robust, but not extending to the end of abdomen. *Wings* similar to the male, but the pterostigma paler yellow and without the pale outer frame, more narrow than in the male. Nodal index similar to the male.

Distribution.—WESTERN INDIA west of the Ghats. The species occurs in sunny swampy localities, weedy ponds

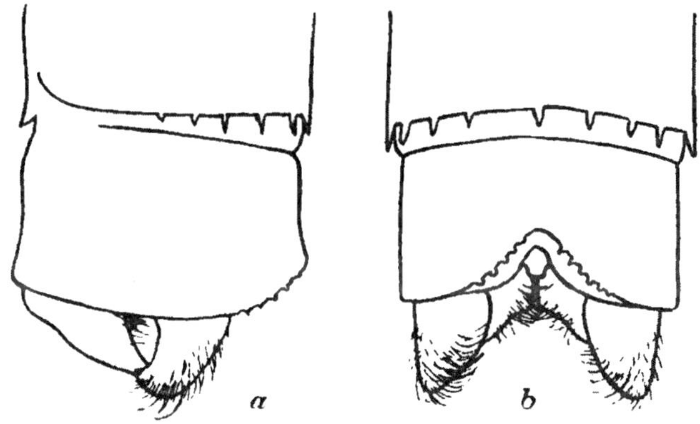

Fig. 135.—Anal appendages of *Ceriagrion rubiæ* Laid., male. *a*. Left lateral view. *b*. Dorsal view.

being preferred. I have taken it as far north as Bombay, in marshy spots on the Bhor Ghat rising to Khandala, and also found it common in Coorg, South Kanara, and parts of Malabar. It occurs before and after the South-West monsoon.

Type in the Indian Museum ; paratypes in the British Museum and most of the better-known private collections.

128. **Ceriagrion prætermissum** Lieftinck. (Fig. 136.)

Ceriagrion prætermissum Lieft., Tijdschr. Ent. vol. lxx, pp. 88, 91 (1927).

Male.—Abdomen 23–24 mm. Hind-wing 15 mm.

Head : labium pale yellow ; labrum, genæ, bases of mandibles and anteclypeus yellow ; postclypeus, frons, and vertex reddish-brown ; occiput dark ochreous ; eyes olivaceous, paler beneath; head pale yellow beneath. *Prothorax* and *thorax* olivaceous, with a golden sheen on dorsum and paling to yellow laterally and to pale bluish-green beneath. *Legs* yellow with very short black spines. *Wings* hyaline; pterostigma small, shorter than in *C. rubiæ*, the inner border very oblique and not parallel with the outer, golden olivaceous, framed in dark brown nervures, covering one cell; *ab* arising at *ac*; 9 to 10 postnodal nervures in fore-wings, 8 to 9 in the hind. *Abdomen* bright red throughout, the sides of segment 1 and the ventral surface of all yellowish. *Anal appendages* ferruginous; superiors as long as segment 10, ungulate, apex rounded and ending below in a minute black tooth; inferiors directed very obliquely upward to meet the superiors, tapering to a point ending in a minute black tooth; on the hinder border a broad robust tooth directed straight back,

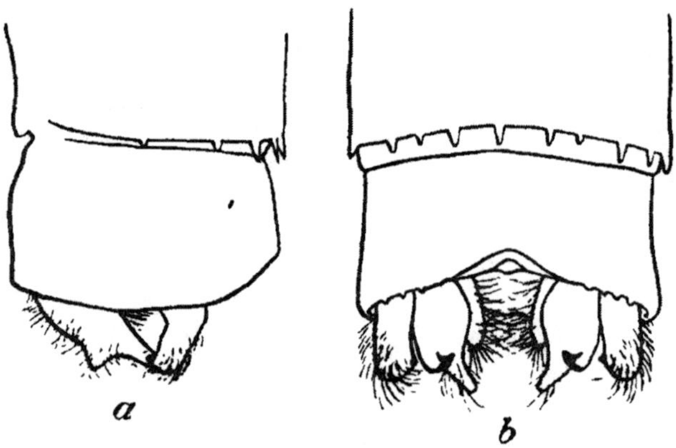

Fig. 136.—Anal appendages of *Ceriagrion prætermissum* Lieft., male. *a*. Left lateral view. *b*. Dorsal view.

a feature unique in the genus, and simulated only by the hinder angulation of the same appendage in *C. melanurum* from Japan.

Female.—Abdomen 22–24 mm. Hind-wing 16 mm.

Very similar to the male, the *head* and *thorax* identical, as also the *legs*. *Wings* with 10 to 11 postnodal nervures in fore-wings, 8 to 9 in the hind; pterostigma similar to the male in shape and colour. *Abdomen* yellow, taking on a slight reddish tint from segment 5 onwards, the intersegmental joints darker, and as distinct narrow apical annules from segments 5 to 8, this latter segment with its mid-dorsum diffusely black; segment 9 with its dorsum black at base, and in some specimens also at apex of segment; 10 with a minute spot of black on dorsum. *Anal appendages* short, conical, olivaceous yellow; vulvar scale robust, extending to the end of body or even beyond it.

Distribution.—Known only from LOWER BURMA and Java, but probably widely distributed over Malaysia.

Distinguished from all other species of the genus by the shape of the inferior anal appendages, by its slight delicate form, and by the shape of the pterostigma. The female closely similar to *C. rubiæ* and *C. erubescens*, but distinguished from both by its smaller size and by the black markings on the terminal segments of abdomen. It is closely related to these two species, and is the smallest of the Indian species. The Javan form has the dorsal spine on the inferior appendages slenderer and more acute, longer, and the apex slightly hooked, so that the Burmese form may be a local race. The female of this form also has no black markings on segments 8 to 10.

Type in the Leiden Museum collection; paratypes in the Laidlaw and Fraser collections.

129. **Ceriagrion fallax** Ris. (Fig. 137.)

Ceriagrion fallax Ris, Ent. Mitteil. vol. iii (2), pp. 47–8, fig. 2 (1914); Laid., Rec. Ind. Mus. vol. xvi, p. 189 (1919); Fras., J. Bombay Nat. Hist. Soc. vol. xxix, p. 748 (1923); Needham, Man. Dragonflies of China, pp. 264, 265 (1930).

Ceriagrion melanurum Selys, Ann. Mus. Civ. Genova, (2) vol. x (xxx), p. 517 (1891) (nec Selys, 1876); Ris, Abh. Senckenb. Ges. vol. xxxiv, p. 520 (1913).

Ceriagrion cerinomelas Lieft., Tijdschr. Ent. vol. lxx, pp. 88–90 (1927).

Male.—Abdomen 30–32 mm. Hind-wing 20–21 mm.

Head: labium palest brown; labrum citron-yellow; clypeus, frons, bases of mandibles, and genæ pale yellow, with a faint greenish tint; vertex dark olivaceous green, taking on a ferruginous tint posteriorly; occiput bright ochreous, this colour sharply delimited from the olivaceous of vertex; eyes dark olivaceous above, paling below; beneath head pale greenish-white. *Prothorax* and *thorax* olivaceous green on dorsum, yellow on the sides, the former with a mid-dorsal stripe of black on the middle lobe and the base of the posterior lobe narrowly, the latter with a small black point on the upper part of humeral suture and another similar spot on the upper part of postero-lateral suture; citron-yellow beneath thorax. *Legs* pale yellow, with short black spines. *Wings* hyaline; pterostigma very narrow and rather elongate, covering one cell, brown, framed narrowly in yellow and thick black nervures; 13 to 15 postnodal nervures in fore-wings, 12 to 13 in the hind; *ab* arising distinctly before the level of *ac* in all wings of all specimens. *Abdomen* pale lemon-yellow, the end segments marked with black on dorsum as follows:—Segment 6 with a narrow apical annule; segment 7 nearly as far as its base or just meeting the black annule on 6; segments 8 to 10 entirely black on dorsum and greater part of sides. *Anal appendages* black, the inferiors yellow at base; superiors slightly shorter than segment 10, conically triangular as seen in profile, with an obtuse apex, hood-shaped as seen from above, the outer side straight, the inner slightly sinuous and ending below apex in a small tooth, the apex thickly coated with hairs; inferiors of the same length, directed obliquely upwards and backwards, tapering to an acute apex, which bears a small inwardly directed tooth.

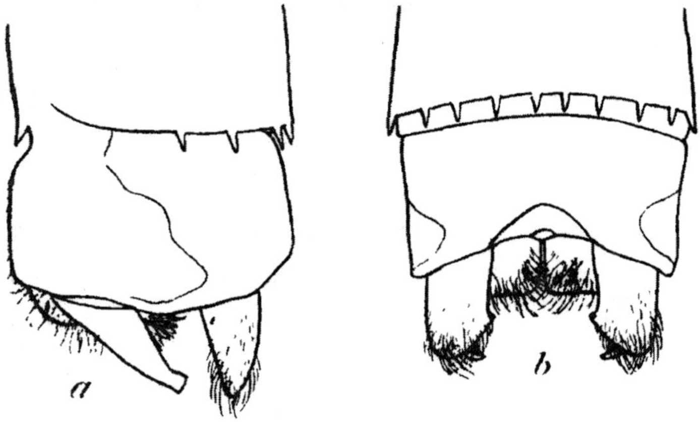

Fig. 137.—Anal appendages of *Ceriagrion fallax* Ris, male.
a. Left lateral view. *b*. Dorsal view.

Female.—Abdomen 28–34 mm. Hind-wing 21–24 mm.

Slightly more robust and more dully coloured than the male. *Head*: labium pale yellow; vertex sometimes reddish-brown: *Thorax* similar, or the dorsum dark olivaceous with a golden sheen, whilst beneath thorax and the first two segments of abdomen often pruinosed white. *Wings* similar or palely enfumed; pterostigma pale amber, narrowly framed in yellow and thick black nervures; 14 postnodal nervures to fore-wings, 12 to 13 in the hind; *ab* arising still more proximal to *ac*. *Legs* as for the male. *Abdomen* olivaceous brown on dorsum, or with a reddish or carneous tint on segments 1 to 4 and basal half of 5, from which point the dorsum changes rather abruptly to black as far as segment 10; intersegmental joints narrowly black. *Anal appendages* very short, black; vulvar scale yellow to brownish-black, extending to the end of abdomen.

Distribution.—The hilly tracts of ASSAM, SIMLA HILLS, BENGAL, SIKKIM, and Tibet.

The form *C. cerinomelas* was compared by its author with *C. melanurum*, and had he compared it with *C. fallax* the error would not have arisen. The colour differences noted are due to post-mortem changes, abundant artefacts being found in preserved specimens of this species. Distinguished from other species by the shape of its anal appendages; from *C. melanurum*, which has the same colouring, by the absence of angulation of the inferior anal appendages and by the absence of the unique latero-apical angulations on the hind border of segment 10 seen in that insect; from *C. nigroflavum* by the much longer superior anal appendages, etc.

Type in the Königsberg Museum; paratypes in the British Museum, Pusa collection, Indian Museum, and several private collections.

130. **Ceriagrion nigroflavum**, sp. nov. (Fig. 138.)

Male.—Abdomen 28 mm. Hind-wing 18 mm.

Head: labium palest yellow; labrum, clypeus, bases of mandibles, genæ, and frons citron-yellow, deepening slightly

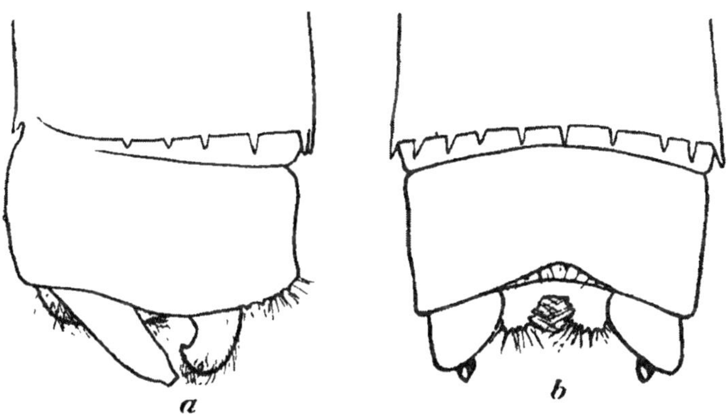

Fig. 138.—Anal appendages of *Ceriagrion nigroflavum* Fras., male. *a.* Left lateral view. *b.* Dorsal view.

on the latter; vertex dark olivaceous; occiput ochreous, rather sharply delimited from the darker vertex; eyes olivaceous, paling to greenish below; beneath head pale yellow. *Prothorax* and *thorax* golden olivaceous on dorsum, paling to greenish-yellow on the sides and pruinosed white beneath. *Legs* yellow, with short black spines. *Wings* hyaline; pterostigma golden yellow, paler at the borders and framed in dark brown nervures, very narrow, oblique, covering one cell; 11 to 12 postnodal nervures in fore-wings,

10 in the hind ; *ab* arising at the level of *ac*. *Abdomen* pale citron-yellow, dorsum of segments 7 to 10 black, this not quite extending to base of the first. *Anal appendages* very short, reddish-brown, the inferiors paler at bases ; superiors barely half the length of segment 10, ungulate, curled slightly down and with a minute tooth beneath apex, which is coated with long hairs ; inferiors broad at base, tapering to a point, sloping strongly up to slightly overlap the superiors ; segment 10 broadly and shallowly notched at apical border, the border here being minutely denticulate, the outer angles simple, not projecting and angulated as in *C. melanurum*.

Female unknown.

Distribution.—Lower Burma and Siam.

Distinguished from *C. fallax* and *C. melanurum* by *ab* rising definitely at the level of *ac* instead of proximal to it ; further distinguished from *C. fallax* by the much shorter superior anal appendage. These two species are again distinguished from *C. melanurum* by the shape of the hinder border of the 10th segment, which is notched and simple, whereas in the latter species it presents a sharp angulated abutment on each side, which bears minute teeth. This is well shown in Dr. Ris's figures (*l. c.*), and is described by Selys as " bout du 10e dont l'échancrure arrondie terminale est bordée de dentelures noires dont l'externe forte."

Type in the author's collection : a paratype will be deposited in the British Museum.

131. **Ceriagrion olivaceum** Laidlaw. (Fig. 139.)

Ceriagrion olivaceum Laid., Rec. Ind. Mus. vol. viii, pp. 345–6 (1914) ; id., ibid. vol. xii, p. 133 (1916) ; id., ibid. vol. xvi, p. 188 (1919) ; Fras., J. Bombay Nat. Hist. Soc. vol. xxix, p. 748 (1923) ; id., Rec. Ind. Mus. vol. xxvi, pp. 428, 492 (1924) ; id., ibid. vol. xxxiii, p. 448 (1931).

Ceriagrion aurantiacum Fras., J. Bombay Nat. Hist. Soc. vol. xxix, p. 748 (1923) ; id., Rec. Ind. Mus. vol. xxvi, pp. 428, 492 (1924) ; id., ibid. vol. xxxiii, p. 448 (1931).

Male.—Abdomen 34–38 mm. Hind-wing 22–23 mm.

Head : labium dirty brown or palest yellow ; labrum and whole of head pale olivaceous brown, a lighter ochreous tint behind eyes and across occiput ; eyes olivaceous brown, darker above than below ; beneath head whitish-brown. *Prothorax* and *thorax* pale olivaceous, paler on the sides and beneath, where it is almost white. *Legs* yellow, with black spines. *Wings* hyaline, palely enfumed in adults ; pterostigma palest brown, pale yellow around the circumference and framed in darker brown nervures, oblique, diamond-shaped, covering one cell ; 15 postnodal nervures in fore-wings, 13 in the hind ; *ab* arises at the level of *ac*. *Abdomen* uniform olivaceous brown, paling to yellowish beneath. *Anal appendages* brown; superiors shorter than segment 10, triangular as seen from the dorsum, and hollowed out on the inner side, the apex rather broadly rounded and coated with long curly hairs; seen from the side shaped like a short pruning-knife, the apex bent under and ending in a small tooth; inferiors sloping strongly up to meet the superiors, broad at base and tapering to a point which bears a small crochet-like hook.

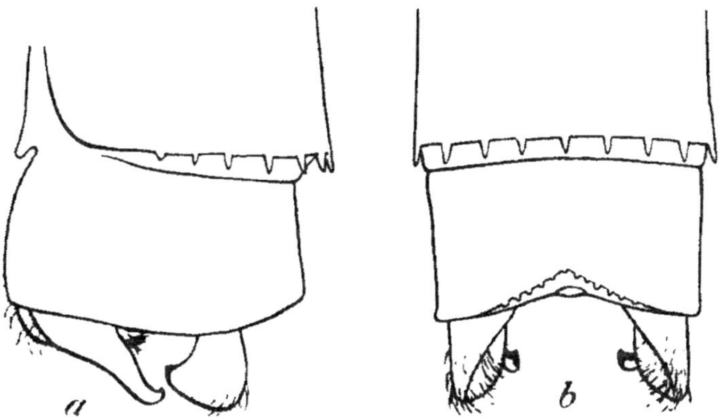

Fig. 139.—Anal appendages of *Ceriagrion olivaceum* Laid., male. *a*. Left lateral view. *b*. Dorsal view.

Female.—Abdomen 33–35 mm. Hind-wing 22 mm.

Exactly similar in colour to the male; nodal index slightly lower; pterostigma paler. *Anal appendages* brownish-red, short, conical; vulvar scale not extending to the end of abdomen, reddish-brown.

Race AURANTIACUM.

The specimens taken in the Nilgiri Wynaad and elsewhere on the Western Ghats which I named *C. aurantiacum* are here treated as a race of *C. olivaceum*. The general build, the uniform coloration, and the anal appendages of the same shape are sufficient justification for this. The following differences are, however, to be noted:—

The ground-colour is a dull mahogany red or reddish-brown rather than olivaceous, the reddish colour being most evident on the abdomen. The labrum is citron-yellow, or even orange in some adults, whilst the anteclypeus and front of frons are pale brown, the intervening postclypeus dark brown; vertex warm brown or dark olivaceous brown. *Thorax* with a golden sheen suffusing the olivaceous. *Legs* yellow, all femora with a black stripe running along the extensor surface. *Wings* with only 12 to 13 postnodal nervures in fore-wings, 10 to 11 in the hind. *Abdomen* olivaceous, the dorsum ferruginous as far as segments 6 or 7, dark brown thereafter. *Anal appendages* similar to those of *C. olivaceum*.

Female not differing from the male in coloration.

Distribution.—Locally very common in many parts of India. The *type* comes from the Kachin Hills, N. Shan States, BURMA, and is in the Indian Museum; I possess specimens from ASSAM and BENGAL. It is a common insect in the Khasia Hills. The southern race *aurantiacum* occurs in many parts of the WESTERN GHATS in large colonies, inhabiting wet jungle and usually well away from water. I noted immense numbers at Devarashola, Nilgiri Wynaad, during August, and also found it swarming in long dry grass in the Empress Gardens, Poona, this colony more closely resembling the type in appearance.

Contrary to the general habits of the genus, they appear to breed in streams. It is to be noted that the "darker, rather green antehumeral stripe on either side of thorax" and the "brown metepisternal area" on the side of the thorax, noted by the author of the original description, are mere artefacts due to decomposition and drying, and are not present in the living insect. Distinguished from all others by the uniform olivaceous colouring.

132. **Ceriagrion cerinorubellum** (Brauer). (Fig. 140.)

Pyrrhosoma cerinorubellum Brauer, Verh. zool.-bot. Ges. Wien, vol. xv, p. 511 (1865).

Agrion cerinorubellum Brauer, Reise d. 'Novara' (Neur.) p. 59 (1866).

Ceriagrion cerinorubellum Selys, Bull. Acad. Belg. (2) vol. xli, p. 526 (1876); Selys, Ann. Mus. Civ. Genova (2) vol. vii (xxvii) (1889); Kirby, Cat. Odon. p. 154 (1890); id., J. Linn. Soc. Lond., Zool. vol. xxiv, p. 564 (1894); Kruger, Stett. Ent. Zeit. pp. 119–20 (1898); Karsch, Ent. Nach. vol. xvii, no. 16 (1891); Laid., Skeat Exped. pt. ii (Odonata), p. 388 (1902); Ris, Abh. Senckenb. naturf. Ges. vol. xxxiv, p. 519 (1913); id., Tijdschr. Ent. vol. lviii, p. 13 (sep.) (1915); Laid., Rec. Ind. Mus. vol. xii, pp. 133–5 (1916); id., ibid. vol. xvi, p. 188 (1919); id., Proc. Zool. Soc. Lond. p. 335 (1920); Fras., J. Bombay Nat. Hist. Soc. vol. xxix, p. 748 (1923); Laid., Spolia Zeylan. vol. xii, p. 369 (1924); Fras., Rec. Ind. Mus. vol. xxvi, pp. 428, 492–3 (1924); Laid., J. Malay Br. Roy. As. Soc. vol. iv, pt. ii, p. 231 (1926); Fras. Rec. Ind. Mus. vol. xxxiii, p. 448 (1931).

Male.—Abdomen 31–33 mm. Hind-wing 20–21 mm.

Head: labium whitish; labrum dark olivaceous, bordered broadly and diffusely with ochreous; anteclypeus and post-clypeus dark greenish olivaceous; base of mandibles, genæ, and front of frons pale greenish-yellow ; frons above and vertex dark reddish-brown ; occiput bright ochreous, sharply defined from the dark vertex ; eyes dark olivaceous above, paler below ; beneath head pale whitish-green. *Prothorax* and *thorax* green, changing to blue on the sides, yellow beneath. *Wings* hyaline ; pterostigma amber-tinted, paler around the circumferenc : and framed in reddish-brown nervures, elongate, diamond-shaped, oblique, covering one cell ; 11 to 12 postnodal nervures to fore-wings, 9 to 10 in the hind ; *ab* arising at the level of *ac* or, in the hind-wings, a shade proximal to that level. *Abdomen* multicoloured, segments 1 and 2 and the basal half of 3, or, in Assam specimens, segments 1 and 2

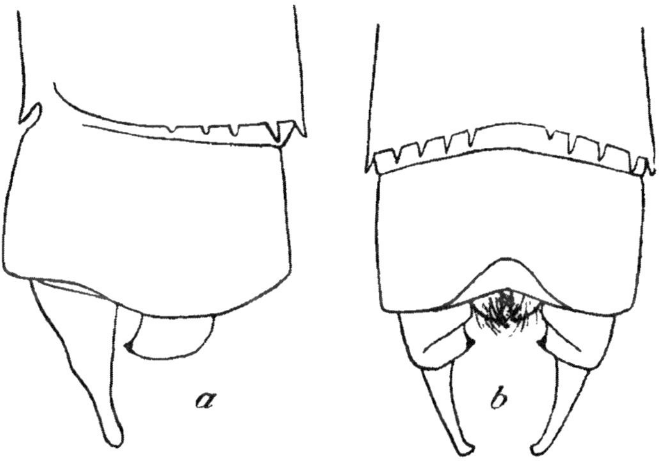

Fig. 140.—Anal appendages of *Ceriagrion cerinorubellum* (Brauer), male. *a*. Left lateral view. *b*. Dorsal view.

only, bright brick-red, the apical half of segment 3 and segments 4 to 6 on the dorsum black : segments 7 to 10 again bright brick-red, but the base of 7 sometimes black on the dorsum ; the sides of segments 3 to 7 blue. *Legs* yellow or bright ochreous, with short black spines. *Anal appendages* reddish-brown, or the inferiors ochreous, tipped with black ; superiors less than half the length of segment 10, short, beaked structures as seen in profile, resembling the outline of a bird's head, subquadrate as seen from above, with rounded outer angle and a small black tooth on the inner ; inferiors much longer, slightly longer than segment 10, broad at base, tapering sinuously to a fine point, and directed rather steeply upwards, the apices curving a little inward and ending in a small black tooth as seen from above.

Female.—Abdomen 31–35 mm. Hind-wing 20–21 mm.

Very similar to the male in colouring; the labrum sometimes entirely ochreous, the upper surface of *head* a darker reddish-brown, the dorsum of *thorax* often suffused with ochreous or golden brown, whilst beneath the thorax in adults is pruinosed white. *Abdomen* similar to the male, but the end segments usually a duller brownish-red. *Wings* and *legs* similar to the male.

Distribution.—Occurs in colonies widely distributed throughout INDIA, CEYLON, BURMA, Malaysia, Sumatra, Java, and Borneo. The whereabouts of the type unknown; paratypes in most national collections. The author found it swarming on a grassy tank in North Coorg during October, and has specimens from Gauhati and Sibsagar, in Assam. The latter specimens are more definitely green on the thorax and the sides are more blue than in southern specimens. The beautiful combination of colours of the abdomen will serve to distinguish it from all other species of the genus.

133. **Ceriagrion azureum** (Selys). (Fig. 141.)

Pseudagrion azureum Selys, Ann. Mus. Civ. Genova, (2) vol. x (xxx), pp. 513–15 (1891).

Ceriagrion cœruleum Laid., Rec. Ind. Mus. vol. xvi, pp. 188–90 (1919); Fras., J. Bombay Nat. Hist. Soc. vol. xxix, p. 748 (1923).

Male.—Abdomen 34–38 mm. Hind-wing 23–25 mm.

Head: labium pale brownish-white; labrum pale turquoise-blue; anteclypeus and genæ a deeper glossy turquoise-blue; postclypeus, frons, and anterior part of vertex dull turquoise-blue, changing to olivaceous on the posterior part of vertex and occiput, but the latter sometimes paler and tinted with ochreous; eyes deep blue above, paler below; beneath head palest blue. *Prothorax* and *thorax* uniform pale blue, paling to white beneath; a small black point on the upper part of humeral suture, and occasionally an obscure point on the upper part of second lateral suture. *Legs* yellow, with short black spines. *Wings* hyaline; pterostigma elongate, inner and outer sides very oblique, diamond-shaped, braced, covering one cell, amber tinted, paler around the circumference, and framed in dark reddish-brown nervures; 12 to 15 postnodal nervures in fore-wings, 12 to 13 in the hind; *ab* arising at the level of *ac*. *Abdomen* uniform glossy turquoise-blue, segment 9 with a dorsal shield-like black spot, the corners of which are continued basally and laterally as far as the base of segment to enclose dorsal spots of the ground-colour; segment 10 with its dorsum black. *Anal appendages* reddish-brown; superiors about half the length of segment 10, conical, coated with long hairs at apex and on inner side, the apex expanding and turned inwards and downwards, with a twist on the long axis of appendage, and ending below in a minute tooth; inferiors of the same length, directed very steeply up towards the superiors, broad at base, tapering to a button-like point turned slightly in as viewed from above.

Female.—Abdomen 36 mm. Hind-wing 23 mm.

Closely similar to the male in colouring, but more robust and with the wings palely enfumed. *Head*: labrum and genæ pale greenish-yellow; anteclypeus, postclypeus, frons, and front of vertex olivaceous, rest of vertex plumbeous, but blue against the eyes; occiput pale ochreous: dorsum of thorax a deeper blue than in the male or even olivaceous; beneath prothorax and thorax pruinosed white in adults. *Wings* with 13 or 14 postnodal nervures in the fore-wing, 12 in the

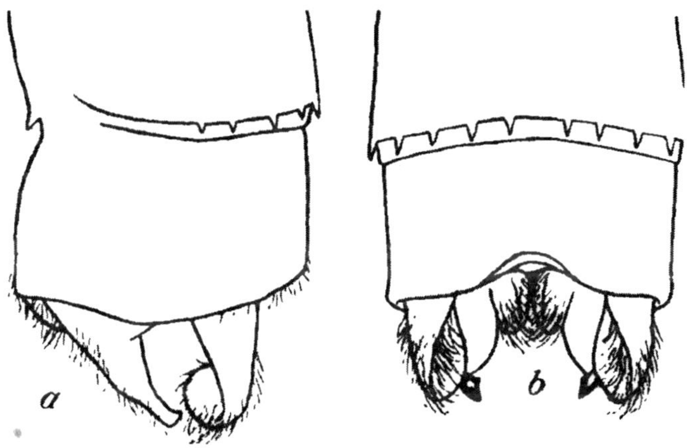

Fig. 141.—Anal appendages of *Ceriagrion azureum* (Selys), male. *a*. Left lateral view. *b*. Dorsal view.

hind: pterostigma brown, paler at the circumference, framed in reddish-brown nervures. *Abdomen* blue, but deepening in tint towards the hinder segments, and without any black marks on segments 9 and 10. *Anal appendages* very short, carneous; vulvar scale robust, extending to the end of abdomen, ochreous or pale reddish-brown.

Distribution.—Bengal (Pashok), Assam (Shillong, in the Khasia Hills), and Burma (Maymyo during May and June, Copabo at the end of September). Mr. Bainbrigge Fletcher found this insect not uncommon on the Ward Lake, Shillong, and Col. F. Wall sent me a series from Maymyo, taken during May and June.

Type in the Genoa Museum: paratypes in the British, Indian and Pusa Museums.

The un-*Ceriagrion*-like colour of this insect, coupled with the fact that the frontal ridge is not quite so well developed as in other species of the genus, evidently led Selys to place this species in the genus *Pseudagrion* on account of its venation; there is no doubt, however, as to where it really belongs, and, as Dr. Laidlaw remarks, it is closely related to *C. olivaceum*. Its beautiful blue colouring will serve to distinguish it from all other species of *Ceriagrion*.

Genus **HIMALAGRION** Fraser (1919). (Fig. 142.)

Himalagrion Fras., Rec. Ind. Mus. vol. xvi, pp. 452, 453 (1919); id., J. Bombay Nat. Hist. Soc. vol. xxix, p. 743 (1923).

Dragonflies of moderately small size and of slender build, coloured black, marked with blue: wings hyaline, long, and narrow, subacute at apices; pterostigma similar in shape in fore- and hind-wings, rather longer than broad, proximal and distal sides oblique, unbraced, covering slightly less than

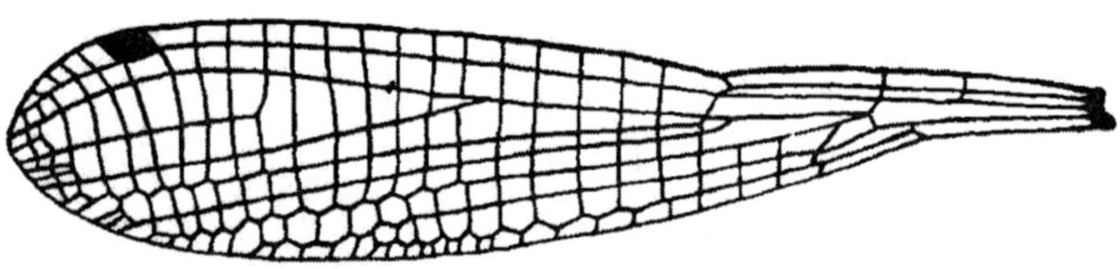

Fig. 142.—Hind-wing of *Himalagrion exclamationis* Fras., male.

one cell; 15 to 16 postnodal nervures in fore-wings; discoidal cell acutely pointed at distal end, with costal side less than half the length of posterior in fore-wing, rather more than half that length in the hind, basal side equal to costal in fore-wing, only half that length in the hind: arc situated at the level of the distal antenodal nervure; *ab* arising very slightly proximal to the level of *ac*, which lies about midway between the two antenodals, complete and continued on as *IA*, but a distinct angulation (medio-anal link) at the junction of the two; *Riv+v* arising well proximal of the oblique nervure descending from the subnode; *IRiii* in continuation of that nervure.

Head moderately narrow, bearing large triangular post-ocular coloured spots; thorax rather small and slim; posterior lobe of prothorax rounded, simple: abdomen cylindrical, rather long and robust; legs medium in length, tibial spines slim, numerous, about 10 to 12 on the hind tibiæ.

Anal appendages short; superiors obtuse, irregular: inferiors broad at base, acutely pointed, forcipate. Genitalia: lamina deeply and narrowly cleft; hamules foliate, quadrate structures; lobe short, rounded, flask-shaped; vulvar scales robust; no ventral spine on segment 8 in the female.

Genotype, *Himalagrion exclamationis* Fraser.

Distribution.—A monotypic genus known only from BENGAL and SIKKIM: life-history unknown, but, from the nature of the country in which it is found, is evidently a riverine breeder. The origin of *ab*, the postocular spots, and the anal appendages suggest a near relationship to the genera *Pseudagrion* and *Ceriagrion*.

134. **Himalagrion exclamationis** Fraser. (Figs. 142 & 143.)

Himalagrion exclamationis Fras., Rec. Ind. Mus. vol. xvi, p. 453 (1919); id., J. Bombay Nat. Hist. Soc. vol. xxix, p. 743 (1923).

Male.—Abdomen 32 mm. Hind-wing 21 mm.

Head: labium white; labrum, genæ, bases of mandibles, anteclypeus, and frons, including the basal joint of antennæ, dull blue, a tiny median basal point and a fine basal line on the labrum and the whole of postclypeus black; vertex and occiput black, marked with large triangular blue postocular spots; eyes dark brown for the upper half, pale olivaceous for the lower, the change from one colour to the other abrupt at the equator of eyes. *Prothorax* black, with the sides broadly, the anterior lobe and the border of the posterior lobe finely, turquoise-blue; posterior lobe broadly rounded, simple. *Thorax* black on dorsum to a little beyond the humeral suture, marked with moderately broad blue antehumeral stripes which are broadened and a little bent outwards above; laterally blue, with a narrow black stripe on the postero-lateral suture broadening below, and a small point of black on the anterior suture; white beneath. *Legs* pale blue, femora broadly black on extensor surface, and the tibiæ equally so on the flexor surface. *Wings* hyaline; pterostigma blackish-brown, framed in thick black nervures; 15 to 16 postnodal nervures to fore-wings, 14 in the hind. *Abdomen* black, broadly marked with blue, or vice versa, as follows:—Segment 1 azure blue with a small dorsal basal black spot, and a short lateral oblique streak; segment 2 blue with a dorsal subapical triangular black spot, with its point directed towards base of segment, an oblique thick lateral streak and the ventral borders narrowly black; segments 2 to 4 with narrow apical black rings at the joints; segment 3 with its ventral borders narrowly black and a rather thick subapical black ring, the rest azure blue; segment 4 similar, but the fine apical and broad subapical rings fused together, enclosing a mere spot of the ground blue on each side; segment 5 similar, but the apical black showing signs of extending basalwards laterally; segment 6 entirely black save for a narrow basal blue ring interrupted on the dorsum; segment 7 black, its apical two-thirds, or slightly more, pale azure blue, this tapering rapidly from apex to base of segment, so that the sides at the apical end are more broadly blue, the black indenting the blue at the sides here; segments 8 and 9 entirely pale azure blue, the former, in some specimens, with a fine stippled or broken ventral black line; segment 10 black, with a small ventro-lateral blue spot and the dorsal apical notch finely outlined in the same colour. *Anal appendages* black; superiors about half the length of segment 10, digitate, the apical border emarginate, the upper end of notch prolonged back slightly; inferiors slightly longer, broad at the base, tapering to a rather obtuse point, which is directed

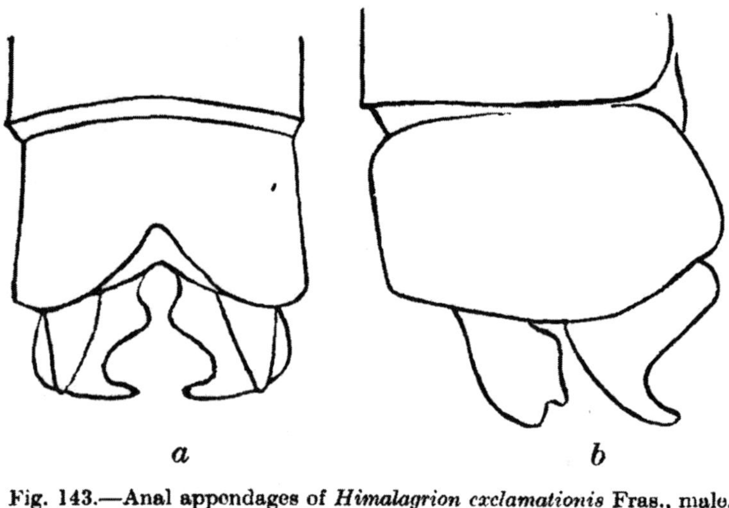

Fig. 143.—Anal appendages of *Himalagrion exclamationis* Fras., male. *a*. Dorsal view. *b*. Right lateral view.

backward and inward, the tapered portion directed obliquely upward as seen from the side, and forcipated and nearly meeting as viewed from above.

Female.—Abdomen 31 mm. Hind-wing 32 mm.

Closely similar to the male, the abdominal markings alone differing as follows:—Segment 1 with the lateral oblique stripe confluent with the dorsal basal spot; segment 2 with the apical dorsal triangular spot confluent dorsally with the apical ring and extended as far as the base of segment, enclosing between itself and the lateral oblique black stripe a narrow hook-shaped blue one; segments 3 to 6 entirely black dorsally, and with a lateral apical spot of blue and a lateral stripe of the same colour which tapers from the base towards the spot, thus forming lateral "notes of exclamation" on each segment; segment 7 with or without a similar marking and, when present, the lateral stripe very fine and even throughout; segment 8 black, with a narrow blue apical ring confluent with a large apico-lateral blue spot; segment 9 blue, marked basally with a black spot shaped like a coronet, and a tiny mid-dorsal apical black triangle; segment 10 entirely blue. *Anal appendages* black, small, conical; vulvar scales robust, marked laterally with whitish blue.

Distribution.—Darjeeling District, BENGAL. Four pairs from Sitong, near Mangpu, taken in July, at about 4,000 ft.

Type and paratypes in the Indian Museum, one male in the Fraser collection.

The generic characters, with the bold and beautiful blue markings so strongly contrasted against the black background, will serve to distinguish this species from any *Pseudagrion* or *Ceriagrion*.

Genus **ACIAGRION** Selys (1891). (Fig. 144.)

Pseudagrion, pars, Selys, Bull. Acad. Belg. (2) vol. xlii, p. 509 (1876).
Aciagrion Selys, Ann. Mus. Civ. Genova, (2) vol. x (xxx), pp. 509–11 (1891); Laid., Rec. Ind. Mus. vol. xvi, p. 184 (1919); Fras., J. Bombay Nat. Hist. Soc. vol xxix, p. 748 (1923); Laid., Proc. U.S. Nat. Mus. vol. lxvi, Art. 10, pp. 1–5 (1924).

Dragonflies of small size and usually of very slender build, coloured blue or violaceous, marked with black, less commonly pale and inconspicuously coloured, non-metallic; wings very narrow and subacute at apex, hyaline; pterostigma differing in the fore- and hind-wings, that of fore-wing nearly double the size of the hind, narrow, diamond-shaped, distal and proximal sides very oblique in fore-wing, but the hind with the distal side much less oblique than the proximal, poorly braced, covering less than one cell; postnodal nervures numbering 10 to 13 in fore-wings; discoidal cell acutely pointed at distal end, with costal side less than half the length of posterior in fore-wing, rather more than half the length in the hind, basal side equal to the costal in fore-wing, only half the length of costal in the hind, distal side almost in line with costal; arc situated at the level of the distal antenodal nervure; *ab* taking origin at the point where *ac* meets the hinder border of wing, and continued on as the nervure *IA*, but their junction strongly angulated; *Riv+v* arising well before, or just before, the level of the nervure descending from the subnode; *IRiii* in continuation of that nervure; *ac* situated about midway between the two nervures.

Head narrow, usually bearing triangular or elongate postocular areas or spots; eyes globate; thorax rather slim; posterior lobe of prothorax rounded, simple, without hooks or spines in the female; abdomen variable, of moderate or of great length, and moderately stout or exceedingly slim and cylindrical; legs short, tibial spines moderately long, about five in number on the hind pair; anal appendages variable, very short and insignificant, occasionally notched at apex or squared, inferiors very minute. Genitalia: lamina deeply but narrowly cleft; hamules small quadrate plates meeting across the middle line; lobe small, depressed, slipper-shaped; penis broad at apex, where it bears a short spine as seen in profile, the end curling strongly over dorsum to embrace the stem of the organ by a shallow or deep bifurcation at the extreme end; a short spine on each side and another on the dorsum of stem beneath the arch formed by the overhanging end segment; vulvar scales robust; a short spine beneath at apex of segment 8.

Genotype, *Aciagrion hisopa* Selys.

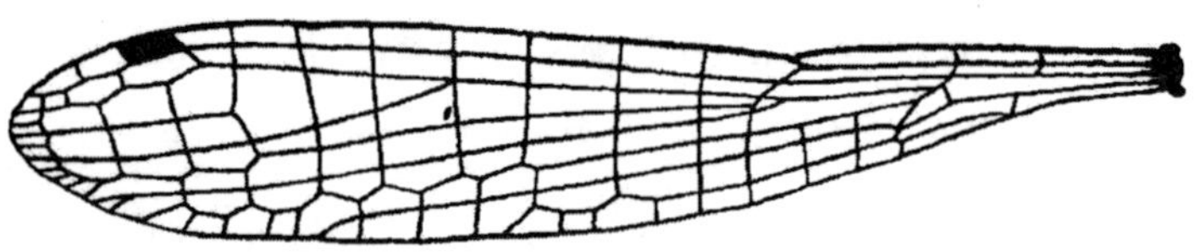

Fig. 144.—Hind-wing of *Aciagrion hisopa* Selys, male.

Distribution.—Continental Africa, CEYLON, CONTINENTAL INDIA save in the desert areas, BURMA, Malaysia, Borneo, and Australia.

Species of the genus are locally common and, where found, are usually exceedingly abundant. Primarily they are marsh-breeders, but many are found breeding in small tanks and shallow lakes, especially if these have a marshy border. Their breeding places suggest a primitive origin; Laidlaw places them as a specialized end branch of the *Enallagma* series, but the long petiolation of the wings hardly accords with such a theory.

Key to the Species of Aciagrion.

1.	Ground-colour pale brown, without black markings on head and thorax; species found in dry grassy lands or dry deciduous forests or scrub jungle	*pallidum* Selys, p. 344.
	Ground-colour blue, with black markings on head, thorax, and abdomen; species found in wet evergreen forests or marshes	2.

2. Very small, very slender species; abdominal segment 8 with a black elongate dorsal triangular mark, its base at apex of segment, apex nearly reaching base of segment .... *occidentale*, Laid., [p. 335.
Slightly larger, more robust species; abdominal segment 8 without any black marking or with a short lateral black stripe on each side .... 3.

3. Large species, with abdomen over 30 mm. in length .... 4.
Smaller species, with abdomen not more than 25 mm. in length .... 5.

4. Superior anal appendages bifid at apex; postocular spots connected across middle line of occiput .... *olympicum* Laid., [p. 337.
Superior anal appendages not bifid at apex; postocular spots not connected across occiput .... *azureum* Fras., p. 339.

5. Posterior lobe of prothorax not crenate behind; black markings on dorsum of segments 2 and 10 restricted .... *hisopa* (Selys), p. 340.
Posterior lobe of prothorax trilobate behind; black markings on dorsum of segments 2 and 10 very broad .... *approximans* (Selys), [p. 342.

135. Aciagrion occidentale Laidlaw. (Fig. 145.)

Aciagrion hisopa (Selys) ? race *occidentalis* Laid., Rec. Ind. Mus. vol. xvi, p. 186 (1919).
Aciagrion paludensis Fras., J. Bombay Nat. Hist. Soc. vol. xxviii, pp. 698. 699 (1922).
Aciagrion occidentalis Fras., J. Bombay Nat. Hist. Soc. vol. xxix, p. 749 (1923); id., Rec. Ind. Mus. vol. xxvi, pp. 428, 491, fig. 3, iii, p. 496 (1924); id., ibid. vol. xxxiii, p. 448 (1931).
Aciagrion hisopa Laid., Spolia Zeylan. vol. xii, p. 368 (1924).

Male.—Abdomen 23–24 mm. Hind-wing 15–16 mm.

Head: labium white; labrum pale blue, with a median spot and its base narrowly black; genæ, anteclypeus, and frons pale azure blue: postclypeus glossy black, rest of head black, marked on occiput with transversely elongate postocular blue spots, narrowly confluent across the middle of occiput or joined by a narrow blue stripe; eyes bottle-green above, with a small black cap, palest blue beneath, changing through pale greenish-yellow to the darker shade above. *Prothorax* black on dorsum, azure blue laterally, anterior lobe entirely blue, whilst the posterior is very finely lined with yellow and its border subcrenate. *Thorax* pale azure blue, the dorsum steely black as far as midway between the humeral and first lateral sutures, marked with a narrow greenish-yellow antehumeral stripe on each side, extending the whole length of dorsum; laterally a tiny point of black on the upper part of the second lateral suture: beneath white. *Legs* palest blue, femora broadly black on extensor surface, whilst the tibiæ have a short streak of the same colour on their proximal third or half. *Wings* hyaline, very narrow; pterostigma of fore-wing nearly double the size of that of hind-wing, but coloured similarly in the two wings, blackish-grey; the costal and distal borders framed broadly in white, covering about three-fourths of a cell; 9 postnodal nervures in fore-wings, 8 in the hind. *Abdomen* extremely slender and comparatively long, pale azure blue at sides of segments 1 to 3, pale yellow from 4 to 7, blue from 8 to 10, broadly marked with black on segments 1 to 8 as follows:—Segment 1 with a broad dorsal marking from base to apex; segment 2 with a broad stripe, slightly expanded subapically, and then slightly constricted to join a narrow apical ring; segments 3 to 7 broadly bronzed black on dorsum, but the pale colour of sides

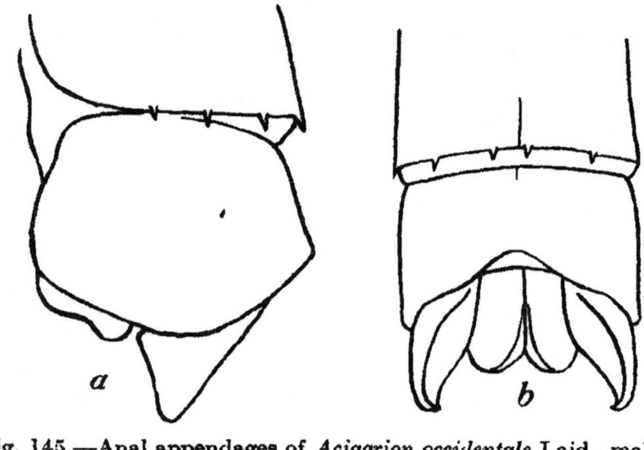

Fig. 145.—Anal appendages of *Aciagrion occidentale* Laid., male. *a*. Left lateral view. *b*. Dorsal view.

invading the base narrowly to form small paired spots as viewed dorsally, whilst on 7 the dorsal band expands continuously as far as apex of segment; on the others it expands abruptly subapically, and then contracts again to join a narrow apical ring; segment 8 with a narrow dorsal triangle of black, the base of the triangle on apical border of segment, its point extending variably nearly or quite up to base of segment; segment 9 entirely blue; 10 with a small X-shaped black dorsal spot, broadly blue on the sides. *Anal appendages* black, superiors about half the length of segment 10, conical as seen from the side, triangular as seen from above, the outer side straight, the inner deeply concave, the apex acute and very slightly turned in; inferiors small rounded tubercles less than half the length of superiors and barely visible from above.

Female.—Abdomen 24 mm. Hind-wing 16 mm.

Somewhat similar to the male, but much more robust, and with a stouter abdomen; markings entirely similar to those of the male, except on the three terminal abdominal segments; segment 8 with the dorsal marking much broader and of almost even width throughout; 9 with a basal dorsal spot extending for one-third the length of segment; 10 entirely blue, or with a small subdorsal apical black spot on each side; the ground-colour of face pale yellow, as also the occipital stripe and antehumeral stripes; the eyes olivaceous, paling to yellow beneath, and with a cap of black ringed outwardly with pale brown. *Anal appendages* very small, conical, black; vulvar scales not conspicuous, pale blue.

Distribution.—Throughout S. INDIA and CEYLON, very common on the foot-hills to east of the WESTERN GHATS from Coorg to Travancore, where in the early months of the year marshes and the jungle in their neighbourhood swarm with countless numbers of these insects. In spite of its delicate build, it takes to migration readily, rising high in the air and taking advantage of its light weight and the use of air-currents. Specimens have been taken over 40 miles out at sea off the western shores of India and Ceylon. The male is easily distinguished by its attenuated abdomen and the triangular black spot on the dorsum of segment 8.

Type in the Indian Museum; paratypes in the British Museum and author's collections.

136. **Aciagrion olympicum** Laidlaw. (Fig. 146.)

Aciagrion olympicum Laid., Rec. Ind. Mus. vol. xvi, pp. 171, 184–6, fig. 1 (1919); id., Proc. U.S. Nat. Mus. vol. lxvi, Art. 10, pp. 4, 8 (1924); Fras., J. Bombay Nat. Hist. Soc vol. xxix, p. 749 (1923).

Male.—Abdomen 32–34 mm. Hind-wing 20–22 mm.

Head: labium palest brown; labrum carneous, its base with a small medial black spot or, more rarely, a fine black line; clypeus, bases of mandibles, genæ, and frons as far posterior as the roots of antennæ carneous or palest blue, anteclypeus finely black at base, postclypeus broadly black along its anterior border, this sometimes confluent with a medial basal black point; vertex and occiput black, the latter with large transversely oval bluish-green postocular spots connected by a narrow isthmus of blue across the middle line; beneath head palest brown; eyes brown above, pale below. *Prothorax* broadly black on dorsum, pale yellow or carneous laterally and across the anterior lobe. *Thorax* black on dorsum, marked with rather broad pale blue antehumeral stripes, these latter, and the margins of the alar sinus and mid-dorsal carina, lying in a narrow halo of ferruginous or ochreous; laterally pale blue, changing to pale yellow below and beneath; upper end of postero-lateral suture finely black. *Legs* creamy white, femora black on external surface, the two anterior pairs of tibiæ on the outer surface. *Wings* hyaline; pterostigma pale brown, framed in yellow and darker bordering nervures, that of fore-wing larger than that in the hind, covering less than one cell; 12 to 13 postnodal nervures in fore-wings, 11 in the hind. *Abdomen* pale blue at basal and anal segments, pale yellow on the medial, marked with black as follows:—Segment 1 with a large dorsal spot not quite extending to the apical end; segment 2 with a small subapical dorsal triangle of black lying in a halo of ochreous, its apex sometimes prolonged as a narrow mid-dorsal stripe towards

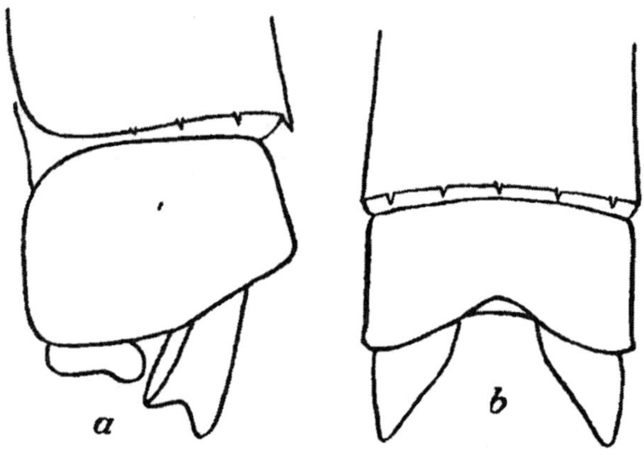

Fig. 146.—Anal appendages of *Aciagrion olympicum* Laid., male. *a.* Left lateral view. *b.* Dorsal view.

the base of the segment; segments 3 to 7 with broad dorsal black stripes, which broaden to encircle the apical ends of segments but do not quite reach the base, so that the ground-colour here forms narrow basal rings; segments 8 to 10 azure blue, the latter with a narrow X-shaped dorsal black marking. *Anal appendages* carneous or palest blue tipped with black; superiors as long as segment 10, triangular as viewed from above, the external border in line with the abdomen, the inner border very oblique, base broad, apex pointed; viewed from the side it is digitate in outline, but rather deeply emarginate at apex, the inner surface excavate; inferiors considerably shorter than superiors, broad at base, ending in an angulate process directed obliquely up towards the superiors, bluntly rounded at apex.

Female.—Abdomen 32 mm. Hind-wing 22 mm.

Very similar to the male in colour and markings, the latter differing only on the *abdomen* as follows :—Segment 1 with dorsal marking broader and extending to apical border of segment ; segment 2 with the dorsal marking extending from base to apex of segment, the apical end narrowing and then becoming broadly confluent with an apical black annule ; segment 8 with a black dorsal spot extending from base along mid-dorsum to bifurcate into two processes at the middle of segment or as an elongate triangle with its base at the middle of segment and apex at base ; segment 9 with two black, dorsal, basal triangles confluent or not at a point at base of segment ; segment 10 unmarked. *Anal appendages* short, conical, pointed at apex, black ; vulvar scales moderately robust, carneous, extending to end of abdomen.

Distribution.—E. HIMALAYAS, Darjeeling District and Sikkim. The *type*, in the Indian Museum collection, comes from Sureil, above Mangpu, Sikkim, 5,500 ft. I took this insect sparingly at Mangpu, during May, haunting swampy ground at the foot of jungly slopes. It is a marsh-breeder, like *A. occidentalis.*

It differs from other species by its great length and by the characteristic markings on segment 2 of the male and segments 8 and 9 of the female. It is closely related to the next species, *A. azureum*, which, however, has large isolated postocular spots and a lower nodal index.

137. **Aciagrion azureum** Fraser.

Aciagrion azureum Fras., Mem. Dept. Agric. India (Ent.), vol. vii, no. 7, p. 51 (1922) ; id., J. Bombay Nat. Hist. Soc. vol. xxix, p. 749 (1923) ; Laid., Proc. U.S. Nat. Mus. vol. xxv, Art. 10, pp. 5, 8 (1924).

Male.—Abdomen 30 mm. Hind-wing 20 mm.

Head : labium whitish ; labrum, bases of mandibles, genæ, anteclypeus, and frons pale yellow ; rest of the head black, except for very large isolated blue postocular spots ; eyes bluish above, pale yellow below. *Prothorax* black on dorsum, pale blue laterally. *Thorax* broadly black on dorsum, marked with narrow pale blue antehumeral stripes ; laterally pale blue, changing to pale yellow on the sides and beneath. *Wings* hyaline ; pterostigma reddish-brown, covering about one cell, that of fore-wing considerably larger than that of the hind ; 10 to 11 postnodal nervures in fore-wings, 9 in the hind. *Abdomen* : segment 1 azure blue, with a large quadrate black spot covering the basal two-thirds dorsally ; segment 2 broadly blue at the sides, with a broad black marking on dorsum shaped like a thistle-head ; segments 3 to 7 with broad dorsal black stripes, which end in a point at base of segments and expand subapically to contract again and become confluent with narrow annules at the apex; on segment 7, however, the band expands gradually as far as the apex of segment; segments 8 to 10 azure blue, without any black markings. *Anal appendages* short, very similar to those of *A. olympicum*, but more conical as seen from the dorsum, and not bifid at apex as seen from the side.

Female unknown.

Distribution.—ASSAM only. The *type*, from Margherita, Assam, taken in May, is now in the British Museum collection. Closely related to *A. olympicum*, from which it is distinguished by having the large postocular spots isolated and not connected by a mid-occipital stripe, by the lower nodal index, and by the superior anal appendages being non-bifid at apex.

138. **Aciagrion hisopa** (Selys). (Figs. 144 & 147.)

Pseudagrion hisopa Selys, Bull. Acad. Belg. (2) vol. xlii, p. 509 (1876); Kirby, Cat. Odon. p. 153 (1890).

Aciagrion hisopa Selys, Ann. Mus. Civ. Genova, (2) vol. x (xxx), pp. 511, 512 (1891); Laid., Rec. Ind. Mus. vol. xvi, p. 172 (1919); Fras., J. Bombay Nat. Hist. Soc. vol. xxix, p. 749 (1923); Laid., Spolia Zeylan. vol. xii, p. 368 (1924); id., Proc. U.S. Nat. Mus. vol. lxvi, Art. 10, pp. 5, 6 (1924); Fras., Rec. Ind. Mus. vol. xxxiii, pp. 448, 464 (1931).

Aciagrion hisopa race *krishna* Fras., J. Bombay Nat. Hist. Soc. vol. xxvii, p. 543 (1921).

Male.—Abdomen 24–26 mm. Hind-wing 15–16 mm.

Head: labium palest yellow; labrum, bases of mandibles, genæ, anteclypeus, and frons broadly pale blue or greenish-blue; postclypeus black; vertex and occiput black, the latter with a broad blue postocular area confluent across the mid-occiput, its anterior border forming a straight line from eye to eye; eyes capped with black above, pale greenish at sides and yellow below; beneath head whitish. *Prothorax* broadly black on dorsum, the anterior lobe and sides broadly azure blue; posterior lobe rounded, simple. *Thorax* broadly black on dorsum to about half-way between humeral and antero-lateral sutures, marked with narrow lilaceous violet antehumeral stripes; laterally lilaceous violet, fading to pinkish beneath; the lower part of sides and beneath pulverulent white in old specimens. *Legs* creamy white, with a few very short slim black spines; femora and anterior pair of tibiæ black on the outer side. *Wings* hyaline; pterostigma considerably larger in the fore-wing than in the hind, its posterior side shorter than costal, and outer side very oblique, black, framed finely in yellow and enclosed in thick black nervures, covering about half a cell; 12 postnodal nervures in fore-wings, 10 in the hind. *Abdomen* pale blue at the sides, marked broadly on dorsum with black; segment 1 with dorsum broadly black; segment 2 with the dorsal marking rounded basally, not quite extending to base of segment, contracted laterally, and then expanded again near apex of segment, where it joins a narrow apical black ring ; segments 3 to 7 with similar but elongated stripes, that on 7 expanding continuously as far as apex of segment ; segments 8 to 10 azure blue, 8 with a short narrow black stripe on each side not extending as far as apical end of segment, 9 unmarked, 10 with a small cross-shaped dorsal marking. *Anal appendages* black ; superiors roughly triangular as seen from above, the outer side in line with the abdomen, inner oblique, apex covered with long curly hairs, deeply bifid as viewed in profile ; inferiors broad at base, then narrowed, cylindrical, obtuse at apex, much shorter than superiors, which are nearly as long as segment 10.

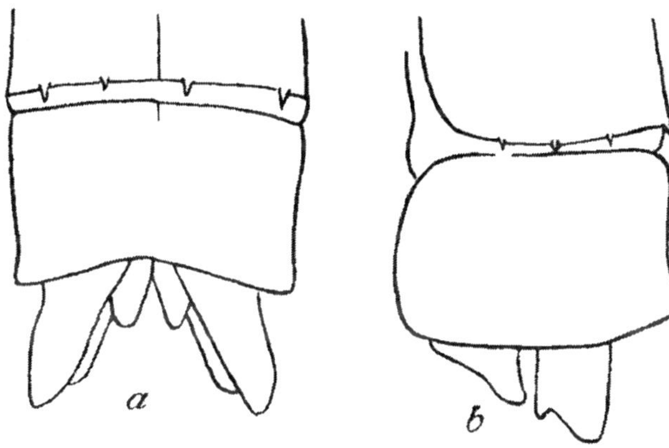

Fig. 147.—Anal appendages of *Aciagrion hisopa* (Selys), male. *a.* Dorsal view. *b.* Left lateral view.

Female.—Abdomen 26 mm. Hind-wing 17 mm.

Closely similar to the male in colouring and markings ; the labrum narrowly black at base ; postocular spots narrow, transversely elongate, hardly broader than the isthmus connecting them across occiput ; eyes greenish, broadly capped with black above. *Thorax* greenish-blue at sides, as well as the narrow antehumeral stripes ; abdominal segments 1 to 7 similar to the male, segments 8 and 9 broadly black on dorsum, the latter with a narrow azure blue apical ring ; segment 10 azure blue, narrowly black at base. *Wings* with pterostigma paler or olivaceous, more nearly equal in the two wings ; 10 postnodal nervures in fore-wings, 9 in the hind. *Anal appendages* short, barely as long as segment 10, bluish, conical, pointed at apex ; vulvar scales yellow, extending beyond end of abdomen.

Distribution.—INDIA, BURMA, and CEYLON ; Malaysia. The *type*, in the Selysian collection, is from Pulo Besoar,

Malay States. Occurs in large colonies in widely separated districts and at all altitudes up to over 7,000 ft. It is common on all streams in the Nilgiris and Ceylon Hills, and is equally common throughout the Deccan. I found it very common in paddy-fields around Bangalore.

The above description is made from specimens from Mahableshwar, Satara, and Poona districts, and is a description of the race named *krishna*. A number of such races exist, and differ only in the markings of the hinder segments of the abdomen, so that it does not seem worth while to separate them; in the type the end segments are unmarked with black. The labrum is narrowly black at base and has a tiny median black spot; this is visible only in the female of Mahableshwar specimens. Breeds alike in rivers and ponds.

139. **Aciagrion approximans** (Selys). (Fig. 148.)

Pseudagrion microcephalum race *approximans* Selys, Bull. Acad. Belg. (2) vol. xlii, pp. 507, 508 (1876).

Pseudagrion approximans Kirby, Cat. Odon. p. 153 (1890).

Aciagrion approximans Selys, Ann. Mus. Civ. Genova, (2) vol. x (xxx), p. 512 (1891); Laid., Rec. Ind. Mus. vol. xvi, pp. 172, 187 (1919); id., Proc. U.S. Nat. Mus. vol. lxvi, pp. 4, 8 (1924).

Aciagrion tillyardi Laid., Rec. Ind. Mus. vol. xvi, pp. 172, 187 (1919); Fras., J. Bombay Nat. Hist. Soc. vol. xxix, p. 749 (1923).

Enallagma assamica Fras., J. Bombay Nat. Hist. Soc. vol. xxvi, pp. 877, 888 (1919).

Male.—Abdomen 22–24 mm. Hind-wing 17 mm.

Head: labium pale yellow; labrum pale blue, narrowly black at base; anteclypeus, bases of mandibles, genæ, and frons palest blue; postclypeus black; vertex and occiput black, the latter with a small postocular blue spot on each side, narrowly connected across the mid-line by a stripe of the same colour; eyes capped with black above, bottle-green at sides and below, sometimes marked with a narrow equatorial band of black; beneath head palest green or white, black against the eyes. *Prothorax* black on dorsum, blue obscured with pruinescence at the sides. *Thorax* broadly black on dorsum to nearly as far lateral as the antero-lateral sutures, marked with narrow blue antehumeral stripes, laterally pale blue, obscured in old specimens with pruinescence, as also below. *Legs* white, femora black on extensor surface. *Wings* hyaline; pterostigma brownish, paler at circumference, and framed in black nervures, covering less than one cell, of practically the same shape, size, and colour in the fore- and hind-wings; 10 postnodal nervures in fore-wings, 9 in the hind. *Abdomen*: segment 1 broadly black on dorsum, blue obscured with pruinescence at the sides; segment 2 broadly black on dorsum, this band expanding subapically, pale blue or bluish-green laterally; segments 3 to 7 broadly black on dorsum and connected with narrow apical black rings on each segment, pale blue laterally; segments 8 and 9 azure blue; segment 10 black, emarginate at its apical border. *Anal appendages* black; superiors nearly as long as segment 10, upper and lower borders parallel, apical border deeply notched as seen in profile, deeply hollowed on the inner apical side, and furnished with long hairs, triangular in outline as viewed from above; inferiors shorter, broad and rounded, broadly notched, and with the inner border of notch forming a robust tooth, directed inward and upward.

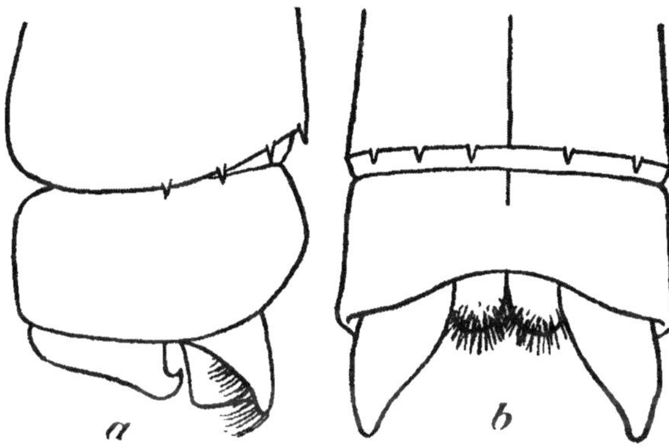

Fig. 148.—Anal appendages of *Aciagrion approximans* (Selys), male. *a*. Left lateral view. *b*. Dorsal view.

Female.—Abdomen 22–24 mm. Hind-wing 15·5–17·5 mm. Closely similar to the male in markings, but the ground-colour pale bluish-green, less often pruinosed; antehumeral stripes pale greenish-yellow, as also the postocular marking; eyes bluish-green at the sides, greenish-yellow below, the equatorial band more marked than in the male. *Abdomen* with segments 8 and 9 broadly black on dorsum, the sides only azure blue, whilst segment 10 is entirely blue. *Wings* similar, but the pterostigma more brownish-yellow; 11 to 12 postnodal nervures to fore-wings, 9 to 10 in the hind. *Anal appendages* bluish, short, conical, pointed at apex; vulvar scales pale, extending to a little beyond end of abdomen; spine at apex of segment 8 very robust.

Distribution.—*Type* in the Dresden Museum, without mention of locality, but indicated by Selys, with some doubt, as being from Malaysia. A very common species in the Khasia Hills, Assam, at 5,000–6,000 ft. Paratypes in the British, Indian, and Pusa Museums, and also in the author's collection. Type of *E. assamica* in the British Museum, of *A. tillyardi* in the Indian Museum. Differs from *A. olympicum* and *A. azureum* by its shorter and more robust abdomen, and by having segment 10 black above. From *A. hisopa*, to which it is very closely related, the broader markings on segments 2 and 10 and the shape of the hinder border of the posterior lobe of the prothorax, which is trilobate in *A. approximans* and simple and rounded in *A. hisopa*, will serve to separate it.

140. **Aciagrion pallidum** Selys. (Fig. 149.)

Aciagrion pallidum Selys, Ann. Mus. Civ. Genova, (2) vol. x (xxx), pp. 512, 513 (1891); Laid., Rec. Ind. Mus. vol. xvi, pp. 172, 186 (1919); Fras., J. Bombay Nat. Hist. Soc. vol. xxvii, p. 543 (1921); id., ibid. vol. xxix, p. 749 (1923); id., Rec. Ind. Mus. vol. xxvi, p. 491 (1924); Laid., Proc. U.S. Nat. Mus. vol. lxvi, Art. 10, pp. 4, 7, 8, pl. i, figs. 2–7 & 9 (1924); Fras., Rec. Ind. Mus. vol. xxxiii, p. 448 (1931).

Male.—Abdomen 31 mm. Hind-wing 18 mm.

Head: labium white; labrum and anteclypeus pale brown; bases of mandibles, genæ, postclypeus, and frons pale carneous, almost white; vertex and occiput palest brown, with large

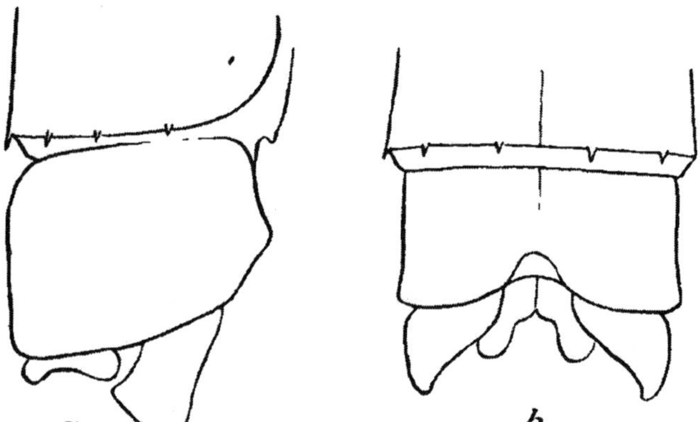

a b

Fig. 149.—Anal appendages of *Aciagrion pallidum* Selys, male. *a*. Right lateral view. *b*. Dorsal view.

triangular pale blue postocular spots narrowly connected by a median stripe of the same colour; eyes brown above, creamy white at sides and below, and with a pale brown equatorial stripe lying a little below the brown polar cap; beneath head whitish. *Prothorax* brown on dorsum except anterior lobe, which is pale blue, pale whitish-brown laterally; posterior border of posterior lobe sinuous. *Thorax* warm brown on dorsum nearly as far lateral as postero-lateral suture, with a narrow ill-defined pale blue antehumeral stripe on dorsum and a similar one lying midway between the lateral sutures; white beneath. *Legs* white, the distal ends of anterior two pairs of femora striped with black; five short fine black spines on hind tibiæ. *Wings* hyaline, very narrow and pointed at apices; pterostigma pale brown, covering less than one cell, very small; 10 to 12 postnodal nervures to fore-wings, 9 to 10 in the hind. *Abdomen* very long and very slender, dilating somewhat at the anal end; segment 1 very pale blue, almost white, with a large dorsal black spot, not extending to apical end of segment; segment 2 white, palest blue near the base and broadly black on dorsum except at extreme base, the black confluent with a narrow apical black ring; segments 3 to 7 similar to 2 but white laterally, and with narrow black apical rings; segments 8 to 10 pale azure blue, unmarked. *Anal appendages* carneous; superiors rather shorter than segment 10, triangular as viewed from above and pointed at apex, truncate and squared at apex as seen from the side, hollowed out beneath; inferiors considerably shorter, hardly visible from the side, broad at base, broadly and very shallowly notched, the end sloping upward, obtuse.

Female.—Abdomen 30 mm. Hind-wing 20 mm.

Closely similar to the male in colouring, the only differences being found in the dorsal markings of the abdomen; segments 1 and 2 more blue at the sides, segment 2 with an arrow-head marking on dorsum, black lying in a halo of reddish-brown, the point of the arrow directed apically and just touching an apical black annule; segments 3 to 7 warm reddish-brown on dorsum, this colour finely bordered with black on segments 3 and 4, becoming more black at apical end of 5 and almost completely black on segments 6 and 7, the marking a little reddish at the base on segment 6, and changing from black to reddish apically on 7; segments 8 to 10 almost white, sometimes pinkish or palest blue; segment 9 with two elongated narrow, dorsal, basal black points not quite extending as far as apex of segment. *Anal appendages* very short, white or pinkish, conical, pointed at apex. *Wings* similar to the male.

Distribution.—The species is peculiarly scattered, colonies cropping up in widely separated areas. It is essentially an insect of dry areas, however, and is especially abundant in parts of the DECCAN and in the dry zones of ASSAM and BURMA, whilst it is unknown, save as an occasional migrant, in wet areas such as the Western Ghats, where I have only once taken a specimen. It occurs in vast numbers in parts of Poona and Satara Districts, and may be seen to rise in scores from the long dry grass in which it shelters, and in which it is remarkably inconspicuous on account of its cryptic colouring. Its colour will serve to distinguish it from all other species of the genus.

Type in the Genoa Museum; paratypes in most national and private museums.

Genus **ISCHNURA** Charpentier (1840). (Fig. 150.)

Ischnura Charp., Lib. Europ. p. 20 (1840); Selys, Bull. Acad. Belg. (2) vol. xli, p. 259 (1876); Kirby, Cat. Odon. p. 140 (1890); Laid., Rec. Ind. Mus. vol. xii, pp. 124–32 (1916); id., ibid. vol. xvi, pp. 172–4 (1919); Ris, Ann. S. African Mus. vol. xviii, p. 333 (1920); Fras., J. Bombay Nat. Hist. Soc. vol. xxix, p. 745 (1923); Needham, Zool. Sinica, vol. xi, ser. A, p. 281 (1930).
Micronympha Kirby, Cat. Odon. p. 140 (1890).

Dragonflies of small size and slender build, body rather more robust and abdomen rather shorter than in the genus *Ceriagrion*; colours non-metallic, usually bright reddish-orange marked more or less with black, or blue or green marked with black; females polychromatic; wings hyaline; pterostigma of males shaped and coloured differently in fore- and hind-wings, that of the fore-wings very variable in the species, braced, covering one cell or less; postnodal nervures numbering from 8 to 9 or 10 in the fore-wings, 6 or 7 in the hind; discoidal cell acutely pointed at the distal end, with costal side about half the length of posterior side in fore-wing and two-thirds the length of that side in the hind, basal side about one and

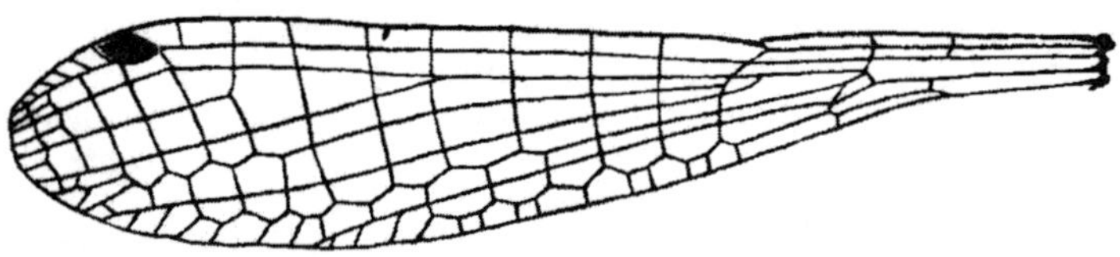

Fig. 150.—Hind-wing of *Ischnura senegalensis* (Ramb.), male.

a half times the length of costal in fore-wing, less than half the length in the hind, distal side very oblique; sectors of arc arising from the lower end of arc, divergent from origins; arc situated at, or a little distal to, the level of the distal antenodal nervure; *ab* always present and complete, arising well proximal to the level of *ac*, which usually lies midway between the two antenodal nervures; *ab* continued on as the nervure *IA*, but an angulation (medio-anal link) present at the junction of the two; *Riv+v* arising well before the oblique nervure descending from the node; *IRiii* in continuation of that nervure.

Head narrow, without any frontal ridge; postocular spots always present; thorax short and robust; posterior lobe of prothorax usually simple, but sometimes prolonged at its centre; anterior border of thorax usually with a pair of small horseshoe-shaped hooks. Abdomen moderately short and robust, tumid at base in many species, abruptly narrowed and cylindrical from segment 3, then dilating again from the apical end of segment 7 to the end; segment 10 with two prominent tubercles on dorso-apical border; legs very short, femoral and tibial spines few and short; anal appendages very variable in the species; segment 8 in the female with a ventral apical spine.

Genitalia: lamina narrowly and very deeply cleft; hamules broad quadrate plates, notched anteriorly and meeting across the middle line; lobe small, partially hidden; penis with its end curled over dorsum, its apex ending in two long curled branches which embrace the stem of the organ; beneath the overarching portion two acute black spines directed very obliquely forwards.

Genotype, *Ischnura pumilio* Charp.

Distribution.—Cosmopolitan; widely distributed throughout both hemispheres, except in South America. The genus has undergone a remarkable development in the islands of the Pacific, where several bizarre forms occur. Eight species occur within Indian limits, and are widely distributed throughout the continent. These fall naturally into two groups, one in which an orange ground-colour predominates in both sexes, and with black markings extremely restricted, the other in which this colour is restricted to heterochromatic females, the males being blue or green, marked broadly with black. Most teneral forms, on first emergence, are orange in colour, and many of the females remain so, so that we may assume that this is the primitive colouring of the genus; the *rufostigma* group will then be the most primitive; this group is linked up to the second, the *elegans* group, by *I. mildredæ*, which first exhibits blue markings on the terminal abdominal segments. These markings are more in evidence, and are combined with bright yellow colouring in an intermediate group—that of *I. inarmata.*

Key to the Species of Ischnura.

1.	Species with the ground-colour of both sexes bright orange-red	2.
	Species with ground-colour of males and isochrome females blue or pale grass-green	4.
2.	Abdominal segments 8 to 10 only black..	*rufostigma* Selys, [p. 362.
	Abdominal segments 7 to 10 black	3.
3.	Segment 8 in both sexes marked with blue.	*mildredæ* Fras., p. 366.
	Segment 8 unmarked with blue	*annandalei* Laid., [p. 364.
4.	Abdominal segments 3 to 6 citron-yellow.	5.
	Abdominal segments 3 to 6 black on dorsum	6.
5.	Inferior anal appendages longer than superiors, and about as long as segment 10; segment 6 with some apical dorsal black markings	*inarmata* Calv., p. 357.
	Inferior anal appendages shorter than superiors and segment 10; segment 6 without any black dorsal markings ...	*delicata* (Hagen), p. 360.

6.	Segment 10 with blue spot on dorsum; pterostigma of fore-wing with costal side shortest........................	*forcipata* Mort., p. 354.
	Segment 10 unmarked with blue; costal side of pterostigma not markedly shorter than the other sides	7.
7.	Posterior lobe of prothorax greatly prolonged at the middle; 2nd abdominal segment non-metallic	*elegans* (Vander Lind.), [p. 351.
	Posterior lobe of prothorax not prolonged; 2nd abdominal segment steely metallic blue on dorsum	*senegalensis* (Ramb.), [p. 348.

141. **Ischnura senegalensis** (Rambur). (Figs. 150 & 151.)

Agrion senegalensis Ramb., Ins. Névrop. p. 276 (1842); Selys, Rev. Odon. p. 186 (1850).

Ischnura senegalensis Selys, Bull. Acad. Belg. (2) vol. xli, p. 273 (1876); id., Ann. Soc. Esp. Hist. Nat. vol. xi, no. 60 (1882); id., Ann. Soc. Ent. Belg. vol. xxvii, p. 55 (1883); id., Ann. Mus Civ. Genova, (2) vol. vii (xxvii), no. 64 (1889); Karsch, Ent. Nach. vol. xvii, no. 16 (1891); Kruger, Stett. Ent. Zeit. vol. xxi (1898); Campion, Percy Sladen Exp. no. xxvii (Odonata), p. 443 (1913); Ris, Suppl. Ent. no. 5, Deut. Ent. Mus. pp. 78, 79 (1916); Laid., Rec. Ind. Mus. vol. xii, p. 130 (1916); Ris, Zool. Meded. Deel. vol. x, p. 21 (1917); id., Ann. S. African Mus. vol. xviii, pp. 333, 335 (1918); id., Jena. Denkschr. vol. xiii, Schultze, Forschungsriese Sudafr I. pp. 303, 310 (1919); Fras., J. Nat. Hist. Siam, vol. iii, no. 4, p. 46 (1919); id., Rec. Ind. Mus. vol. xxiv, p. 310 (1922); id., J. Bombay Nat. Hist. Soc. vol. xxix, p. 746 (1923); id., Rec. Ind. Mus. vol. xxvi, pp. 428, 489–90 (1924); Laid., Spolia Zeylan. vol. xii, p. 369 (1924); id., J. Malay. Br. Roy. As. Soc. vol. iv, pt. ii, p. 219 (1926); Fras., Treubia, vol. iii, livr. 3–4, p. 469 (1926); id., J. Nat. Hist. Soc. Siam, vol. vii, p. 90 (1927); Ris, Wien. Ent. Zeit. vol. xliv, pp. 159–60 (1928); id., Treubia, vol. vii, Suppl. p. 145 (1929); Fras., Rec. Ind. Mus. vol. xxxiii, p. 448 (1931).

Micronympha senegalensis Kirby, Cat Odon. p. 141 (1890); id., J. Linn. Soc. Lond., Zool. vol. xxiv, p. 564 (1894).

Male.—Abdomen 21–23 mm. Hind-wing 13–15 mm.

Head: labium pale yellow; labrum pale blue, narrowly black along the base; genæ, bases of mandibles, anteclypeus, and a broad fascia traversing the frons pale azure blue; postclypeus steely blue-black; vertex and occiput black, with a small round blue postocular spot on each side; beneath head whitish; eyes black on the upper half, pale green to yellow on the lower, these two zones abruptly separated. *Prothorax* black, the anterior collar pale blue, the sides pale green; posterior lobe shallow, rounded, not produced at the middle. *Thorax* bronzed black on dorsum as far back as a little beyond the humeral suture, the sides palest green, pale yellow beneath; narrow citron-yellow or pale green antehumeral stripes extending the whole length of dorsum. *Legs* black, the flexor surfaces of femora, extensor surfaces of tibiæ, and the tarsi yellow; spines very numerous, very short, black. *Wings* hyaline; pterostigma of fore-wings very oblique, diamond-shaped, black, but the outer angle and costal border narrowly white, tinted with blue on the upper surface of wing, covering less than one cell; 8 post-nodal nervures to fore-wings, 6 in the hind; pterostigma of hind-wings smaller than that of fore, uniform pale brown or carneous, framed in heavy black nervures. *Abdomen* black, marked with yellow and blue as follows:—Segment 1 broadly azure blue at the sides and along apical border, the dorsum broadly metallic blue-black; segment 2 metallic blue-black on dorsum, this extending broadly on to sides on the basal half of segment, laterally azure blue, fading to pale green along the ventral border. In some specimens there

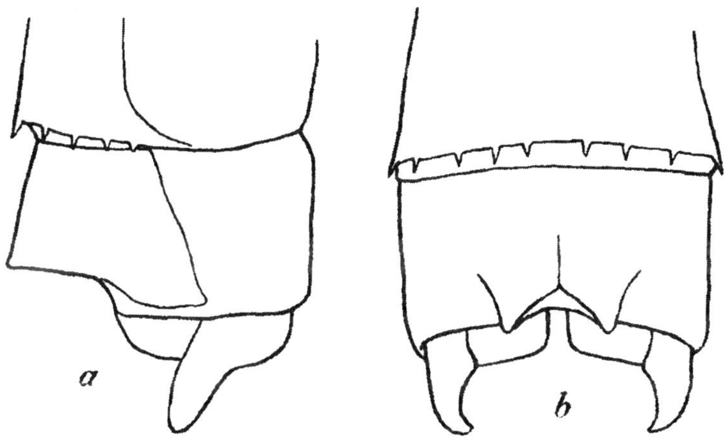

Fig. 151.—Anal appendages of *Ischnura senegalensis* (Ramb.), male. *a.* Right lateral view. *b.* Dorsal view.

is a narrow lateral longitudinal black stripe dividing the blue from the green, and further subdividing the blue into two spots; segments 3 to 7 citron-yellow at the sides, black on the dorsum, this black band on each segment ending in a point basally and expanding subapically except on segment 7, where the band broadens gradually as far as the apical end; all inter-segmental joints narrowly black; segment 8 azure blue, but often narrowly black at the base; segment 9 broadly black on dorsum, azure blue laterally, segment 10 black on dorsum, broadly yellow on the sides, its apical border raised into two tubercles, beneath and behind which the segment is blue. *Anal appendages* black and yellow; superiors barely one-third the length of segment 10, squared as seen from above, rounded as seen from the side; inferiors about three-fourths the length of segment 10, very broad and rounded at the base, then abruptly tapered, sloping strongly upwards and backwards, the apices acute and turned inwards, somewhat forcipated.

Female.—Abdomen 20–24 mm. Hind-wing 14–16 mm.

Three forms are found, one of which is the teneral condition of a heterochromatic form, and the third an isochrome.

Forma 1.—*Head* bright orange, the postclypeus and a broad transverse band on vertex at level of ocelli black; eyes emerald green, suffused with bright orange above; no postocular spots present. *Thorax* bright orange except the dorsum between the humeral sutures, which is black. *Prothorax* similar, with a broad black X-shaped mark on dorsum. *Legs* yellow striped with black. *Abdomen* with segments 1 and 2 bright orange, the latter segment with its apical border narrowly black and confluent with a broad black mark extending along dorsum half-way to base; segments 3 to 6 and the basal three-fourths of 7 pale blue at the sides, the dorsum marked as in the male; segments 8 to 10 and the apical third or fourth of 7 orange, fading to yellow beneath, all the segments broadly black on dorsum, but the apical borders ringed narrowly with orange. *Anal appendages* short, pointed, orange; vulvar scales yellow.

Forma 2.—This is the adult stage of forma 1. *Head* with black markings similar to the male, but the postocular blue spots poorly defined and merging into an orange or ochreous background on the occiput; the labrum olivaceous suffused with black, whilst the pale colour is greenish rather than blue; eyes olivaceous on the lower half, traversed by a narrow equatorial black ring. *Prothorax* broadly black on dorsum, pale greenish laterally; posterior lobe simple, rounded. *Thorax* broadly black on dorsum, and with or without a narrow antehumeral carneous stripe, and, when without, then the dorsal black extending outwards only halfway to humeral suture; the sides pale carneous or suffused with ochreous. *Legs* as for the male. *Abdomen* with the dorsum broadly marked with black from segment 1 to the end, the apical borders of segments 1, 8, 9, and 10 narrowly pale blue, the sides of all grey or pale carneous.

Forma 3.—An isochromatic form barely distinguishable from the male except by sexual characters. Segments 1 and 2 are non-metallic, whilst segment 9 is green laterally instead of blue; segment 2 often with the lateral blue present as two spots.

All forms, including the isochrome, with the wings similar; pterostigma in fore- and hind-wings similar, pale pinkish or greyish, with the proximal, distal, and posterior sides black, costal side yellow, covering less than one cell; 8 postnodal nervures in fore-wings, 7 in the hind.

Distribution.—Throughout INDIA, BURMA, and CEYLON,

and extending beyond these limits eastwards as far as Japan and the Philippines, and westwards throughout the greater part of the African continent. This small, dominant insect is found in odd colonies from sea-level to over 7,000 ft.; it has populated the Ootacamund Lake, 7,250 ft., to the exclusion of all other Zygoptera, and literally swarms there. Apart from this colony I have not found it a common insect, and in some years I never once came across it.

The whereabouts of the type is unknown, but paratypes are present in most national and private collections.

142. **Ischnura elegans** (Vander Linden). (Fig. 152.)

Agrion elegans Vander Lind., Opusc. Sci. vol. iv, p. 104, no. 6, t. 4, fig. 5 (1823); id., Mon. Lib. Eur. p. 40, no. 39 (1825); Steph., Ill. Brit. Ent., Mand. vol. vi, p. 72 (1836); Ramb., Ins. Névrop. p. 274 t. 7, figs, 6, 7, *f*, *g* (1842); Evans, Brit. Lib. p. 13, t. 3, figs. 3, 4 (1845); Selys, Rev. Odon. p. 188 (1850); Rost., Bull. Ent. Ital. vol. xvii, p. 246, t. 2 (i), t. 3 (1885).

Agrion pupilla Hansem, Wied. Zool. Mag. vol. ii (1), p. 156 (1823); Selys, Mon. Lib. Europ. p. 157 (1840).

Agrion tuberculatum Charp., Hor. Ent. p. 21 (1825); Burm., Handb. Ent. vol. ii, p. 819, no. 11 (1839); Charp., Lib. Europ. p. 152, t. 38, fig. 2 (1845).

Agrion ezonatum Steph., Ill. Brit. Ent., Mand. vol. vi, p. 72 (1836); Evans, Brit. Lib. p. 14, t. 5, fig. 5 (1845).

Agrion rufescens Steph., Ill. Brit. Ent., Mand. vol. vi, p. 74 (1836); Evans, Brit. Lib. p. 15, t. 5, fig. 2 (1845).

Agrion aglæa Fons., Ann. Soc. Ent. France, vol. vii, p. 567, t. 15, figs. 1, 2 (1838).

Agrion hastulatum Burm., Handb. Ent. vol. ii, p. 820, no. 13 (1839).

Agrion rubens Evans, Ill. Brit. Ent., Mand. vol. vi, p. 14, t. 3, fig. 6 (1845).

Ischnura elegans Selys, Bull. Acad. Belg. (2) vol. xli, p. 277 (1876); Kirby, Cat. Odon. p. 142 (1890); Lucas, British Dragonflies, pp. 271–6, pl. xxiv, fig. 2 (larva, p. 44) (1900); Campion, Entomologist, pp. 2, 6 (Dec. 1906); Morton, Trans. Ent. Soc. Lond. p. 306 (1907); Campion, Entomologist, p. 2 (Jan. 1909); id., ibid. p. 177 (1909); id., ibid. p. 294 (Nov. 1909); id., ibid. p. 77 (1913); id., ibid. p. 115 (April 1919); Fras., Rec. Ind. Mus. vol. xviii, pp. 80–1 (1919); id., J. Bombay Nat. Hist. Soc. vol. xxix, p. 746 (1923); Morton, Entomologist, vol. lix, p. 237 (Sept. 1926); Schmidt, Die Tierwelt Mitteleurop. iv Band, Lief. i, 6, pp. 15–18, fig. 16, *l–n* (1929); Needham, Zoologia Sinaica, vol. xi, ser. A, pp. 282, 283 (1930).

Var. *a*. *Agrion elegans* race *magna* Rost, Bull. Ent. Ital. vol. xvii, p. 247 (1885).

Var. *b*. *Agrion elegans* race *excelsa* Rost, Bull. Ent. Ital. vol. xvii, p. 248, pl. 6, fig. 1, *a–c* (1885).

Var. *c*. *Agrion elegans* race *exigua* Rost, Bull. Ent. Ital. vol. xvii, p. 248, pl. 6, fig. 2, *a–c* (1885).

Var. *d*. *Agrion elegans* var. *aurantiaca* Rost, Bull. Ent. Ital. vol. xvii, p. 249, pl. 6, fig. 3, *a–c* (1885).

Micronympha elegans Kirby, Cat. Odon. p. 142 (1890).

Male.—Abdomen 23–25 mm. Hind-wing 15–16·5 mm.

Head: labium white; labrum pale blue, narrowly black at base; genæ, bases of mandibles, and a stripe across the frons pale greenish-blue; anteclypeus bluish, postclypeus steely blue-black; vertex and occiput black, the latter with a large round blue postocular spot on each side; eyes black above, greenish-yellow in the lower half; beneath head palest blue, black against the neck. *Prothorax* black on dorsum, broadly pale bluish-green on the sides; posterior lobe produced at its middle into a prominent lobule directed up and back. *Thorax* black on dorsum to well beyond humeral sutures, and marked with narrow antehumeral bluish-green stripes; laterally grass-green, with the upper part of the antero-lateral suture and the whole length of the others finely mapped out in black; beneath pale greenish-yellow. *Legs* pale yellow, broadly marked on the outer sides with black; spines few and short. *Wings* hyaline; pterostigma of fore-wing diamond-shaped, with the inner nervures very thick and black, the

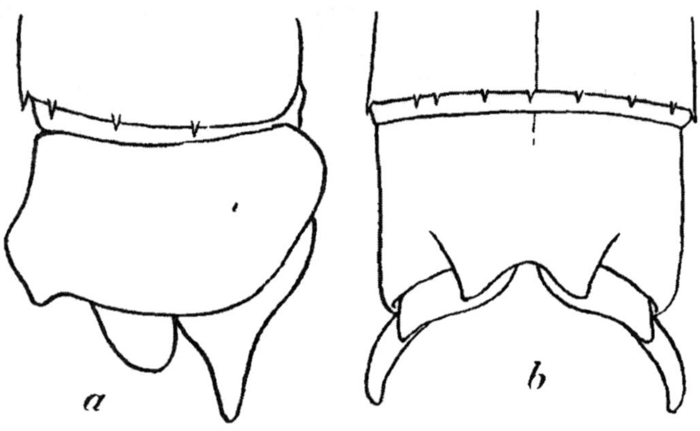

Fig. 152.—Anal appendages of *Ischnura elegans* Vander Lind., male. *a*. Right lateral view. *b*. Dorsal view.

outer finer, the membrane pure white or tinted with pale blue, its inner half suffused with black, which is vignetted outwardly, covering slightly less than one cell; in the hind-wing pterostigma but slightly smaller, its inner half but palely blackish; 8 to 10 postnodal nervures in fore-wings, 7 in the hind. *Abdomen*: segment 1 black on dorsum, green or bluish-green on the sides and apical border; segment 2 broadly black, with a steely blue reflex on dorsum, the sides pale grass-green; segments 3 to 7 broadly black on dorsum, bright lemon-yellow laterally, the black bands expanding subapically on each segment, narrowed to a point basally; segment 8 azure blue, finely bordered with black at base and apex; segments 9 and 10 broadly black on dorsum, blue laterally; the latter segment with two prominent tubercles on the dorsum at apical border, between and below which is a small bluish area. *Anal appendages* black; superiors short, one-fourth the length of segment 10, tumid, curled on themselves as seen from behind, rounded as seen in profile and above; inferiors slightly more than half the length of segment 10, tumid at base, then tapering to a long nipple-like process directed straight back as seen from the side, but curved and slightly forcipated as viewed from the dorsum.

Female.—Abdomen 24–25 mm. Hind-wing 18 mm.

Isochromatic and heterochromatic forms occur, and a number of varieties have also been described.

Forma 1.—Isochromatic form. Differs only from the male by its sexual characters, by the colour of the pterostigma, which resembles in all wings that of the hind-wing of the male, but with paler centre, and lastly, by the ground-colour, which is violaceous or pale brown, except on the terminal segments, which are coloured exactly the same as in the male; nodal index usually higher, 10 to 12 postnodal nervures in fore-wings, 9 in the hind; antehumeral stripe and sides of *thorax* carneous or greyish. *Anal appendages* short, conical, black; vulvar scales carneous.

Forma 2.—Heterochromatic form. Pale markings of *head* carneous or violaceous, the postocular spots brick-red. *Prothorax* black on dorsum, with a small round red spot on each side of middle lobe, the sides reddish, as also the anterior collar. *Thorax* brick-red, with a broad dorsal black stripe extending out from mid-dorsal carina about half-way to humeral suture; in some adult specimens the red limited to the juxta-humeral regions only, and the sides olivaceous; beneath pale yellow or carneous. *Legs* reddish or carneous, black markings less extensive than in the male, and often broken up into spots. *Wings* similar to forma 1. *Abdomen* with segments 1 and 2 reddish, 3 to 6 pale green at the sides, 7 to 10 similar to the male, or 9 and 10 somewhat reddish; all segments with dorsal black markings similar to the male except on segment 2, where the marking is broad at the apical half and then contracted to half its width, expanding again slightly towards the base, so that the stripe is vase-shaped; segment 8 invariably blue, but the apical and basal rings are often broader than in the male; lastly, segment 7 may bear small apical blue spots.

Distribution.—Extending from the British Isles across Europe and mid-Asia. Within our limits confined to the N.W. FRONTIER PROVINCE and BALUCHISTAN. I have seen specimens from Baluchistan and Seistan, and possess a male from Chakdara (N.W.F.P.). Morton has reported the species from Quetta.

Very closely related to *Ischnura senegalensis*, from which the male may be distinguished by the truncate lobule of the middle of the posterior lobe of prothorax. Heterochromatic females may be distinguished by having segment 8 blue instead of black on dorsum, whilst both isochromatic and heterochromatic forms may be determined by the prolonged posterior lobe of prothorax.

Whereabouts of type unknown, but paratypes in most public and private collections. The author possesses a pair *in cop.*, and it may be seen that the superior anal appendages rest on the summit of the middle lobe of the prothorax of the female, whilst the inferiors are opened widely, like a grappling-iron, and grip the sides of the same lobe ; thus the posterior lobe, in spite of its curious development, plays no part in the primary copulation.

143. **Ischnura forcipata** Morton. (Fig. 153.)

Ischnura forcipata Mort., Trans. Ent. Soc. Lond. p. 306, pl. xxiv figs. 1–3 (1907) ; Laid., Rec. Ind. Mus. vol. xii, pp. 129–130 (1916) ; id., ibid. vol. xvi, pp. 171, 173, 174 (1919) ; Mort., Ent. Month. Mag. (3) vol. v, p. 149 (1919) ; Fras., J. Bombay Nat. Hist. Soc. vol. xxix, p. 746 (1923) ; Lieft., Tijdschr. Ent. vol. lxx, pp. 91, 92 (1927).
Ischnura musa Barten, Rev. Russ. Ent. vol. xiii, no. 1, pp 187–9 (1913)
Ischnura gangetica Laid., Entomologist, p. 235 (1913).

Male.—Abdomen 22–23 mm. Hind-wing 13–14 mm.

Head : labium white ; labrum, bases of mandibles, genæ, and a broad stripe traversing frons pale blue, the former with its base narrowly black and a small median spot of the same colour ; vertex and occiput black, the latter with a pair of very small round blue postocular spots ; behind and beneath head palest blue ; eyes black above, pale bluish-green in the lower half. *Prothorax* black, the anterior collar and a moderately large pyriform spot on each side of middle lobe pale blue, as also the lower part of sides ; posterior lobe of prothorax simple, rounded, narrowly lined with blue. *Thorax* black on dorsum to well beyond the humeral sutures, and marked with a pair of complete narrow blue antehumeral stripes ; laterally pale blue, white beneath ; the upper half of antero-lateral suture finely black. *Legs* pale yellow, with a few very short black spines ; femora broadly black on outer surface, tibiæ with the extreme proximal end black on the inner surface and a fine line on the outer surface of anterior pair. *Wings* hyaline ; pterostigma of fore-wings more than twice the size of that of the hind, its costal side only half the length of the proximal and posterior, the distal side slightly shorter than the posterior, the latter rounded, inner half black, outer half blue ; the same organ in hind-wings colourless, or with its middle faintly grey, framed in thick black nervures ; 8 postnodal nervures in fore-wings, 6 in the hind ; arc situated distal to the 2nd antenodal nervure. *Abdomen* with the basal and apical segments blue, segments 4 to 6 bright chrome-yellow, segments 7 to 10 azure blue, the dorsum of all except 8 and 9 broadly black ; the apex of segment 1 narrowly blue, the black dorsal marking of segment 2 vase-shaped, with short narrow stem at the apex of segment ; on segments 3 to 6 each dorsal stripe tapers to a point at the base of the segment and expands subapically to contract again and become confluent with a narrow apical black ring; on 7 the dorsal band ends at the apex of segment in three points, whilst the black dorsal

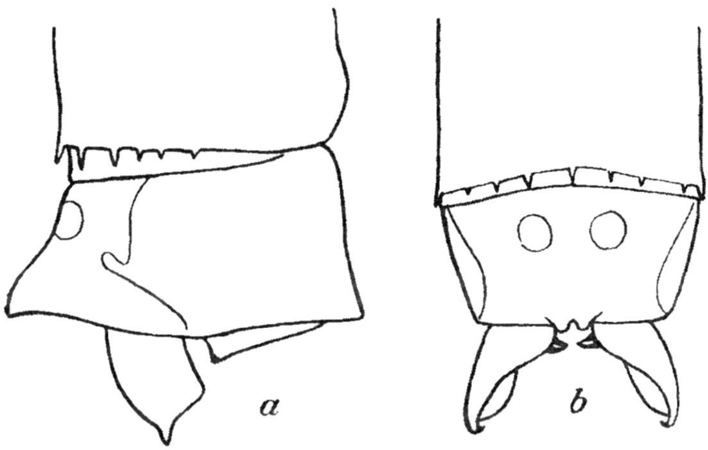

Fig. 153.—Anal appendages of *Ischnura forcipata* Mort., male. *a.* Right lateral view. *b.* Dorsal view.

marking on 10 has either an oval median basal blue point or a pair of small blue spots in the same place. *Anal appendages* black ; superiors not quite as long as segment 10, moderately broad and angulate as seen from the side, and ending in a minute point, whilst seen from the dorsum they are triangular, divaricate, and deeply hollowed out on the inner side, which is white ; inferiors much shorter, only just visible from the side, curled shell-like processes, their convexities opposed and ending in small black spines ; segment 10 strongly keeled and rising steeply as a robust minutely notched spine at apex of segment.

Female.—Abdomen 23 mm. Hind-wing 13 mm.

Only one form, a heterochrome, known, which is red in the subadult stage, but takes on blues and other colours in the final. *Head*: labrum, bases of mandibles, genæ, clypeus, and frons as far as just posterior to the roots of antennæ ochreous, the labrum often brighter orange; a broad stripe traversing the vertex from eye to eye; occiput orange, but with a diffuse blue postocular spot at its outer ends which blends through green to the orange; eyes orange above, changing to olivaceous green below: in the adult stage each eye acquires a small cap of black, and the genæ and frons are distinctly bluish-green. *Prothorax* orange in tenerals, bluish-green in adults, with a broad black dorsal marking extending over the middle and posterior lobes, the former marked with a small medial orange spot; posterior lobe low, convex, simple. *Thorax* with a broad black bar on mid-dorsum extending half-way out to humeral sutures; tenerals bright orange on dorsum, changing to pale greenish-yellow laterally and bright yellow beneath, whilst adults are olivaceous at the humeral region, changing to pale blue on the sides and greenish beneath. *Legs* yellow, with minute black spines; femora of the two anterior pairs of legs with a blackish-brown stripe on the outer sides. *Wings* with pterostigma similar to that in the hind-wings of male, almost hyaline, pale yellow, with a suspicion of a darker centre; nodal index similar to that of the male; arc at or very slightly distal to the level of distal antenodal nervure. *Abdomen* with segments 1 and 2 bright orange, changing to pale yellow or greenish in the adult stage; segment 2 with the mid-dorsal carina finely black and with a small basal triangle of black and another subapically connected with this line; in adults this dorsal marking has a halo of bright reddish-orange, whilst segment 1 has also a black basal spot similarly surrounded with orange; segments 3 to 8, and also 9 and 10 in the adult stage, with the dorsum very broadly black, the sides yellow in tenerals, changing to orange on the end segments, and blue throughout in adults; segment 9 in tenerals with a basal black dorsal spot forking strongly on the apical side, the prongs of the fork extending to the apical border of segment. *Anal appendages* very tiny, yellowish or blue; vulvar scales robust, yellow or orange.

Distribution.—Northern India in the hilly tracts of the PUNJAB and BENGAL, BALUCHISTAN. I have specimens from Mussoorie and the Simla Hills taken during April and May. The type of Laidlaw's *Ischnura gangetica* comes from Sat Tal and Shamket, Kumaon, whilst Morton's type, now in the British Museum collection, was taken in June at Quetta.

The curious shape of the pterostigma in the fore-wings will serve to distinguish this species from all others of the genus, whilst the shape of the anal appendages, the blue dorsal spot on segment 10, and the arc distal to the distal antenodal are other marked features. The female is distinguished by the dorsal marking on segment 2. The females mentioned by me in the Rec. Ind. Mus. (*l. c.*) under the name of *Ischnura elegans* belong really to this species; one of these has the ground-colour entirely pale blue save for the face, hinder part of occiput, and the halo around the dorsal marking on segment 2, which are pale ochreous; the postocular spots are nearly encircled with black and are definitely blue. One orange female of this lot, from Baluchistan, has no black spot on segment 1, but a broad vase-shaped dorsal stripe on segment 2, whilst segments 9 and 10 have also black dorsal markings; the anteclypeus is black in this specimen, and the postocular orange spot is entirely enclosed in black. There are probably many varieties of the female, but more material is needed to settle this point.

144. **Ischnura inarmata** Calvert. (Fig. 154.)

Ischnura inarmata Calv., Proc. Acad. Nat. Sci. Phil. pp. 147, 148, figs 1, 2 (1898); Morton, Trans. Ent. Soc. Lond. p. 307 (1907); Laid., Rec. Ind. Mus. vol. xii, pp. 130–1 (1916); id., ibid. vol. xvi, pp. 171, 173, 175 (1919); Fras., J. Bombay Nat. Hist. Soc. vol. xxix, p. 746 (1923).

Male.—Abdomen 23–25 mm. Hind-wing 16 mm.

Head: labium pale yellow; labrum, bases of mandibles, genæ, anteclypeus, and a broad stripe traversing frons, including the basal joints of antennæ, pale blue, the labrum narrowly black at base; postclypeus black, as also vertex and occiput, the latter with a pair of very large ovoid azure blue postocular spots; eyes black on extreme summit, blue to green at the sides and beneath; beneath head pale blue, but black adjacent to neck. *Prothorax* broadly black on dorsum, pale blue at the sides and anterior lobe; posterior lobe rounded, slightly elevated. *Thorax* broadly black on dorsum as far out as half-way between humeral and antero-lateral sutures, marked by narrow azure blue antehumeral stripes which expand slightly anteriorly; laterally blue, with the upper ends of both sutures marked with black; beneath pale bluish-green. *Legs* pale blue or bluish-green, the extensor surfaces of both tibiæ and femora marked with glossy black; spines few, black. *Wings* hyaline; pterostigma of fore-wings diamond-shaped, the costal side but a shade shorter than the posterior, ochreous framed in hyaline and dark brown nervures; pterostigma of hind-wing considerably smaller, palest yellow, with centre broadly greyish, covering one cell; 8 to 9 postnodal nervures in fore-wings, 8 in the hind; arc at the level of the distal antenodal nervure. *Abdomen* with segment 1 blue laterally, changing to pale green and then bright citron-yellow on segment 2 and onwards to segment 7, at the apical end of which it again takes on blue; segments 8 and 9 azure blue, with a row of fine black apical spines, and on segment 8 a very narrow black basal line; segment 10 entirely black, except on the lower parts of the sides, which are blue; segment 1 with a broad basal black spot confluent with a narrow apical black line, which is separated from segment 2 by a blue apical ring; segment 2 with a broad dorsal black marking connected to a narrow black apical ring, this marking tapering from the base of segment to apex, and often with a slight constriction subapically; segments 3, 4, and 5 with narrow black apical rings, the latter also with a small subapical black spot; segment 6 with rather less than its apical half on dorsum black, this marking rapidly tapering to a fine

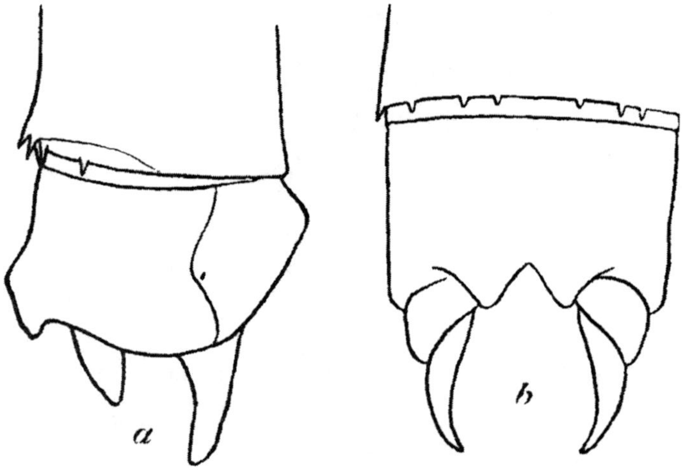

Fig. 154.—Anal appendages of *Ischnura inarmata* Calv., male. *a*. Right lateral view. *b*. Dorsal view.

point on dorsum basally and showing some evidence of the subapical dilatation seen in other species; segment 7 almost entirely black, only a narrow area of yellow on the sides near ventral borders. *Anal appendages* black, the inferiors bluish at the base; superiors triangular in shape as seen both from above and the sides, less than half the length of segment 10, slightly divaricate; inferiors about as long as segment 10, directed straight back, tapering to a somewhat obtuse point, curved and slightly forcipate as seen from above; segment 10 with the apical dorsal tubercle deeply notched.

Female.—Abdomen 23 mm. Hind-wing 16–17 mm.

Two forms occur, one of which is a partial isochrome, the other heterochromatic.

Forma 1.—Isochromatic, with enclosed antehumeral stripes on thorax. *Head* similar to the male, but the blue ground-colour bluish-green or green on the face and frons, and the postocular spots larger and not enclosed posteriorly, where they are bordered with orange; eyes capped above with warm brown, pale green otherwise. *Prothorax* and *thorax* similar to the male, but the antehumeral stripes and sides pale green. *Legs* marked with reddish-brown, not black. *Wings* with 8 to 10 postnodal nervures in fore, 6 to 7 in the hind; pterostigma similar to that of the hind-wing of male, that of the fore-wing decidedly larger than that of hind. *Abdomen* differing markedly as follows:—Segments 2 and 3 reddish-brown on dorsum, changing to ochreous on the base of segment 2 and dorsum of segment 1, the sides of which are greenish; remaining segments, 4 to 10, with the dorsum broadly black, the sides yellow, changing to bluish-green on segment 7 and dull orange on the end segments; segments 8 to 10 with narrow apical blue annules. *Anal appendages* short, black; vulvar scales extending to end of abdomen, dull ochreous.

Forma 2.—Orange form without any dark markings on thorax. Labrum, bases of mandibles, genæ, anteclypeus, and frons variably bright orange, the former with its base narrowly black, the latter with or without the hinder border of the orange irregular; vertex traversed by a broad black stripe extending from eye to eye, its anterior border straight or with two indentations just inside the roots of the antennæ; occiput broadly bright orange, bordered outwardly with black and without coloured postocular spots; postclypeus glossy black; eyes bluish, capped with orange above. *Prothorax* bright orange, the middle lobe with two large lateral black spots, posterior lobe narrowly black at base, its posterior border crenate, the middle portion projecting back as a broad convex lobe. *Thorax* bright orange, without markings, or with the mid-dorsal carina a darker orange. (A rare form has a broad mid-dorsal black band similar to the heterochrome of *I. elegans*, and in this form the bright orange colouring is replaced by pale carneous, and on the abdomen by pale bluish.) *Wings* similar to the isochrome form, but the pterostigma more nearly equal in size. *Legs* bright orange, usually unmarked. *Abdomen* with segments 1 to 3 bright orange, the latter with a squared apical dorsal spot confluent with an apical black annule; segments 4 to 10 similar to the isochromatic form, but 4 often with the black dorsal stripe broken or not extending as far as the base of the segment.

Distribution.—KASHMIR, in June.

Distinguished by the shape of its anal appendages and by the bright yellow segments 3 to 5 of abdomen without black markings. The heterochromatic form of the female is easily distinguished from other species by its thorax being without any black markings, save in the rare form mentioned above, which, however, has the 2nd segment orange or carneous, with a small black diamond-shaped apical dorsal spot not found in other species.

Type in the Philadelphia Museum ; paratypes in the Morton, Fraser, Pusa, and British Museum collections.

145. **Ischnura delicata** (Hagen). (Fig. 155.)

Agrion delicatum Hagen, Verh. zool.-bot. Ges. Wien, vol. viii, p. 479 (1858).

Agrion aurora Brauer, Verh. zool.-bot. Ges. Wien, vol. xv, p. 510 (1865) ; id., Reise 'Novara,' p. 52 (1866).

Ischnura delicata Selys, Bull. Acad. Belg. (2) vol. xli, p. 281 (1876) ; Mart., Mém. Soc. Zool. France, vol. xix, p. 246 (1901) ; Tillyard, Proc. Linn. Soc. N.S. Wales, vol. xxxii, (2) pp. 384, 389 (1907).

Micronympha aurora Kirby, Cat. Odon. p. 143 (1890) ; id., J. Linn. Soc. Lond., Zool. vol. xxiv, p. 564 (1894).

Ischnura aurora Ris, Nova Caledonia, Zool. vol. ii, 4, p. 67 (1915) ; Laid., Rec. Ind. Mus. vol. xii, p. 131 (1916) ; id., ibid. vol. xvi, p. 175 (1919) ; Fras., J. Bombay Nat. Hist. Soc. vol. xxvii, pp. 542, 543 (1921) ; id., Rec. Ind. Mus. vol. xxiv, p. 310 (1922) ; id., J. Bombay Nat. Hist. Soc. vol. xxix, p. 746 (1923) ; id., Rec. Ind. Mus. vol. xxvi, pp. 428, 489 (1924) ; Laid., Spolia Zeylan. vol. xii, p. 369 (1924) ; Fras., Rec. Ind. Mus. vol. xxxiii, p. 448 (1931).

Nanosura aurora Kennedy, J. Sci. Ohio Univ. vol. xxi, no. 2, p. 88 (1920).

Male.—Abdomen 16–20 mm. Hind-wing 10–12 mm.

Head : labium white ; labrum citron-yellow, with the base narrowly bordered with black ; anteclypeus pale olive-green ; postclypeus bronzed black ; bases of mandibles, genæ, frons, and the three basal joints of antennæ pale grass-green ; vertex bronzed black, as also occiput, except along its hinder border, where it becomes abruptly pale greenish-yellow ; a small postocular spot on each side of occiput, rounded, deep azure blue ; eyes a beautiful olive-green, dark olive above, where is found a small semilunar spot of black bordering the eye, pale olive beneath. *Prothorax* bronzed black on dorsum, the sides and anterior lobe blue. *Thorax* bronzed black on dorsum as far posterior as midway between the humeral and antero-lateral sutures, marked with narrow grass-green antehumeral stripes ; laterally grass-green, with a small streak of black on upper part of postero-lateral suture, beneath white. *Legs* pale citron-yellow ; anterior and middle pairs of femora with a short black stripe on the outer side of the distal half ; anterior pair of tibiæ with a similar stripe. *Wings* hyaline ; pterostigma differing in the fore- and hind-wings, that of fore-wings slightly broader than long, kite-shaped, rose-red for its proximal half, hyaline for the distal, inner and posterior borders thick, black, anterior border white, outer border yellow; in the hind-wing much smaller, about half the size, uniform pale grey, almost colourless, all bordering nervures except the costal thick and black, costal border white. A few of the nervures below and external to pterostigma bright yellow; 7 postnodal nervures to fore-wings, 6 in the hind; *ac* situated nearly half-way between the two antenodal nervures, *ab* beginning slightly proximal to the level of basal antenodal nervure. *Abdomen* citron-yellow, except segments 8 to 10, which are azure blue; in some specimens, or races, the basal segments 1 to 3 are more or less red on dorsum (Ootacamund, Nilgiri Hills); segment 1 with a large dorsal bronzed black spot extending the whole length of

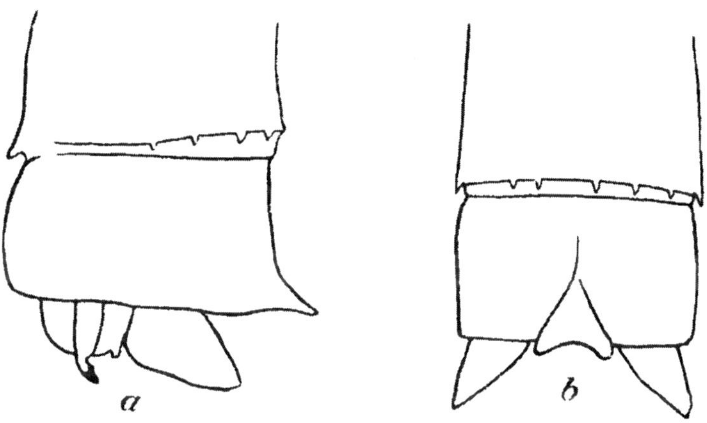

Fig. 155.—Anal appendages of *Ischnura delicata* (Hagen), male.
a. Left lateral view. *b*. Dorsal view.

segment; segment 2 with its base on dorsum narrowly black, a fine mid-dorsal carinal line running from this nearly as far as the apical border of segment; segment 6 with its apical border finely black, and a large subapical diamond-shaped spot on dorsum; segment 7 broadly bronzed black except at sides, and partially so at extreme base; 8 to 10 entirely azure blue save the last segment, which has a broad quadrate black dorsal spot, the bifid dorsal apical tubercle white at borders. *Anal appendages* pale ochreous, tipped with black or suffused with reddish-brown; superiors about as long as segment 10, triangular as seen both from above and the side, hollowed out within; inferiors broad at extreme base, then ungulate, the apex turned slightly up and tipped with black, ending in a small black point; of almost equal length to the superiors; penis with its end curled strongly over dorsum, dividing into two branches at the end which embrace the dorsum and turn abruptly back towards the apex of the organ, nearly completing a circle; in the lower arc of this, two robust, very long chitinous spines spring, and are directed towards and beyond the apex of the organ.

Female.—Abdomen 18–20 mm. Hind-wing 14–15 mm.

A stouter and less conspicuously marked insect than the male. Pale markings of head palest green; postocular spots small, blue; behind occiput often ochreous. *Prothorax* and *thorax* palest yellow on the sides, almost white beneath; antehumeral stripe citron-yellow. *Wings* hyaline, 8 postnodal nervures to fore-wings, 6 in the hind; pterostigma palest pink, shaped similarly to that of the male, and differing similarly in size. *Abdomen* with a broad black dorsal stripe extending the whole length, but interrupted by narrow yellow annules on segment 1, base of segment 8, and apical borders of segments 9 and 10; the sides broadly yellow, this colour paling to greenish beneath; vulvar scales palest green, as also *anal appendages*, which are very short and conical.

Distribution.—Widely distributed throughout southern Asia, INDIA, CEYLON, and BURMA. Malaysia, the Sondaic Archipelago, Borneo, New Guinea, Australasia, Philippines, and Samoa. *I. delicata* is one of those insects which has made full use of the upper air currents, crossing seemingly impossible barriers of land and ocean. Only the slightest variations are found; specimens from Ootacamund, Nilgiris, 7,500 ft., have the basal segments definitely red, and this feature is again met with in Australian specimens: Samoan specimens differ in no way from Indian. The only other species of the genus at all resembling it in colour is *I. inarmata* Calvert, but this is a much larger insect, and has very differently shaped anal appendages. I have retained Hagen's name as, although no detailed description accompanied it, we have the undoubted authority of Selys that it was conspecific with Brauer's *Agrion aurora*, published at a later date.

146. **Ischnura rufostigma** Selys. (Fig. 156.)

Ischnura rufostigma Selys. Bull. Acad. Belg. (2) vol. xli, p. 283 (1876); Morton, Trans. Ent. Soc. Lond. p. 307 (1907); Laid., Rec. Ind. Mus. vol. viii, p. 344, pl. xvi, fig. 5 (1914); id., ibid. vol. xii, p. 130 (1916); id., ibid. vol. xvi, pp. 171, 173, 175 (1919); Fras., J. Bombay Nat. Hist. Soc. vol. xxix, p. 746 (1923); Needham (*I. annandalei*, nec *I. rufostigma*). Zool. Sinica, vol. xi, pp. 277, 292 (1930).

Micronympha rufostigma Kirby, Cat. Odon. p. 143 (1890).

Male.—Abdomen 23 mm. Hind-wing 14–15 mm.

Head: labium pale yellow; labrum pale blue or pale ochreous, with the base narrowly black; bases of mandibles, genæ, anteclypeus, and the frons broadly pale ochreous or pale blue; postclypeus glossy steely black; vertex and occiput black, the latter with very small azure blue post-ocular spots; eyes dark olivaceous, paler below; beneath head bluish, black against the neck. *Prothorax* black on dorsum, pale blue at sides as well as the whole of anterior lobe; posterior lobe shallow, rounded. *Thorax* steely black on dorsum to nearly as far as the antero-lateral suture, marked with narrow pale blue antehumeral stripes; laterally pale blue, pale yellow beneath; upper part of postero-lateral suture with a minute spot of black. *Legs* pale yellow, with a few short black spines; femora striped posteriorly with black. *Wings* hyaline; pterostigma of fore-wing very narrow, elongate, diamond-shaped, its outer angle very acute, bright

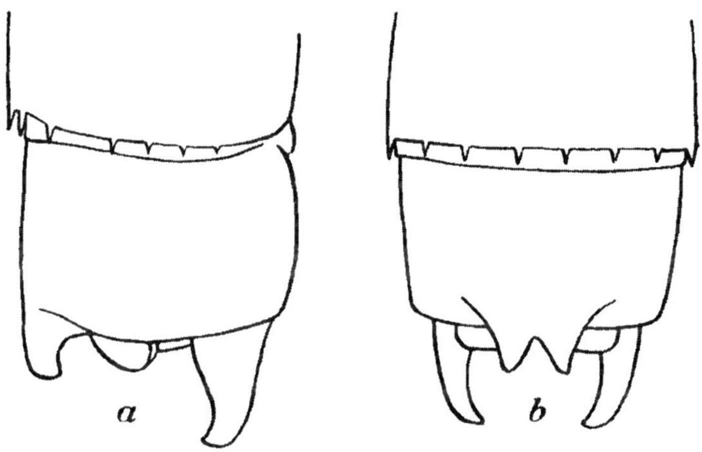

Fig. 156.—Anal appendages of *Ischnura rufostigma* Selys, male. *a*. Right lateral view. *b*. Dorsal view.

brick-red, the costa adjacent to it distally sharing in this colour; pterostigma of hind-wing much smaller, but of the same shape, dirty yellow, covering about half a cell; 7 to 8 postnodal nervures to fore-wings, 6 in the hind; arc distinctly distal to the level of outer antenodal nervure. *Abdomen* bright reddish-orange, marked with a quadrate dorso-basal black spot on segment 1, and broadly black on dorsum of segments 8 to 10, the intervening segments unmarked; segment 10 prolonged into two apical blunt spines, which represent the usual tubercle very deeply notched. *Anal appendages* ochreous, the inferiors tipped with black; superiors very short, rounded, deeply excavate on the inner under-sides, ending in a robust spine which is directed in and slightly down; inferiors about half the length of segment 10, broad and tumid at base, then abruptly narrowed, cylindrical, ending in a small point which is curled slightly in.

Female.—Abdomen 22 mm. Hind-wing 16–17 mm.

Closely similar to the male; only one form, which is isochromatic. Differs in the following points:—Postocular spots enclosed only in very aged specimens, usually opening on to an orange fascia which traverses the occiput, or entirely absent and replaced by a broad orange area connected by an isthmus across the middle line; antehumeral stripes greenish-blue, the bordering black on the outer side often incomplete above; pterostigma of both *wings* pinkish, the centres darker, covering nearly one cell, subequal; 9 postnodal nervures in fore-wings, 7 to 8 in the hind; position of arc variable. *Abdomen* bright reddish-orange, marked with black as follows:—Segment 1 with a dorsal black spot shaped like a trident, its handle connected with a narrow apical black ring; segments 2 to 6 with the mid-dorsal carina finely black, expanding into a small spot or cross subapically, these markings very variable, entirely absent in some adults on segments 3 to 6, and only present as a small apical dorsal spot on segment 7, which is, however, usually broadly black from base to apex; segments 8 to 10 broadly black on dorsum, with narrow apical orange rings. *Anal appendages* very short, conical, pointed, yellow; vulvar scales orange, extending just to end of abdomen.

Distribution.—This is peculiar. The type is merely labelled "India," and most specimens so far taken have come from Bengal, Assam, and Bihar; I found it quite common on the river encircling Mhow in the Central Provinces during February and March. Specimens in the Indian Museum are labelled "Calcutta," although I never came across it in that neighbourhood. Other localities are Pusa, Bihar in February, and Tezpur, Assam, during May; from nowhere else in India or Burma has it been so far reported. Needham mentions it from China, but his description is more in accord with that of *I. annandalei*, and the distribution of the latter would also suggest this.

Type in the Selysian collection; paratypes in the Indian, Pusa, and British Museums, and the Fraser, Morton, and Laidlaw collections.

147. **Ischnura annandalei** Laidlaw. (Fig. 157.)

Ischnura annandalei Laid., Rec. Ind. Mus. vol. xvi, pp. 175, 177 (1919); Fras., J. Bombay Nat. Hist. Soc. vol. xxix, p. 746 (1923).

Male.—Abdomen 24 mm. Hind-wing 14 mm.

Head: labium pale yellow; labrum, bases of mandibles, genæ, anteclypeus, and a broad stripe across frons palest blue or pale yellow in subadults; base of labrum finely and postclypeus wholly black; vertex and occiput black, with a pair of very small blue postocular spots on the latter; beneath head creamy white, black against the neck; eyes dark olivaceous, capped with black above. *Prothorax* broadly black on dorsum, the anterior lobe and the sides pale blue; posterior lobe narrow, rounded, simple. *Thorax* broadly black on dorsum to about midway between the humeral and antero-lateral sutures, marked with narrow pale blue antehumeral stripes; laterally pale blue to pale greenish below and beneath, the lateral sutures marked finely with black in their upper parts. *Legs* pale bluish or carneous, with a few short black spines; femora black on extensor surface. *Wings* hyaline:

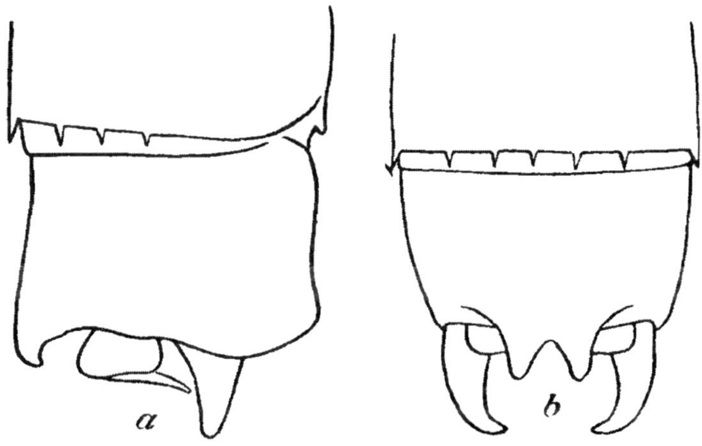

Fig. 157.—Anal appendages of *Ischnura annandalei* Laid., male. *a*. Right lateral view. *b*. Dorsal view.

pterostigma of fore-wings reddish-orange, its inner two borders black, outer two yellow, this colour being prolonged for a short distance on to costa as in *I. rufostigma*, very narrow, elongate, very acute at outer angle, covering one cell or less; in the hind-wing considerably smaller, but of the same shape, pale yellow stippled with grey and framed in black; 8 postnodal nervures in fore-wings, 6 in the hind; arc usually slightly distal to the level of the second or distal antenodal nervure. *Abdomen* bright reddish-orange; segment 1 with the dorsum broadly black, blue at the sides; segment 2 greenish-yellow, changing to orange laterally, and with a broad black mark on dorsum shaped like a goblet with a very short stem, which is connected to a narrow apical black ring; segments 3 to 6 unmarked, save for articular fine black rings; segments 7 to 10 black; segment 10 with a pair of small robust spines on its dorsal apical border which represent a deeply bifid ochreous tubercle. *Anal appendages* almost identical with those of *I. rufostigma*, but less robust, and the inferiors decidedly shorter and finer.

Female.—Abdomen 23 mm. Hind-wing 16 mm.

Two distinct forms, isochromatic and heterochromatic, occur, which differ by the presence or absence of enclosed antehumeral stripes.

Forma 1.—Isochrome. Identical in colouring and markings with the male; segment 2 with the dorsal goblet-shaped marking not usually connected with its stem, or with the stem divorced from the apical black ring; intersegmental rings more pronounced; pterostigma of fore-wings pale crimson, that of hind yellowish, subequal; 9 postnodal nervures in fore-wings, 7 in the hind.

Forma 2.—Heterochrome. *Head* differing from the male in that the postocular spots are replaced by a broad orange area connected by a narrow isthmus of the same colour across the middle line. *Thorax* differs by the black being restricted to the middle part of dorsum, so that there are no antehumeral stripes; humeral region a rich reddish-orange; laterally greenish, the sutures unmarked. *Legs* similar to those of the male. *Wings* with pterostigma similar in colour in fore and hind and almost so in size, pale ochreous or yellow, framed in black. *Abdomen* with segments 1 and 2 and 7 to 10 similar to the male, but 3 to 6 with the mid-dorsal carina very finely black, crossed at right angles subapically by an equally fine black line; some specimens, although belonging to this form, have the whole of the abdominal segments broadly black on dorsum.

Distribution.—North Shan States, BURMA. The *type*, in the Indian Museum, is from the Inle Lake; I have specimens from Maymyo, Upper Burma, collected in July, and others, collected in March, from a more northern locality. Needham reports it from China, but has determined it, in error, as *I. rufostigma*.

Very closely related to *I. rufostigma*, from which it differs by having segments 2 and 7 marked broadly with black (unmarked in *I. rufostigma*). It is interesting to note that in the isochrome female the pterostigma shares in the general resemblance to the male.

148. **Ischnura mildredæ** Fraser.

Ischnura mildredæ Fras., Rec. Ind. Mus. vol. xxix, pp. 87, 88 (1927).

Male.—Abdomen 24 mm. Hind-wing 16 mm.

Head: labium white; labrum blue, its base narrowly black; bases of mandibles, genæ, anteclypeus, and a broad stripe across frons bluish or greenish-blue; vertex and occiput black, the latter marked on each side by a small round blue postocular spot; eyes olivaceous, capped above with black; beneath head bluish-grey, black against the neck; postclypeus glossy steely black. *Prothorax* black on dorsum, ochreous on the sides. *Thorax* broadly black on dorsum, this extending outwards to about half-way to the antero-lateral suture, marked with narrow blue antehumeral stripes; laterally blue to bluish-green, the upper ends of the lateral sutures finely black. *Legs* yellow; femora black on extensor surfaces. *Wings* hyaline; pterostigma narrow, diamond-shaped, acutely pointed at outer end especially, bright tawny, with the proximal borders black, distal yellowish; 8 postnodal nervures to fore-wings, 7 in the hind. *Abdomen* with segment 1 pale green, marked dorsally with a large quadrate black spot; segment 2 greenish basally, changing to bright reddish-orange apically, the dorsum with a large round black spot on the basal two-thirds and a smaller round spot on the apical third, the two narrowly confluent and the latter with a narrow black apical ring; segments 3 to 6 bright reddish-orange, with narrow apical black rings; segments 7 to 10 black, 8 bearing a small round spot of azure blue at the centre of dorsum; segment 10 with a bifid tubercle at its apical end. *Anal appendages* carneous, inferiors tipped with black; superiors very short, curled on themselves, about half the length of segment 10, with a fine downwardly directed spine on inner side; inferiors tumid at base, abruptly tapered, cylindrical thereafter, obtusely pointed, the apices curling strongly inwards and nearly meeting.

Female.—Abdomen 24 mm. Hind-wing 16 mm.

Closely similar in colour and markings to the male. *Abdomen* more robust; segment 2 marked quite differently on dorsum, this mark shaped like a small wineglass, its rim apposed to the base of segment, its base connected by a fine stem, situated a short distance from the apical border of segment, but connected by a fine black line on mid-dorsal carina to a narrow black ring at apical border; segment 8 with a very large azure blue spot which covers the whole of dorsum of segment except narrow basal and apical black rings; laterally the blue spot finely bordered with black; segments 7 to 10 with the ventral borders, as well as vulvar scales, ochreous. *Wings* similar to the male, but the pterostigma of similar size and colouring in both pairs of wings, pale yellow framed in dark brown. *Anal appendages* small, pointed, ochreous.

Distribution.—Known only from Maymyo, UPPER BURMA, during June.

Type in the author's collection. Distinguished from *I. annandalei*, to which it is very closely related, by the dorsal marking on segment 2, and also by the blue dorsal spot on segment 8. The latter is of great interest as presaging the evolution of the more primitive orange forms into the modern blue and black species. Only one form of the female is known —the isochrome.

Genus **RHODISCHNURA** Laidlaw. (Fig. 158.)

Rhodischnura Laid., Rec. Ind. Mus. vol. xvi, pp. 171, 177 (1919); Fras., J. Bombay Nat. Hist. Soc. vol. xxix, p. 747 (1923).

Dragonflies of very small size and moderately stout build; colours non-metallic, bright red and yellow, or green with black markings; females polychromatic; wings hyaline; pterostigma shaped and coloured differently in fore- and hind-wings, that of fore-wings coloured; only six postnodal nervures in fore-wings; discoidal cell acutely pointed at the distal end, with costal side about half the length of posterior side in fore-wing and two-thirds the length in the hind, basal side

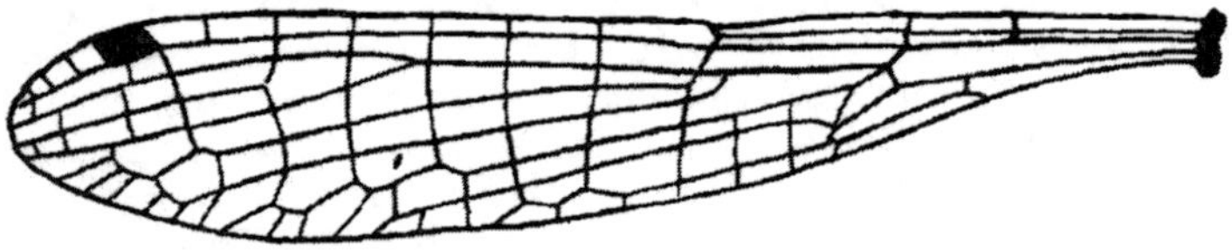

Fig. 158.—Hind-wing of *Rhodischnura nursei* (Mort.), male.

about one and a half times the length of costal in fore-wings, less than half the length in hind, distal side very oblique; sectors of arc arising from the lower end of arc, divergent from origins; arc situated at the level of distal antenodal nervure; *ab* always present and complete, arising well proximal of the level of *ac*, which lies about midway between the two antenodal nervures; *ab* continued on as the nervure *IA*, but an angulated medio-anal link present at the junction of the two; *Riv+v* arising well before the oblique nervure descending from the node; *IRiii* in continuation of that nervure.

Head narrow, without any frontal ridge; postocular spots always absent in the adult stage; thorax short and robust; posterior lobe of prothorax simple; abdomen short and robust, of nearly even width throughout, cylindrical; legs very robust, spines short and few in number; anal appendages short, inconspicuous, inferiors the more robust and sub-forcipate; segment 8 of female with a ventral apical spine.

Tubercles on apical border of segment 10 in the male widely separated by a deep notch. Genitalia: similar to those of *Ischnura*, the penis of the only known species ending in two long branches which form a close spiral on each side of the stem, the dorsal spines lying closely apposed to the stem and directed straight forwards

Genotype, *Rhodischnura nursei* (Morton)

149 Rhodischnura nursei (Morton) (Figs 158 & 159)

Ischnura nursei Mort, Trans Ent Soc Lond pp 306, 307, pl xxiv, figs 4–6 (1907)

Rhodischnura nursei Laid, Rec Ind Mus vol xvi, pp 171, 177, 178 (1919), Fras, Rec Ind Mus vol xix, pp 31–32 (1920), id, J Bombay Nat Hist Soc vol xxix, p 747 (1923)

Male —Abdomen 14 mm Hind-wing 9 5 mm

Head labium bright yellow, labrum, bases of mandibles, genæ, anteclypeus, and frons pale greenish-yellow, post-clypeus, vertex, and occiput black, eyes green above, pale greenish-yellow below, beneath the head and behind occiput

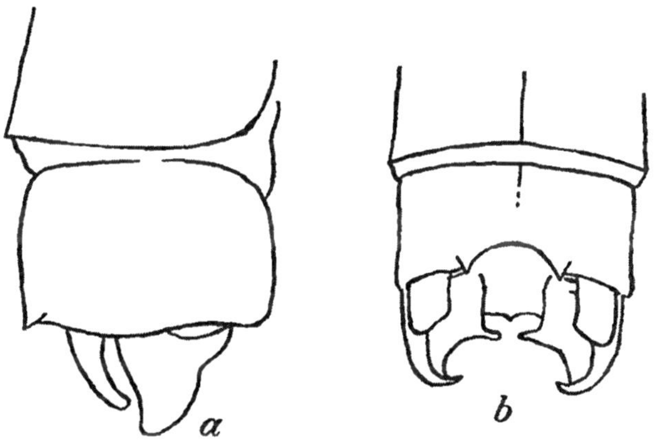

Fig 159 —Anal appendages of *Rhodischnura nursei* (Mort), male. *a* Right lateral view *b* Dorsal view

white or palest yellow *Prothorax* broadly black on dorsum, pale greenish-yellow on sides and anterior lobe, posterior lobe narrow, simple, rounded *Thorax* broadly black on dorsum as far lateral as half-way between the humeral and antero-lateral sutures, marked with narrow green antehumeral stripes, laterally green, paling to yellow below and beneath, this part in adults obscured with pruinescence *Legs* pale yellow, with a few fine black spines on tibiæ and femora *Wings* hyaline, pterostigma of fore-wing diamond-shaped, proximal and posterior borders thickened, black, distal thinner, black, costal pale yellow, membrane pale, red on the inner half changing to pink and then hyaline at the distal angle, covering less than one cell, pterostigma in hind-wing much smaller, hyaline, uncoloured, borders similar to that of fore-wing; 6 postnodal nervures in fore-wings, 5 in the hind. *Abdomen*: segments 1 to 4 bright crimson, 1 with two small black basal dorsal marks, 2 occasionally presenting a short transverse blackish spot subapically; segment 5 and the basal two-thirds of 6 pale citron-yellow, but in some specimens the latter segment becoming red again near its apical end; segments 7 to 10 and the apical third of 6 black, the intersegmental joints finely reddish. *Anal appendages* reddish or ochreous; superiors very short, curved strongly downward, hollowed out on the inner side, rounded as seen from above; inferiors half as long again, about as long as segment 10, very broad at base, deeply hollowed out on the inner side, the apex curling in and acutely pointed; apical border of segment 10 presenting two small tubercles separated by a broad deep notch.

Female.—Abdomen 15–17 mm. Hind-wing 11 mm.

Heterochromes and an isochromatic form are met with.

Isochromatic form.—Similar to the male, except for the following characters:—Pale parts of face yellow; postocular areas and occiput broadly yellow; antehumeral stripes and sides of *thorax* yellow, the stripes in the upper parts changing to rich ochreous, bordered finely with black; basal spots on segment 1 larger, and a fine mid-dorsal carinal black line on segment 2; segment 6 nearly entirely ochreous, changing to reddish at apical end, dorsum only slightly enfumed at apical end; ventral borders of segments 7 to 10 and the vulvar scales bright ochreous; pterostigma of all wings pale, with thick black proximal and posterior borders as in the hind-wing of male, that of fore-wing nearly twice the size of that of hind; 7 postnodal nervures in fore-wings, 6 in the hind. *Legs* with distal extensor surface of all femora black.

Heterochromatic form A.—Ground-colour similar to isochromatic form, but markings differing as follows:—Postocular areas and occiput bright reddish; eyes bluish above, paling to greenish below; antehumeral stripes pale bluish-green, unenclosed with black on the outer side, the dorsal black limited outwardly by the antehumeral stripes; segment 1 with the basal spots confluent at the base; segment 2 with a narrow thistle-shaped marking on dorsum finely confluent by its stalk to a narrow apical black ring; segments 3 to 6 with the mid-dorsal carina very finely black, this expanding subapically into a tiny cross-like spot; segments 7 to 10 with the dorsum narrowly black, this not extending to the base of 7 or 8, and not to apical border on 9 and 10; *anal appendages* short, pinkish, conical.

Heterochromatic form B.—Similar in many respects to the last, but the black markings much more extensive. *Head* similar to the male, postocular area black; antehumeral stripes not enclosed on the outer side, but margined here by a diffuse dark ochreous band; sides of *thorax* pale greenish; whole of abdominal segments broadly black on dorsum, only the sides and the basal ends narrowly of segments 3 to 7 pale greenish-yellow; *anal appendages* black.

Distribution.—Mainly confined to dry zones in N.W. INDIA, but occurring also in parts of BENGAL, BIHAR, UNITED and CENTRAL PROVINCES. Being very small and inconspicuous, it has probably escaped notice in many parts, and its distribution will probably be found in the future to be more extensive. I found it moderately common around Karachi and near tanks in Hyderabad, Sind. The *type* comes from Deesa, and is in the British Museum; other localities cited are Agra, Pusa, Nagpur, and Baghwonie, Duars (Bengal), from which latter locality I have several examples. It appears from September to November.

The bright crimson citron-yellow and black abdomen will serve to distinguish this species from any other; *Ischnura aurora*, of the same size, has the abdomen citron-yellow and black, and sometimes the basal segment red, but the terminal segments are azure blue.

Genus **ENALLAGMA** Charpentier (1840). (Fig. 160.)

Enallagma Charp., Lib. Europ. p. 21 (1840); Selys, Bull. Acad. Belg. (2) vol. xli, p. 496 (1876); Kirby, Cat. Odon. p. 145 (1890); Ris, Denkschr. med.-naturw. Ges. Jena, Bd. xii, pp. 310–13 (1908); id., Ann. S. African Mus. vol. xviii, pp. 317–19 (1918); Munz, Mem. Amer. Ent. Soc. no. 3, p. 61 (1919); Laid., Rec. Ind. Mus. vol. xvi, p. 183 (1919); Fras., J. Bombay Nat. Hist. Soc. vol. xxix, p. 747 (1923).

Dragonflies of small size and slender build, body rather more robust and the abdomen rather shorter than in the genus *Ceriagrion*; colours non-metallic, usually black, marked with

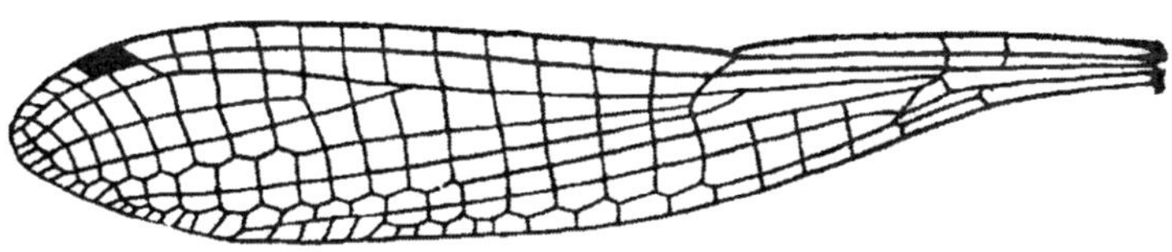

Fig. 160.—Hind-wing of *Enallagma cyathigerum* Charp., male.

blue or green; females not usually polychromatic; wings hyaline; pterostigma shaped and coloured the same in fore- and hind-wings, braced, covering about one cell or less; about 8 to 15 postnodal nervures in fore-wings, 6 to 10 in the hind; discoidal cell acutely pointed at the distal end, with costal side less than half the length of posterior in fore-wing and about half the length in the hind, basal side of the same length as, or shorter than, costal side in fore-wing, of the same length or only two-thirds the length in the hind; arc situated at the level of the distal antenodal nervure; *ab* always present and complete, arising well proximal to the level of *ac*, which usually lies about midway between the two antenodal nervures or rather nearer the proximal one; *ab* continued on as *IA*, but an angulation (medio-anal link) present at the junction of the two: *Riv+v* arising well before the oblique nervure descending from the node, *IRiii* in continuation of that nervure.

Head narrow, without any frontal ridge; postocular coloured spots always present, but often elongate and confluent across the occiput; thorax short and robust; posterior lobe of prothorax usually simple, rounded; anterior border of thorax without hooks. Abdomen moderately short and slender, slightly tumid at base, and again at terminal segments; legs variable, short, or moderately long, with few or numerous fine spines; segment 8 in the female with a ventral apical spine. Genitalia: lamina narrowly and deeply cleft; hamules broad and foliate, opposing borders slightly divergent; lobe depressed, scrotal-shaped.

Genotype, *Enallagma cyathigerum* Charp.

Distribution.—Throughout Europe, Asia, America, and Africa.

The genus is richest in North American species, and has only a single representative in Europe, whilst a distinct group is confined to Africa. Within our limits only four species have been described, one of which, a cosmopolitan species, has invaded Kashmir from the north; another, a Javan species, occurs in Ceylon, and sporadically in parts of India; whilst a third, *E. parvum*, occurs widely in the Oriental Region, and has been placed by Kennedy in a separate genus, *Amphiallagma*; the fourth species is known only from the female.

Key to the Species of Enallagma.

1.	Very small species, with abdomen not more than 17 mm. in length; postnodal nervures in fore-wings only 6 to 7 in number.	*parvum* Selys, p. 376.
	Larger species, with abdomen more than 20 mm. in length; postnodal nervures in fore-wing more than 7 in number....	2.
2.	Dorsum of segment 2 with a small subapical cordate black spot isolated or narrowly confluent with an apical annule	*cyathigerum* Charp., [p. 373.
	Dorsum of segment 2 with a broad black stripe extending from base to apex of segment	3.
3.	Dorsal marking on segment 1 extending from base to apex of segment; fore-wing with 10 postnodal nervures	*insulæ* Fras., p. 378.
	Dorsal marking of segment 1 not extending to apex of segment; only 8 postnodals.	*malayanum* Selys, [p. 375.

150. Enallagma cyathigerum Charpentier. (Figs. 160 & 161.)

Agrion aereus Harris, Expos. Eng. Ins. pl. xxix, figs. 3, 4 (1782).
Agrion hastulatum Steph. Ill. Brit. Ent. Mand. vol. vi, p. 74 (1836); Selys, Lib. Europ. p. 219 (1840); Ramb., Ins. Névrop. p. 273, pl. vii, fig. 3 c (1842).
Enallagma cyathigerum Charp., Lib. Europ. p. 163 (1840); Selys, Bull. Acad. Belg. (2) vol. xli, p. 505 (1876); Kirby, Cat. Odon. p. 145 (1890); Lucas, Brit. Drag. pp. 297–307 (1900); MacL., Ent. Month. Mag. (2) vol. ii, pp. 110–11 (1900); Campion, Entomologist, p. 3 (Dec. 1906); Morton, Trans. Ent. Soc. Lond. p. 307 (1907); Campion, Entomologist, p. 177 (Jan. 1909); id., ibid. p. 295 (1910); Kennedy, Proc. U.S. Nat. Mus. vol. lii, no. 2192, pp. 599, 600, 603, 607, 608, 612, 613, 615, 629, 630, 633 (1917); Campion, Entomologist, pp. 116, 204 (1919); Laidlaw, Rec. Ind. Mus. vol. xvi, p. 183 (1919); Fras., J. Bombay Nat. Hist. Soc. vol. xxix, p. 747 (1923); Needham, Zool. Sinica, ser. A, vol. xi, pp. 265, 266 (1930).
Agrion cyathigerum Charp., Lib. Europ. p. 163, pl. xlii, fig. 1 (1840); Selys, Rev. Odon. p. 205 (1850); MacL., Cat. Brit. Neur. vol. xviii (1870).
Agrion charpentieri Selys, Rev. Zool. p. 214 (1840); id., Bull. Acad. Belg. (2) p. 95 (1840).
Agrion pulchrum Hagen, Syn. Lib. Europ. p. 80 (1840).
Agrion brunnea Evans, Brit. Libel. p. 15, pl. iv, fig. 8 (1845).

Male.—Abdomen 25–28 mm. Hind-wing 18–21 mm.

Head: labium white or pale blue laterally; labrum, bases of mandibles, genæ, anteclypeus, and frons pale azure blue, as well as the basal joint of antennæ, a basal median small point of black on the labrum; vertex and occiput black, marked with large pyriform postocular pale blue spots, and a small streak of the same colour lying between but not joining them; eyes dark blue above, pale blue beneath; beneath head pale blue, black centrally. *Prothorax* black, the sides broadly pale azure blue as well as the anterior lobe, a large pyriform subdorsal spot on the middle lobe, and the lateral borders of the posterior lobe finely; the latter broad, rounded, trilobate, the middle portion slightly projecting and rounded. *Thorax* bronzed black on dorsum as far lateral as the humeral suture, marked with very broad, even, pale azure blue antehumeral stripes; laterally pale azure blue, with a small black spot on the upper part of the postero-lateral suture; beneath pale, often thinly pruinosed. *Legs* black, flexor surface of femora and extensor of tibiæ yellow, the former often white with pruinescence; 8 to 10 fine spines on hind femora and tibiæ. *Wings* hyaline; pterostigma black, framed finely in yellow and thick black nervures, the middle often bluish from pruinescence in adults, covering one cell; 14 to 15 postnodal nervures in fore-wings, 11 to 13 in the hind. *Abdomen* bright glossy azure blue, marked with black as follows:—Segment 1 with a broad basal spot not nearly extending to apex of segment; segment 2 with an apical spot, variable in shape, connected to a fine black apical ring by a short stalk, and extending basally for nearly half the length of segment, this spot sometimes detached from the apical ring and shaped like a bird in flight, or cordate or hatchet-shaped: segments 3 to 5 with broad apical black rings extending basally for about one-fourth the length of segments, but also variable and pointed on the mid-dorsum or extending to half the length of segment 5; segment 6 with this apical marking extending for more than half the length of segment; segment 7 entirely black on dorsum, whilst 8 and 9 are entirely blue, but sometimes with a small lateral black spot; segment 10 broadly black on dorsum. *Anal appendages* black; superiors about one-third the length of segment 10, obtuse, quadrate, hollowed out within, the

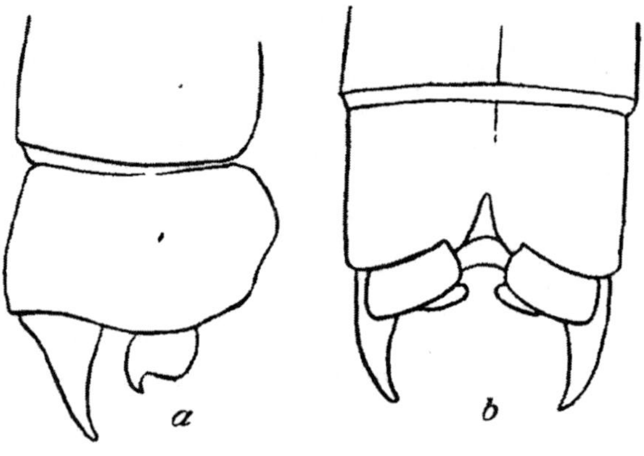

Fig. 161.—Anal appendages of *Enallagma cyathigerum* Charp., male. *a.* Left lateral view. *b.* Dorsal view.

inner inferior angle turned down and inwards; inferiors nearly as long as segment 10, broad at base, tapering rapidly to a fine point, strongly convex as viewed from below, and with the apices turned slightly in.

Female.—Abdomen 22–28 mm. Hind-wing 18–22 mm.

Closely similar to the male in markings, but the ground-colour pale greenish-yellow instead of blue. The mid-dorsal carina of *thorax* finely yellow; the pterostigma yellow, slightly darkening at the middle and pruinosed in adults; abdominal markings differing as follows:—Segment 1 with a larger spot; segment 2 with the apex of apical spot prolonged broadly as far as base of segment, the marking thus shaped like the head of a thistle; segments 3 to 6 with the apical spots prolonged as a thick stripe nearly to base of segments, where they end in a dorsal point; segment 7 very similar to 6; segment 8 with a broad dorsal spot shaped like a bishop's mitre, the point just reaching base of segment; segments 9 and 10 with broad dorsal marking, broadest at base, narrowing gradually towards apical borders of segments. *Anal appendages* small, pointed, yellow; vulvar scale yellow, robust, extending to end of abdomen; spine beneath segment 8 very robust and long.

Distribution.—Europe, Central Asia, including KASHMIR and Tibet, North America, and the British Isles. I have specimens from Kashmir and Tibet which do not differ markedly from those from Europe and America. Breeds in ponds and marshes.

151. **Enallagma malayanum** Selys.

Enallagma malayanum Selys, Bull. Acad. Belg. (2) vol. xli, p. 536 (1876); Kirby, Cat. Odon. p. 147 (1890); Ris, Denkschr. med.-naturw. Ges. Jena, Bd. xii, pp. 311–13 (1908); Laid., Rec. Ind. Mus. vol. xvi, p. 183 (1919); Fras., J. Bombay Nat. Hist. Soc. vol. xxix, p. 748 (1923).

Male.—Abdomen 22 mm. Hind-wing 15 mm.

Head: labrum pale blue or white; labrum, bases of mandibles, genæ, clypeus, and frons pale azure blue; vertex and occiput black, the latter marked with bluish-green oval postocular spots and a narrow line between them of the same colour; eyes olivaceous, paler below; beneath head bluish-white. *Prothorax* black, anterior lobe and a large spot on each side azure blue; posterior lobe rounded. *Thorax* bronzed black on dorsum, marked with a broad ante-humeral stripe of even width, greenish-yellow in colour; laterally bluish-green, with a short black streak on the upper part of each lateral suture; whitish beneath. *Legs* bluish-white, marked broadly with black on extensor surface of femora and flexor surface of tibiæ. *Wings* hyaline; pterostigma yellowish-white, framed in thick black nervures, covering less than one cell, distal side oblique; 7 to 8 postnodal nervures in the fore-wing, 6 to 7 in the hind. *Abdomen* pale blue at the sides, changing to creamy yellow beneath, marked with black as follows:—Segment 1 with a quadrate spot at base; segment 2 with a broad dorsal band not extending to base, but confluent by a narrow stalk with a narrow apical ring, the band bilobate; segments 3 to 6 with broad dorsal bands tapered to a point at the base of segments and expanding subapically, before finally becoming confluent with narrow apical rings, this expanded portion on the last two segments enclosing a lateral blue spot; segment 7 entirely black on dorsum except a narrow basal blue annule; segments 8 and 9 azure blue, unmarked; segment 10 blue, with a narrow dorsal black band, its sides also black, the anterior border emarginate. *Anal appendages*: superiors slightly shorter than segment 10, broadly conical and slightly notched at the apex, which is black; inferiors nearly as long as superiors, simple, sloping up towards superiors.

Female.—Abdomen 20 mm. Hind-wing 15 mm.

Closely resembling the male, but the ground-colour greenish-yellow instead of blue, and the markings on abdomen somewhat different, as follows:—Segment 2 with a broader dorsal black band; segments 8 and 9 with broad dorsal black bands extending their whole length; segment 10 blue at the apical border only; vulvar scale robust, extending to end of abdomen; a robust ventral spine at the apex of segment 8.

Distribution.—The *type*, a male in the Selysian collection, comes from Java. Laidlaw (*l. c.*) has reported it from Nagpur, CENTRAL PROV., 1,000 ft., during September. I have seen a few specimens from ASSAM, and Ris has reported it from Colombo, CEYLON. The species appears to have a wide and scattered distribution, and is common nowhere. Distinguished from *E. cyathigerum* by the broad complete black bands on abdomen, and from *E. parvum* by the broad oval post-ocular spots.

152. **Enallagma parvum** Selys. (Fig. 162.)

Enallagma parvum Selys, Bull. Acad. Belg. (2) vol. xli, p. 537 (1876); Kirby, Cat. Odon. p. 147 (1890); Morton, Trans. Ent. Soc. Lond. p. 307, pl. xxiv, figs. 13, 14 (1907); Laid., Rec. Ind. Mus. vol. xvi, p. 184 (1919); Fras., ibid. vol. xxiv, p. 310 (1922); id., J. Bombay Nat. Hist. Soc. vol. xxix, p. 747 (1923); id., Rec. Ind. Mus. vol. xxvi, p. 491 (1924); id., ibid. vol. xxxiii, p. 448 (1931).

Ischnura immsi Laid., Entomologist, pp. 236–7, text-fig. (1913); id., Rec. Ind. Mus. vol. xvi, p. 184 (1919).

Male.—17 mm. Hind-wing 11 mm.

Head: labium white; labrum, bases of mandibles, genæ, and anteclypeus pale sky-blue; postclypeus black or broadly black at base; frons pale blue; vertex and occiput black, the latter with small postocular spots confluent across the middle line by a narrow occipital stripe; eyes deep sky-blue, paling below; behind head black. *Prothorax* broadly black on dorsum, the sides pale blue. *Thorax* black on dorsum, with very broad antehumeral azure blue stripes narrowly bordered outwardly with black; laterally pale blue, fading to white below and beneath. *Legs* white, with a black line on the extensor surface of femora, spines short, 4 to 5 on hind tibiæ. *Wings* hyaline; pterostigma small, covering less than one cell, very oblique distally, yellow in tenerals, blackish in adults, with a fine pale frame and thick black enclosing nervures; 7 postnodal nervures in fore-wings, 6 in the hind. *Abdomen* pale sky-blue, marked with black as follows:—Segment 1 with a broad quadrate dorsal spot and narrow blue apical annule; segment 2 with a broad thistle-shaped dorsal spot extending the whole length of segment; segments 3 to 7 with fine apical annules confluent with a rounded spot, the apex of which is continued on as a narrow stripe as far as the base of segments; segments 8 to 10 azure blue, the latter with a mid-dorsal narrow black stripe. *Anal appendages*: superiors slightly shorter than segment 10, curled plates, the concavities facing one another, the inferior ends quadrate and slightly truncate; inferiors of the same length, conical, broad at base, sloping strongly upward, and ending in a moderately acute point.

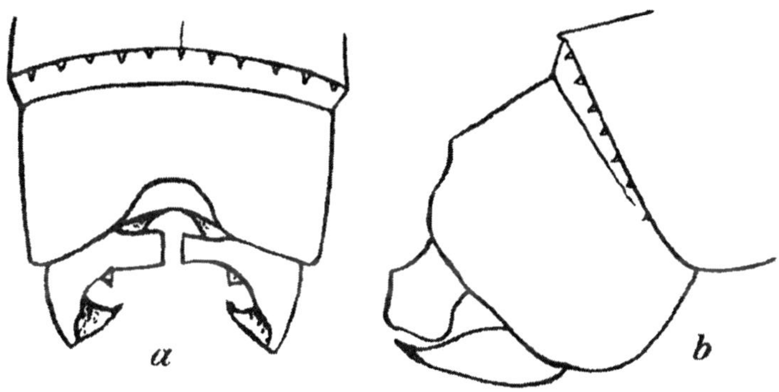

Fig. 162.—Anal appendages of *Enallagma parvum* Selys, male.
a. Dorsal view. *b*. Right lateral view.

Isochrome female.—Abdomen 17 mm. Hind-wing 11 mm. *Head* similar to the male, but the middle portion of the occipital stripe yellowish, the postocular spots greenish; eyes paler azure blue. Ground-colour of *thorax* greenish-yellow or definitely yellow, paling to pale blue below and beneath, the antehumeral stripe bordered rather more broadly with black outwardly. *Abdomen* with the dorsal stripes broader, so that the apical dilatation scarcely shows, and the confluence with the apical rings is much broader; segments 8 to 10 broadly black on dorsum, but the border of the latter blue. *Anal appendages* very small, blue, conical processes; vulvar scales pale blue, extending to end of abdomen.

Heterochrome female.—Differs from the above as follows, but the ground-colour similar:—Postclypeus entirely blue, unmarked; postocular spots and stripe unenclosed, occiput black posteriorly and blue at centre, as well as outwardly;

prothorax unmarked with black, dorsum pale yellow, blue at the sides; *thorax* with the antehumeral stripes unenclosed with black.

In some specimens of the male the postocular spots are bordered narrowly with black behind, but the central stripe not so, so that the spots are merely narrow gulfs opening on to the blue at the back of the head. In these specimens, also, the black on the postclypeus is restricted to its base, and the anterior collar of the prothorax is blue. The description of the type, written, apparently, from a teneral specimen, approaches this condition, which is similar to that of the heterochromatic female, as the postclypeus is unmarked with black and the postocular spots and stripes are unenclosed with black behind.

Distribution.—Widely distributed through South Asia, and abundant in most parts of INDIA, BURMA, and CEYLON. Breeds in weedy tanks during the greater part of the year. It is one of the smallest Indian dragonflies, a character which serves to distinguish it from other species of the same genus; the shape of the anal appendages and the postocular spots will also distinguish it from others.

Type in the Selysian collection; paratypes in the Indian, Pusa, and British Museums and most national and private collections.

153. **Enallagma insulæ** Fraser.

Enallagma insula Fras., Rec. Ind. Mus. vol. xix, pp. 32, 33 (1920); id., ibid. vol. xxiv, p. 309 (1922); id., J. Bombay Nat. Hist. Soc. vol. xix, p. 748 (1923).

Female.—Abdomen 24 mm. Hind-wing 17 mm.

Head: labium white; labrum palest green, finely black at base; bases of mandibles, genæ, frons, and anteclypeus pale green, postclypeus black; vertex black; occiput pale blue, a short streak of black projecting into the blue from border of eyes so as nearly to enclose a large postocular blue spot; eyes dark brown above, greenish at the sides and beneath. *Prothorax* broadly black on dorsum, broadly blue on the sides, the anterior collar, two tiny points on mid-dorsum of middle lobe, and the rounded posterior border of the posterior lobe finely blue. *Thorax* broadly black on dorsum, marked with a narrow pale blue antehumeral stripe on each side; laterally palest blue, without markings; whitish beneath. *Wings* hyaline; pterostigma black, finely framed in white and thick black nervures, covering nearly one cell; 10 post-nodal nervures in fore-wings, 9 in the hind. *Abdomen* pale blue, broadly marked with black as follows:—Segment 1 with a broad quadrate spot confluent with an apical ring; segment 2 with a broad dorsal streak not quite reaching the base of the segment, but expanded subapically and then constricted again to join a narrow apical ring; segments 3 to 8 similar, but the streaks longer and narrower, that on the last of nearly even width throughout; segment 9 with two small subdorsal basal projecting black points which extend for about half the length of the segment; segment 10 entirely blue. *Anal appendages* pale blue, conical, very short; vulvar scale bluish-white, moderately robust.

Male unknown.

Distribution.—The *type*, in the Indian Museum, the only specimen known, is a female from the Chilka Lake, Ganjam district, MADRAS PRES., collected in October. Distinguished from *E. parvum* by its much larger size, from *E. cyathigerum* by the complete black band on dorsum of segment 2, from *E. malayanum* by the larger number of postnodal nervures and the dorsal spot on segment 1 extending from base to apex of segment.

Genus **AGRIOCNEMIS** Selys (1869). (Fig. 163.)

Agriocnemis Selys, Pollen & Van Dam, Faune Mad., Ins. p. 24 (1869); id., Bull. Acad. Belg. (2) vol. xliii, p. 142 (1877); Kirby, Cat. Odon. p. 158 (1890); Ris, Ann. S. Africa Mus. vol. xviii, p. 335 (1918); Laid., Rec. Ind. Mus. vol. xvi, p. 178 (1919); Munz, Mem. Amer. Ent. Soc. no. 3, p. 60 (1919); Fras., J. Bombay Nat. Hist. Soc. vol. xxix, p. 746 (1923); Ris, Ark. Zool. Uppsala, Bd. 21 A, no. 31, p. 5 (1930).

Dragonflies of the smallest size and slender build; colours non-metallic, usually black marked with blue, or with the

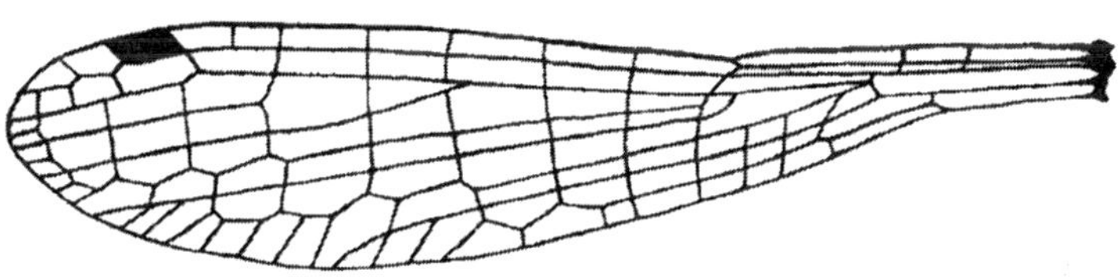

Fig. 163.—Hind-wing of *Agriocnemis pygmæa* (Ramb.), male.

abdomen terminating with bright ochreous; females polychromatic; wings hyaline; pterostigma similar or dissimilar in fore- and hind-wings, very small, covering less than one cell, diamond-shaped, but distal side rather more oblique than proximal, usually braced; postnodal nervures numbering from 5 to 6 or, more rarely, 9; discoidal cell acutely pointed at the distal end, with costal side rather more than half the length of posterior in fore-wing, and about three-fourths the length in the hind; basal side considerably shorter than costal side in fore-wing, and only half the length of that side in the hind, distal side moderately oblique; sectors of arc arising from lower end of arc, divergent from their origins; *arc situated distal to the level of the distal antenodal nervure*; *ab* always present and complete, arising well proximal of the level of *ac*, which usually lies slightly nearer the level of the distal antenodal nervure; *ab* continued on as the nervure *IA*, but an angulation (medio-anal link) present at the junction of the two; *Riv+v* arising well before the oblique nervure descending from the subnode; *IRiii* in continuation of that nervure.

Head narrow, without any frontal ridge; postocular coloured spots invariably present; thorax short and robust; posterior lobe of prothorax simple, but slightly variable in the species; no hooks on anterior border of thorax. Abdomen slim and cylindrical, dilating somewhat towards the anal segments; legs very short, 4 to 5 spines on the hind pair of tibiæ and usually about 7 on the corresponding femora; anal appendages very variable in the species; segment 8 in the female without any ventral spine.

Genitalia: lamina very deeply and very narrowly cleft; hamules quadrate plates, the free borders of which almost meet across the middle line; penis simple, its apex curling over dorsum and ending in two short points; lobe small, rounded, hidden.

Genotype, *Agriocnemis lacteola* Selys.

Distribution.—Throughout the tropical zones of the Old World, from Africa to Australia and Oceania. The genus, however, is essentially an Oriental one, the bulk of the species coming from southern Asia. They rank among the smallest known dragonflies, and yet, in spite of their diminutive size, by utilizing the air-currents have extended over an enormous area. Like *Ischnura*, teneral specimens are coloured red, and pass through a series of shades until the adult colours are finally reached. Adults, especially the males, are inclined to become more or less densely pruinosed white on the thorax and head. They breed in weedy ponds and marshes or in the shallows of sluggish rivers.

Species of the genus fall rather naturally into two groups, at least in so far as the Oriental members are concerned.

Group 1 is characterized by the labrum being metallic blue and the anal appendages variable.

Group 2 has the labrum non-metallic and the anal appendages more or less homogeneous, the superiors being considerably longer than the inferiors. In this group the majority of the species have a broad black marking on the dorsum of segment 2 of the abdomen along with a pair of small blue spots, not unlike the spectacles on the hood of a cobra.

Key to the Species of Agriocnemis.

1.	Labrum metallic blue	2.
	Labrum non-metallic	3.
2.	Superior anal appendages longer than inferiors	*pygmæa* (Ramb.), p. 398.
	Superior anal appendages shorter than inferiors	*femina* (Brauer), p. 402.
3.	Abdomen palest blue, with the black dorsal markings very restricted	4.
	Abdomen darker blue or the terminal segments ochreous, with the black dorsal markings largely obscuring the blue ground-colour	5.
4.	Abdominal segments 4 to 10 unmarked with black	*lacteola* Selys, p. 381.
	Abdominal segments 4 to 6 with black apical rings confluent with a subapical spot shaped like an arrow-head; segment 7 with a dorsal stripe on its basal three-fourths	*pieris* Laid., p. 384.
5.	Black dorsal marking on segment 2, without blue eye-spots	*splendidissima* Laid., p. 392.
	Black dorsal marking on segment 2 with a pair of blue eye-spots	6.
6.	Fore-wings with 8 or 9 postnodal nervures	*aborensis* (Laid.), p. 394.
	Fore-wings with only 6 or 7 postnodal nervures	7.
7.	Abdominal segment 8 ochreous, with a black anchor-shaped marking on dorsum.	*d'abreui* Fras., p. 396.
	Abdominal segment 8 entirely black	*clauseni* Fras., p. 390.
	Abdominal segment 8 blue, with black markings	8.
8.	Segment 8 blue, with a mid-dorsal carinal black stripe which tapers away towards the apex of the segment, and with or without a small black spot on each side of its end, which may or may not be confluent with it	*naia* Fras., p. 387.
	Segment 8 blue, with a mid-dorsal carinal black stripe which is very broad at and encircles the anal end of the segment and tapers away towards the base	*nana* Laid., p. 386.

154. **Agriocnemis lacteola** Selys. (Fig. 164.)

Agriocnemis lacteola Selys, Bull. Acad. Belg. (2) vol. xliii, p. 144 (1877); Kirby, Cat. Odon. p. 158 (1890); Laid., Rec. Ind. Mus. vol. viii, pp. 347, 348 (1914); id., ibid. vol. xvi, pp. 178, 179 (1919); Fras., J. Bombay Nat. Hist. Soc. vol. xxix, p. 747 (1923).

Male.—Abdomen 16–18 mm. Hind-wing 9·5–10·5 mm.

Head: labium white; labrum creamy white; anteclypeus and postclypeus, bases of mandibles, genæ, and the frons broadly, as far as base of antennæ, putty-coloured; vertex and occiput black as well as the base of postclypeus narrowly ; a narrow stripe across occiput, slightly interrupted on both sides of the centre of occiput, and very sinuous external to this interruption, pinky white in colour ; eyes black above, palest blue beneath and at sides. *Prothorax* black on dorsum, palest blue or bluish-white at the sides ; posterior lobe produced squarely back at its central part, and finely margined with white ; anterior lobe creamy white. *Thorax* black on dorsum as far lateral as the antero-lateral suture, with two very narrow antehumeral stripes ; laterally pale blue, marked with a small spot of black at the upper part of postero-lateral suture ; beneath white. *Legs* white, with some obscure mottling at extreme distal ends ; spines black, to the number of four, on the hind tibiæ. *Wings* hyaline ; pterostigma pale blue

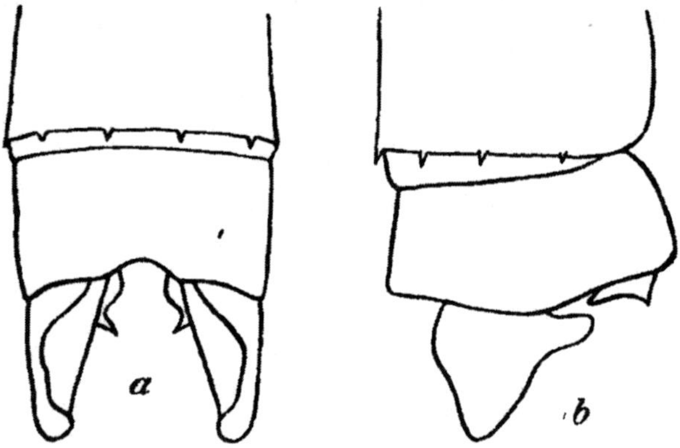

Fig. 164.—Anal appendages of *Agriocnemis lacteola* Selys, male.
a. Dorsal view. *b.* Right lateral view.

in the hind-wings on the upper surface, but with the hinder three-fourths black beneath, black in the fore-wings, with a fine blue line along the costal border above, but entirely blackish beneath ; in some specimens all pterostigmas alike black, with the costal border finely blue, covering less than one cell, unbraced ; 6 to 7 postnodal nervures in fore-wings, 5 in the hind ; the arc very far distal to the level of outer antenodal nervure. *Abdomen* pale blue, almost white on the proximal segments, marked with black as follows :—Segment 1 with a dorsal spot expanding subapically and confluent with a fine apical black ring ; segment 2 with a mid-dorsal stripe, thick for the basal three-fourths, expanded somewhat at base and again squarely subapically, but the apical fourth very fine ; in other specimens this dorsal marking rather broader and shaped like a vase, the apical stem confluent with a narrow apical annule; segment 3 with a narrow mid-dorsal stripe tapering finely to near apical end of segment or else extending so as to become confluent with an apical ring, and with a small triangular subapical expansion; remaining segments unmarked or segments 4 and 5 with the small subapical triangular spot only. *Anal appendages* pale blue; superiors as long as segment 10, broadly triangular as seen in profile and obtusely pointed, whilst the ventral border is produced into a long, slightly curled, hollow spine, which is usually more or less hidden beneath segment 10; inferiors very much shorter, and almost entirely hidden within segment 10, broad and stunted, with a robust spine directed upward, and another, slightly shorter, directed inward.

Female.—Abdomen 16–18 mm. Hind-wing 15·5–17 mm.

Differs rather widely from the male—a much stouter insect, with the black markings more in evidence. *Head* with face definitely blue, except the labrum, which is as in the male; postocular spots a beautiful azure blue, and usually rather broader than in the male, forming large comma-shaped spots on each side of occiput; eyes olive-green, capped with black. *Prothorax*, including the shape of the posterior lobe, similar to the male. *Thorax* also similar to the male, but the ante-humeral stripes often bluish-green and the sides a deeper blue. *Legs* with the extensor surface of all femora black, this expanding at distal ends to form an incomplete ring. *Wings* similar, but the pterostigma pale yellow, centred with greyish and framed in thick black nervures, the anterior border paler, more than twice as long as broad. *Abdomen* azure blue, the dorsum of all segments except 10 rather broadly black, this expanding subapically on all segments except 9, and again narrowing to become confluent with a narrow apical black ring; segment 9 with a broad black basal triangle on dorsum, which may extend as far as the apical border of segment on mid-dorsum and at the sides to enclose a triangular subdorsal blue spot: segment 8 has occasionally a narrow lateral longitudinal stripe, which is confluent at a point with the dorsal black to enclose an apical lateral blue spot on segment 9. *Anal appendages* small, conical, blue; vulvar scales blue.

Distribution.—BENGAL, ASSAM, and SIKKIM. I have examined specimens from Sikkim, from Hasimara and Jalpaiguri (Bengal), and from Margherita (Assam). More recently a number have reached me from Nowgong, Assam; Dr. Laidlaw reports the species from Dibrugarh, Assam.

The *type*, in the Selysian collection, is from Bengal. The female has not hitherto been described.

155. Agriocnemis pieris Laidlaw.

Agriocnemis pieris Laid., Rec. Ind. Mus. vol. xvi, pp. 179, 180 (1919); Fras., J. Bombay Nat. Hist. Soc. vol. xxix, p. 747 (1923); id., Rec. Ind. Mus. vol. xxvi, p. 490 (1924); id., ibid. vol. xxxiii, p. 448 (1931).

Male.—Abdomen 16–18 mm. Hind-wing 9–10 mm.

Head: labium white; labrum, bases of mandibles, genæ, clypeus, and a stripe across the frons palest azure blue; postclypeus narrowly black at base, from which, at the middle, runs a fine black line to interrupt the frontal stripe by becoming confluent with the black on vertex; the latter and occiput and a broad cap on upper surface of eyes black; occiput with a narrow transverse stripe made up of two narrow comma-like postocular pinkish spots not quite connected by an equally narrow median stripe; eyes at sides and below pale blue. *Prothorax* black on dorsum, palest blue at the sides; posterior lobe produced squarely back at its middle and finely pinkish along its borders, as also is the anterior lobe. *Thorax* black on dorsum as far lateral as the antero-lateral suture, but rather more extensively so above than below, marked by pinkish narrow antehumeral stripes; laterally blue, with a small black spot on the upper part of postero-lateral suture; beneath white. *Legs* white, broadly marked with black on the extensor surface of all femora, the black expanding at extreme distal ends of limbs; spines black, 4 on hind tibiæ, 5 on the femora. *Wings* hyaline; pterostigma very narrow and elongate, distal side rounded off into posterior and very oblique, pale yellow or pinkish, with the centre blackish or grey, especially in the hind-wings, enclosed in thick black nervures covering nearly one cell; 6 or 7 postnodal nervures in fore-wings, 5 in the hind; arc very far distal to the outer antenodal nervure. *Abdomen* pale blue, deepening on the distal four segments (china-blue or the pale blue of the early morning sky), marked with black as follows:—Segment 1 broadly black on dorsum, and with a narrow apical annule; segment 2 with an anchor-shaped dorsal marking, the flukes of the anchor sometimes curled round until they become confluent with the shaft to enclose two small spots of the ground-colour; rarely the anchor spot finely connected in the middle line to an apical annule; segment 3, and often 4, but there more finely, black along the mid-dorsal carina, this black expanding subapically to form a small triangular spot; segments 2 to 6 with well-defined apical annules; segments 4 to 6 with subapical dorsal black fleur-de-lis spots narrowly confluent with the apical annules; on segment 6 this spot extends basally along the mid-dorsal carina as a fine stripe which, however, falls short of the base of segment; segment 7 somewhat similar, but the dorsal stripe much thicker and reaching the base at a point, and with the fleur-de-lis spot not connected to an apical annule, whilst in some this spot is entirely missing; segment 8 unmarked, or with a fine basal annule prolonged for a variable distance along the mid-dorsal carina; segments 9 and 10 unmarked, but the former in old specimens sometimes with a fine basal annule. *Anal appendages* pale blue or pinkish; superiors slightly longer than segment 10, shaped very much as in *A. lacteola*; inferiors similar to those of this species, the spines tipped with black.

Female.—Abdomen 18 mm. Hind-wing 11·5 mm.

A stouter, more robust insect than the male, and with the black markings much more extensive on the *abdomen*. Differs as follows:—Labrum rather more broadly black at base; postclypeus with three black spots along its anterior margin as as well as the base more broadly black; the pale band on frons broadly interrupted at the middle by confluence of the black on postclypeus and vertex. *Prothorax* as in the male. *Legs* more broadly black on extensor surface of femora, and the flexor surfaces of tibiæ in addition black for their proximal three-fourths. *Wings* with 7 to 8 postnodal nervures in the fore, 6 in the hind; pterostigma golden yellow, suffused with greyish at the centre. *Abdomen* broadly marked with black on dorsum from segments 1 to 9; segments 3 to 7 with the dorsal band meeting base of segments at a point and broadly confluent with an apical ring distally, the band showing an abrupt expansion subapically; segment 2 with the band extending broadly to the base and with the subapical expansion; segments 8 and 9 entirely black, save for a narrow bordering of yellow along the venter, and on segment 8 a narrow longitudinal lateral stripe of blue on the basal two-thirds; segment 10 black at the base only, this black showing a square projection on each side. *Anal appendages* short, conical, blue; vulvar scales robust, extending to beyond the end of abdomen, carneous. Ground-colour of thorax and abdomen much darker blue than in the male.

Distribution.—Western Ghats of India, south of Bombay. *Type* in the Indian Museum, from Castle Rock, N. Kanara. I have found it common in S. Kanara, Malabar, Nilgiri Wynaad, and Coorg.

The species frequents grassy marsh-lands, in which it breeds. Very conspicuous on the wing in spite of its small size, owing to its pearly blue colouring. Closely allied to *A. lacteola*, from which it differs principally by abdominal segments 4 to 10 also bearing black markings, these segments being immaculate in the former.

156. Agriocnemis nana (Laidlaw). (Fig. 165.)

Argiocnemis nana Laid., Rec. Ind. Mus. vol. viii, p. 348 (1914); id., ibid. vol. xvi, p. 182 (1919); Fras., J. Bombay Nat. Hist. Soc. vol. xxix, p. 747 (1923).

Male.—Abdomen 18 mm. Hind-wing 10 mm.

Head: labium white; labrum pale blue, broadly black at base; bases of mandibles, genæ, ante- and post-clypeus, and frons pale blue, the postclypeus black at its base; vertex and occiput black, the latter with a comma-shaped postocular blue spot on each side slightly separated from a narrow stripe of the same colour in the middle of occiput; eyes black above, blue laterally and below. *Prothorax* broadly black on dorsum, blue laterally; posterior lobe produced at its centre into a slightly emarginate lobe. *Thorax* black on dorsum as far lateral as the antero-lateral suture, but more so above than

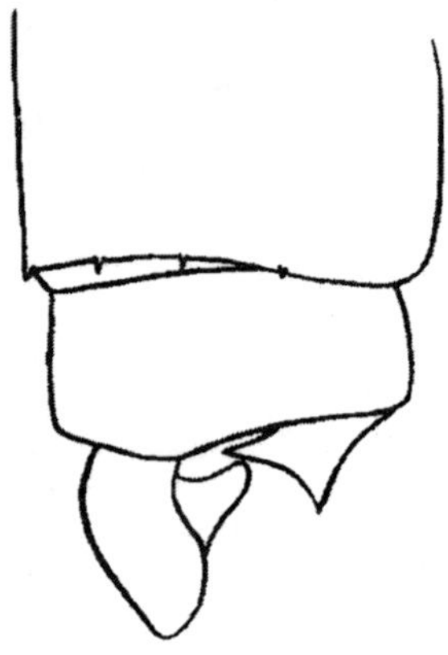

Fig. 165.—Anal appendages of *Agriocnemis nana* (Laid.), male, right side.

below, marked with narrow blue antehumeral stripes; laterally blue, paling to white beneath, with a small black spot on the upper end of the postero-lateral suture. *Legs* white, extensor surface of femora and flexor surface of all tibiæ black; about four black spines on hind pair of tibiæ. *Wings* hyaline; pterostigma greyish-black, framed in thick black nervures, and with its costal border pale bluish-white, covering less than one cell, distal side rounded off with the posterior and very oblique, as in *A. lacteola* and *A. pieris*; arc well distal to the outer antenodal nervure; 6 postnodal nervures in fore-wings, 4 to 5 in the hind. *Abdomen* azure blue, marked with black as follows:—Segment 1 with a broad quadrate dorsal spot, its apical joint blue ; segment 2 with a broad, squared, black dorsal spot constricted abruptly near apical border of segment to become confluent with a narrow apical annule, the sub-basal portion bearing a pair of small oval blue spots, the whole marking strikingly like the "spectacles" of the cobra's hood ; segments 3 to 7 with a broad dorsal stripe constricted subapically and expanded immediately proximal to this, meeting the base of segments at a point and confluent with narrow apical rings ; segment 8 with a broad dorsal stripe, which on the apical half of segment expands as a broad annule, and may partially enclose on each side a spot of the ground-colour ; segments 9 and 10 entirely black, but 9 with or without a small lateral blue elongate spot. *Anal appendages* : superiors pale blue on dorsum, black beneath, shaped very similarly to those of *A. lacteola*, about as long as segment 10, roughly triangular in profile, and with a stout, obtusely pointed ventral spine, hollowed out within ; inferiors broad and very short, hardly visible in profile, and almost hidden beneath segment 10, with a robust upwardly directed spine and a stouter shorter one directed backwards and inwards.

Distribution.—UPPER BURMA, the *type* coming from the Kachin Hills, and now in the Indian Museum (a spirit-specimen in a dilapidated condition).

The species is closely related to *A. lacteola* and *A. pieris* (the *naia* group) by the shape of its appendages and by the cobra-hood marking on dorsum of segment 2 (this marking is rudimentary in the two species mentioned, or only occasionally complete in the second, but in a subspecies of *A. pieris* from Siam the hood-mark is fully developed). Laidlaw remarks that the species is clearly related to the group (genus ?) which includes *A. pygmæa* ; this is an evident error, and, as he had lumped a whole lot of species under *Agriocnemis*, he probably meant that this particular one belonged to *Agriocnemis*. I have re-examined the type and am satisfied that the latter solution is correct, and that it belongs to the *naia* group and not to that of *pygmæa*.

157. **Agriocnemis naia** Fraser. (Fig. 166.)

Agriocnemis naia Fras., J. Bombay Nat. Hist. Soc. vol. xxix p. 747 (1923).

Male.—Abdomen 18–20 mm. Hind-wing 13·5 mm.

Head : labium white ; labrum azure blue, the base narrowly black ; bases of mandibles, genæ, ante- and postclypeus, and frons azure blue, the postclypeus black at base ; vertex and occiput black, the latter with azure blue postocular spots shaped like an elongate comma, and narrowly connected across the hinder border of occiput; eyes black above, blue at sides and beneath. *Prothorax* black on dorsum, blue laterally; anterior lobe also blue; posterior lobe produced squarely at its middle as in *A. lacteola*, and with its margin finely outlined in pale blue. *Thorax* black on dorsum as far lateral as the antero-lateral suture, marked with narrow dark blue antehumeral stripes; laterally deep azure blue, with a small black spot on the upper end of the postero-lateral suture. *Legs* palest blue, all femora black on the extensor surface, tibiæ on the flexor surface. *Wings* hyaline; pterostigma black, framed in thick black nervures, the anterior border narrowly bluish, distal and posterior sides in a continuous curve, covering less than one cell. Seven postnodal nervures in fore-wings, 5 in the hind. *Abdomen* deep azure blue, marked with black as follows:—Segment 1 with a broad

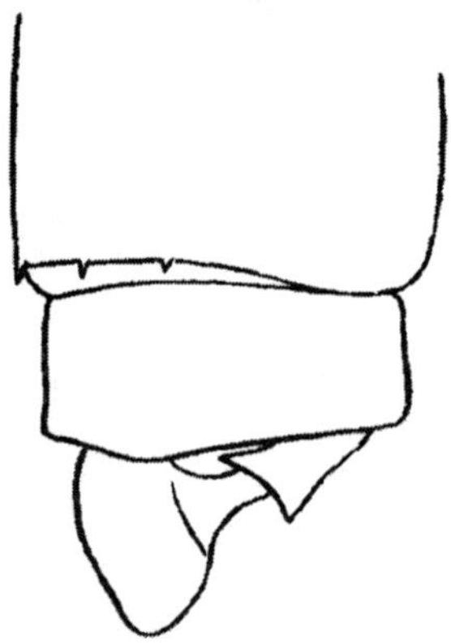

Fig. 166.—Anal appendages of *Agriocnemis naia* Fras., male, right side.

triangle on the dorsum, its base applied to apical border of segment, which is blue; segment 2 with a cobra-hood mark similar to that seen in *A. nana*; segments 3 to 6 with broad dorsal stripes not quite extending to the base of the segments, but confluent with narrow apical rings, showing a constriction just before this junction and a dilatation just prior to this; segment 7 variable, either somewhat similar to 6 or more often with a narrow apical ring followed by a small fleur-de-lis spot such as is found in *A. pieris*, which is narrowly confluent at its middle leaf with a broad dorsal stripe which extends nearly to base of segment; segment 8 also a little variable, usually with a small basal triangle, the dorsal point of which is prolonged along the mid-dorsal carina for about three-fourths the length of the segment, to amalgamate finally with a small fleur-de-lis which is always isolated from a narrow apical ring (in one specimen only it just touches the ring); in some specimens the fleur-de-lis is represented by a small spot on each side of the mid-dorsal stripe just at its termination; segments 9 and 10 entirely black, with narrow blue basal joints. *Anal appendages*: superiors blue on dorsum, black beneath, shaped almost exactly as in *A. pieris*; inferiors with the upper spine longer, the hinder one shorter than in *A. pieris*.

Female.—Abdomen 18–20 mm. Hind-wing 14 mm.

A more robust insect than the male, and with the black markings more extensive. The *wings* with an extra postnodal nervure, the pterostigma golden yellow, centred with greyish. *Head*, *prothorax*, and *thorax* similar to the male. *Abdomen* differing as follows:—Segment 1 with the dorsal marking quadrate; segment 2 with a dorsal marking shaped like the head of a thistle, and without the pair of blue spots seen in the male; segments 3 to 6 similar to the male, but the dorsal stripes broader; a second narrower stripe runs along the ventral border of each of these segments to enclose a narrow lateral blue stripe, very sinuous towards the apical end of the segments on account of the subapical dilatation of the dorsal band; segment 7 with the lateral stripe quite straight, and interrupted at the apical end so as to leave an apical isolated dot of blue; segment 8 with but a short lateral stripe, whilst segments 9 and 10 are unmarked. *Anal appendages* short, conical, black; vulvar scales blue, robust, extending to end of abdomen.

Distribution.—Lower Burma.

A certain amount of variability in the markings of both sexes is to be noted. In the male the pair of blue spots on the dorsum of segment 2 is sometimes absent: the variability in the markings of segment 8 has already been noticed, but the marking is always present, even in teneral specimens, and never extends as far as the apical border of the segment. The female in the teneral state is without the ventral black stripes, so that the whole of the sides are blue. A comparison of tenerals with adults shows that variations of the dorsal markings are true varieties, and not age developments. Closely related to *A. nana*, from which it is distinguished by the dorsal marking on segment 8 tapering away towards the apex of the segment instead of broadening and covering it as in *A. nana*.

Type in the British Museum collection, from King Island, Mergui, Lower Burma; paratypes in the Pusa collection, from the same locality.

158. **Agriocnemis clauseni** Fraser. (Fig. 167.)

Agriocnemis clauseni Fras., Mem. Dept. Agric. India (Ent.), vol. vii, no. 7, pp. 53–5 (1922); id., J. Bombay Nat. Hist. Soc. vol. xxix, p. 747 (1923).

Male.—Abdomen 20–22·5 mm. Hind-wing 11–13·5 mm.

Head: labium yellow; labrum turquoise-blue, with the base narrowly black; anteclypeus, postclypeus, genæ, and bases of mandibles turquoise-blue, with a broad black band at base of postclypeus: vertex and occiput black, as also frons, except at the sides, where the blue of the genæ extends up for some distance; occiput with deep azure blue postocular comma-like spots, a fine line of the same colour almost connecting them: eyes black above, blue at sides and beneath. *Prothorax* black on dorsum and sides, anterior lobe narrowly pale blue, as also an angulated spot on the postero-lateral

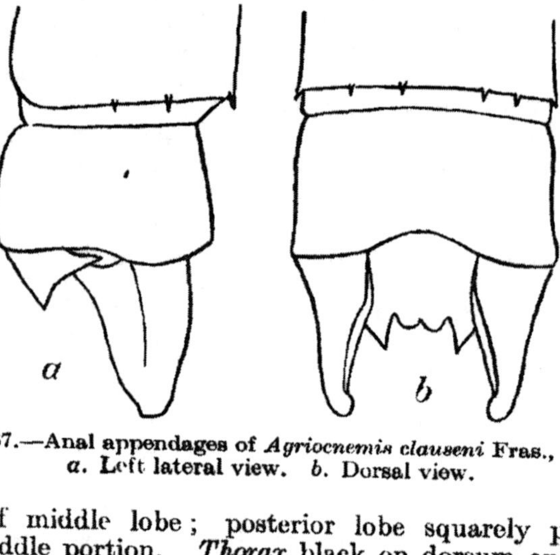

Fig. 167.—Anal appendages of *Agriocnemis clauseni* Fras., male. *a*. Left lateral view. *b*. Dorsal view.

aspect of middle lobe; posterior lobe squarely produced at its middle portion. *Thorax* black on dorsum and as far lateral as midway between the humeral and antero-lateral sutures, but meeting this last suture above; narrow azure blue antehumeral stripes tapering somewhat above and nearly confluent below with two small spots of the same colour; laterally pale azure blue, with a small black spot on the upper part of postero-lateral suture, and another larger spot just above the coxæ of the middle legs. *Legs* pale blue, femora marked broadly with black on extensor surface and tibiæ on the flexor surface. *Wings* hyaline; pterostigma rather shorter than in *A. lacteola*, pale yellow, centred with greyish, and framed in thick black nervures; 8 to 10 postnodal nervures in fore-wings, 6 to 7 in the hind; arc much less distal to the level of outer antenodal nervure than in *A. nana* or its relatives. *Abdomen* azure blue, marked with black as follows :—Segment 1 with a broad quadrate spot on dorsum extending to apical border ; segment 2 with the cobra-hood marking on dorsum, the paired blue spots moderately large, the border of the marking very ragged, extending out on to sides of segment at the middle, and then narrowing apically in a series of steps, the stem at apex being very fine and confluent with a narrow apical ring ; segments 3 to 6 with broad dorsal stripes which fall just short of the base of segments or meet the base at a point only, and which show an expansion subapically and a marked constriction just before they become confluent with a narrow apical ring ; in old specimens these segments also have an additional lateral black stripe which, by becoming confluent with the dorsal black in points, encloses a lateral blue stripe and apical spot ; segment 7 blue, with rather less than the apical fourth ringed with black, and with a quadrate mid-dorsal basal black spot which is connected to the base and apical black by a fine carinal black line ; segments 8 to 10 black, the intersegmental joints blue. *Anal appendages* blue ; superiors slightly longer than segment 10, ungulate, and rather obtusely pointed as seen in profile, narrow and bluntly pointed as seen from above, the apex here slightly turned inwards ; below and near base usually hidden under segment 10, a robust ventral spine curved downwards and inwards towards the anus ; inferiors much shorter, shaped similarly to the same organs in *A. lacteola*, but only the backward robust spine at all visible in profile.

Female.—Abdomen 21–22 mm. Hind-wing 15 mm.

A larger and more robust insect than the male, but coloured and marked very similarly. The face pale greenish-yellow, without any black at base of labrum ; the postocular spots larger ; eyes dark brown above. *Wings* with a deeper yellow pterostigma, the centres paler ; 10 to 12 postnodal nervures in fore-wings, 9 in the hind. *Abdomen* with segments 1 and 2 deep azure blue, 3 to 7 bluish-green and 8 to 10 entirely black ; segment 1 marked very broadly on dorsum with black ; segment 2 with a dorsal thistle-shaped mark without the pair of blue eye-spots on it ; segments 3 to 6 similar to the male, but the stripes thicker ; segment 8 marked somewhat similarly to the male, but the basal portion of dorsal stripe of even thickness from base until it joins the broad apical ring. *Anal appendages* short, conical, carneous ; vulvar scales blue or yellowish, extending to end of abdomen.

Distribution.—Assam, Bengal, Burma, and Siam, from May to November. *Type* in the British Museum, from

Shillong, Khasia Hills, Assam; paratypes in the Pusa and author's collections.

Rather closely resembles *A. nana* and *A. naia*, but is a larger insect, with higher nodal index; the characteristic dorsal marking on segment 7, as well as the longer superior anal appendages, will also aid in distinguishing it from these species. Siamese examples are somewhat smaller, the legs are very broadly marked with black, and the eye-spots on the dorsal marking on segment 2 are tiny round dots, not oval spots.

159. **Agriocnemis splendidissima** Laidlaw. (Fig. 168.)

Agriocnemis splendidissima Laid., Rec. Ind. Mus. vol. xvi, pp. 180–2 (1919); Fras., J. Bombay Nat. Hist. Soc. vol. xxix, p. 747 (1923); id., Rec. Ind. Mus. vol. xxvi, pp. 470, 491 (1924); id., ibid. vol. xxxiii, p. 448 (1931).

Male.—Abdomen 18 mm. Hind-wing 10 mm.

Head: labium yellow; labrum pale blue; bases of mandibles, genæ, and anteclypeus blue; postclypeus, frons, vertex, and

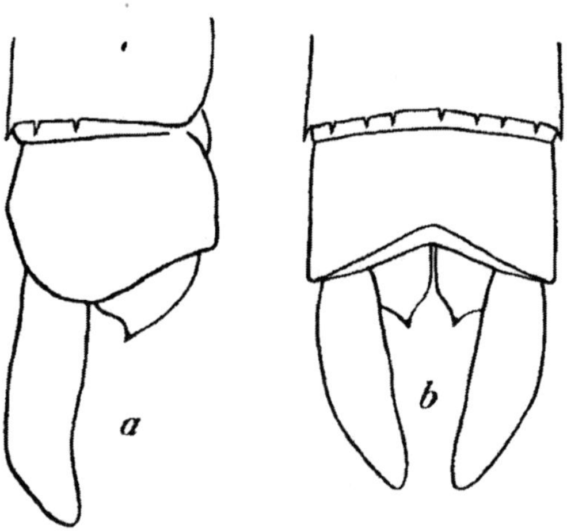

Fig. 168.—Anal appendages of *Agriocnemis splendidissima* Laid., male, *a*. Right lateral view. *b*. Dorsal view.

occiput black, as well as upper surface of eyes; comma-shaped blue postocular spots on occiput, with a fine line of the same colour lying between them; a fine brown equatorial line on circumference of eyes, which are bluish-green below. *Prothorax* black on dorsum, palest blue on the lower part of sides; posterior lobe produced quadrately at its centre, and very slightly emarginate. *Thorax* black on dorsum as far out as the antero-lateral suture, marked with narrow blue or pinkish antehumeral stripes ; blue on the sides, with a small black spot on the upper part of the postero-lateral suture ; the sides and beneath pruinosed white in adults. *Legs* carneous, black on the extensor surface of femora and flexor surface of tibiæ, but this often obscured by pruinescence ; 3 or 4 black spines on hind pair of tibiæ. *Wings* hyaline ; pterostigma shorter than in *A. lacteola,* black, framed finely in pale brown or yellow and thick black nervures, covering less than one cell ; 7 postnodal nervures in fore-wings, 5 to 6 in the hind. *Abdomen* blue, heavily marked with black as follows :—Segment 1 with the whole of dorsum black ; segment 2 with a broad dorsal marking shaped like the head of a thistle, or this expanded laterally to enclose an apical and a basal blue spot or a ventral stripe ; segments 3 to 7 with a broad dorsal black stripe which expands subapically and again contracts to join a narrow apical ring ; in adults the subapical expansion spreads well on to sides and then curves back towards base of abdomen for a variable distance along the segments, partially to enclose a narrow stripe of the ground-colour, this latter stripe expanding at base of segments to form blue basal rings ; segments 8 to 10 black, but 8 and 9 usually with a small basal lateral blue spot, whilst the sides of all are blue ventrally. *Anal appendages* black, blue internally, nearly half as long again as segment 10, narrow and elongate and curved slightly downward as seen from the side, the apices pointed, hollowed out within, curved gently in towards one another as viewed from the dorsum, without the ventral spine seen in the foregoing species ; inferiors less than half the length of superiors, pale carneous, conical, with rudimentary upper and posterior obtuse spines.

Female.—Abdomen 17–18·5 mm. Hind-wing 12–13 mm.

Very similar to the male in colour and markings, but more robust. Differs as follows :—Labrum and genæ pale green ; eyes beneath olivaceous ; anterior lobe and a lateral spot on middle lobe of *prothorax* greenish-blue ; black on dorsum of *thorax* not extending as far laterally, bordered with pale brownish, changing to bluish-green on the sides, but the colour here, as well as beneath the thorax, usually obscured by pruinescence ; pterostigma golden yellow, only slightly darker at centre ; dorsal marking on segment 2 much narrower and always shaped like a narrow thistle-head. *Anal appendages* and vulvar scale yellowish.

Distribution.—WESTERN GHATS, CENTRAL PROV., and DECCAN. I have taken this species at Mhow, Poona, Khandala (Bombay Pres.), Malabar and the Nilgiri Wynaad, and in Coorg.

A very common insect on some of the Malabar streams at the end of the N.E. monsoon. It frequents the beds of streams in shallows where the herbage and grass are thrusting up above the water, and is only found during the dry seasons. Distinguished from the previously described species by the absence of the eye-spots on dorsum of segment 2, as well as by the conspicuous alternating black and blue stripes on the sides of the abdominal segments; the anal appendages without ventral spine also distinguish it from other described species, except *A. d'abreui*, *A. pygmæa*, and *A. femina*, all of which, however, differ in many other respects.

Type and cotype in the Indian Museum, from Talewadi, N. Kanara; paratypes from Chalakudi, Cochin, in the same museum, whilst others are to be found in the British Museum, Morton, Laidlaw, and author's collections.

160. **Agriocnemis aborensis** (Laidlaw). (Fig. 169.)

Argiocnemis aborense Laid., Rec. Ind. Mus. vol. viii, p. 347 (1914); id., ibid. vol. xvi, p. 182 (1919); Fras., J. Bombay Nat. Hist. Soc. vol. xxix, p. 747 (1923).

Male.—Abdomen 23–24 mm. Hind-wing 14–16 mm.

Head: labium pale yellow; labrum pale blue, with two large black spots at base which may be confluent or replaced by one large spot which covers the greater part of labrum; anteclypeus, bases of mandibles, genæ, and postclypeus pale blue, the latter narrowly black at base; frons blue outwardly, black at its centre; vertex and occiput black, the latter with azure blue rounded postocular spots; eyes black above, blue at sides and below; beneath head black, but an elongate spot of yellow anteriorly. *Prothorax* black on dorsum except the anterior lobe, which is very pale blue; posterior lobe with its central portion produced into a rounded lobe, the lateral portions narrow, pale blue; a large or small blue spot low down on the sides. *Thorax* black on dorsum to as far as the antero-lateral suture, marked with a narrow pale blue antehumeral stripe on each side; laterally azure blue, with a thick black oblique stripe on the postero-lateral suture; beneath yellowish, the central part darker. *Legs* creamy white, extensor surface of femora and flexor of tibiæ black; 3 to 4 black spines on hind tibiæ. *Wings* hyaline; pterostigma about twice as long as broad, blackish-brown, framed in pale yellow and thick black nervures, covering about one cell; 8 to 9 postnodal nervures in fore-wings, 6 to 7 in the hind; arc situated well beyond the distal antenodal nervure. *Abdomen*: segment 1 with dorsum black, sides and apical border azure blue; segment 2 with the dorsum broadly black except near apical end, where the black is greatly constricted, and at base, where are a pair of small oval blue spots narrowly separated medially ; segments 3 to 6 black, with narrow greenish-blue lateral stripes nearly confluent across the dorsum at base, and again approaching each other subapically on each segment (these markings, except at base, nearly obsolete on segment 6) ; segment 7 with a pair of basal blue dorsal spots only ; segment 8 with its ventral border azure blue ; segment 9 entirely blue save for an apical row of black spines ; segment 10 azure blue, with its apical border and the mid-dorsal line narrowly black. *Anal appendages* black, the inferiors paler below ; superiors rather longer than segment 10, curved a little downwards as seen from the side, broad and hollowed out on the inner side, the apex hooked inwards as seen from above ; inferiors considerably shorter, broad at base as seen

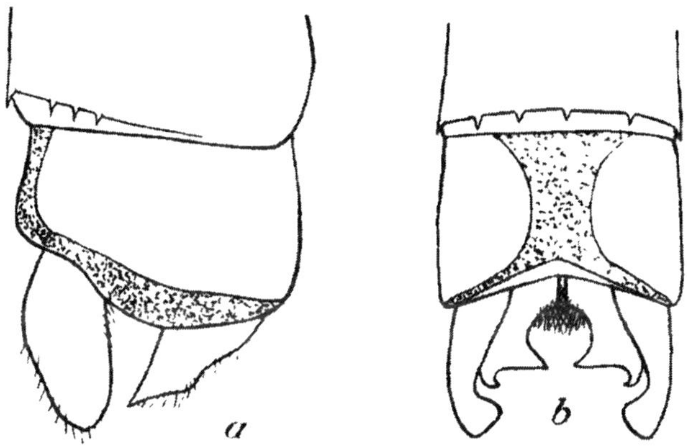

Fig. 169.—Anal appendages of *Agriocnemis aborensis* (Laid.), male. *a.* Right lateral view. *b.* Dorsal view.

from the side, tapering, the apex bevelled upwards and ending in a sharp spine ; this spine turned slightly in as seen from above and a second and more robust spine seen on the inner border of appendage.

Female unknown.

Distribution.—The *type*, which appears to have been lost, was taken at Dibrugarh, Assam ; the only other specimens known are in the author's collection, two males from Hasimara, Duars, Bengal, collected by Mr. H. V. O'Donel, and a male from Bangkok, Siam, collected in August and September.

The species has many distinctive features, the two large black spots on labrum, segment 9 entirely blue, 10 with its apical border and mid-dorsal ridge narrowly black, the shape of the anal appendages, etc., all serving to distinguish it from other species of the genus. The dorsal marking of segment 2 shows a clear relation to the *naia* group. A specimen will be lodged in the British Museum.

161. **Agriocnemis d'abreui** Fraser. (Fig. 170.)

Agriocnemis d'abreui Fras., Rec. Ind. Mus. vol. xvi, pp. 454, 455 (1919); id., Mem. Dept. Agric. India (Ent.), vol. vii, no. 7, pp. 52, 53 (1922): id., J. Bombay Nat. Hist. Soc. vol. xxix, p. 747 (1923).

Male.—Abdomen 13–14 mm. Hind-wing 9–9·5 mm.

Head: labium pale yellow; labrum blue (yellow in Siamese specimens), with the base broadly black (a pair of tiny blue spots enclosed in this black band in some specimens); bases of mandibles, genæ, anteclypeus, and a narrow stripe traversing the frons pale blue; vertex and occiput black, the latter with narrow elongate, pyriform, blue postocular spots narrowly connected by a yellow band across the middle line; eyes black above, blue laterally, changing to olive-green below; beneath head pale bluish-green. *Prothorax* black on dorsum, the lower part of sides, the anterior lobe, and the borders of the posterior lobe finely blue; the latter produced squarely at its centre much as in *A. lacteola*. *Thorax* black on dorsum as far lateral as the antero-lateral suture, marked by narrow azure blue antehumeral stripes; laterally blue, changing to bluish-green on the lower part of sides and beneath, and

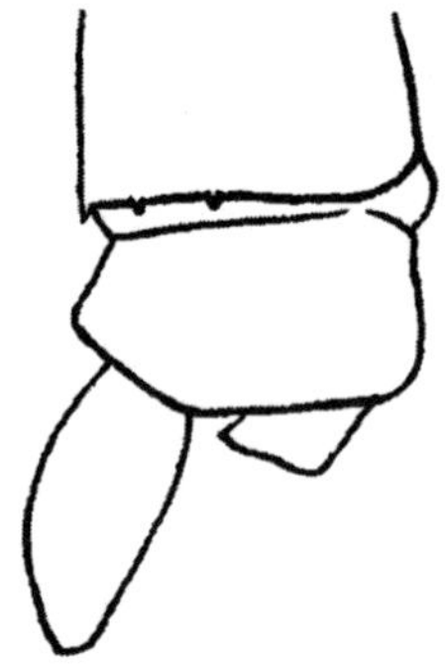

Fig. 170.—Anal appendages of *Agriocnemis d'abreui* Fras., male, seen from the right side.

with a small spot of black on the upper part of the postero-lateral suture. *Legs* yellow or creamy white, with the extensor surface of femora black. *Wings* hyaline; pterostigma brownish-black, framed in paler brown and thick black nervures, covering slightly less than one cell; 5 to 6 postnodal nervures in fore-wings, 4 in the hind. *Abdomen* with the ground-colour blue from segments 1 to 6, changing through blue to greenish-blue and then yellow, and finally bright ochreous on segment 7, the final segments bright ochreous; segment 1 with a broad black dorsal marking, its apical border blue; segment 2 with the "spectacles" (cobra's hood) marking typically developed, the marking constricted slightly apical to the pair of blue spots, and then showing a further constriction which becomes confluent with a narrow apical black ring; segment 3 to 6 with a narrow dorsal stripe not quite reaching base of segments, but expanding subapically and then constricted to join narrow apical rings; segments 7 and 8 with similar markings, but the subapical expansion throwing out an arm on each side which extends basalwards along the side of the segments for about half their length, the whole marking thus shaped like an anchor; segment 9 with an abbreviation of this marking, the stem of the anchor shortened to an equilateral triangle, the apex of which just joins the arms or flukes of the anchor; segment 10 with a dorsal triangle of black, the apex just falling short of the apical border of the segment. In a male from Siam these markings on segments 7 to 10 are but poorly developed, and the segments are a darker ochreous. *Anal appendages*: superiors about half as long again as segment 10, broad and spatulate as seen from the side, hollowed out on the inner side, narrow and obtusely pointed as seen from above, and very similar to those of *A. aborensis*; inferiors ochreous or yellow, much shorter than superiors, squared and blunt as seen from the side, the upper angle slightly produced to form a short stout spine.

Female.—Abdomen 15–16 mm. Hind-wing 10–12 mm.

Differs considerably from the male in colour and markings, and a number of what appear to be age-varieties exist, which differ among themselves in the amount of melanism present. Labrum dark or paler blue, with the base narrowly black; postclypeus blue, black at the base; the blue band across frons narrowly interrupted in the middle line, where the black of vertex becomes confluent with that at base of postclypeus; postocular spots similar to the male or, in very old specimens, entirely isolated and small. *Thorax* in young specimens very similar to the male, but the antehumeral stripe pale yellow instead of blue, this colour developing only in very old specimens; the sides of the thorax in old specimens black except for a lower irregular area, the black encroaching on this along the lateral sutures; antehumeral stripes in these absent. *Wings* with 6 to 8 postnodal nervures in the fore, 5 to 6 in the hind; pterostigma similar to that of the male. *Abdomen* with the ground-colour azure blue throughout, but the basal segments 1 and 2 may be tinged with greenish; marked with black as follows:—Segment 1 similar to the male, but in old specimens entirely black, except the apical border; segment 2 with a dorsal thistle-shaped marking narrowly confluent by its stalk to a narrow black apical ring, or, in very old specimens, the whole of the segment black except for a small latero-apical blue spot on each side; segments 3 to 6 similar to the male, or with an extra longitudinal stripe along each side near the ventral border ; segment 7 similar, but the additional stripe confluent with the dorsal black apically, so as partially to enclose a blue stripe ; segments 8 to 10 broadly black on the dorsum and sides except along the ventral borders, which are blue ; segment 8 has a lateral basal oval blue spot, and sometimes a second smaller lateral apical spot ; segment 9 has usually a small rounded lateral spot, whilst the apices of all the segments 7 to 10 are narrowly blue ; vulvar scales blue, robust, extending to end of abdomen. *Anal appendages* blue above, black beneath, pointed, conical, very short.

Distribution.—The *type* is a female in the British Museum, formerly in the Pusa collection, from Lamta, Balaghat, 1,130 ft., CENTRAL PROVINCES, taken in March. Other specimens in the Pusa collection are from Gauhati, and from Tokhlai in the Jorhat District, both in ASSAM, taken in November. A pair in my collection come from Siam, and these differ from the type only by not having the anchor-shaped markings on the terminal segments of abdomen ; the female differs only by having the labrum glossy black except for the anterior border, the postclypeus entirely black, but the transverse band on the frons not interrupted ; its abdomen resembles that of the more melanotic form of the female. The species appears to be widely spread throughout southern Asia ; by the character of its anal appendages, without ventral spine, and by the character of the markings of the female it is evidently closely allied to *A. splendidissima*. The male is distinguished from others by the ochreous end of abdomen with black anchor-shaped dorsal markings ; the female by the black end-segments of the abdomen marked laterally with blue spots, or, when these are absent, by the extreme development of melanism on the sides of the thorax.

162. **Agriocnemis pygmæa** (Rambur). (Figs. 163 & 171.)

Agrion pygmæum Ramb., Ins. Névrop. p. 278 (1842) ; Brauer, Verh. zool.-bot. Ges. Wien, vol. XIV, p. 161 (1864) ; id., Reise, 'Novara,' p. 103 (1866).

Agrion velare Hagen, Verh. zool.-bot. Ges. Wien. vol. viii, p. 479 (1858) (*nom. nudum*).

Agriocnemis pygmæa Selys, Syn. Agr. 5me Légion : Agrion (suite et fin), p. 52 (1877) ; Kirby, Cat. Odon. p. 158 (1890) ; Selys, Ann. Mus. Civ. Genova (2), vol. x (xxx), p. 518 (1891) ; Kirby, J. Linn. Soc. Lond., Zool. vol. xxiv, p. 564 (1893) ; Ris, Tijdschr. Ent. vol. lv, p. 159 (1912) ; Campion, Trans. Linn. Soc. Lond. vol. xv, 4, p. 443 (1913) ; Laid., Rec. Ind. Mus. vol. viii, p. 348 (1914) ; Ris, Suppl. Ent. no. 5, pp. 20–22 (1916) ; Laid., Rec. Ind. Mus. vol. xvi, pp. 171, 178 (1919) ; Fras., ibid. vol. xxiv, pp. 309, 310 (1922) ; id., J. Bombay Nat. Hist. Soc. vol. xxix, p. 747 (1923) ; Laid., Spolia Zeylan. vol. xii, p. 372 (1924) ; Fras., Rec. Ind. Mus. vol. xxvi, p. 490 (1924) ; id., ibid. vol. xxxiii, p. 448 (1931).

Agriocnemis velaris Selys, Ann. Soc. Españ. Hist. Nat. vol. xi (p. 29, sep.), (1882); Kirby, Cat. Odon. p. 158 (1890); id., J. Linn. Soc. Lond., Zool. vol. xxiv, p. 564 (1893); Martin, Mém. Soc. Zool. France, vol. xix, p. 248 (1901); Tillyard, Proc. Linn. Soc. N.S. Wales, vol. xxxii, p. 388 (1907).

Agriocnemis hyacinthus Tillyard, Proc. Linn. Soc. N.S. Wales, vol. xxxvii, p. 457, tab. 48, figs. 15, 16 (1913).

Male.—Abdomen 16–17 mm. Hind-wing 9·5–10 mm.

Head: labium palest yellow; labrum entirely brilliant metallic blue; anteclypeus, bases of mandibles, genæ, and the frons narrowly pale apple-green; postclypeus, vertex, and occiput black, the former narrowly confluent with the black of vertex, thus narrowly interrupting the pale green on frons; occiput with very small rounded pale green post-ocular spots; eyes black above, pale apple-green on the sides and beneath. *Prothorax* black on dorsum, the anterior lobe,

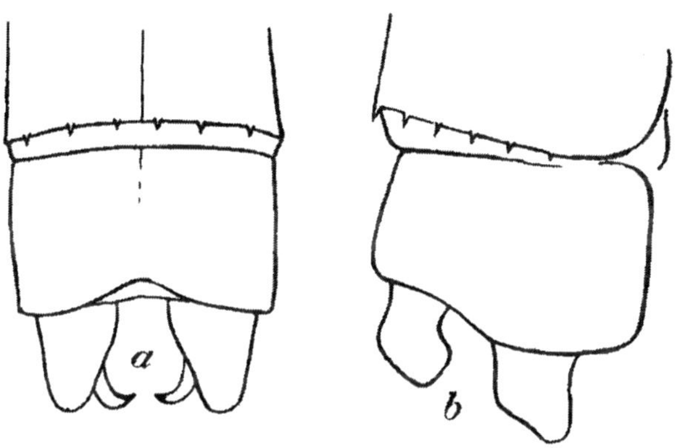

Fig. 171.—Anal appendages of *Agriocnemis pygmæa* (Ramb.), male. *a*. Dorsal view. *b*. Right lateral view.

lower part of sides, and the extreme edge of posterior lobe apple-green; posterior lobe trilobate, the middle lobe produced backwards. *Thorax* black on dorsum as far lateral as the antero-lateral suture, marked with narrow antehumeral apple-green stripes; laterally apple-green, paling to yellow below and with a very small spot of black on the upper part of postero-lateral suture. *Legs* yellow, extensor surface of femora black. *Wings* hyaline; pterostigma pale yellow in the fore-wings, black in the hind, the latter very finely paler at its circumference outwardly, covering less than one cell, that of hind-wing slightly shorter than in the fore; 6 to 7 postnodal nervures in fore-wings, 5 to 6 in the hind, usually the lesser number. *Abdomen*: segments 1 to 6 with the ground-colour pale greenish-yellow, the terminal segments brick-red, marked with bronzed black as follows :—Segment 1 with the dorsum broadly black, the apical border green; segment 2 with a broad dorsal thistle-shaped marking extending the whole length of the segment and confluent at apex with a narrow apical ring; segments 3 to 6 with a broad dorsal stripe expanding subapically, then again contracting to become confluent with a narrow apical black ring; segment 7 somewhat similar, but the stripe expanding gradually as far as the apical border, where it becomes somewhat diffused and lost; segments 8 to 10 with the dorsum clouded or not with black. *Anal appendages* brick-red; superiors longer than inferiors, not quite as long as segment 10, subquadrate as seen in profile, and with the lower angle produced slightly as a short obtuse spine, triangular as seen from the dorsum; inferiors half the length, tiny quadrate processes with a small black upper spine.

Female.—Abdomen 18 mm. Hind-wing 11–12 mm.

A more robust insect than the male, and exhibiting a number of polychromatic forms which seem to depend largely on the age of the specimens, although not entirely so. These extend through a red form with unenclosed antehumeral stripes (heterochromes) to one in which these stripes are enclosed by a darker outer stripe, a condition approaching that of the male (isochromes).

Red form.—Front and back of *head*, *prothorax*, and *thorax*, except a broad black band on dorsum, and segments 1 to 7 dark salmon-pink or a dark brick-red, the terminal segments of *abdomen* suffused with black, and a narrow black stripe across the vertex, the blue postocular spots being unenclosed behind.

Subadult forms.—Labrum and face unmarked, yellowish or pale green; a broad black stripe across vertex limited anteriorly by the bases of antennæ, and behind by small rounded postocular blue spots which are connected by a narrow green stripe across occiput; the black of vertex continued on to the eyes as a polar cap, beneath which the eyes are apple-green to greenish-yellow. *Prothorax* brownish on dorsum, broadly blue at the sides, the posterior lobe less produced than in the male. *Thorax* with a broad black stripe on mid-dorsum, bounded externally by a pale blue antehumeral stripe (pink or violaceous in tenerals); laterally pale green, changing to yellowish-green low down on the sides and beneath, with a small black spot on the upper part of the postero-lateral suture. *Legs* similar to the male, but the tibiæ black on flexor surface. *Wings* with 7 postnodal nervures in the fore, 5 or, rarely, 6 in the hind; pterostigma similar in all wings, yellow, framed in darker coloured nervures, covering less than one cell. *Abdomen* pale apple-green, changing to yellow on segments 8 to 10, marked similarly to the male, but the stripes broader and the end segments very broadly black on dorsum, the joints here bluish or yellow. *Anal appendages* small, conical, yellow; vulvar scale robust, extending to end of body, yellow.

Fully adult or isochrome form.—Face, frons, and the eyes, except for a black polar cap, definitely bright yellow, with indications of a black basal line to labrum; postocular spots blue, connected by a mid-occipital narrow green stripe, and partially enclosed by black behind. *Prothorax* brick-red, unmarked. *Thorax* similar to subadults, but the antehumeral stripe narrower and enclosed or bordered outwardly by a reddish-brown or dark violaceous stripe; laterally green, as for the subadult. *Abdomen* similar to the male in every respect except for the ground-colour of segments 2 to 6, which is bright yellow instead of apple-green; the end-segments bright brick-red, as in the male.

Distribution.—Throughout the Oriental Region, Australia, and the Pacific Islands; throughout INDIA, CEYLON, and BURMA except in desert tracts. Ris gives the following localities and countries:—Singapore, Ceylon, Calcutta, Assam, Nicobars, Java, Manila, Burma, Seychelles, Queensland, Sydney, Formosa, China, N. Celebes, New Guinea, Philippines, etc., so that it will be seen that, although it is the smallest species of the Indian fauna, it has, perhaps, the widest distribution, except *P. flavescens* and a few other Libellulines.

The male, when very adult, becomes pruinescent on the dorsum of head and thorax, and in the dry season of the year specimens are met with where these parts of the body are snowy white, all markings beneath being quite obscured; the female does not appear to share in this character, but may change to very dull brown. The isochrome is the form most commonly met with in the Deccan, whereas the heterochrome is more typical of the coastal area, *e. g.*, around Madras, and Malabar. The male is easily distinguished by the pterostigma differing in colour in the fore- and hind-wings, by the reddish coloured end-segments of abdomen, and by the shape of the anal appendages. Teneral females are more difficult to distinguish from others, as the whole genus shares in the same colouring in the young or newly emerged form; the unenclosed antehumeral stripes will, however, help to distinguish the adult.

Type in the Serville collection, a female almost totally destroyed. Paratype male in the Selysian collection; paratypes in most national and private collections.

163. **Agriocnemis femina** (Brauer). (Fig. 172.)

Ischnura femina Brauer, Verh. zool.-bot. Ges. Wien, vol. xviii, p. 554 (1868).

Agriocnemis incisa Selys, Syn. Agr. 5me Légion, p. 55 (1877); id., Ann. Mus. Civ. Genova, vol. xiv, p. 290 (1879); Albarda, Veth's Midden Sumatra, Neur. p. 12 (1881); Selys, Ann. Soc. Españ. Hist. Nat. vol. xi (p. 29, sep.), (1882); id., Ann. Mus. Civ. Genova, vol. xxvii, p. 218 (1889); id., Ann. Soc. Españ. Hist. Nat. vol. xx, p. 218 (1891); Kruger, Stett. Ent. Zeit. vol. lxix, p. 127 (1898); Ris, Arch. Naturgesch. Bd. i, p. 200, tab. 10, fig. 19 (1900); Laid., Proc. Zool. Soc. Lond. p. 388 (1902); Martin, Mission Pavie (p. 18, sep.) (1904); Laid., Rec. Ind. Mus. vol. viii, p. 349 (1914); id., ibid. vol. xvi, p. 178 (1919); Fras., J. Bombay Nat. Hist. Soc. vol. xxix, p. 747 (1923).

Agriocnemis pulverulans Selys, Syn. Agr. 5me Légion, p. 56 (1877); id., Mitt. Mus. Dresd. vol. iii, p. 297 (1878); Kirby, Cat. Odon. p. 159 (1890); Kruger, Stett. Ent. Zeit. vol. lxix, p. 127 (1898); Ris, Arch. Naturgesch. Bd. i, p. 200, tab. 10, fig. 18 (1900); Laid., Proc. Zool. Soc. Lond. p. 388 (1902); Needham, Proc. U.S. Nat. Mus. vol. xxvi, tab. 53, fig. 7 (1903); Tillyard, Proc. Linn. Soc. N.S. Wales, vol. xxxvii, p. 461 (1913); Ris, Nova Guinea, vol. xiii, Zool. p. 121 (1915).

Agriocnemis materna Selys-Hagen, Syn. Agr., 5me Légion, p. 57 (1877); id., Ann. Mus. Civ. Genova, vol. xxvii, p. 484 (1889); Kirby, Cat. Odon. p. 159 (1890); Karsch, Ent. Nachr. vol. xvii, p. 243 (1891); Kruger, Stett. Ent. Zeit. vol. lxix, p. 127 (1898); Ris, Arch. Naturgesch. Bd. i, p. 200, tab. 10, fig. 18 (1900); Tillyard, Proc. Linn. Soc. N.S. Wales, vol. xxxvii, p. 461 (1913); Ris, Nova Guinea, vol. xiii, Zool. p. 121 (1915).

Agriocnemis femina Kirby, Cat. Odon. p. 158 (1890); Ris, Tijdschr. Ent. p. 9 (1915); id., Nova Guinea, vol. xiii, Zool. p. 121 (1915); id., Suppl. Ent. no. 5, pp. 22–26, text-figs. 3–7 (1916); id., Zool. Meded. Leiden, vol. x, p. 22 (1927); id., Treubia, vol. vii, Suppl. p. 145 (1929); id., Ark. Zool. Uppsala, Bd. xxi, A, no. 31, p. 7, fig. 2 (1930).

Male.—Abdomen 16–17 mm. Hind-wing 10·5–11 mm.

Head: labium pale yellow; labrum brilliant metallic Prussian blue; bases of mandibles, genæ, anteclypeus, and a stripe across frons apple-green; postclypeus, vertex, and occiput black, the latter enclosing rounded azure postocular spots; eyes capped with dark brown, but apple-green at the sides and beneath. *Prothorax* black on dorsum, blue or bluish-green on the lower part of sides, and blue on the anterior lobe and border of posterior lobe; the latter with middle portion produced slightly. *Thorax* black on dorsum as far lateral as the antero-lateral suture, marked with pale blue narrow antehumeral stripes; laterally apple-green, paling below to yellow. *Legs* pale yellow, femora black on extensor surface. *Wings* hyaline; pterostigma yellow, slightly darker at its centre, and with darker surrounding nervures, similarly coloured in all wings, covering less than one cell; 5 to 6 postnodal nervures in fore-wings, 4 in the hind. *Abdomen* with the ground-colour of segments 1 to 6 blue or pale green, of segments 7 to 10 bright chrome-yellow, marked with black as follows :—Segment 1 with a broad dorsal spot, the sides and apical border blue to bluish-green ; segment 2 with a broad thistle-shaped spot on dorsum, its stem confluent with a narrow apical ring ; segments 3 to 7 with moderately broad dorsal stripes which do not quite extend to base of segments, but show an expansion subapically, after which the bands contract again to become confluent with a narrow apical ring ; segment 8 with the mid-dorsum finely black for the basal three-fourths of the segment ; segments 9 and 10 unmarked. *Anal appendages* bright yellow ; superiors very short, showing some slight variations in local races, nearly as long as segment 10, rather narrow at base, expanding towards apex,

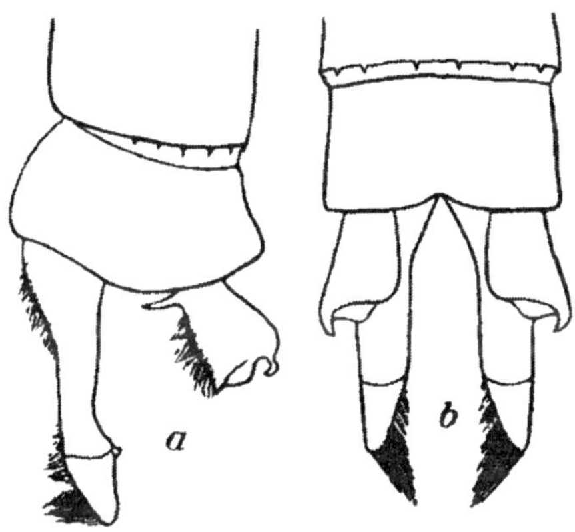

Fig. 172.—Anal appendages of *Agriocnemis femina* (Brauer), male. *a*. Left lateral view. *b*. Dorsal view.

which is very obtuse, grooved, furnished at the upper outer angle with a small curved blackish spine, and with a short spine at the lower angle and a long thin spine below at base ; inferiors twice the length of superiors, rather broad at base, sinuous and cylindrical, ending in an obtuse apex, which is furnished with an inner lower tuft of stiff bristles and shows a short blunt spine below near the apex.

Female.—Abdomen 18 mm. Hind-wing 11 mm.

More robust than the male, and showing a series of colour varieties as in *A. pygmæa*, but which appear to be entirely due to age stages. Only heterochromes are known.

Teneral form.—*Head* : labrum purplish metallic, bordered with yellow, rest of face apple-green as far as base of antennæ ;

vertex black, occiput cherry-red, enclosing pale lilaceous postocular rounded spots; eyes with a black cap above, apple-green elsewhere. *Prothorax*, *thorax*, and *abdomen* lemon-yellow, segments 7 to 10 alone enfumed or blackish above.

Subadult form.—Head very similar to the last, but the postclypeus black and the postocular spots better defined. *Thorax* cherry-red, with a broad medial black band on dorsum. *Prothorax* similarly coloured, and with a small black anterior dorsal spot; posterior lobe very large, elevated markedly at its centre, its sides oblique, the central portion widely notched. *Abdomen* with segments 1 to 6 cherry-red, the latter segment only marked on dorsum with a diffuse black band, which falls short of the base of the segment but broadens apically; segments 7 to 10 with broad dorsal stripes, the apical borders and sides yellow or ochreous.

Adult form.—Head similar to the last, but the red on occiput a dark blood-red, and the postocular spots blue and well defined. *Prothorax* pale pinkish-brown, with a small black dorsal spot. *Thorax* similar, but the black on dorsum broader and the sides pinkish-brown or fawn, with a small black spot at the upper end of postero-lateral suture. *Wings* similar to the male in all respects. *Legs* pale yellow, the femora black on extensor surface at distal ends only. *Abdomen* with ground-colour blue from segments 1 to 6, changing to reddish on segment 7, or ochreous marked broadly with black on dorsum, as in the subadult, but the markings more sharply defined. *Anal appendages* small, conical, pointed, ochreous; vulvar scales ochreous, robust, extending to end of abdomen.

Distribution.—Within our limits confined to BENGAL, BURMA, and ASSAM. Ris gives Ceylon, but I have seen many collections from that island and have never once met with it from there; I can only think that a mistake has crept in, as Fruhstorfer, who collected Ris's specimens, also collected in Java, and may thus have mixed up the localities. The distribution beyond Indian limits is almost the same as for *A. pygmœa*; Ris gives Formosa, South China, Philippines, Annam, Sumatra and Java, Borneo, Singapore, Lombok, N. Celebes, Amboina, and Darnley Island in the Torres Straits.

Distinguished from all other species by the remarkable shape of its anal appendages. The whereabouts of the type is doubtful; Ris was of opinion that the female type of *A. pygmœa* was really the type female of *A. femina*; paratypes in most national and private collections. The author's collection contains examples from Gauhati, Assam, collected in September by Mr. T. B. Fletcher.

Genus ARGIOCNEMIS Selys (1877). (Fig. 173.)

Argiocnemis Selys, Bull. Acad. Belg. (2) vol. xliii, p. 135 (1877); Kirby, Cat. Odon. p. 158 (1890); Ris, Abh. Senckenb. naturf. Ges. Bd. xxxiv, p. 517 (1913); Munz, Mem. Amer. Ent. Soc. no. 3, p. 60 (1919); Laid., Rec. Ind. Mus. vol. xvi, p. 182 (1919); Fras., J. Bombay Nat. Hist. Soc. vol. xxix, p. 746 (1923); Ris, Ark. Zool. Uppsala, Bd. xxi, A, no. 31, p. 5 (1930).

Dragonflies of rather larger size than the last; colours bright red and yellow in the subadult stage, black, marked with blue in the full adult; wings hyaline; pterostigma similar in fore- and hind-wings, small, covering about one cell, nearly twice as long as broad, distal side oblique, poorly braced; 10 to 12 postnodal nervures; discoidal cell acutely pointed at the distal end, with costal side about half the length of posterior in the fore-wing and about three-fourths in the hind; basal side equal to costal in the fore-wing, nearly half the length in the hind; distal side moderately oblique; sectors of arc arising from the lower end of arc, divergent from origin; arc situated distal to the level of the distal antenodal

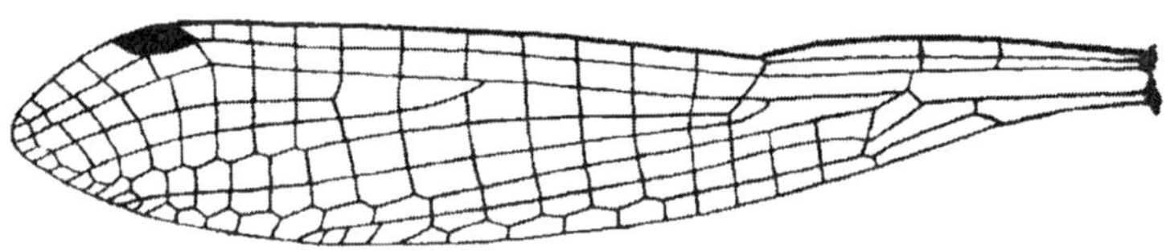

Fig. 173.—Hind-wing of *Argiocnemis rubescens* Selys, male.

nervure; *ab'* always present and complete, arising well proximal of the level of *ac*, which lies about midway between the two antenodal nervures; *ab* continued on as *IA*, but an angulation (medio-anal link) present at the junction of the two; *Riv+v* arising well proximal to the level of the oblique nervure descending from the subnode; *IRiii* in continuation of that nervure.

Head narrow, without any frontal ridge; postocular coloured spots invariably present; thorax short and robust; posterior lobe of prothorax simple; no hooks on anterior border of thorax; abdomen slim and cylindrical, dilating slightly towards the anal segments; legs very short, about five spines on the hind tibiæ and femora. Anal appendages: superiors about as long as segment 10, rather broad and obtuse structures, curving slightly downwards as viewed in profile; inferiors shorter, very small and aborted. Genitalia: lamina very deeply and very narrowly cleft; hamules square plate-like structures almost meeting across the middle line; lobe small, scrotum-shaped; penis similar to the last genus.

Genotype, *Argiocnemis rubescens* Selys.

Distribution.—Throughout N.E. India, Burma, Malaysia and Indo-China; the Sondaic islands, Borneo, and as far as

Australia. Very little is known about their habits, but they breed in running waters and pass through red teneral stages as in the last genus. It is on account of these latter stages that so much synonymy has arisen, the genus being represented in fact by very few forms.

164. **Argiocnemis rubescens** Selys. (Figs. 173 & 174.)

Argiocnemis rubescens Selys, Bull. Acad. Belg. (2) vol. xliii, p. 136 (1877); Kirby, Cat. Odon. p. 158 (1890); Ris, Abh. Senckenb. naturf. Ges. Bd. xxxiv, p. 517 (1913); Laid., Rec. Ind. Mus. vol. xvi, p. 182 (1919); Fras., J. Bombay Nat. Hist. Soc. vol. xxix, p. 746 (1923); Ris, Zool. Meded. Leiden, vol. x, p. 22 (1927); id., Treubia, vol. vii, Suppl. p. 145 (1929).

Argiocnemis rubeola Selys, Bull. Acad. Belg. (2) vol. xliii, p. 137 (1877); Kirby, Cat. Odon. p. 158 (1890); Ris, Abh. Senck. naturf. Ges. Bd. xxxiv, p. 517 (1913); Laid., Rec. Ind. Mus. vol. xvi, p. 182 (1919).

Agriocnemis rubescens var. *intermedia* Selys, Bull. Acad. Belg. (2) vol. xliii. p. 138 (1877); Kirby, Cat. Odom. p. 158 (1890); Ris, Abh. Senck. naturf. Bd. xxxiv. p. 517 (1913); Laid., Rec. Ind. Mus. vol. xvi. p. 182 (1919).

Agriocnemis lunulata Selys, Bull. Acad. Belg. (2) vol. xliii, p. 139 (1877); Kirby, Cat. Odon. p. 158 (1890); Ris, Abh. Senck. naturf. Ges. Bd. xxxiv, p. 517 (1913); Laid., Rec. Ind. Mus. vol. xvi, p. 182 (1919).

Agriocnemis nigricans Selys, Bull. Acad. Belg. vol. xliii, p. 140 (1877); Kirby, Cat. Odon. p. 158 (1890); Krug., Stett. Ent. Zeit. vol. lix, p. 290 (1898); Ris, Abh. Senck. naturf. Ges. Bd. xxxiv, p. 517 (1913); Laid., Rec. Ind. Mus. vol. xvi, p. 182 (1919).

Argiocnemis rubescens race *sumatrana* Krug., Stett. Ent. Zeit. vol. lix. p. 290 (1898).

Argiocnemis obscura Laid., Rec. Ind. Mus. vol. viii, p. 346 (1914); id., ibid. vol. xvi, p. 182 (1919).

Male.—Abdomen 28–30 mm. Hind-wing 17–20 mm.

Head: labium pale yellow or straw-coloured; labrum pale turquoise-blue, unmarked with black; bases of mandibles, genæ as far up as bases of antennæ, and clypeus pale blue; rest of head mat black, marked posteriorly with small rounded pale blue postocular spots; eyes deep blue above, capped with black, paler blue beneath. (In a note on the living insect made by Col. F. Wall the clypeus and cheeks are said to be apple-green and the postocular spots dull yellow.) *Prothorax* black on dorsum and sides; the anterior lobe, a narrow streak on the lower part of each side of middle lobe, and the outer ends of the posterior lobe pale blue; the latter structure prolonged at the centre as a truncated plate squared at the end. *Thorax* black on dorsum as far laterally as the antero-lateral suture, marked with narrow pale greenish-blue ante-humeral stripes; laterally pale blue, with an oblique narrow black stripe bordering the postero-lateral suture anteriorly; beneath pale blue. *Wings* hyaline; pterostigma narrow, with the outer angle prolonged somewhat, blackish-brown with a bright yellow frame on the inner sides of the thick black bordering nervures, covering one cell; 10 to 12 postnodal nervures in fore-wings, 8 to 9 in the hind. *Legs* pale bluish-green to yellowish, hinder femora almost entirely black, middle femora heavily marked with black on the extensor surface, the anterior pair less so, tibiæ black on flexor surface proximally. *Abdomen* pale blue laterally, heavily marked with black on dorsum as follows:—Segment 1 with a small quadrate spot on mid-dorsum; segment 2 with a broad dorsal spot extending from end to end, its lateral borders deeply crenate; segments 3 and 4 with a narrow dorsal stripe, which tapers abruptly at the base of each and expands broadly at the apical ends of the segments, sending a lateral prolongation forwards on each side nearly as far as the base which encloses partly a streak of the ground-colour, and a small round spot of the same at the sides at apical

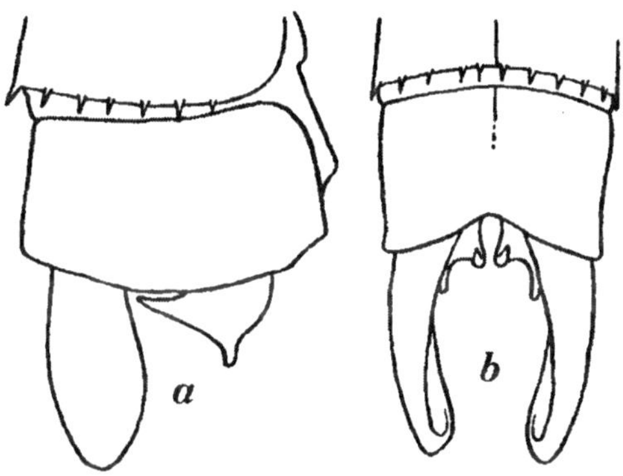

Fig. 174.—Anal appendages of *Argiocnemis rubescens* Selys, male. *a*. Right lateral view. *b*. Dorsal view.

ends of segments; segments 3 to 7 similar, but the sides entirely black and without any apico-lateral spots; segments 8 and 9 pale azure blue, the former with the apical border very narrowly black, but, in some specimens, extending basalwards as an apical dorsal triangle of black, with a small black spot on each side of its apex; segment 9 with the apical border narrowly black, and confluent with an apical dorsal spot shaped like a crown or a fleur-de-lis; segment 10 entirely black. *Anal appendages* black, the superiors pale brown on the inner side, slightly longer than segment 10, convex outwardly, deeply excavate on the inner side, obtuse at apex, curved very gently downwards; inferiors much shorter, less than half the length of superiors, conical as seen in profile, broad at base, and with a minute black point at apex; a more robust dorsal spine directed sharply upward, not visible as a rule, but hidden beneath the superiors.

Female.—Abdomen 28–29 mm. Hind-wing 18–20 mm.

The sub-adult and teneral form exactly similar to the same stage in the male, the first five segments of abdomen bright cherry-red and with fine apical black rings, the remaining segments dull brown to black.

Adult similar to the adult male except for the terminal segments of abdomen, which are entirely black, save for a small basal lateral blue spot on segment 8 and a narrow lateral stripe on segment 7. The under surface of abdomen warm brown, continuous with the lateral apical prolongations of black on segments 3 to 7.

Distribution.—Bengal, Assam, Burma, Malaysia, and Indo-China. The type, a female, comes from Queensland, Australia. The species is distinguished at a glance from *A. rubeola* by the total absence of black on the labrum. In Java specimens this structure is entirely glossy black, and in strong contrast to the pale turquoise colouring of *A. rubescens.* (*A. lunulata* Selys and *A. nigricans* Selys are merely age-varieties of *A. rubeola* Selys ; *A. solitaria* is probably a distinct species, but is not found within Indian limits.)

Type in the Selysian collection ; paratypes in the author's collection and in the Indian Museum and Pusa collections. I possess specimens from Hasimara, Duars, Bengal, taken on 19.viii.23 ; from Maymyo, Burma, on 19.v.25 ; a pair from Kachar, Assam, and several tenerals from Siam taken on 18.xi.23 at Nang Khor, near Surache.

Genus **MORTONAGRION** Fraser (1920). (Fig. 175.)

Mortonagrion Fras., J. Bombay Nat. Hist. Soc. vol. xxvii, p. 148 (1920) ; Ris, Ark. Zool. Uppsala, Bd. 21, A, no. 31, pp. 1–7 (1930).
Indagrion Fras., Mem. Dept. Agric. India (Ent.), vol. vii, no. 7, p. 50 (1922).

Dragonflies of very small size and slender build ; colours non-metallic, usually black marked with green or blue, or palest brown marked with very pale blue, females isochromatic; wings hyaline ; pterostigma shaped like an orange-pip, the posterior and distal borders forming a continuous convexity, the costal and proximal borders only slightly angulated at their junction, usually braced ; postnodal nervures numbering from 8 to 10 in fore-wings, 7 to 8 in the hind ; discoidal cell acutely pointed at distal end, with costal side nearly half the length of posterior in the fore-wing and three-fourths the length in the hind, basal side slightly or considerably shorter than the costal in fore-wing, and less than half the length in the hind ; sectors of arc arising from the lower end of arc and divergent from origin ; arc situated well distal to the level of distal antenodal nervure ; *ab* always present and complete, arising well proximal to the level of *ac*, which lies slightly nearer the level of the distal antenodal nervure; *ab* continued on as *Ia in the same straight line, no angulated medio-anal link intervening at the junction*; *Riv+v* arising slightly before the level of the oblique nervure descending from the subnode in the fore-wing, but well proximal to that level in the hind; *IRiii* in continuation of that nervure.

Head narrow, without any frontal ridge; postocular coloured spots present or absent; thorax short and robust; posterior lobe of prothorax with the middle portion produced in a rounded arch. Abdomen slim and cylindrical, very slightly dilated at end-segments; legs short, two rows of spines on tibiæ and femora, a long and a short row, 4 to 5 long spines on hind tibiæ, and 3 to 4 long spines on the femora. Anal appendages variable in the species, the superiors usually digitate, the inferiors tapering to a point and sloping upwards; segment 8 in the female without a

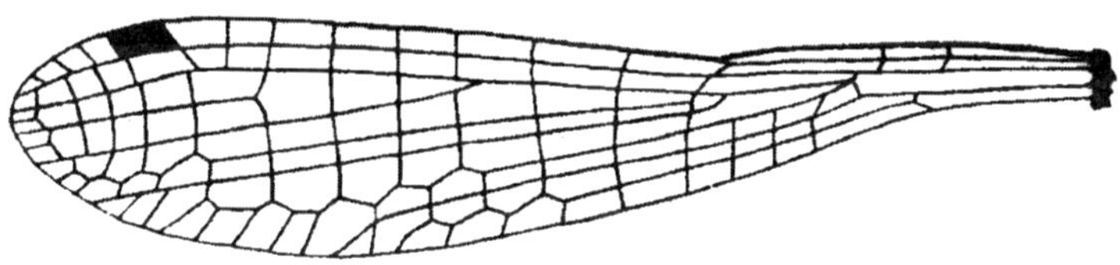

Fig. 175.—Hind-wing of *Mortonagrion varralli* Fras., male.

ventral spine. Genitalia: lamina very narrowly and very deeply cleft; hamules quadrate plates similar to those of *Agriocnemis*; lobe slim and rounded, scrotum-shaped.

Genotype, *Mortonagrion varralli* Fras.

Distribution.—WESTERN INDIA, Sumatra, and Japan. Probably more species remain to be discovered in intervening areas.

Key to the Species of Mortonagrion.

Ground-colour of head and thorax black......	*gautama* (Fras.), p. 411.
Ground-colour of head and thorax palest brown, without any black markings	*varralli* Fras., p. 409.

165. **Mortonagrion varralli** Fraser. (Figs. 175 & 176.)

Mortonagrion varralli Fras., J. Bombay Nat. Hist. Soc. vol. xxvii, p. 148 (1920); id., Rec. Ind. Mus. vol xxvi, p. 490 (1924); Ris, Ark. Zool. Uppsala, Bd. xxi, A, no. 31, pp. 7, 13–14 (1930); Fras., Rec. Ind. Mus. vol. xxxiii, pp. 448, 464 (1931).

Male.—Abdomen 23–25 mm. Hind-wing 14–15 mm.

Head: labium straw-coloured; labrum, genæ, bases of mandibles, and clypeus pale morning-sky blue, vertex pale reddish-brown as well as occiput, except for a pale blue postocular oval spot bordering the eye inwardly; eyes dove-grey, capped with reddish-brown above; head bluish-white beneath. *Prothorax* pale reddish-brown, the anterior lobe and the sides pale blue; posterior lobe trilobate, the median portion the larger and strongly convex. *Thorax* pale reddish-brown on dorsum and sides as far as the antero-lateral suture, marked with a narrow antehumeral pale blue stripe on each side; beneath and the sides palest blue, marked with a diffuse pale brown stripe on each postero-lateral suture. *Legs* pale brown, spines blackish. *Wings* hyaline; pterostigma covering less than one cell, braced, pale golden brown, framed finely in yellow and thick black nervures; 9 to 10 postnodal nervures in fore-wings, 8 in the hind. *Abdomen* pale reddish-brown, this colour deepening subapically, leaving a narrow paler annule between it and narrow black apical rings; the sides of segments 1 and 2 diffusely pale blue, and segments 3 to 7 with narrow pale blue basal annules; segment 8 with a broader pale sky-blue basal annule which extends apically on each side, but is deeply indented by the ground-colour

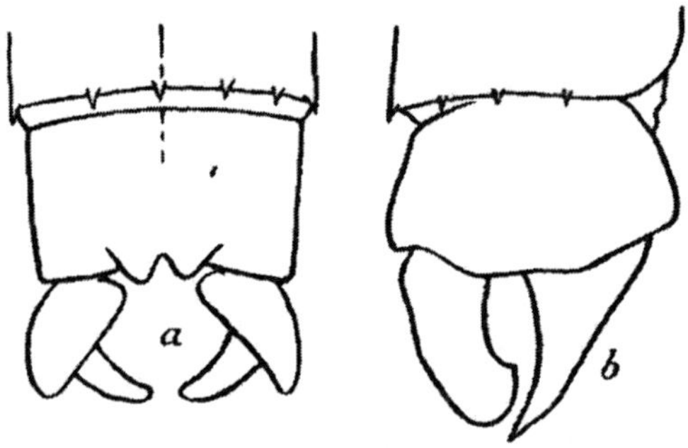

Fig. 176.—Anal appendages of *Mortonagrion varralli* Fras., male. *a*. Dorsal view. *b*. Right lateral view.

on the mid-dorsum; segments 9 and 10 unmarked, but the latter paler laterally, and presenting a small apical dorsal bifid tubercle not unlike that seen in species of *Ischnura*. *Anal appendages*: superiors nearly as long as segment 10, broad and subquadrate as seen from the side, obtusely ended, sloping downward, divergent as seen from above; inferiors slightly longer, very broad at base, then abruptly truncated and ending in a fine spine, directed strongly upward and somewhat inward between the superiors, resembling somewhat the appendages of *C. coromandelianum*.

Female.—Abdomen 24 mm. Hind-wing 14 mm.

Coloured exactly as the male except for segment 8, on which the basal annule is of even width, covering about one-third of the segment.

Distribution.—Western India at sea-level or on the foot-hills. A large colony exists in the scrub jungle alongside the

Vihar and Pawai lakes near Bombay. I took a single female on a hill-stream near the foot of the Vayitri Ghat, South Malabar; this was evidently a wanderer, as no others were ever subsequently found near there. The habits of the species are closely similar to those of *Copera*, the insect threading its way in jerks through the dense gloomy undergrowth, and visible only by the pale blue identification marks, which are seen as tiny moving white points.

166. **Mortonagrion gautama** (Fraser).

Indagrion gautama Fras., Mem. Dept. Agric. India (Ent.), vol. iii, no. 7, p. 50 (1922); id., J. Bombay Nat. Hist. Soc. vol. xxix, p. 746 (1923).

Female.—Abdomen 24·5 mm. Hind-wing 16·5 mm.

Head: labium whitish; labrum, bases of mandibles, genæ, and clypeus pale bluish-green, the former broadly black except for its base; vertex, occiput, and summits of eyes black, the occiput marked with large pale blue post-ocular spots; beneath eyes pale blue. *Prothorax* black on dorsum, the anterior lobe and the sides broadly blue; posterior lobe rounded, narrowly edged with blue. *Thorax* black on dorsum as far lateral as the antero-lateral suture, with a narrow pale blue antehumeral stripe on each side; laterally pale blue, with a narrow black stripe on the postero-lateral suture; palest blue beneath. *Legs* white; femora broadly black on extensor surface, tibiæ on the flexor surface. *Wings* hyaline; pterostigma black, finely framed in pale yellow and thick black nervures; 9 postnodal nervures in fore-wings, 7 in the hind. *Abdomen* bluish-green, broadly marked with black as follows:—Segment 1 with a broad subtriangular spot on dorsum separated from the apical border by a narrow blue apical ring; segment 2 with a broad thistle-shaped dorsal marking confluent behind with a narrow apical black ring; segments 3 to 7 all broadly marked with black on dorsum, this marking not quite extending to base of segments, and presenting a subapical dilatation before contracting to become confluent with a narrow apical ring; segments 8 to 10 broadly black on dorsum, blue at the sides, segment 8 with a narrow blue apical ring. *Anal appendages* small, conical, black; vulvar scale blue, robust, extending to end of abdomen.

Distribution.—The *type*, a female in the British Museum, is the only specimen known, and comes from Sadiya, ASSAM, May 23rd, 1920. Its venation agrees with the genotype, except for the presence of only two cells between the discoidal cell and the nervure descending from the subnode (three in the genotype), but we find the same condition in *M. simile* Ris, so that the genus *Indagrion* Fraser can no longer stand.

Genus **CŒNAGRION** Kirby (1890). (Fig. 177.)

Agrion Fabr., Syst. Ent. p. 425 (1775); Leach, Edinb. Encycl. vol. ix, p. 137 (1819); Steph., Ill. Brit. Ent., Mand. vol. vi, p. 71 (1836); Charp., Lib. Europ. p. 21 (1840); Selys, Mon. Lib. Europ. p. 146 (1840); id., Rev. Odon. p. 171 (1850); id., Bull. Acad. Belg. (2) vol. xli, p. 1246 (1876); Lucas, Brit. Drag. p. 61 (1900).

Cœnagrion Kirby, Cat. Odon. p. 148 (1890); Munz, Mem. Amer. Ent. Soc. no. 3, p. 61 (1919); Fras., J. Bombay Nat. Hist. Soc. vol. xxix, p. 748 (1923); Needham, Zool. Sinica, ser. A, vol. xi, pp. 265, 267 (1930).

Dragonflies of small or moderate size and slender build; colours non-metallic, usually black marked with blue, or blue marked with black; females always coloured differently from the males; wings hyaline; pterostigma similar in fore- and hind-wings, slightly longer than broad, the distal and posterior borders in a single rounded nervure, poorly braced, covering less than one cell; 8 to 11 postnodal nervures in fore-wings; discoidal cell acutely pointed at distal end, with costal

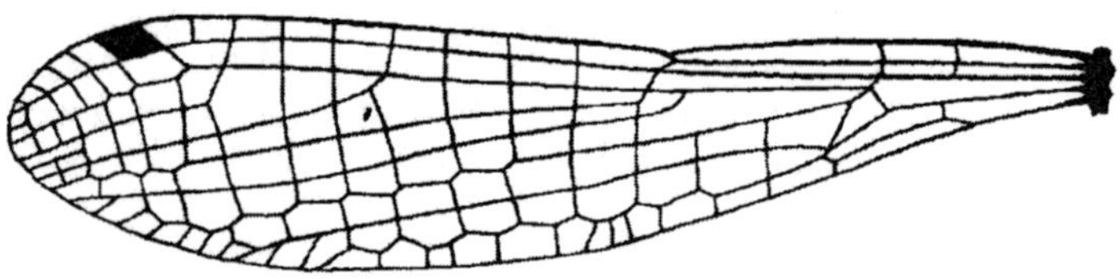

Fig. 177.—Hind-wing of *Cœnagrion dyeri* Fras., male.

side about half the length of posterior in fore-wing, two-thirds that length in the hind, basal side shorter than costal side in fore-wing, about half the length in the hind; distal side very oblique; sectors of arc arising from lower end of arc, divergent from their origins; arc situated at the level of the distal antenodal nervure; *ab* present and complete, arising well proximal to the level of *ac*, which lies midway between the two antenodal nervures or rather nearer the proximal one; *ab* continued on as *IA*, but an angulation (medio-anal link) present at the junction of the two; *Riv+v* arising well proximal to the oblique nervure descending from the subnode, *IRiii* in continuation of that nervure.

Head narrow, without any frontal ridge; postocular spots usually present, but occasionally absent; thorax short and robust; posterior lobe of prothorax simple. Abdomen slim and cylindrical, dilated but very slightly at the base and again at anal end; legs short, spines rather short and slim, 5 to 7 on the hind tibiæ; anal appendages variable in the species, usually very short, the inferiors shorter than the superiors;

segment 8 of the female without a ventral spine. Genitalia: lamina deeply cleft; hamules broad quadrate plates with borders closely apposed in the middle line; penis with end curled strongly over stem, embracing the latter with its branches; lobe depressed, scrotum-shaped.

Distribution.—A large holoarctic genus with a number of species distributed through Europe and Central Asia, but only one known from within Indian limits. Species of the genus generally breed in ponds, tanks, or still waters, and often rest flat on floating objects such as lotus-leaves, an uncommon habit with dragonflies when over water.

Genotype, *Cœnagrion puella* (Linn.).

167. **Cœnagrion dyeri** Fraser. (Figs. 177 & 178.)

Argiocnemis dyeri Fras., Rec. Ind. Mus. vol. xvi, pp. 451, 452 (1919).
Argiocnemis gravelyi Fras., Rec. Ind. Mus. vol. xvi, p. 451 (1919).
Cœnagrion dyeri Fras., J. Bombay Nat. Hist. Soc. vol. xxix, p. 748 (1923); id., Rec. Ind. Mus. vol. xxvi, pp. 428, 498 (1924).
Cœnagrion gravelyi Fras., J. Bombay Nat. Hist. Soc. vol. xxix, p. 748 (1923).
Cœnagrion violacea Fras., Mem. Dept. Agric. India (Ent.), vol. viii, no. 8, pp. 84, 85 (1924); id., J. Bombay Nat. Hist. Soc. vol. xxix, p. 748 (1923); id., Rec. Ind. Mus. vol. xxxiii, p. 448 (1931).

Male.—Abdomen 22–23 mm. Hind-wing 16 mm.

Head: labium white; labrum, genæ, bases of mandibles, clypeus, and frons pale azure blue, the labrum with a minute medial basal spot of black, the postclypeus very narrowly black at base, the frons with a median transverse black stripe not extending to eyes; vertex and occiput black, the former with a small round spot on the outer side of the middle ocellus, the latter with pyriform postocular blue spots. In old adults the blue markings on frons and vertex are obscured by pruinescence which is snowy white on the frons; the postocular spots are also partially or wholly hidden by pruinescence, which thickens in intensity on the back of the occiput; eyes deep blue above, paler below; beneath the head black, with a broad streak of pale blue against the eyes. *Prothorax* black on dorsum, the anterior lobe, the hinder border of the posterior lobe, a round spot on each side, and the lower part of sides azure blue; occasionally a small geminate blue spot on the mid-dorsum of middle lobe; in old adults the whole prothorax densely pruinescent, obscuring all markings; posterior lobe broad and shallow and very minutely emarginate at the middle. *Thorax* black on dorsum to a little beyond the level of humeral suture, marked with broad azure blue antehumeral stripes of even width throughout; laterally azure blue, with a fine black line on the upper part of each lateral suture. In old adults the whole of thorax densely white or violaceous with pruinescence. *Legs* white, black on extensor surface of femora, which are pruinosed in old adults; hind tibiæ with five spines. *Wings* hyaline; pterostigma covering a little more than half a cell, usually braced, yellow, framed in heavy black nervures; 9 to 10 postnodal nervures in fore-wings, 7 to 8 in the hind; *ac* situated nearest the proximal antenodal nervure, *ab* beginning widely distal to the level of this, so that the petiolation of the wing begins proximal to the level of the proximal antenodal. *Abdomen* azure blue, marked with black as follows:—Segment 1 with a large dorsal black spot and a fine apical ring; segment 2 with a broad dorsal spot shaped like a thistle-head connected narrowly to a fine apical ring; segments 3 to 7 with broad dorsal stripes not quite extending to base of segments, expanding subapically, and

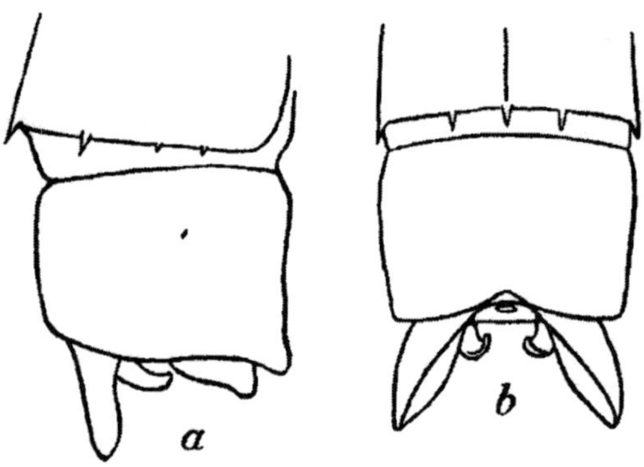

Fig. 178.—Anal appendages of *Cœnagrion dyeri* Fras., male. *a.* Right lateral view. *b.* Dorsal view.

then contracting again to join narrow apical rings; segments 8 and 9 entirely blue, but with an apical row of fine black spines; segment 10 blue, with a narrow mid-dorsal streak of black. *Anal appendages*: superiors black, as long as segment 10, widely divaricate, truncate, ending in a rather acute apex, the inner border prolonged vertically downwards as a long robust black spine; inferiors much shorter, only just visible in profile, white, small, tumid processes.

Female.—Abdomen 21–23 mm. Hind-wing 15–16 mm.

Differs considerably from the male, and there are again differences between young and adult forms.

Subadult.—Markings of *head* similar to those of the male, but the ground-colour of face greenish-yellow and the postclypeus black, more or less enclosing two large yellowish spots; postocular spots bluish-green or green; head not pruinosed. Ground-colour of *prothorax* and *thorax* dove-grey or pale olivaceous, with the mid-dorsal carina and humeral sutures very finely delineated in black; a very narrow black line running parallel and close to the mid-dorsal carina; markings on lateral sutures similar to the male. *Legs* yellow, femora black on extensor surface; tibiæ with six spines. *Wings* similar to the male, but only 9 postnodal nervures in fore-wings and 7 in the hind. *Abdomen* similar to the male, except that the ground-colour is greenish-yellow and segments 8 and 9 are broadly black on dorsum, whilst 10 is bluish-green and with the base only finely black. *Anal appendages* short, conical processes, shorter than segment 10, yellow; vulvar scales robust, but not quite extending to end of abdomen, yellow.

Adult form.—Labrum often dark brown at centre; post-clypeus entirely black, or with a small median yellow spot; vertex and occiput entirely black, without postocular spots, or these very obscure. *Prothorax* broadly black on dorsum, obscuring the dorsal marks, pruinosed white at sides. *Thorax* broadly black on dorsum as far lateral as half-way between the humeral and antero-lateral sutures, marked with narrow pale green antehumeral stripes and narrower interrupted humeral stripes, the two confluent anteriorly; laterally bluish, with a more or less broken stripe on the antero-lateral suture and a black spot on the upper part of the postero-lateral suture; the sides beneath densely pruinosed white, largely obscuring the markings. *Abdomen* similar to the subadult, but segments 1 and 2 and the sides of 8 and 9 pruinosed white.

Distribution.—Widely distributed throughout PENINSULAR INDIA. The type is from the Central Provinces, but the insect is commonest along the Western Ghats, especially in Coorg and Kanara, where at certain times of the year the vegetation on the sides of tanks is blue with their numbers. It is a sub-montane insect, breeding in stagnant and weedy ponds, resting flat on floating grasses and lotus-leaves.

Type in the Indian Museum; paratypes in the Pusa and British Museums and in the author's collection. I possess a female specimen from Maymyo, UPPER BURMA, in which the abdominal markings are bronzed black, segments 9 and 10, as well as a broad apical ring on segment 8, pale sky-blue; there are 10 postnodal nervures in the fore-wings, 9 in the hind; the labrum has a small basal median black spot, and the post-clypeus is black, with two small yellow spots included; the postocular spots are present, small, nearly round and green; the thoracic and prothoracic markings are similar to the adult form described above. The female specimen described as *C. gravelyi* by myself is evidently the old adult form of *C. dyeri*, and *C. violacea* Fras. is the old adult pruinosed form of the male of the same species.

Genus ONYCHARGIA Selys (1865). (Fig. 179.)

Onychargia Selys, Bull. Acad. Belg. (2) vol. xx, p. 416 (1865); Kirby, Cat. Odon. p. 139 (1890); Ris, Nova Guinea, vol. xiii (Zool.), pp. 94, 121 (1915); Munz, Mem. Amer. Ent. Soc. no. 3, p. 56 (1919); Laid., Rec. Ind. Mus. vol. xvi, p. 172 (1919); Fras., J. Bombay Nat. Hist. Soc. vol. xxix, p. 745 (1923).

Dragonflies of small size and moderately robust build; colours black or bronzed purple, marked with bright citron-yellow in the subadult, but immaculate in the full adult; wings hyaline, rather short and broad; pterostigma nearly half as long again as broad, costal and posterior sides equal or subequal, distal side more oblique than proximal, braced, covering less than one cell; 11 to 18 postnodal nervures in fore-wings, 11 to 15 in the hind; discoidal cell acutely pointed at distal end, costal side one-third to one-half the length of posterior in fore-wings, about half or more than half the length in the hind; basal side slightly shorter than or half as long again as the costal in fore-wings, equal to or only about half the length in the hind; sectors of arc arising from

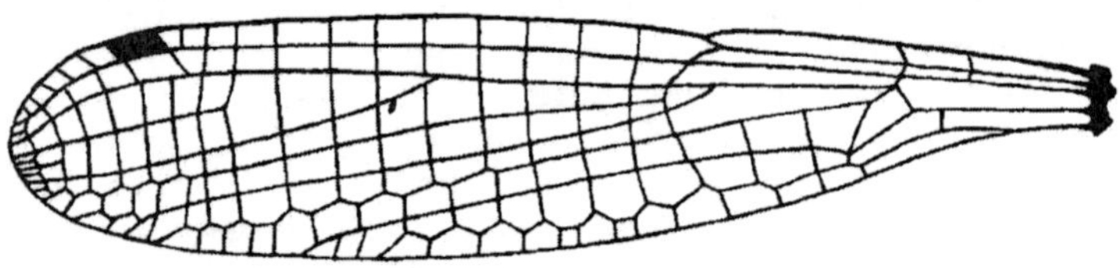

Fig. 179.—Hind-wing of *Onychargia atrocyana* Selys, male.

lower end of arc; arc situated opposite the distal antenodal nervure; *ab* always present and complete, arising far proximal to the level of *ac* or even the level of the proximal antenodal nervure, and continued on as *IA*, but with a marked angulation at the junction of the two; *ac* situated considerably nearer the proximal antenodal nervure; *Riv+v* arising well proximal to the level of the oblique nervure descending from the subnode; *IRiii* in continuation of that nervure.

Head narrow, frons rounded and rather prominent, face and frons very hirsute; postocular coloured spots absent; thorax short and robust; posterior lobe of prothorax rounded in the male, but the middle portion greatly prolonged in the female. Abdomen comparatively short and robust, cylindrical, but slightly dilated at the base and anal ends; segment 10 deeply emarginate on the mid-dorsum, and with two tubercles at the apical border; legs moderately long and robust, hind tibiæ with about 9 or 10 long slim spines, claw-hooks at the extreme end of claws, which thus appear bifid; anal appendages variable in the species, but superiors and inferiors equal in length; segment 8 in the female without a ventral spine.

Genitalia : lamina deeply and rather widely cleft ; hamules quadrate prominent plates with the inner posterior corner acutely pointed ; penis with the end curled strongly over, the apical portion greatly expanded and embracing the stem of the organ, but without branching processes ; lobe scrotum-shaped, depressed, deeply grooved on its naked surface ; vulvar scales robust, extending to end of abdomen.

Genotype, *Onychargia atrocyana* Selys.

Distribution.—Southern Asia from the Western Ghats of India to the Philippin s and New Guinea. Within our limits from the WESTERN GHATS of India, ASSAM, BENGAL, and BURMA. The genus is a small one, and only a single species is found within Indian limits ; by its extremely short petiolation it is related to *Argia* and *Palaiargia*.

168. **Onychargia atrocyana** Selys. (Figs. 179 & 180.)

Onychargia atrocyana Selys, Bull. Acad. Belg. (2) vol. xx, p. 416 (1865) ; Kirby, Cat. Odon. p. 139 (1890) ; id., J. Linn. Soc. Lond., Zool. vol. xxiv, p. 563 (1893) ; Kruger, Stett. Ent. Zeit. p. 118 (1898) ; Ris, Nova Guinea (Zool.), vol. xiii, 2, pp. 94, 95, fig. 13 (1914) ; Laid., Rec. Ind. Mus. vol. xvi, pp. 171, 172 (1919) ; id., Proc. Zool. Soc. Lond. p. 335 (1920) ; Fras., J. Bombay Nat. Hist. Soc. vol. xxix, p. 745 (1923) ; id., Rec. Ind. Mus. vol. xxvi, pp. 428, 489 (1924) ; id., ibid. vol. xxxiii, p. 448 (1931).

Onychargia vittigera Selys, Bull. Acad. Belg. (2) vol. xx, p. 416 (1865) ; Laid., Rec. Ind. Mus. vol. viii, p. 343 (1914).

Male.—Abdomen 23 mm. Hind-wing 17 mm.

Head : labium mat black ; labrum and clypeus glossy black ; rest of head velvety black, unmarked in the full adult, but with a narrow stripe of citron-yellow extending from eye to eye through the interocellar space and the genæ, bases of mandibles and labrum citron-yellow in subadults, the latter finely bordered with black and with a small median basal black point ; eyes black above, brown below. *Prothorax* and *thorax* velvety black, the latter with a deep purplish reflex, especially on the dorsum. In the subadult and teneral stage the dorsum marked with narrow citron-yellow antehumeral stripes, the sides bright citron-yellow, marked with a broad oblique black stripe over the postero-lateral suture ; beneath thorax black, often pruinosed. *Legs* black. *Wings* hyaline ; pterostigma olivaceous yellow, paler around the borders, and framed in thick black nervures, braced, covering slightly less than one cell ; 10 to 11 postnodal nervures in fore-wings, 10 in the hind. *Abdomen* black, unmarked in the adult, but with the sides of segment 1, a lateral stripe on segment 2, and a less evident one on segment 3 bright citron-yellow ; segments 3 to 6 have also narrow bluish basal rings broadly interrupted on the dorsum. *Anal appendages* black ;

superiors ungulate, slightly shorter than segment 10, broad at base, curled slightly in and down and rounded at apex; inferiors of the same length, but more massive, very broad at base, curled slightly in, and subacute at apex.

Female.—Abdomen 23 mm. Hind-wing 18 mm.

Coloured exactly similar to the subadult male; the anteclypeus brownish, postclypeus black at its centre, bright yellow laterally. *Prothorax* with a small subdorsal and a very large lateral spot citron-yellow; antehumeral stripes linear, the ground-colour of *thorax* always velvety black, without the purple reflex. *Wings* with an extra postnodal nervure. *Abdomen* marked with yellow as in the subadult male, but

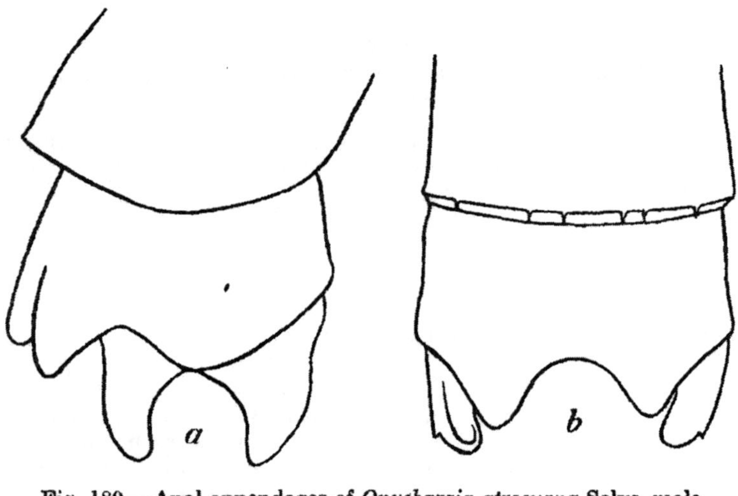

Fig. 180.—Anal appendages of *Onychargia atrocyana* Selys, male. *a.* Right dorso-lateral view. *b.* Dorsal view.

in addition a yellow spot on the pleural surface of segment 8 and another on the sides of the vulvar scales. *Anal appendages* short, pointed, conical, black.

Distribution.—Scattered colonies occur throughout the wet sub-montane areas of India and Burma. The insect is rarely plentiful, but occasional large colonies are found, two of such being known to me in N. Coorg and Malabar, and a third, but smaller one, at Paralai, Anaimalai Hills. The species occurs also in Bengal, Assam, and Burma, and in the hilly tracts of Ceylon.

Type in the Selysian collection; paratypes in most national collections.

ALPHABETICAL INDEX.

[All names printed in *italics* are synonyms. The page on which the description occurs is indicated by thickened numerals.]